T0135070

Smart Innovation, Systems and Technologies

Volume 324

Series Editors

Robert J. Howlett, Bournemouth University and KES International, Shoreham-by-Sea, UK

Lakhmi C. Jain, KES International, Shoreham-by-Sea, UK

The Smart Innovation, Systems and Technologies book series encompasses the topics of knowledge, intelligence, innovation and sustainability. The aim of the series is to make available a platform for the publication of books on all aspects of single and multi-disciplinary research on these themes in order to make the latest results available in a readily-accessible form. Volumes on interdisciplinary research combining two or more of these areas is particularly sought.

The series covers systems and paradigms that employ knowledge and intelligence in a broad sense. Its scope is systems having embedded knowledge and intelligence, which may be applied to the solution of world problems in industry, the environment and the community. It also focusses on the knowledge-transfer methodologies and innovation strategies employed to make this happen effectively. The combination of intelligent systems tools and a broad range of applications introduces a need for a synergy of disciplines from science, technology, business and the humanities. The series will include conference proceedings, edited collections, monographs, handbooks, reference books, and other relevant types of book in areas of science and technology where smart systems and technologies can offer innovative solutions.

High quality content is an essential feature for all book proposals accepted for the series. It is expected that editors of all accepted volumes will ensure that contributions are subjected to an appropriate level of reviewing process and adhere to KES quality principles.

Indexed by SCOPUS, EI Compendex, INSPEC, WTI Frankfurt eG, zbMATH, Japanese Science and Technology Agency (JST), SCImago, DBLP.

All books published in the series are submitted for consideration in Web of Science.

Chakchai So-In · Narendra D. Londhe ·
Nityesh Bhatt · Meelis Kitsing
Editors

Information Systems for Intelligent Systems

Proceedings of ISBM 2022

Editors
Chakchai So-In
Khon Kaen University
Khon Kaen, Thailand

Narendra D. Londhe
National Institute of Technology
Raipur, Chhattisgarh, India

Nityesh Bhatt
Nirma University
Ahmedabad, Gujarat, India

Meelis Kitsing
Estonian Business School
Tallinn, Estonia

ISSN 2190-3018   ISSN 2190-3026 (electronic)
Smart Innovation, Systems and Technologies
ISBN 978-981-19-7449-6   ISBN 978-981-19-7447-2 (eBook)
https://doi.org/10.1007/978-981-19-7447-2

This Springer imprint is published by the registered company Springer Nature Singapore Pte Ltd.
The registered company address is: 152 Beach Road, #21-01/04 Gateway East, Singapore 189721,
Singapore

Preface

The first series of World Conference on Information Systems for Business Management ISBM 2022 is an extension to our ICT4SD conference series which is now in 7th Edition.

The first edition of ISBM conference will serve a potential to focus on the avenues and issues related to governance involved with information systems and business management in different forms and would be attracted by delegates from more than 50 nations which will present their latest research and its practical implications for business, society, academia, and governance. It will be an excellent platform to deliberate upon global themes and dimensions of Information Management.

The conference will be held on September 8–9, 2022, at Physical at Hotel Novotel Bangkok Siam Square. Bangkok, Thailand, and digitally via Zoom. The conference is organized by Global Knowledge Research Foundation, Supporting Partner Springer, Springer Nature, InterYIT, International Federation for Information Processing, and Knowledge Chamber of Commerce and Industry.

Research submissions in various advanced technology areas were received, and after a rigorous peer-review process with the help of program committee members and 78 external reviewers for 357+ papers from 17 different countries out of which 57 were accepted. These will be presented in nine parallel sessions in two days organized physically and virtually including two inaugural and four keynote sessions.

The conference is anticipated to attract a large number of high-quality submissions and stimulate the cutting-edge research discussions among many strategists, managers, academic pioneering researchers, scientists, industrial engineers, students, directors, scientists, lawyers, policy experts, and information technology professionals from all around the world and provide a forum to discuss more on the theme.

Propose new technologies, share their experiences, and discuss future solutions for design infrastructure. Discuss new case studies which haven't come across, share their experiences, and discuss future impacts and possible solutions for Good Technology Governance. Support and share greater and more impactful recommendation for Global GOOD Governance Policies.

Provide a common platform for academic pioneering researchers, scientists, engineers, managers, and students to share their views and achievements. Enrich technocrats, management professionals, and academicians by presenting their innovative and constructive ideas. Focus on innovative issues at international level by bringing together the experts from different countries.

Khon Kaen, Thailand   Chakchai So-In
Raipur, India   Narendra D. Londhe
Ahmedabad, India   Nityesh Bhatt
Tallinn, Estonia   Meelis Kitsing

Contents

About the Editors

Dr. Chakchai So-In (SM, IEEE(14); SM, ACM(15)) is Professor of Computer Science in the Department of Computer Science at Khon Kaen University. He received B.Eng. and M.Eng. from KU (TH) in 1999 and 2001, respectively, and M.S. and Ph.D. from WUSTL (MO, USA) in 2006 and 2010, all in computer engineering. In 2003, he was Internet Working Trainee in a CNAP program at NTU (SG) and obtained Cisco/Microsoft Certifications. From 2006 to 2010, he was Intern at Mobile IP division, Cisco Systems, WiMAX Forum, and Bell Labs (USA). His research interests include mobile computing/sensor networks, Internet of Things, computer/wireless/distributed networks, cybersecurity, and intelligent systems and future Internet. He is/was Editor/Guest Member in *IEEE Access*, *PLOS One*, *PeerJ (CS)*, *Wireless and Mobile Computing*, and *ECTI-CIT*. He has authored/co-authored over 100 publications and 10 books including *IEEE JSAC*, *IEEE Communications/Wireless Communications Magazine*, *Computer Networks*, and *(Advanced) Android Application Development*, *Windows Phone Application Development*, *Computer Network Lab*, *Network Security Lab*. He has served as Committee Member and Reviewer for many prestigious conferences and journals such as *ICNP*, *WCNC*, Globecom, ICC, ICNC, PIMRC; and *IEEE Transactions (Wireless Communications, Computers, Vehicle Technology, Mobile Computing, Industrial Informatics)*; *IEEE Communications Magazine, Letter, System Journal, Access*; and *Computer Communications, Computer Networks, Mobile and Network Applications, Wireless Network*.

Dr. Narendra D. Londhe is presently working as Associate Professor in the Department of Electrical Engineering of National Institute of Technology Raipur, Chhattisgarh, India. He completed his B.E. from Amravati University in 2000 followed by M.Tech. and Ph.D. from Indian Institute of Technology Roorkee in the years 2006 and 2011, respectively. He has 14 years of rich experience in academics and research. He has published more than 150 articles in recognized journals, conferences, and books. His main areas of research include medical signal and image processing, biomedical instrumentation, speech signal processing, biometrics, intelligent healthcare, brain–computer interface, artificial intelligence, and pattern recognition. He has been awarded by organizations like Taiwan Society of Ultrasound in Medicine, Ultrasonics Society of India, and NIT Raipur. He is an active member of different recognized societies from his areas of research including senior membership of IEEE.

Dr. Nityesh Bhatt is Professor and Chairperson of Information Management Area at Institute of Management, Nirma University, Ahmedabad. He holds M.B.A. (Marketing) and Ph.D. (e-Governance) degrees from M. L. Sukhadia University, India. He has also completed a four-month Faculty Development Programme (FDP) of IIM-Ahmedabad and an online 'Internet Governance Capacity Building Programme' of Diplo Foundation, Malta, in 2006. He has more than 22 years of experience in academia, corporate training, and research. In 1998, he was awarded as the best faculty of NIIT in North India. Credited with 52 research papers and management cases, he has also co-edited 13 co-edited books. In 2006, he was invited to participate in the first Internet Governance Forum (IGF) meeting jointly organized by the United Nations—IGF Secretariat and Government of Greece at Athens. He is the recipient of 'Dewang Mehta Best Teacher of Information Technology Award' in 2009. Four of his students have been awarded Ph.D. Another three students are pursuing their Ph.D. with him. He has completed four consultancy/research assignments for ISRO and Government of Kerala. Currently, he is involved in a major research project with ISRO on Government–Academia Interface. He has traveled to 10 countries for different academic initiatives. Since 2007, he is Member of the National Executive Committee of Special Interest Group on e-Governance (SIGeGov) set up by the Computer Society of India. He is Life Member of the Computer Society of India and Indian Society for Training and Development.

Dr. Meelis Kitsing is Author of *The Political Economy of Digital Ecosystems* (Routledge 2021). He is Rector of and Professor of Political Economy at the Estonian Business School. Previously, he worked as Head of Research at Foresight Center, Think-Tank at the Estonian Parliament, Adviser at the Strategy Unit of the Estonian Government Office, and Head of Economic Analysis at the Estonian Ministry of Economic Affairs and Communication. He also served as President of Estonian Economic Association. His current research interests focus on the political economy of digital platform ecosystems. His research has been published by *Transnational Corporations Review*, *Journal of Politics*, *Policy and Internet*, *Journal of Information Technology and Politics* as well as by Springer, IEEE, ACM, Leuven University, and MIT Press. He has conducted research and taught at the University of Massachusetts, National Center for Digital Government (US), George Mason University, Central European University, Harvard University, University of Connecticut, Stockholm School of Economics, and numerous other universities in Europe and in the USA. He has been quoted in the Financial Times, Economist, Reader's Digest, El Comercio, and Postimees. He earned his Ph.D. from the University of Massachusetts Amherst (US), his M.A.L.D. as Fulbright Scholar from the Fletcher School at Tufts University (US) and his M.Sc. as Peacock Scholar from London School of Economics (UK).

Chapter 1
Social Media as Communication–Transformation Tools

Waralak V. Siricharoen ⓘ

Abstract Social media is rapidly being used for a range of purposes, including communication, education, collaboration, and a variety of other factors that have had an impact on society throughout time. Even while it has become widespread, it is also a double-edged sword that may lead to negative repercussions such as social media addiction if not managed properly. Despite this, little attention has been dedicated to explaining the components that lead to social media addiction and dependence. This study will investigate the interaction between social media and people by investigating the influences and consequences of social media use on the behaviour of social media users, as well as interpretations of the beliefs that individuals hold about social media addiction and addiction to other forms of media. It is based on a survey on the topic that enabled people to express their opinions on social media use and responses. The survey was conducted online. People aged 18 and above who lived in Bangkok and the Metropolitans of 150 individuals were surveyed regarding their social media usage habits, including how they responded to which social media resources they used, and about their social media usage behaviours that were negatively correlated with the balance of their social media usage habits.

1.1 Introduction

The usage of social media is getting more widespread, and it is progressively invading our everyday routines. Some of the applications may be advantageous in that they may assist persons in building relationships and acquiring important bits of information that can aid in the betterment of the individual's situation. Individuals may now communicate with one another via social media, and they can also have access to other people's lives from a variety of angles, ranging from their personal lives to their ideas on life as a whole. It is the goal of this research to better understand how social media has influenced people and how it has altered individuals' lives in a variety of

---

W. V. Siricharoen (✉)
Faculty of Information and Communication Technology, Silpakorn University, Nonthaburi 11120, Thailand
e-mail: siricharoen_w2@su.ac.th

ways, given the fact that social media has played a significant role in contemporary society [6].

However, because of the widespread use of social media platforms and the ease with which people can access the Internet, the risk of developing social media addiction has increased in individuals. Social media addiction is defined as the unreasonable and excessive use of social media to the point where it interferes with other aspects of one's daily life [11, 13]. Addiction to social media has been connected to a number of emotional, relationship, health, and performance concerns in the recent years. Because of this, it is vital to understand the causes, consequences, and therapies of social media dependence. In this study, we looked at social media relationships and perceptions through the eyes of individuals, taking into consideration the wide range of ages that have been associated with social media behaviours of the users, what individuals think about social media as it has transformed into everyday life, and what relationships they do have with social media. We also looked at the effectiveness of the intervention in terms of reducing the number of social media platforms and the potential negative consequences of the circumstance.

1.2 Literature Review

Social Media Usage and Characteristics

As a blanket phrase referring to a vast range of online platforms with a variety of different qualities, communication methods, and sociability purposes, all social media apps share some core traits. Social media blurs the barrier between media and audience by inviting contributions and comments from anybody interested in contributing to the discourse [1]. When highlighting social media capabilities, it is critical to remember that they enable users to share and interact with one another, resulting in the distribution of more democratising material than ever before [5, 7]. While involvement levels vary, social media has been used to foster a participatory culture, albeit with different degrees of effectiveness. Additionally, it has been associated with participation, which is defined as the extent to which senders and receivers actively participate in the interaction when it comes to sending and receiving messages. Participation in this might be regarded as action-oriented contact in some ways, since it has shown a range of behaviours in communities, from the providing of critical information to the provision of expertise to assist others [16].

Apart from providing two-way communication rather than one-way transmissions or distributions of information to audiences, social media supports bidirectional communication in conventional media. While conventional communication established a communication pattern, the Internet established a more bidirectional communication pattern than conventional communication. While this has enhanced social media's pace, it is also concerning for the communication aspect of the involvement, as previously mentioned [12]. The relevance factor has had a large effect on the potential for social media and associated businesses. For example, Facebook provides a plethora of communication components for conversations, whereas Twitter utilises the microblogging tool, and the amount of two-way communication that occurs on these social media platforms is frequently dependent on the amount of two-way communication that occurs on these social media platforms using the fundamental and structural elements inherent in these social media platforms [14].

While physical presence in social interactions is ideal, interpersonal ties may be sustained via a mix of face-to-face contact and mediated engagement through communication technology. By offering connections to other websites, services, and people, social media enables media users to travel between virtual locales and fosters a sense of belonging and connection with their peers [3]. Interpersonal, communal, and larger societal links may all be categorised as such in terms of social connectivity. It has been defined as social connectedness and is often linked to interpersonal actions and self-identification. As a result, people with a high degree of connection often feel more connected and pleasant towards others in their social group, whilst others may feel more remote in social situations. Individuals who are more linked to the outside world and more eager to expand their experience in mediated communication situations benefit from being more connected [20].

When it comes to social media usage, women are more likely to use it than males, implying that women prefer online connections over male encounters. On the other hand, male impulses are predominantly driven by social recompenses. Males are more likely to use media to make new acquaintances, learn about current events, and search, while females are more likely to use media to retain current or existing relationships and are more likely to use media for communication, amusement, and time-killing activities than men are [2].

1.2.1   Influences of Social Media

Social Media Addiction

A kind of Internet addiction, social media addiction may be thought of as a condition in which individuals have an overpowering need to use social media. Social media addiction is characterised by individuals being overly preoccupied with it and being propelled to use it by an uncontrollable need to log on and use the platform. As of right now, it is estimated that a significant number of users across social media platforms have begun to suffer from the addiction. According to several research on the association between social media usage and mental health [8, 17], extended use of social media platforms is detrimental to mental health. Examples include Facebook, which has been linked to the long-term well-being of people as a result of its relationship with mental health difficulties. Additionally, they may be associated with the amount of time spent on social media, which has been shown to have an impact on depressive symptoms in individuals, particularly among young people [4].

Further research has found that while using social media for academic purposes did not predict academic success as measured by the cumulative grade average, using social media for non-academic purposes, and multitasking on social media did predict academic success as measured by the cumulative grade average. Studies have also indicated that the overall GPAs of students have been negatively impacted by their use of social media, and tests have revealed a negative correlation between social media use and the academic performance of people. When it comes to learning, multitasking via different social media activities such as texting, emailing, and posting demonstrated a high level of aptitude.

Notably, frequent social media usage does not necessarily imply social media addiction, and as a result, it does not always have negative repercussions on people's mental well-being or academic achievement. Regular social media overuse, which many individuals experience from time to time, and social media addiction, on the other hand, is markedly different in that the latter is associated with negative consequences when online social networking becomes uncontrolled and compulsive.

In the vast majority of studies on social media addiction, it has always been found to be connected with anxiety, sadness, and sleeplessness, but it has also become subjective when it comes to life satisfaction, well-being, and vitality, as has been shown in this study. Furthermore, research has shown that social media addiction has a detrimental impact on academic achievement as well.

1.2.2   Social Media Value

Several people have found convenience in social media's use, as it allows them to maintain a smooth and simple connection with one another. Social media is regarded to be one of the most essential communication channels in the world because of its widespread use. However, it has changed throughout the years, not only in terms of the relationships between people, but also in the business sector, making it more accessible to individuals than it was in the past as a result of the transformation. As a result, technology has taken over and controlled various facets of life, particularly in many businesses and has had an influence on how communication is conducted in the corporate world. It not only facilitates learning and makes business contacts more profitable, but it also serves as a significant benefit to the business sector by making it simpler to reach out to audiences and by increasing the number of customer bases that are valuable to the company [18].

Besides gaining prominence in the market, it has also found ways to integrate into a variety of other fields, such as science and technology. Examples include Osterrieder [9]'s literature, which is based on the photo and pose contributions that depicted plant content and which primarily displayed the information by making use of hashtags to start conversations, share information, make information more accessible and raise awareness about information while also encouraging participation in the field of science. In addition to being able to exchange information, it was determined that the material could be published and developed on a big scale. Social media has served as a tool to enable professionals such as scientists, designers and programmers to communicate more effectively and to produce new ideas that have the potential to benefit society in the future, particularly in the areas of scientific content and educational objectives.

We must recognise, however, that the development of social media has begun to demonstrate a shift away from traditional academic publishing or information such as papers to the development of online scientific publications that help capture and track the information from social media, which individuals can also use as a professional means to explore and learn from in more positive purposes such as user satisfaction surveys.

Satisfaction of Social Media Usage

Individuals of all ages, not only adolescents but also older adults, according to studies conducted by Ractham et al. (2022), can benefit from social media in terms of their lives and satisfaction as they begin to integrate the use of social media into their daily activities across a variety of domains. According to the findings of the study, technological advancements among older adults have resulted in an improvement in communication tools for them because it aids in social connectivity, the development of new skills and the enhancement of existing skills that have been linked to the understanding of the uses of social media. It is preferred by the majority of older folks to exchange information and create more relationships in order to keep up with their social lives online rather than in person.

According to the findings of the studies, older adults were interviewed about how social media has been utilised in their lives and how it has resulted in various aspects of their lives. It was discovered that social media usage among adults had positive impacts on the development of satisfaction in several domains and the reduction of stress, confusion and tension of the individuals when they begin to use the social media platforms. In this study, it was discovered that older adults find social media to be more pleasant than adolescents because of the purposes for which they use it and because they have more control over how they use it. It was also discovered that individuals have more insight when engaging with social media than adolescents. When compared to other elements of life, older persons frequently get a greater variety of motives, which seems to have been one of the most significant things for most older adults in terms of fulfilment, amusement and the well-being of individuals within society rather than other considerations.

1.2.3 Methodology

Participants

The participants are individuals aged between 18 and above, of all genders with at least a Bachelor's degree in their educational background, who are active on social media, and who have resided in Bangkok and the surrounding areas were included in the study, which had a total sample size of 150 participants.

Procedure and Materials

The survey will be centred on online platforms from both within and outside of the classroom, with a series of questions concentrating on the interpretations of social media tools as they are used in class. They would be alerted about the survey and would be required to complete the whole survey using the Google Form platform to be considered.

Social Media Behaviours

Individuals would be asked based on the duration of social media usage based on the scales of how often and the purposes which individuals have perceived regarding the usages of social media platforms.

Social Media Perceptions

The responses to the questions about the significance of social media may be provided by people based on the alternatives that can be picked once or several times in order to get the perspectives that individuals have about social media use.

Social Media Addiction

The set of questions based on the social media addiction symptoms and the duration of social media usage of individuals which pertain to recent experiences of the individuals based on the scale ranging from more often to east often such as the question based on how many hours they use social media platforms and what applications have been used the most.

1.2.4   Findings

First, the result displayed the respondents' patterns of social media use, as well as their basic behaviours and the goals for why they use social media platforms, in addition to the results of the survey. The inquiry's findings are presented as an examination into the connection between social media usage as a tool in everyday life and the behaviour of social media usage in order to get a better knowledge of the insights offered on the understanding of social media use. The results of the survey, which were completed by persons between the ages of 18 and 29, revealed that the largest age difference is between 18 and 24 years old and that 40.9% of the participants were girls and 59.1% were men. The study showed that the vast majority of individuals (90.9%) often checked social media as the first thing they did after waking up in the morning. The majority of people believe that social networking is the most important use of social media (90.9%), with gaming coming in second place at 59.1% as the primary reason for using social media. According to the findings of the study, the minority group uses it for podcasts and YouTube (4.5%). Social networking and learning are the two most popular uses of social media, accounting for 77.3% of all use and 72.7% of all learning, with the Facebook application accounting for the majority of all usage among other applications (90.9%).

For the behavioural aspect, it was discovered that several individuals scored similarly in different areas as the majority of individuals spent approximately 4–5 h on average per day online, whereas other groups are seen to be almost identical at 22.7%, with the exception of those who spent less than an hour, which is made up of 3%. Individuals have indicated that social media has made their lives much simpler, with the majority of responses falling within the moderate, high, and very high categories. It is also often used to execute the most common everyday duties. Individuals, according to the results of the survey, like to believe that social media has given them the greatest amount of escape (4 out of 5) from reality, as well as neutral benefits (54.5%) and more advantages than negatives (45.5%).

Moreover, the survey also highlighted the urge to disconnect from reality and the loss of control over everyday activities as the first two key causes of social media addiction, with habituation being regarded as the least significant component (4.5%). Personality and mental well-being, among other aspects, have been demonstrated to have the largest effect on users on social media, which shows that it has been one of the strongest influencers out of all difficulties.

The percentages in the table represent the results of the survey responses from the sample group, giving a clear conclusion based on the social media behaviours and usages that individuals engage in see in Table 1.1.

1.3 Discussions

It is shown that the objective of this study is to assemble information based on people's social media use and purpose in order to explore the relationship between social media satisfaction and the attitudes and behaviours of persons towards social media usage. The findings suggest that social media has played a significant role in people's lives, particularly among young people, from a variety of viewpoints, and that this has resulted in the development of new habits and behaviours among individuals [19]. The use of social media has a positive impact on a variety of areas, ranging from one's own identity to social interactions. Individuals' pleasure and positive attributes when they use social media have been proven to be influenced by some components of their social media lifestyle, which has been shown in the research on social media lifestyle. Although social media is used for the purposes of relaxation and meeting new people, it also expresses the use of social media in the context of building connections and assisting with job advancement, as well as the purposes of understanding and learning new things from online platforms for self-improvement. This lends credence to studies that have focused on the benefits that social media has offered to individuals when used carefully and within reasonable limits, rather than on the development of social media addiction, which is common among young people.

Table 1.1 Percentages of the survey responses

Answers	Responses (%)
Is social media the first thing you check after waking up?	
Yes	90.9
No	9.1
What social media tool(s) do you use the most? (you can choose more than one)	
Social networking	90.9
News and blogs	36.4
Search engine/Wikis	36.4
Games	59.1
Podcasts	4.5
YouTube	4.5
What do you use social media for? (you can choose more than one)	
Networking	77.3
Online shopping	63.6
Online learning	72.7
Work/business-related	31.8
Watch movies or cartoons	4.5
Nothing	4.5
Entertainment	4.5
What social network applications do you use? (you can choose more than one)	
Facebook	90.9
Twitter	50
Instagram	81.8
WhatsApp	4.5
Snapchat	0
TikTok	4.5
Line, Discord	4.5
How much does social media make your life easier?	
Very high	40.9
High	31.8
Moderate	27.3
Low	0
Very low	0
How often would you use social media to complete daily activities (e.g. routine tracking, shopping, reminder, alarm, etc.)?	

(continued)

Table 1.1 (continued)

Answers	Responses (%)
Very often	45.5
Often	27.3
Neutral	22.7
Not often	4.5
Rarely	0
From scale 1 to 5: how likely would you use social media as an escape from reality?	
1	4.5
2	0
3	22.7
4	54.5
5	18.2
What is the most possible factor(s) for social media addiction among users? (you can choose more than one)	
Want to separate themselves from reality	59.1
Lose control over daily activities	59.1
Compulsive craving for the behaviour	27.3
Sustain the need to develop interpersonal skills	45.5
Habituation	4.5
Which factor social media has impacted users the most?	
Personality and mental well-being	40.9
Cultural perspectives	9.1
Opportunities to grow social presence/business	31.8
Barrier elimination of communication	18.2

Moreover, in order to create a connection and utilise social media platforms to their full potential, it is necessary to understand the behaviour and routine of persons in order to be able to regulate and use social media platforms responsibly [15]. A comparison of social media use across each age group revealed that social media is largely utilised for engagement, expression and enjoyment rather than for any other purpose. The findings revealed that, in addition to supporting literature review, young users have relatively high technological demands, and they use technology for a variety of reasons, including learning new skills, mind exercising, relaxation, leisure, self-presentation, information-seeking, purchasing and marketing attributes [10].

In our technologically advanced culture, it has demonstrated the growth of social media from all perspectives, with the result implying that social media platforms have now dominated society as well as playing a significant role in increasing the number of new activities, as well as serving a variety of functions that are beneficial to one's well-being and way of life.

1.4 Conclusions

In conclusion, social media has influenced almost every aspect of life and may have transformed and changed the lifestyles of individuals. It has constantly changed with each passing year, but with the same underlying concept that social media will continue to be the central concept of our existence. The use of social media to prevent addiction may thus become a difficult matter rather than a simple one, and it may have a negative rather than positive effect, as social media does not only provide negative outcomes but also positive ones that individuals can take advantage of within a certain time limit.

Despite the fact that social media continues to have an impact on a variety of sectors, such as the emergence of new apps that may cause future problems for people, it may be the right moment to implement new developments. Even if people are beginning to form certain norms for diverse social situations, society is still assessing the full consequences of these standards for others. Being able to learn more about social media and be able to understand how to use it securely, particularly in terms of growth and development or professional elements, would assist to make social media use one of the most productive ways to use social media.

References

1. Boahene, K.O., Fang, J., Sampong, F.: Social media usage and tertiary students' academic performance: examining the influences of academic self-efficacy and innovation characteristics. Sustainability **11**(8), 2431 (2019). https://doi.org/10.3390/su11082431
2. Chan-Olmsted, S.M., Cho, M., Lee, S.: User perceptions of social media: a comparative study of perceived characteristics and user profiles by social media. Online J. Commun. Media Technol. **3**(4) (2013). https://doi.org/10.29333/ojcmt/2451
3. Chou, W., Ying S., Hunt, Y., Moser, R., Hesse, B.: Social media use in the United States: implications for health communication. PsycEXTRA Dataset (2009). https://doi.org/10.1037/e521582014-115
4. Choudhury, M., Counts, S., Horvitz, E.: Social media as a measurement tool of depression in populations. In: *Proceedings of the 5th Annual ACM Web Science Conference on WebSci '13*, 2013. https://doi.org/10.1145/2464464.2464480
5. Fu, J., Shang, R.-A., Jeyaraj, A., Sun, Y., Hu, F.: Interaction between task characteristics and technology affordances. J. Enterp. Inf. Manag. **33**(1), 1–22 (2019). https://doi.org/10.1108/jeim-04-2019-0105
6. Kavanaugh, A.L., Fox, E.A., Sheetz, S.D., Yang, S., Li, L.T., Shoemaker, D.J., Natsev, A., Xie, L.: Social media use by government: from the routine to the critical. Gov. Inf. Q. **29**(4), 480–491 (2012). https://doi.org/10.1016/j.giq.2012.06.002
7. Kung, Y.I.N.G.M.A.I., Oh, S.A.N.G.H.E.E.: Characteristics of nurses who use social media. CIN: Comput. Inf. Nurs. **1** (2014). https://doi.org/10.1097/cin.0000000000000033
8. Liu, C., Ma, J.: Development and validation of the Chinese social media addiction scale. Personality Individ. Differ. **134**, 55–59 (2018). https://doi.org/10.1016/j.paid.2018.05.046
9. Osterrieder, A.: The value and use of social media as communication tool in the plant sciences. Plant Methods **9**(1), 26 (2013). https://doi.org/10.1186/1746-4811-9-26
10. Oxley, A. (2013). Security threats to social media technologies. *Security Risks in Social Media Technologies*, 89–115. https://doi.org/10.1016/b978-1-84334-714-9.50003-6

11. Quan-Haase, A., Young, A.L.: Uses and gratifications of social media: a comparison of Facebook and instant messaging. Bull. Sci. Technol. Soc. **30**(5), 350–361 (2010). https://doi.org/10.1177/0270467610380009

12. Rodgers, R.F., Rousseau, A.: Social media and body image: modulating effects of social identities and user characteristics. Body Image **41**, 284–291 (2022). https://doi.org/10.1016/j.bodyim.2022.02.009

13. Salem, J., Borgmann, H., Baunacke, M., Boehm, K., Hanske, J., MacNeily, A., Meyer, C., Nestler, T., Schmid, M., Huber, J.:. Widespread use of internet, applications, and social media in the professional life of urology residents. Can. Urol. Assoc. J. **11**(9) (2017). https://doi.org/10.5489/cuaj.4267

14. Seo, E.J., Park, J.-W., Choi, Y.J.: The effect of social media usage characteristics on e-WOM, trust, and brand equity: focusing on users of airline social media. Sustainability **12**(4), 1691 (2020). https://doi.org/10.3390/su12041691

15. Shabir, G., Hameed, Y.M., Safdar, G., Gilani, S.: The impact of social media on youth: a case study of Bahawalpur City. Asian J. Soc. Sci. Human. **3**(4), 132–151 (2014). http://www.ajssh.leena-luna.co.jp/AJSSHPDFs/Vol.3(4)/AJSSH2014(3.4-13).pdf

16. Smith, E.E.:. Social media in undergraduate learning: categories and characteristics. Int. J. Edu. Technol. Higher Edu. **14**(1) (2017). https://doi.org/10.1186/s41239-017-0049-y

17. Sun, Y., Zhang, Y.: A review of theories and models applied in studies of social media addiction and implications for future research. Addict. Behav. **114**, 106699 (2021). https://doi.org/10.1016/j.addbeh.2020.106699

18. UCW.: (n.d.). How has social media emerged as a powerful communication medium? Retrieved April 20, 2022, from https://www.ucanwest.ca/blog/media-communication/how-has-social-media-emerged-as-a-powerful-communication-medium

19. Xiao, G., Lee, H.R., Tessema, K., Wang, S.: The examination of cultural values and social media usages in China. Rev. Mark. Sci. **19**(1), 101–120 (2020). https://doi.org/10.1515/roms-2020-0044

20. Zolkepli, I.A., Kamarulzaman, Y.: Social media adoption: the role of media needs and innovation characteristics. Comput. Hum. Behav. **43**, 189–209 (2015). https://doi.org/10.1016/j.chb.2014.10.050

Chapter 2
Bidirectional DC–DC Converter-Based Energy Storage System Method for Electric Vehicles

Aditya Aniruddha Tipre and Manisha Ingle

Abstract Hybrid electric cars have the same advantages as hybrid cars, but the main difference is that they use an electric motor that is powered by an energy storage system that gets its energy from a source like batteries or the grid to help with the main source of power. As a bonus, the electric motor can also be used to make electricity. When the car brakes, it converts the energy into electricity that can be stored in the car's energy storage unit. A hybrid control strategy is usually used in an energy-saving study of a vehicle. This strategy divides the load between different modes of operation, such as when the vehicle is running. This thesis talks about how electric vehicles (EVs) came to be and how they can be used with a combined energy storage system. To make electric cars last for a long time and keep costs down, this paper proposes a new hybrid energy storage device for electric cars. This thesis proposes the best way to control the hybrid energy storage device, which is made up of a Li-ion battery and a supercapacitor. The battery's capacity dynamic constraint rule-based control is based on the supercapacitor's state of charge. Use of an ANFIS controller makes hybrid energy storage system outputs more accurate and less distorted when they're used in a system.

2.1 Introduction

India doesn't have enough petroleum to last for a long time. It is also at risk from the supply of crude oil and natural gas. Even so, India is now the third-largest importer of oil in the world, after the United States and China, behind both of them. More than 82% of India's total oil and gas imports are made up of crude oil, and more than 45% are made up of natural gas. There has been a lot of effort to cut back on the use of petroleum products in order to cut down on pollution in the air. It also puts a lot of money into the pockets of Indians because they have to buy so much crude oil from other countries. To reach these goals, we would need to use more renewable energy and nuclear energy, as well as less fossil fuels.

A. A. Tipre (✉) · M. Ingle
VLSI and Embedded Systems, MIT World Peace University (MITWPU), Pune, India
e-mail: adityatipre@gmail.com

More than half of the petroleum products are used in cars. As a result, running these cars causes a lot of air pollution, which has a big impact on our environment. It is because India has a lot of cars that run on petroleum products that aren't electric. New technologies, like the battery-powered PHEV and PEV, are becoming more popular in the fight against greenhouse gas emissions and air pollution. Plug-in hybrid cars are electric cars that can be charged with electricity from an outside source, like the electric grid (PEV).

The economy, electricity, and a lot of other things have made transportation more electric in the recent years. Railways also had the good fortune of having a lot of different types of electric locomotive for a long time. They go from A to B. In order to get electric power from a conductor rail, pantograph slider slides make it simple to do so. Trying to understand more, because electric vehicles (EVs) have so many utility options (UTOPIA), it is more difficult to get power in the same way. Other than that, an electric vehicle battery pack (usually a high power, large-capacity one) is usually used as an energy storage system to let the vehicle go a long way. Until now, electric cars have been out of the reach of most people because of government incentives. To help electric cars take over the market, government incentives and tax credits are a must at the moment. Exaggerate the main problem with an electric car is how to store electricity, which is mostly done by a battery. Batteries, on the other hand, are not very useful because they have a short life cycle, are very expensive, and don't have a lot of power. The following must all be met at the same time when designing a battery for an electric car: a large energy volume, a high power density, an economical price, a long cycle life, good safety, and longevity. Lithium-ion batteries are thought to be the most cost-effective choice for electric car batteries at the moment [1]. When you put a full pack of lithium-ion batteries into an electric car, they have an energy capacity of 90–100 Wh/kg [2]. However, the commercialised lithium-ion battery only has an energy capacity of 90–100 Wh/kg [2]. There isn't as much power in this estimate as there is in gasoline. It has just 300 Wh/kg, which isn't very power-packed. To be able to deal with an ICE car's 300-mile range, a pure electric vehicle (EV) needs a lot of powerful and expensive batteries. Cost of a lithium-ion battery now is about $500 per kWh. Battery electric cars cost less to buy and maintain than gasoline-powered cars, but they cost more to buy and maintain. Keeping a battery electric car saves owners an extra $1000 a year [1].

For one thing, electric vehicle batteries need to be charged often. They also need to be charged for a long time, which makes the EV unfeasible for some people. A single battery can be charged in a half hour to a few hours, depending on the power capacity of the charger that is connected to it. This is a lot longer than the process of filling up a car with gasoline. If the battery is dead, electric vehicles can no longer be ready to go. Businesses will make sure that they still have access to an outlet and a charging cord if they try to get around this. To add to (numbers), this is also one of the problems that comes about when you don't plug in. People run out of battery power when they don't plug in. If there are wires on the floor, they could be dangerous to walk on. There is a chance someone could get stuck. The home's occupants and the property itself could be in danger if old wires, which are more likely to break in cold climates, are not replaced. People also have to deal with bad weather, like wind, rain, ice, or snow, in order to turn on the power. As a result, the chance of getting an electric shock goes up. To be clear: [3] As people who own electric cars, they want wireless power transfer (WPT) technology because it makes charging easier. Wireless charging was as easy as moving power from one electric car to another. People only have to park and leave when they use a WPT in a certain place. It is possible for a vehicle to run while it is moving, which means it can keep going without having to stop. A dynamic WPT system, like an EV, can do this. In addition, electric cars that use wireless charging can reduce their battery capacity by up to 20%, but cars that use conductive charging don't. However, because the energy storage unit determines the electrical drive's energy and control capabilities, it is important that energy conservation starts when parts are being made. When it comes to the reliability of a car, these choices make a big difference. In addition, their start-up and ongoing costs can also have an effect on them.

It is true that there are a lot of different ways to store energy in the world, but this post will only talk about batteries and ultra-capacitors. Batteries can store a lot of energy, but ultra-capacitors have a lot of power in a small space. In this case, a hybrid device takes advantage of both technologies [1].

In general, if an energy-based drivetrain like a hybrid has a lot of different sources of energy, it can be mounted either in a series or parallel way. Using a series arrangement, the electric motor only moves the wheels. When using the parallel arrangement, the electric motor, the internal combustion engine, or both move the wheels. This is called a "parallel arrangement". There are a lot of electric cars on the road today, but they are mostly in parallel rather than series configurations. Electric motor: Because it controls everything in a series configuration, it must be the right size for the vehicle's peak output. This is because it is in charge of everything. It might be possible to use a smaller engine because the electric motor only makes up a small part of the car's total power needs when it works with the rest of the car. However, the series configuration of a motor allows the engine to run at its best speed at all times, but it often requires an expensive process: mechanical energy from the engine is turned into electrical energy that can be stored.

It would be more difficult to drive and maintain a car that is not electric because it doesn't have enough space to store energy. People who drive hybrid electric vehicles, plug-in electric vehicles, long-range electric vehicles, and fuel cell vehicles all have a lot to do with how well they work. A lot of energy must be stored in order for the ESS to have a long service life and a low cost of ownership. This means that the ESS must be able to store a lot of energy. For now, pure battery-based ESS (like power batteries) can't meet both of these needs because it has to make some trade-offs to meet them. A hybrid energy storage system is one that combines two or more sources of energy in order to get the most out of each one's special features, which makes the ESS more efficient (HESS). With this in mind, hybridization of high-energy batteries and ultra-capacitors with complementary properties is now the most common way to get more power in today. This paper talks about how important a HESS is when factors like system power, efficiency, cost, practicality, and temperature requirements are taken into account. A HESS is better than other systems when these things are taken into account. Three main types of battery ultra-capacitor HESS are discussed:

the dormant, the semi-active, and the fully active, which are the three main types. HESS control methods that have been used before are then tested, including rule-based or reference curve-based control and fuzzy logic control, as well as fuzzy logic and closed-loop control methods. It came to an end with a new control method that focused on signal isolation through sparse coding.

It has become more and more important for cars to store energy efficiently. Most people like batteries because they can store a lot of energy, which is needed to extend the range of HEVs and EVs because they have a lot of power. As a result, there isn't a single factor (Battery) that can provide all of the desirable characteristics (low power density) on its own. Weight and cost are both cuts when the battery pack gets bigger. Small batteries are used for low energy consumption (average power), and a supercapacitor is used for very high energy consumption while going downhill or during acceleration and regenerative braking. Hybrid energy storage, also called hybridization, is made up of these two types of batteries. A supercapacitor with a high power density (a powerful source of energy) and a battery with high energy density (a good way to store energy) works together to help the device and its customers [1, 2, 4–6]. This way, HESS's different parts work together in a way that helps the device and its customers.

Because there are so many different ways to store energy, this article will only talk about batteries and ultra-capacitors. People use batteries to store a lot of energy, but ultra-capacitors can also store a lot of electricity. Hybrid systems combine the best parts of both technologies [1] to make them even better.

2.2 Implementation

Several scientific disciplines and businesses, including the automobile industry, have benefited from the usage of artificial intelligence (AI). Intelligent control is a control method that makes use of various artificial intelligence technologies such as genetic algorithms, machine learning, neural networks, and fuzzy logic to achieve its goals. A large number of real-time applications have benefited from the use of fuzzy logic controllers, which have played a significant role in the design and improvement of these applications. A fuzzy model is a system description that includes fuzzy quantities that are expressed in terms of fuzzy numbers or fuzzy sets, as opposed to discrete quantities. The right selection of the number, type, and parameters of the fuzzy membership functions and rules is critical for obtaining the required performance. The following are some examples of good selection: adaptive neuro-fuzzy inference system (ANFIS) is a system that combines the fuzzy qualitative approach with the adaptive capabilities of neural networks in order to achieve the required performance.

In Fig. 2.1, model implementation for dual battery storage for electric vehicle is shown on MATLAB Simulink.

In Fig. 2.2, ANFIS controller and SPWM controller are shown.

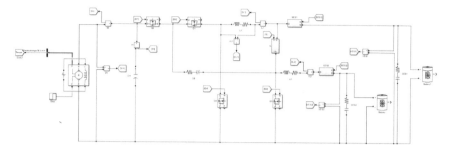

Fig. 2.1 Model implemented for electric vehicle

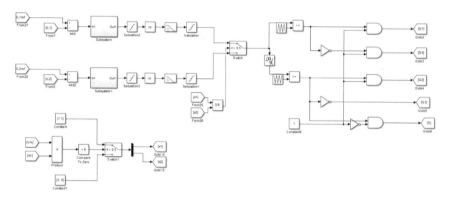

Fig. 2.2 Proposed control system for ANFIS and SPWM

In Fig. 2.3, the error and change in error input are given to fuzzy logic controller for ANFIS rule application.

In Fig. 2.4, fuzzy rule input and outputs are shown.

.

In Fig. 2.5, rules for Sugeno type are shown.

In Fig. 2.6, 5 membership functions for input are shown.

In Fig. 2.7, 5 membership functions for output are shown.

In Fig. 2.7, membership function for output BDC with HESS is shown.

In Fig. 2.8, the final ANFIS structure for bidirectional hybrid storage for electric vehicle is shown. This is proven to improve the distortions found earlier in PI controller.

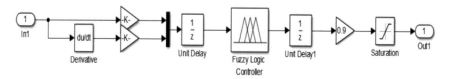

Fig. 2.3 ANFIS part

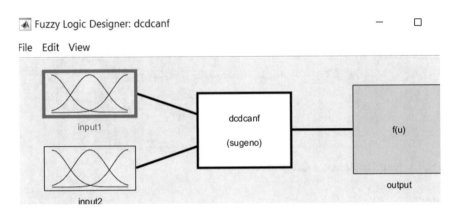

Fig. 2.4 Fuzzy rule input and output

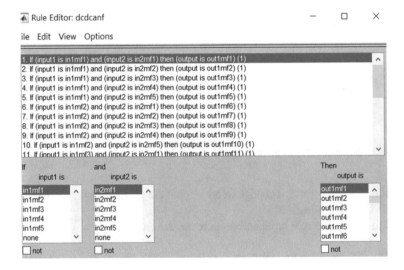

Fig. 2.5 Rules screen

2.3 Results

In Fig. 2.9, switching of gate pulses is shown.

In Fig. 2.10, battery potential and inductor current with low fluctuations are shown. It gives better in case of ANFIS controller.

Figure 2.11 shows the fluctuations are less and smoother transitions in case of mode changes.

In Fig. 2.12, the output is shown which shows better transitions and lesser distortions as compared to PI controller outputs

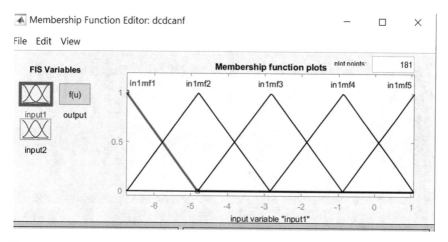

Fig. 2.6 Input membership functions

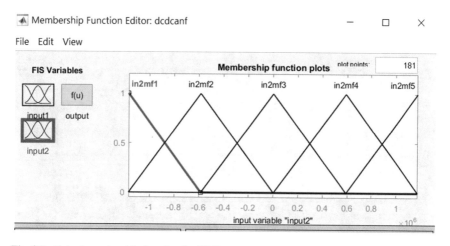

Fig. 2.7 Output membership function for BDC

In Fig. 2.13, the output transitions for inductor current are shown which give less distortions and smoother waveforms.

2.4 Conclusion

Due to their potential to drastically reduce power consumption and pollution, electric automobiles have created enormous enthusiasm. Governments and manufacturers continue to agree on new electric vehicle market objectives, while the cost of manufacturing electric vehicles continues to fall, making them more competitive with

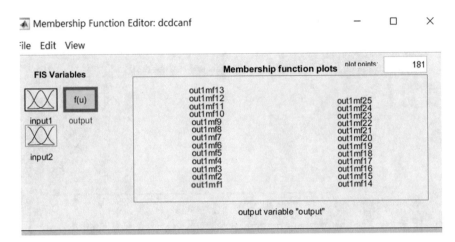

Fig. 2.8 Membership function for output BDC

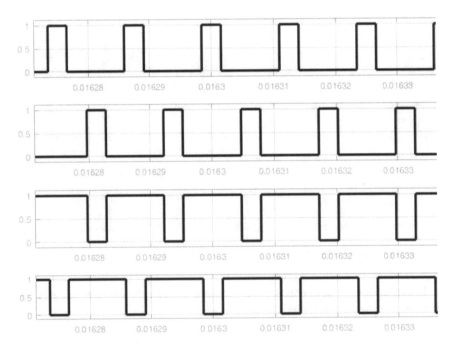

Fig. 2.9 Switching of gate pulses

internal combustion vehicles. Lithium-ion battery technology advancements have been important to the rise of electric cars, and a further move to electric driving would demand greatly improved battery output. The empirical knowledge of the exact environmental implications of electric vehicles is still evolving, and the effects

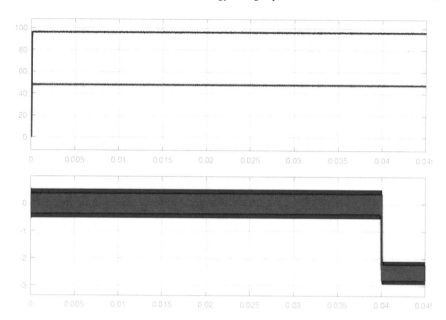

Fig. 2.10 Inductor current and battery potential

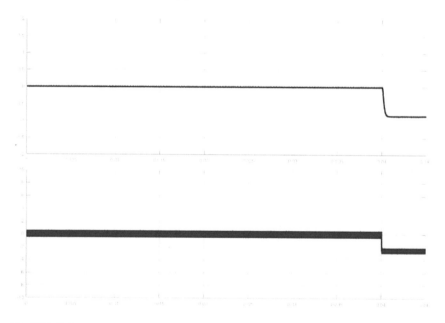

Fig. 2.11 Inductor currents

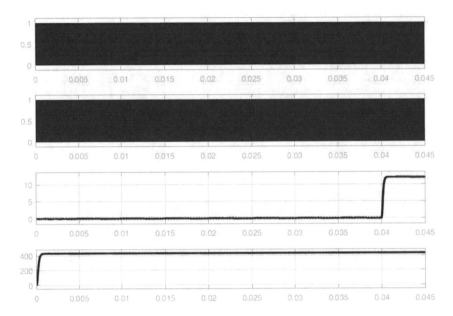

Fig. 2.12 Output waveforms

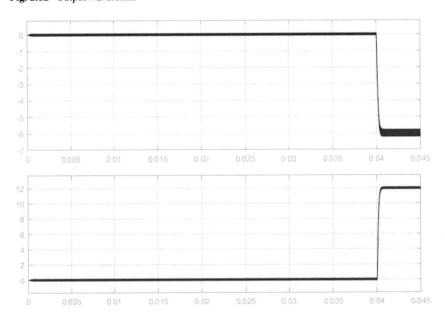

Fig. 2.13 Final inductor currents

of battery output on overall pollution produced by electric vehicles are especially challenging. Recent study on the greenhouse gas emissions related with battery manufacture has produced a wide collection of results and implications. Energy storage devices (ESS) are the brains of electric cars, since they determine their performance, strength, and driving range, among other qualities. Due to the growing demand for new electric cars, the ESS must often maintain a high energy density in addition to a high peak capacity. Nowadays, batteries and supercapacitors are often utilised as the ESS in industrial practise since they are capable of supplying considerable quantities of energy and power. Dual battery energy sources are employed in this thesis, as well as a bidirectional DC–DC converter for hybrid electric cars. Due to the variations associated with the usage of PI controllers, such systems have a shorter lifespan. The suggested study using ANFIS and SPWM demonstrates smoother transitions and less distortion, resulting in a longer lifespan and greater accuracy for such storage devices.

References

1. Singh, K., Bansal, H., Singh, D.: A comprehensive review on hybrid electric vehicles: architectures and components. J. Modern Transp. 27 (2019). https://doi.org/10.1007/s40534-019-0184-3
2. Niu, G., Arribas, A.P., Salameh, M., Krishnamurthy, M., Garcia, J.M.: Hybrid energy storage systems in electric vehicle. In: 2015 IEEE Transportation Electrification Conference and Expo (ITEC), 2015, pp. 1–6. https://doi.org/10.1109/ITEC.2015.7165771
3. Lai, C., Cheng, Y., Hsieh, M., Lin, Y.: Development of a bidirectional DC/DC converter with dual-battery energy storage for hybrid electric vehicle system. IEEE Trans. (SCOPES), 1831–1835 (2016). https://doi.org/10.1109/SCOPES.2016.7955761
4. Lu, S., Corzine, K.A., Ferdowsi, M.: High efficiency energy storage system design for hybrid electric vehicle with motor drive integration. In: Conference Record of the 2006 IEEE Industry Applications Conference Forty-First IAS Annual Meeting, 2006, pp. 2560–2567. https://doi.org/10.1109/IAS.2006.256899
5. Nielson, G., Emadi, A.: Hybrid energy storage systems for high-performance hybrid electric vehicles. In: 2011 IEEE Vehicle Power and Propulsion Conference, 2011, pp. 1–6. https://doi.org/10.1109/VPPC.2011.6043052
6. Krishna, V.V., Kumar, P.A., Chandrakala, K.R.M.V.: Development of hybrid energy storage system for DC motor powered electric vehicles. In: 2019 International Conference on Smart Structures and Systems (ICSSS), 2019, pp. 1–4. https://doi.org/10.1109/ICSSS.2019.8882838. Vehicular Technol. **67**(2), 1036–1052 (2018). https://doi.org/10.1109/TVT.2017.2763157

Chapter 3
Design of Smart Irrigation System in Sone Command Area Bihar for Paddy Crop

Md. Masood Ahmad and Md. Tanwir Uddin Haider

Abstract In Sone Command Area of Bihar, India, 80% of population relies on growing paddy crops which in turn depend mainly on supply of right quantity of water at the right time. In the old conventional system of farming, there is a large wastage of water and also causes damage to the crops due to inadequate or over supply of water. Further, currently there is no provision to handle the flooding of field due to heavy rainfall. Also, farmers are not able to accommodate themselves for changing environmental conditions such as temperature, rainfall and wind speed that dramatically affect the quantity of water needed to sustain the paddy crop. If these elements are monitored, then the crop yield will be maximized with the minimum use of water. All these problems can be solved by the help of smart irrigation system. Hence, this paper proposes an Android App smart irrigation system for paddy crops which provides a promising solution to the paddy growing farmers whose presence in the field is not compulsory during cultivation as required in conventional system. To develop the system, Raspberry Pi 4 microcontroller, water level sensors, relays and DC motors are used. Android mobile App has been developed to send all real-time information to farmers, once the farmer registered themselves through the mobile OTP. The farmer has to enter only the sowing date, after that the entire system operates automatically such as supply of required quantity of water plus checking of flooding of water in field in case of rainfall.

3.1 Introduction

In India, 70–80% of populations rely on agriculture-based economy. This profession is very popular since early civilization but even today, majority of the agricultural processes are still monitored and controlled manually. Further, there are no functional

Md. M. Ahmad (✉)
Maulana Azad College of Engineering and Technology, Patna, Bihar, India
e-mail: masood.macet@gmail.com

Md. T. U. Haider
National Institute of Technology Patna, Patna, Bihar, India

canal network systems available in majority of the areas for supplying irrigation water. Generally, either it is rainfed or groundwater is utilized through bore well. This conventional system not only causes large wastage of water but also requires the physical presence of farmers in the field.

Agriculture consumes approximately 80% of available water resources and day by day, and demands of agriculture products are increasing because of population growth. Hence, there is an urgent need to develop new strategies which optimize the water utilization as well as the agriculture demands.

Today, we live in a smart world where majority of the systems are automated but still, there are sectors like agriculture where smart systems have not been adopted yet. Smart irrigation system is basically intended to ensure the optimum quantities of water at the optimum time throughout the growth period of the crops. In this, the entire process is automated and farmers need not be physically present in the field. They can monitor it through the smart mobile phone. This provides opportunity to farmers to be involved in other income generating activities in addition to farming. Further, wastage of water is very much minimized. Hence, smart irrigation system is recommended for current practices.

This paper presents a smart irrigation system that would solve a number of problems related to irrigation and agriculture of paddy crop such as saving of irrigation water and energy, minimizing dependency on rainfall and manual intervention, protecting the plants from diseases, etc. All the above-described advantages boost to design a smart system which is a sustainable option for the overall improvement of efficiencies of irrigation as well as of agriculture. Hence, the main intention of this study is to design and implement a smart system for paddy crop in Sone Command Area, Bihar, using microcontroller board Raspberry Pie 4.

The Sone Canal Command Area lies at latitude 24° 48' N and longitude 84° 07' E as shown in Fig. 3.1. It is in the Bihar State of India and at present spread over 8 districts—Aurangabad, Patna, Jahanabad, Gaya, Bhojpur, Buxar, Rohtas and Bhabhua [1]. Approximately 1100 mm rainfall on an average occurs over the area, 80% of which is received in between June and September month, i.e., during the Monsoon season [2]. Soils are predominantly alluvial mostly clay loam [3]. In the command area, average population density is 1068 person per sq. km, and the total population was 1.89 crore as per the census of 2011 which would increase manifolds by 2050 [4]. Paddy is the major crop which is grown in the area during the Kharif (Monsoon) season [3]. It is basically a semi-aquatic plant. The production of paddy crop in this area was 844 thousand metric ton [5]. This necessitates mandatory upgradation and introduction of modern technology in the methodology of crop production both in terms of quantity and quality to meet the requirements of staple food. Further, in these areas, paddy growing farmer's incomes are reducing day by day due to many reasons such as heavy dependency on rainfall, manual labors and lack of controlled resources. Further, their potential and abilities in the agriculture sector are dropping as different enterprises attract farmers from the farming activities. Furthermore, manual labor is becoming more and more expensive, so, if no effort is made in optimizing these resources, more money will be involved with a very little return or in some cases losses also. Hence, it is an urgent need of time that

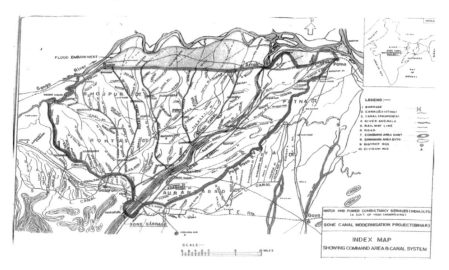

Fig. 3.1 Index map of Sone Command Area [1]

new advanced smart irrigation system must be introduced and implemented in the command area in order to optimize the irrigation system.

3.2   Literature Review

Different crops require different quantity of water during the entire growth period depending upon the type of soil and climatic conditions. Maintenance of optimal level of soil moisture is essential for maximized yield, otherwise, the crop will wilt or in worst case, it might die. Also, presence of excess water will destroy the crop.

Since last decades, large numbers of researchers are working to develop an automated irrigation system for different type of crops and they have suggested various techniques using microcontroller, different types of sensors, Android smart phone, etc. Some of their works are discussed here.

Balaji et al. [6] have developed a smart irrigation system using IOT and image processing techniques. This model uses Arduino microcontroller, Ethernet shield and moisture level and temperature sensor and the data is sent to the BLYNK Android App to monitor the operation of pump. The health of the crop is monitored through image processing technique using Raspberry Pi microcontroller and Webcam. Based on the images captured, the information is sent to the farmer regarding the health of the crop, weather it is growing normally or infected by disease. An Android-based smart irrigation system has been proposed by Hambarde et al. [7] for drip irrigation using Raspberry Pi microcontroller. The main intent of the study is to enhance the yield of the crop with the minimum use of water. Mahesh and Reddy [8] have developed an automatic irrigation system using WSN and GPRS modules through which users get SMS alert and with the information received, users can operate pump ON/OFF from any place. Ata et al. [9] have suggested Web-based model for irrigation system by using wireless sensor network and embedded Linux board. The main aim of this model is to develop a system in which control and monitor can be done from remote place through Web page. A smart irrigation has been presented by Gavali et al. [10] based on wireless soil and temperature sensor, microcontroller and radio receiver. In this, the communication link is established via ZigBee protocol and the irrigation scheduling is monitored through Android application. Abdurrahman et al. [11] have proposed sensor-based automatic irrigation management system for the area where there is a scarcity of water like Ethiopia and for crops which require less quantity of water. In this model, PIC 16F887 microcontroller, water level sensor relay interface board is used. Algorithms were developed by using C programming language. An automated irrigation sensor was developed by Jagüey et al. [12] which was assessed for pumpkin crop. In this, the sensors capture and process the digital images of the soil to estimate the water content in the soil. An Android App in Java was developed to operate directly for computing and providing connectivity of digital camera and Wi-Fi network. Their experimental results have shown that the smart mobile phone can be used as an irrigation sensor and the same could become a future tool for agriculture application. Harishankar et al. [13] have presented a solar powered smart irrigation system. In this, solar power is suggested to be used for all the energy requirements and could be used as an alternative to farmers in the present crisis of energy. This model optimized the uses of water by minimizing the wastage of water and also moderates the human interventions.

In majority of the previous studies, automated irrigation system has been developed without consideration of specific field crop. In most of the cases, it is quite generic in nature. Further, no model has been suggested, how excess water in the field will be managed in case of heavy flood. Also, Arduino-based microcontroller is used in most of the studies. In the present study, a smart irrigation system has been developed for paddy crop in the Sone Command Area (Bihar) which will be a total new concept as the same has not yet designed and implemented. Here, adequate management has been implemented for handling the flooding of field in case of heavy rainfall. For this, a pond has been proposed to take into excess water and also supply the water to the field in case of need. The entire system is fully automated and real-time information will be sent to the farmers for which latest microcontroller Raspberry Pi 4, water sensors and smart mobile have proposed to be used. Hence, this system will give a lot of benefit to the farmers, water resource managers as well as to the society.

3.3 Methodology

3.3.1 Proposed Model

A smart irrigation system having field sensors provides the best alternative solutions to paddy growing farmers as after the plantation, farmer's presence in the field is practically not required. In this smart irrigation system, microcontroller board (Raspberry Pi 4), different types of sensors, relays, DC motors and smart mobile are used. Raspberry Pi 4 microcontroller is a flexible programmable hardware platform that controls the circuit logically, in which automation process is designed by using the inbuilt library in Python programming language. The system will use microcontroller to automate the process of water pumping during both excess or deficit rainfall conditions. Different sensors are used to measure the attributes such as depth of water in the paddy field, rainfall and water level in the storage tank. A threshold value has been preset to fix both minimum and maximum value, so that the motor operates automatically in ON/OFF position as the water level in the field crosses the predefined threshold values. The microcontroller also has an LCD to display the current status of water level in the field. Hence, a sensor-based smart irrigation system is to be designed to maintain the optimum water level in the field during the entire cultivation period. This automated system will always ensure the adequate required quantity of water in the field even though the farmers are away from the site. Further, in addition, it delivers water with maximum water usage efficiency by always keeping the water level in the field at optimum level. Also, a storage pond is to be created in the field to perform dual function. Excess water collected in the field during heavy rainfall is to be directed in this pond through sensor monitored gates provided in the field and field ditches. The same collected water is to be pumped to the field as and when water is required by the crop. The overview of the system to be designed and installed in the Sone Command Area (Bihar) has been shown in Fig. 3.2.

As per the general and current agronomics practices in the command area for the paddy crop, the following conditions are considered for solving the model:

- Total durations of the crop = 120 days
- Initial depth of the water = Nil (Zero mm)
- Bund height = 150 mm
- At the transplanting date, total depth of water applied to the field including rainfall = 150 mm
- From 1 to 44th day, at the beginning of any day, if the depth of water < 10 mm, then on that day irrigation water applied to the field = 50 mm
- All the standing water and rainfall if occurs, should be drained out on 45, 46 and 47th day from the date of transplantation, for efficient supply of nutrients to the crop and effective weed control in the field.
- On 48th day, water applied to the field = 50 mm

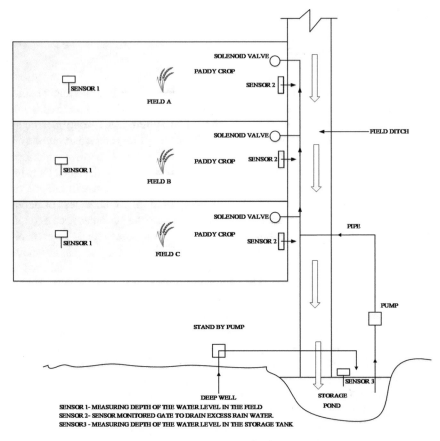

Fig. 3.2 Overview of the system to be installed in the Sone Command Area (Bihar)

- From 49 to 105th day, again at the beginning of any day, if the depth of water <
  10 mm, then on that day irrigation water applied to the field = 50 mm
- From 106 to 120th day, water applied to the field = Nil (Zero mm), as this period
  of the crop is considered as the stage of attaining maturity before harvesting.

3.3.2　Circuit Details of Proposed Model

In this section, the main circuit diagram of the proposed model is illustrated. The
circuit contains the components that are used are Raspberry Pi 4B, e-Tape Liquid
level sensor, 4-channel Relay Module and Brass Solenoid Valve. The main circuit
diagram of the model is shown in Fig. 3.3. Here, Raspberry Pi 4B has been connected
to e-Tape liquid level sensor which is then connected to MCP3008 through a voltage
divider circuit. The sensor measures water level in the field. Since the sensor gives

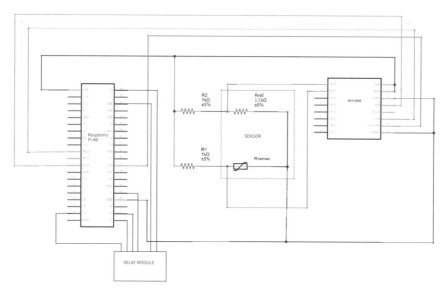

Fig. 3.3 Overall circuit diagram of the proposed model

resistive output, voltage divider circuit has been used in order to convert the resistive output into voltage. Now, since Raspberry Pi cannot read analog (voltage) signals, so, MCP3008 chip has been connected for this purpose. MCP3008 converts the voltage into digital signal which can be read by the Raspberry Pi easily. A 4-channel relay module has also been connected with the Raspberry Pi. The relay module is used here to control the opening and closing of the solenoid valves and the water pump. Here, VCC as + 5 V has been taken and 4 GPIO pins from the Raspberry Pi are used to give inputs to the corresponding 4 relays in the module. First one is connected to water pump, while the rest are connected to the respective solenoid valves in the fields. The valves and the pump are connected to Normally Open (NO) terminals of their respective relays so that they will operate only when their corresponding GPIO input is high.

3.3.3 Operation of Smart Irrigation System

Figure 3.4 illustrates the flowchart of the proposed smart irrigation model where T stands for True and F stands for False. On the first day, i.e., on the day of transplantation, the Raspberry Pi checks the water level in the field. If the sensor senses the water level below 150 mm, then the microcontroller (Raspberry Pi) turns on the pump and the valve with the help of relay module. After that day and up to the 44th day, if at the beginning of any day, the water level in the field is less than 10 mm then the Raspberry Pi microcontroller will turn on the pump until the water level reaches 50 mm, after which, it will turn off the pump. If on any day, the water level in the field is exceeding its maximum limit due to excessive rainfall or any other reasons, then the microcontroller will open the valve using the relay module while keeping the pump off, so that the excessive water drains out to the pond. On 45th, 46th and 47th day, the microcontroller will only open the valve while keeping the motor off until all the water in the field drains out (0 mm). From 48th day till 105th day, the microcontroller will again maintain the water level in the field between 10 and 50 mm. After that for the remaining days, the microcontroller will open the valve while keeping the pump off to drain out all the water in the field. During this whole period of 120 days, the microcontroller will keep on sending the status of the pump, valves and the water level in the field to the server. Thus, the database on the server will keep on getting updated. The app on the farmer's mobile will fetch data from the server and will regularly inform the farmer about the water level in the field, the status of the valves and the pump (open/closed) and will also remind him about the number of days remaining. Thus, the farmer can monitor his field from his home using the mobile app.

3.3.4 Detail of Android App

Android-based App has been developed based on one-time password (OTP). Farmers have to Login by entering their mobile number which will be verified through 6-digit OTP. After verification of OTP, the App will be operative and farmers are registered. Now, they have to enter their name and date of sowing. Once this process is done successfully, the model will start functioning and the entire activities are now controlled through microcontroller Raspberry Pi by processing the data sensed by various sensors provided in the field. Farmers will get continuous information throughout the cultivation period regarding the water level in the field and pond. The interface of App has been shown in Figs. 3.5, 3.6 and 3.7.

Figure 3.5 represents the welcome page of App, on the bottom of which, continue will appear. Farmers have to press it. Then, the second and third page of App will appear as shown in Fig. 3.6. The second page is the Login page where farmers have to enter their mobile number which will be verified through the 6-digit OTP. After successful verification of mobile number, the third page will appear and App will become functional. Now, the farmers have to enter their name and date of sowing of paddy crops through the calendar provided in the App. Once, this information is updated, the App will start receiving real-time information sent by microcontroller Raspberry Pi as shown in Fig. 3.7 regarding the water level in the field as well as in the storage pond. Also, real-time information will be available on the App whether the motor is in ON/OFF position.

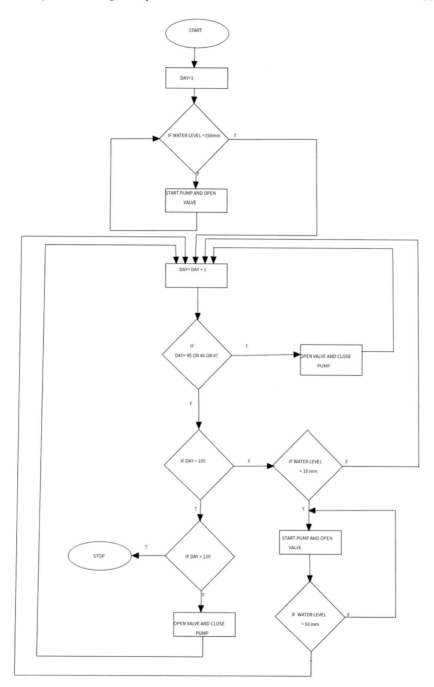

Fig. 3.4 Flow diagram of the proposed model

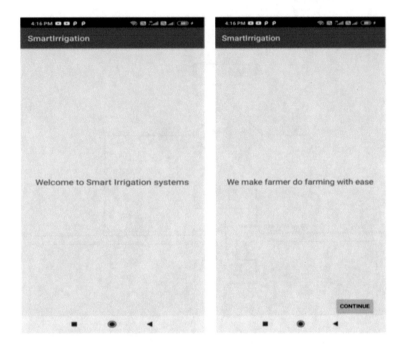

Fig. 3.5 Welcome page of App

Fig. 3.6 App login page

Fig333...7 Displaying real-time information regarding the water level in the field and storage pond

3.4   Conclusion and Future Scope

In this paper, a smart irrigation system model for paddy crop in Sone Canal Command Area, Bihar, India, is proposed and successfully implemented using Raspberry Pi microcontroller. This concept is totally a new one in this command area as the same has not yet been designed and implemented. This smart system automatically controls the entire activities once the farmers will register themselves through the Android App using OTP and entering the starting date of cultivation. After that, model will ensure the optimum supply of water to the field completely ruling out any under irrigation or over irrigation. Farmers can monitor the status of water level and other operating conditions through the mobile App from any place. Hence, this system will give a lot of benefits to the farmers, water resource managers as well as to the society. Further, this will lead to added attraction to farmers toward farming as it will enhance their income because simultaneously, they can engage themselves in other income generating activities. Because of creation of pond in the field, problems of water logging will also be solved as excess water during heavy rainfall will be collected into it and the same will be supplied when it is needed in the field. In this pond, fishing activities can also be added which will fetch additional income to the farmers. Further, spreading of fertilizers can also be introduced by using one more sensor. This sensor will ensure the uniform distribution of fertilizers in the field. Multiple cameras can also be put into the field to monitor the health of the crop through image processing techniques.

Hence, the proposed smart irrigation system will lead to overall prosperity to the farmers in the Sone Canal Command Area of Bihar.

Acknowledgements   This work was supported by World Bank Project 2038–Technical Education Quality Improvement Program (TEQIP) Phase-III Grant under Collaborative Research Scheme (CRS), funded through Aryabhatta Knowledge University, Patna, Bihar, India.

References

1. Bhuvan Homepage: www.bhuvan3.nrsc.gov.in
2. IMD Homepage: www.mausam.imd.gov.in/patna
3. Sone Canal Modernization Project Report by Water and Power Consultancy Services (India) Ltd. (WAPCOS)—1998 report
4. Census 2011 Homepage: www.census2011.co.in
5. DRDPAT Homepage: www.drdpat.bih.nic.in
6. Balaji, V.R.: Smart irrigation system using IOT and image processing. Int. J. Eng. Adv. Technol. (IJEAT) **8**(6S), 115–120 (2019). ISSN 2249-8958
7. Hambarde, H., Jadhav, S.: Android based automated irrigation system using Raspberry Pi. Int. J. Sci. Res. **5**(6) (2016)
8. Mahesh, R., Reddy, A.: An Android based automatic irrigation system using a WSN and GPRS module. Indian J. Sci. Technol. **9**(30), 1–6 (2016)
9. Ata, S.R.: Web based automatic irrigation system using wireless sensor network and embedded linux board. Int. J. Adv. Eng. Technol. Manage. Appl. Sci. **3**(2) (2016)
10. Gavali, M.S., Dhus, B.J., Vitekar, A.B.: A smart irrigation system for agriculture base on wireless sensors. Int. J. Innov. Res. Sci. Eng. Technol. **5**(5) (2016)
11. Abdurrahman, M.A., Gebru, G.M., Bezabih, T.T.: Sensor based automatic irrigation management system. Int. J. Comput. Inf. Technol. **04**(03), 532–535 (2015). ISSN 2279-0764
12. Jagüey, J.G., Francisco, G., Gándara, M.A.: Smartphone irrigation sensor. IEEE Sens. J. **15**(9), 5122–5127 (2015)
13. Harishankar, S., Kumar, R.S., Sudharsan, K.P., Vignesh, U., Viveknath, T.: Solar powered smart irrigation system. Adv. Electr. Electric Eng. **4**, 341–346 (2014)

Chapter 4
A Footstep to Image Deconvolution Technique for the Both Known and Unknown Blur Parameter

Rikita Chokshi⬥, Harshil Joshi⬥, and Mohini Darji⬥

Abstract Deblurring an image is a difficult task in modern times. It is a poorly posed problem that affects a variety of disciplines, including photography, earth space science, geophysical, medical imaging, and lens. Images become distorted as of impacting several factors, including vibration, hand movement, the launch of a vehicle (satellite), the presence of image noise, unfavorable Image or Environment conditions, and the rapid movement of objects. There is no perfect answer for all of these. Therefore, a technique is required to address the abovementioned issues and provide feasible methods for minimizing image distortion. The very first step in image restoration is the identification of blur. Then, based on the deconvolution technique, several steps are carried out. Mainly two types of classification for image deconvolution: Blind and Non-Blind Deconvolution. If restoration uses non-blind deconvolution, then the first image is degraded by any blur from various available sources, and then a technique is applied to restore an image. On the other hand, if the restoration of an image uses blind deconvolution, then blur types and parameters of blur are estimated and then, based on an appropriate estimation technique, is applied for image restoration. Lastly, analysis and comparison of the resultant output image analysis and comparison are carried out, and performance is measured through Structural Similarity Index and Peak Signal-to-Noise Ratio.

4.1 Introduction

Obtaining a clear and good quality image has been a challenging task for decades as the way of capturing an image is not so perfect, and under excellent conditions, the captured image undergoes degradation. When a digital camera is used to capture

R. Chokshi (✉)
ChanduBhai S Patel Institute of Technology and Engineering, Charotar University of Science and Technology, Changa, Anand, India
e-mail: Chokshi.rikita@gmail.com

H. Joshi · M. Darji
Devang Patel Institute of Advance Technology and Research, Charotar University of Science and Technology, Changa, Anand, India

images because of the relative fluctuation between any object and camera or camera defocusing will cause a degraded image, resulting in clarity, reduced contrast, and edge sharpness. Also, it is very strenuous to identify, discover and track the earmark from a blurring image. These causes are mainly due to two factors. One is noise which can be defined as randomness in nature, and the other is a blur which can be defined as some loss in image or corrupting the image. The main objective is to acquire a deblurred image.

The image does not have any noise. The process of retrieving good quality blurred and noisy images is known as Image deblurring. It has several applications such as chromatography, detector, recorder, medical images, astronomical images, or occurring circumstances where multiple images of the similar cannot be captured. In such kinds of images, degradations are unavoidable due to various factors.

Blurring is one of the most common spectacles that degrade the quality of the obtained images. Blurring can be formulated by applying convolution with the latent image and PSF. It is a very challenging task to recuperate this type of degraded image. The methodology of retrieving an image into its primeval form is called the restoration of an image. The main aim behind image restoration is to get improved the quality of the degraded image and restore it as much as the closer original through the prior knowledge of its nature of degradation and accurate image. A resembled version of the original image can be formed.

4.1.1 Degradation of Image and Restoration Process

An Image is a collection of small blocks known as a pixel. It can be represented as a two-dimensional function I like follow

$$I = f(x, y) \tag{4.1}$$

Here, x and y are represented as spatial coordinates and (x, y) as a pixel. I represent the intensities of pixel or gray level value, defined as the amplitude off at any point thus (x, y). Image is defined as a digital image with the defined finite spatial coordinates values and the amplitude.

Based on problem formulation, the blurred image can be represented as $g(x, y)$ as following formula:

$$g(x, y) = h(x, y) * f(x, y) + \eta(x, y) \tag{4.2}$$

Here, degradation function is $h(x, y)$ due to which the image gets blurred or also defined as the point spread function (PSF) in some cases, $f(x, y)$ defining the source image, the * symbol indicates convolution and $\eta(x, y)$ denotes the supplement noise. The same equation can be rewritten as follows:

$$g(x, y) = \text{PSF} * f(x, y) + \eta(x, y) \tag{4.3}$$

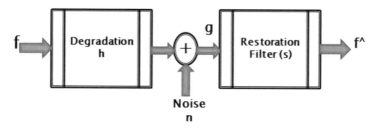

Fig. 4.1 Block diagram of image restoration

The proceeding for degraded image and restoration of the image can be proposed by the block shown in Fig. 4.1 [1]. Where $f(x, y)$ states source image and $g(x, y)$ the blurred one. In the above model, $n(x, y)$ showing an additive noise that is indicated by the system, and $h(x, y)$ is the degradation which causes image blur.

In the matrix vector form, the above equation can be written due to processing the image in digital form. Equation (4.3) in the frequency domain by the Fourier Transform can be given by:

$$G(u, v) = F(u, v)H(u, v) + N(u, v) \tag{4.4}$$

The given image with blur $g(x, y)$, $h(x, y)$ means PSF can be estimated, and the restored image will be obtained. However, it is essential and challenging to identify the true PSF because of the lack of prior knowledge.

4.1.2   Point Spread Function

PSF is known by "point spread function", relation of an imaging system to a point source. In this case, the point spread function may be individualistic of position in the object plane, so-called shift-invariant. It is the responsible factor for distortion in an image. The image of source becomes blurred and degraded due to noise by an imaging system. A point input is represented as a single pixel in the "ideal" image. The output image may then be formulated as a two-dimensional convolution of the "ideal" image having PSF (Fig. 4.2):

$$g = f * \mathrm{PSF} \tag{4.5}$$

Need of Point Spread Function

After detection of the blur type, the main concern is to decide the value of point spread function. There are two techniques listed in the paper for image restoration, so when the blind deconvolution is concerned, no matter about the type of blur, amount of blur is available. Hence, there is the primary role of point spread function.

Fig. 4.2 Blurred image by
PSF

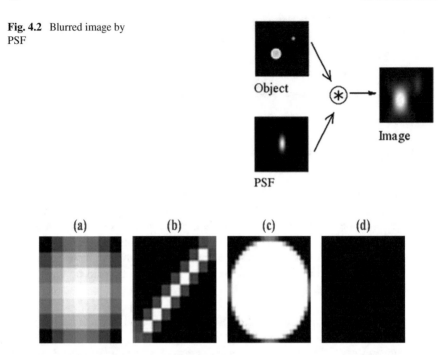

Fig. 4.3 Several types of PSF **a** Gauss ian PSF, **b** Motion PSF, **c** Disk PSF, **d** Average PSF

What are the parameters of blur in the image, and what type of blur is there. The
main task is to decide it and obtain better PSNR for restored images. In the case of
the non-blind deconvolution type of PSF, parameters are known so one can directly
apply the technique to restore an original image. As shown in Fig. 4.3, several types
of PSF are there like Gaussian, motion, disk, and average. They can be with some
parameters due to which image gets degraded [1].

4.1.3 Image Deblurring Techniques Classification

Classification of Image deblurring techniques can be done mainly in two types based
on the knowledge of the point spread function, $H(u, v)$. (i) Non-blind image deconvolution and (ii) Blind image deconvolution as shown in Fig. 4.4. We know the point
spread function and blur image in the first type. We do not have information regarding
PSF and blurred images in the second type. So it is used when we do not have prior
knowledge of PSF, causing blur and process used for degradation [2].

Blind deconvolution is most useful in real-life situations as knowing the PSF is not
possible in most practical cases; for example, in remote sensing and astronomy applications, it is very tough and challenging to estimate the scene, which is something
we have never seen before.

Fig. 4.4 Image deblurring techniques

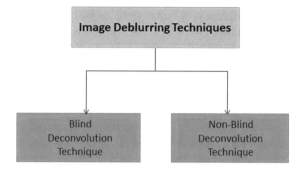

Table 4.1 Blind deconvolution and non-blind deconvolution

Parameter	Blind deconvolution	Non-blind deconvolution
PSF, type of blur	Not known	Known
Approach	Iterative	Non-iterative
Image restoration	First estimation of blur type and parameter is done then restoration technique is applied	Based on the blur present in an image, restoration technique is applied

Table 4.1 shows the difference of the blind and non-blind deconvolution approaches to image restoration.

In this paper, the first survey is carried out regarding image restoration, basic classification, existing algorithms for recovering images, various types of blurs, methodology and algorithms to detach the blur from an input blurred image and various methodology for the same. The overall architecture of image restoration is also shown and discussed.

4.2 Literature Survey

P. Ghugare et al. have proposed various restoration techniques for restoring the original image from the blurred image using a blind deconvolution algorithm for different PSFs. He has shown the performance results by the Gaussian blur, motion blur, and average blur [3]. S. Motohashi et al. have proposed an innovative approach for PSF estimation for blind image deconvolution by utilizing total variation regularization, the gradient reliability map and a shock filter [4]. A. Patel have proposed a method that uses the segmentation of images in different objects. Also, she has used the same class training image data for more accuracy [2]. S. Agarwal have proposed a method for restoring the degraded MRI image to obtain primary undegraded images using both deconvolution methods. They also concluded that the blind deconvolution technique is more practically and experimentally respectively [5]. S. Derin et al.

have proposed a novel total variation-based blind deconvolution methodology that simultaneously estimates the recovered image, the blur, and the hyper-parameters of the Bayesian formulation [6]. O. Whyte et al. have proposed a method for deblurring not many saturated and Shaken images. They have developed an approach to the non-blind deblurring of images that are degraded by camera movements and suffering from saturation. They have analyzed the characteristics and causes of "ringing" artifacts in images without blur as they apply to saturated images [7]. Mane et al. have discussed approaches for image restoration technique and also shown canny edge detector method for deblurring [8]. Xue-fen Wan et al. have shown PSF blur parameter estimating noisy out-of-focus blur image restoration and usage of FIR filter for the same [9]. Bassel Marhaba et al. have proposed the restoration of images using a combination of both approaches for deconvolution. They have also discussed various image restoration techniques [10]. R. Chokshi et al. have proposed a blind deconvolution technique for image restoration by identifying and estimating blur. They also have proposed an approach for the same [1]. P. Patil et al. have compared both blind deconvolution techniques and the flow of them as well [11].

Xuan Mo et al. have e proposed an iterative strategy based blind deconvolution method which is based on PSF constraints. They have concluded that the PSF constraints Iterative blind deconvolution method performs more better when adding more the PSF constraints that is on PSF calculated from it, to ensure that the changed PSF will lead to a particular form [12]. S. A. Bhavani et al. have given an idea about the convolution, what exactly it is and what convolution is and the importance of deconvolution in deblurring an image for removing the noise present in it [13]. M. Kalpana Devi et al. have shown various blind and non-blind deconvolution technique and then made analysis from that. They show that for the blind deconvolution approach, Richardson-Lucy algorithms give good results compared to other algorithms, and blind deconvolution algorithms work better for image restoration than non-blind techniques. They have shown results for deconvolution using the Richardson-Lucy algorithm and regularized filter under non-blind deconvolution and for blind deconvolution restoration using wiener filter [14].

Too much research has been done in this field of image processing and restoration. Based on known and unknown blur parameters, classification is done for image restoration technique: (i) Blind and (ii) Non-Blind Deconvolution as discussed. Restoration can be done quickly in the case of non-blind deconvolution in which blur type is known, but the main challenge is when blur is not known in the degraded image. So, the restoration of an image with unknown blur and parameters is known as Blind Image Restoration (Deconvolution). There are many applications in medical imaging, remote sensing, astronomical imaging, etc.

Further, the restoration of Blind image deconvolution methods are subdivided into two main groups as follows:

- One that estimates PSF, a priori self-sufficient of the correct image. Algorithms that fall in this category are less complicated computationally.
- Those which estimate both the PSF and the correct, accurate image simultaneously.

Algorithms in this computationally more complex category can be used in more general situations.

Blind image deconvolution (BID) is a challenging problem as to obtaining a correct image from a distorted image, using priorly knowledge of the actual original image (called the point spread function (PSF)), which causes the blur is very tough in the entire process. It is tough to calculate the PSF a priori, making Blind image deconvolution. It is predicted that the PSF is priorly known to restoration in traditional restoration techniques. So, the technique is to find the inverse of the process using frequency domain techniques with some regularization to reduce the noise amplification, known as the technique of non-blind deconvolution, whereas in the blind deconvolution technique, guessing of PSF is done. After a random guess, PSF is decided and applied to restore an image based on the trial–error method, thus becoming time-consuming.

4.3 Overall Architecture of Image Restoration

As shown in Fig. 4.5, image restoration takes place in real-time. The first step is to acquire an input image from the source. If the image is not blurred, we must apply blur and make blurred images. The second step is to identify the type of blur, like Gaussian, motion, average, etc., from an input image and estimate the parameter based on the identification of blur type. After that, if everything is known priorly, apply the non-blind deconvolution technique [15]. Otherwise, apply the blind deconvolution approach [3, 4] for image restoration as it is the process of recovering an input image; one needs to compare the results and measure the performance of an output image using PSNR and SSIM. This paper discusses various techniques for both image resto- ration approaches. One can apply the appropriate image restoration technique based on the parameter. Various linear and nonlinear filters may be used to recover the degraded source image as well.

4.4 Conclusion and Future Work

This paper demonstrates several image restoration algorithms using blind deconvolution and non-blind deconvolution approaches and a beneficial overall architecture for the image restoration process. One can get a clear insight into it and follow the same process when doing image restoration. Various methods and algorithms are available to restore blurred images and improve image quality. It is studied that the benefit of using a deblurring image algorithm is removing the blur or recovering the distorted image with priorly available parameters and information of PSF and noise for a very good image restoration process. The difference between blind and non-blind deconvolution based on the various parameter is also shown. According to that,

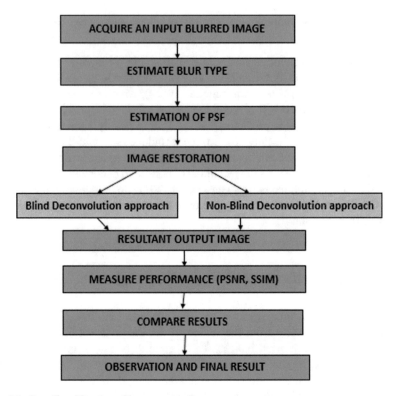

Fig. 4.5 Overall architecture of image restoration

blind deconvolution is an iterative process; each output is fed back to its succeeding input, and this way, a restored image is obtained. It is also that compared to non-blind techniques, the blind deconvolution technique gives good results. Image quality matters a lot when doing image restoration, so it is better to use a blind image deconvolution approach for good quality. With every iteration, it yields a more promising result. From generic techniques discussed in this paper, the non-blind deconvolution technique requires more iteration for yielding better results, while blind deconvolution produces better results. So, in future work, post-processing algorithms can be implemented. Then the number of iterations can be reduced, used for algorithms and estimation of blur parameters can be made for different types of blurs.

References

1. Chokshi, R., Israni, D., Chavda, N.: Proceedings of IEEE International Conference on Recent Trends in Electronics, Information & Communication Technology, RTEICT 2016 (2017)
2. Suthar, A.C., Patel, A.: A Survey on Image Deblurring Techniques Which Uses Blind Image Deconvolution. Int. J. Res. Anal. Rev. **5**, 255 (2018)

3. Jayapriya, P., Chezhian, R.M.:A Study on Image Restoration and its Various Blind Image Deconvolution Algorithms. Int. J. Comput. Sci. Mob. Comput. **2**, 273 (2013)
4. Motohashi, S., Nagata, T., Goto, T., Aoki, R., Chen, H.: 2018. A study on blind image restoration of blurred images using R-map. Int. Work. Adv. Image Technol. IWAIT **2018**, 1 (2018)
5. Agarwal, S., Singh, O.P., Nagaria, D.: Biomed. Pharmacol. J. **10**, 1409 (2017)
6. Babacan, S.D., Molina, R., Katsaggelos, A.K.: Eur. Signal Process. Conf. 2164 (2007)
7. Whyte, O., Sivic, J., Zisserman, A.: Int. J. Comput. Vis. **110**, 185 (2014)
8. Mane, A.S., Pawar, M.M.: Int. J. Innov. Res. Adv. Eng. **1**, 2349 (2014)
9. Wan, X.F., Yang, Y., Lin, X.: Proc. 2010 IEEE Int. Conf. Softw. Eng. Serv. Sci. ICSESS 2010 **2**, 344 (2010)
10. Marhaba, B., Zribi, M., Khodar, W.: Int. J. Eng. Res. Sci. **2**, 225 (2016)
11. Patil, P.U., Lande, D.S.B., Nagalkar, D.V.J., Nikam, S.B.: Int. J. Recent Technol. Eng. **9**, 73 (2020)
12. Bhavani, S.A.: Int. J. Sci. Res. **4**, 194 (2013)
13. Mo, X., Jiao, J., Shen, C.: Lect. Notes Comput. Sci. (Including Subser. Lect. Notes Artif. Intell. Lect. Notes Bioinformatics) **5916 LNCS**, 141 (2009)
14. Dihingia, M.M., Ranadev, M.S., Rani, A.S.: Int. J. Eng. Trends Technol. **10**, 130 (2014)
15. Charu, J., Chugh, A., Yadav, S.: Int. J. Sci. Res. Dev. **7**, 722 (2019)

Chapter 5
Secured Monitoring of Unauthorized UAV by Surveillance Drone Using NS2

Priti Mandal⬿, Lakshi Prosad Roy⬿, and Santos Kumar Das⬿

Abstract Rapid increase in the Unmanned Aerial Vehicles (UAVs) or drones led to its wide application in all the sectors. This makes the situation critical and demands a proper monitoring system to keep an eye on the UAVs in the particular area. In this paper, work is taken up on monitoring/tracking of unauthorized UAV by the surveillance drone using proposed tracking algorithm in NS2 platform. In addition to this, a cryptographic algorithm is proposed to transferred the data of the tracked UAVs to the ground base station with proper routing protocol.

5.1 Introduction

Enormous application of UAVs makes it suitable for both civilian as well non-civilian sector [1]. For any emerging technology, along with its advantages disadvantages are also needed to be dealt. Easy accessibility and low cost create an alarming situation of using UAV inappropriately. So, a proper monitoring system is required towards the unauthorized UAV. In order to do so, swarm of UAVs could be used to determine the presence of unauthorized UAV in particular area.

There are several research works in which network simulator is used to analyze the performance of the Flying Ad hoc Network (FANET). FANET can be deployed in different environment for both air-to-air communication and air-ground communication [2]. While communicating proper routing protocol is to be used. In [3], different routing protocols such as Ad hoc on Demand Distance Vector (AODV) and Destination-Sequenced Distance-Vector (DSDV) are analyzed using NS2 platform.

P. Mandal (✉) · L. P. Roy · S. K. Das
National Institute of Technology, Rourkela, Odisha 769008, India
e-mail: pritimandal2310@gmail.com; 519ec1003@nitrkl.ac.in

L. P. Roy
e-mail: royl@nitrkl.ac.in

S. K. Das
e-mail: dassk@nitrkl.ac.in

© The Author(s), under exclusive license to Springer Nature Singapore Pte Ltd. 2023
C. So-In et al. (eds.), *Information Systems for Intelligent Systems*, Smart Innovation, Systems and Technologies 324, https://doi.org/10.1007/978-981-19-7447-2_5

For the swarm of UAVs, particular topology is to be maintained for proper communication. Different topologies are based on the mobility model of the FANET. In
[4], different mobility model such as random way-point mobility model, random
movements, Gauss-Markov, etc., are explained briefly.

Like other networks, while communicating or exchange of information within
the aerial nodes in FANET security is to be ensured. The data should be secured
from the attacks such as hacking and spoofing. In order to have a proper monitoring
system for unauthorized UAV in a specific area, data is to be encrypted for secured
information transmission. There are different traditional cryptographic techniques
[5] such as AES, DES, Blowfish, and Two-fish. For the high-speed processing of the
UAVs, novel cryptographic technique is required with much more efficiency.

So, here in this paper, work is taken up on continuous tracking of unauthorized
UAV after detection and transmitting the information to the ground base-station
securely in NS2 platform.

The important advantages of the proposed method in secured monitoring of
intruder UAV are summarized as follows:

- After detecting the unauthorized UAV, continuous location of it is tracked using
  proposed tracking algorithm and send to the ground base station for further
  processing.
- Transmitted data includes the location of the unauthorized UAV, which are
  securely transferred with a newly proposed cryptographic technique.

The rest of the paper is arranged as follows. Section 5.2 contains the description
of the system model. In Sect. 5.3, proposed methodology is explained to track the
unauthorized UAV with proper routing protocol along with the proposed encryption
and decryption algorithm for transmitting the packets securely using secret key.
Section 5.4 contains the simulation results of the proposed technique in the NS2
platform. Finally, in Sect. 5.5 concludes the work.

5.2   System Model

Proposed system model is depicted in Fig. 5.1. It comprises of aerial UAV nodes and
ground base station. Proposed algorithm is used to track the unauthorized UAV and
proper routing protocol is used to transfer data among the swarm of UAVs and to
the ground station. After detection of unauthorized UAV, its location is continuously
sent to the base station for further processing and action. The packets are encrypted
with the secret key which is explained in Sect. 5.3.

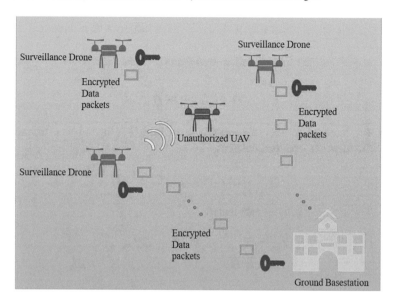

Fig. 5.1 Proposed system model

5.3 Proposed Methodology

The proposed methodology consists the explanation of the proposed tracking method used along with the proposed encryption and decryption techniques with proper routing protocol.

5.3.1 Proposed Tracking Method

The monitoring of UAV in general is done with camera, radar, LiDAR, etc. [6]. In this work, for monitoring an algorithm is proposed by considering the continuous hovering and movement of the UAV. It is assumed in Kalman filter [7] that the present state is evolved from the previous state. The present state could be represented as,

$$Y_t = AY_{t-1} + w_t \tag{5.1}$$

where A is the state transition matrix, t is the current state and $(t-1)$ is the previous state, and w_t is the system noise. The observation state measured from the sensor is represented as,

$$Z_t = GY_t + u_t \tag{5.2}$$

where G is the measurement matrix, and u_t is the measurement noise.

For estimation the priori state and covariance matrix can be represented as,

$$Y_{t/t-1} = AY_{t-1} + w_t \tag{5.3}$$

$$C_{t/t-1} = AC_{t-1} + MQ_{t-1}M^T \tag{5.4}$$

where M is the total number of predicted value, and Q is the noise covariance. The gain of the filter can be represented as,

$$K_t = C_{t/t-1}G^T(GC_{t/t-1}G^T + R)^{(-1)} \tag{5.5}$$

The updated state can be represented as,

$$Y_{t/t} = Y'_{t/t-1} + K_t(Z_t - GY_{t/t-1}) \tag{5.6}$$

where

$$Y'_{t/t-1} = Y_{t/t-1} + \Delta Y_{t/t-1} \tag{5.7}$$

$\Delta Y_{t/t-1}$ determines the direction of the UAVs movement which follows the principle of Dragonfly Algorithm [8].

$$\Delta Y_{t/t-1} = sS_i + vV_i + eE_i + fF_i \tag{5.8}$$

where S represents the separation between the drones while monitoring the unauthorized drone to avoid collision among themselves, V represents the tuned velocity of the swarm of UAVs to work in alignment toward the task, E and F represents the distance which is to be maintained by the UAVs to find the unauthorized UAV. The updated covariance can be represented as,

$$C_{t/t} = (I - Y_tG)C_{t/t-1} \tag{5.9}$$

Algorithm 1 Proposed tracking method

---

1. Initialize Parameters: Y_0, C_0, A, H, Q, R, M
2. **Repeat** each cycle t
3. Estimate current state $Y_{t/t-1}$ at t based on previous state $t-1$
4. Estimate the error covariance $C_{t/t-1}$ based on previous covariance
5. Compute the filter gain
6. Correct the state using (5.6)
7. Update the state using $Y'_{t/t-1}$
8. Update the error in covariance using (5.9)
9. Replace the previous information with the updated information of the unauthorized drone
10. **Until** $t = timeout$

Next sub-section explains about the routing protocol used for entire process.

5.3.2 Routing Protocol

Based on the system model, it could be observed that different kind of interaction between the nodes are to be maintained, interaction among the UAV nodes in the sky and UAV nodes to ground base station. Ground base station is fixed in this scenario which works as a reliable backbone of the entire system. It indicates heterogeneous routing is required to adapt. Le et al. [9] proposed a technique Load Carry and Deliver Routing (LCAD) which is used to enhance the connectivity between the UAV and the ground base station. It uses Disruption Tolerant Network (DTN) in the sky while for the ground base-station Ad hoc On Demand Distance Vector (AODV).

5.3.3 Proposed Encryption and Decryption Algorithm

The proposed encryption algorithm is the improved and more secured version of traditional AES algorithm. In the proposed algorithm, initially the data to be encrypted is considered as plain text of 128-bits. Then, the plain text is divided into four blocks and 4 × 4 state matrix. The add round key is obtained by 128-bits of state XOR with the Round Key which is a transformation of the Cipher Key. Followed by the substitution of the bytes which are arranged randomly and the order is stored in the look-up table, i.e., S_{box}. The next step is to Shift Rows. As 128-bits of plain text is placed in 4 × 4 matrix then the shift operation is performed for the 4 rows. First row remains intact, second row of the matrix moved circularly toward left once, third row is circularly shifted toward left twice, and the fourth row is circularly relocated toward left three times. In the Mix Column state, the 4 columns are combined in a reversible way which could be accessed back. This could be considered as the matrix multiplication. The steps are repeated for $N - 1$ times, i.e., here $N = 10$ as plain text is considered to be 128 bits. This may vary according to the number of bits. In the last round, i.e., Nth round Mix Column step is not considered only three steps are there as shown in the Fig. 5.2. After the AddRound Key in the last stage, the bits are crossover with the random crossover key and the output is further mutate after a constant number of bits. The remaining bits after the crossover of the AddRound Key output and crossover key is consider as the secret key. This secret key makes it more secured and robust. This proposed encryption technique makes it secured than the symmetric encryption method and faster than the asymmetric method.

The decryption method is depicted in Fig. 5.3. In this, the encrypted text is first de-mutate for the particular bits. Then using the secret key crossover is performed. Similar to the encryption technique four stages—Inv Shift Rows, Inv SubByte, Inv Mix Column, and AddRound Key are performed for $N - 1$ times using the round key

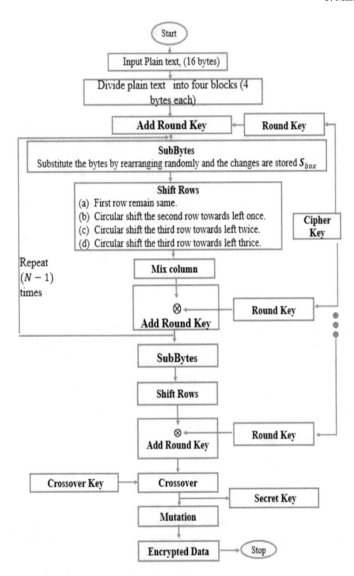

Fig. 5.2 Proposed encryption technique flowchart

as shown in the flowchart. In the last round, Inv Mix Column stage is not considered.
Finally, the plain text is obtained.

Fig. 5.3 Proposed
decryption technique
flowchart

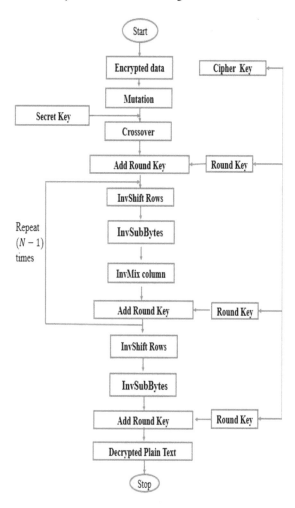

5.4   Simulation Results

In this section, proposed algorithm is analyzed in the NS2 environment. For tracking the unauthorized drone protocol is used. For continuously updating the location of the drones in the base station, information is sent in the packets securely. The parameters for the simulation are as follows: Number of nodes = 4, Wireless channel, LCAD protocol, Directional Antenna, 3 J of node energy, 0.175 W transmission, and reception power with Random-Way point movement model [10].

Monitoring of nodes and data transfer in the NS2 platform can be observed in NAM file as depicted in Fig. 5.4. The surveillance drones are represented in green color and unauthorized/intruder UAV is represented in red color with ground base station in blue. Figure 5.5 depicts the intruder UAV position and speed using proposed

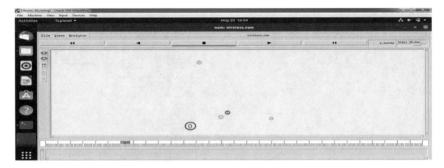

Fig. 5.4 Tracking of unauthorized UAV in NS2 NAM file

tracking algorithm in NS2. The data of the intruder UAV is encrypted and transferred and then decrypted using the proposed algorithm depicted in Fig. 5.6.

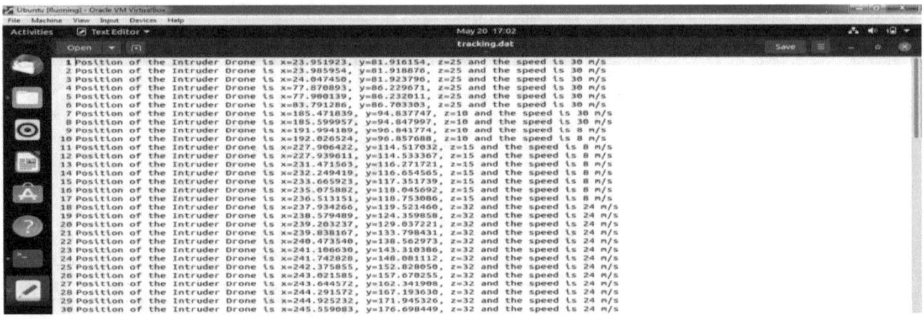

Fig. 5.5 Unauthorized/Intruder UAV position and speed from proposed tracking algorithm

Fig. 5.6 Output of proposed encryption and decryption algorithm

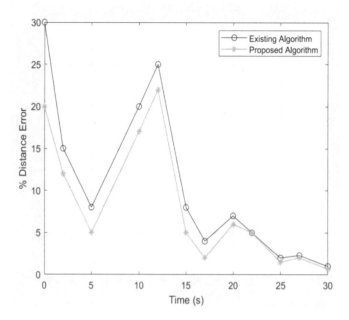

Fig. 5.7 Tracking distance error

Figure 5.7 depicts the distance error computation while tracking unauthorized UAV. Proposed hybrid method performs with much more accuracy as compared to the existing algorithm. The packets are transferred by encrypting it using proposed algorithm. The performance of the proposed algorithm analyzed using throughput.

Figure 5.8 represents the throughput of the proposed encryption algorithm. Greater throughput of the proposed algorithm represents its higher performance. Figures 5.9 and 5.10 depicts the encryption and decryption time for various packet sizes. As the proposed algorithm takes lesser time for encryption and decryption, it is more efficient than other algorithms. Proposed cryptographic algorithm outperforms as compared to the conventional methods.

5.5   Conclusion

In this paper, continuous monitoring of unauthorized drone using surveillance UAV is simulated in NS2 platform. The proposed monitoring algorithm performs better as compared to the existing algorithm which is analyzed in terms of distance error. The information about the unauthorized/intruder drone is securely transmitted using newly proposed cryptographic algorithm which performs better in terms of throughput which makes it more efficient, secured, and faster.

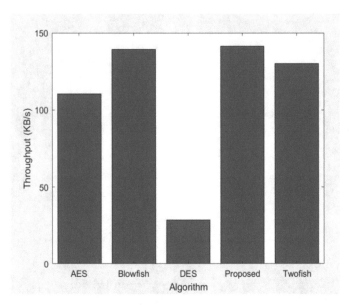

Fig. 5.8 Throughput comparison

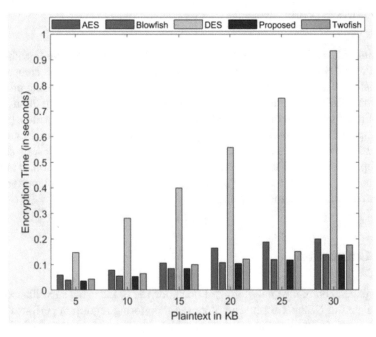

Fig. 5.9 Encryption time comparison

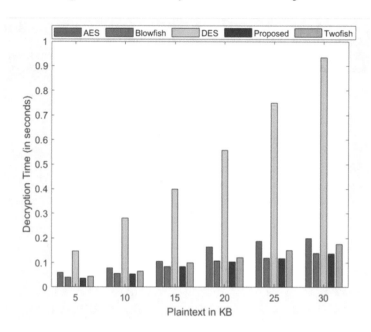

Fig. 5.10 Decryption time comparison

References

1. Mandal, P., Roy, L.P., Das, S. K.: Internet of UAV mounted RFID for various applications using LoRa technology: a comprehensive survey. Internet Things Appl. 369–380 (2022)
2. Azari, M., Sallouha, H., Chiumento, A., Rajendran, S., Vinogradov E., Pollin, S.: Key technologies and system trade-offs for detection and localization of amateur drones. IEEE Commun. Mag. **56**(1), 51–57 (2018)
3. Singh, K., Verma, A.K.: Experimental analysis of AODV, DSDV and OLSR routing protocol for flying adhoc networks (FANETs). In: IEEE International Conference on Electrical, Computer and Communication Technologies (ICECCT), pp. 1–4 (2015)
4. Mowla, M.M., Rahman, M.A., Ahmad, I.: Assessment of mobility models in unmanned aerial vehicle networks. In: 2019 International Conference on Computer, Communication, Chemical, Materials and Electronic Engineering (IC4ME2), pp. 1–4 (2019)
5. Sohal, M., Sharma, S.: BDNA-A DNA inspired symmetric key cryptographic technique to secure cloud computing. J. King Saud Univ.-Comput. Inf. Sci. **34**(1), 1417–1425 (2022)
6. Sie, N.J., Srigrarom, S., Huang, S.: Field test validations of vision based multi-camera multi-drone tracking and 3D localizing with concurrent camera pose estimation. In: 2021 6th International Conference on Control and Robotics Engineering (ICCRE), pp. 139–144 (2021)
7. Nanda, S.K., Bhatia, V., Singh, A.K.: Performance analysis of Cubature rule based Kalman filter for target tracking. In: 2020 IEEE 17th India Council International Conference (INDICON), pp. 1–6 (2020)
8. Amaran, S., Madhan Mohan, R.: An optimal multilayer perceptron with dragonfly algorithm for intrusion detection in wireless sensor networks. In: 2021 5th International Conference on Computing Methodologies and Communication (ICCMC), pp. 1–5 (2021)

9. Le, M., Park, J.S., Gerla, M.: UAV assisted disruption tolerant routing. In: Proceedings of the IEEE Conference on Military Communication (MILCOM), pp. 1–5 (2006)
10. Khan, U.S., Saqib, N.A., Khan, M.A.: Target tracking in wireless sensor networks using NS2. In: Smart Trends in Systems, Security and Sustainability, pp. 21–31, Springer, Singapore (2018)

Chapter 6
Design and Implementation of Machine Learning-Based Hybrid Model for Face Recognition System

Ramesh Chandra Poonia⬤**, Debabrata Samanta**⬤**, and P. Prabu**⬤

Abstract Face recognition technologies must be able to recognize users' faces in a chaotic environment. Facial detection is a different issue from facial recognition in that it requires reporting the position and size of every face in an image, whereas facial recognition does not allow for this. Due to their general similarity in look, the photographs of the same face have several alterations, which makes it a challenging challenge to solve. Face recognition is an extremely challenging process to do in an uncontrolled environment because the lighting, perspective, and quality of the image to be identified all have a significant impact on the process's output. The paper proposed a hybrid model for the face recognition using machine learning. Their performance is calculated on the basis of value derived for the FAR, FRR, TSR, ERR. At the same time their performance is compared with some existing machine learning model. It was found that the proposed hybrid model achieved the accuracy of almost 98%.

6.1 Introduction

A few examples of areas where identification and authentication methods have become essential technologies are building access control, computer access control, day-to-day activities such as withdrawing money from a bank account or dealing with the post office, and the well-known field of criminal investigation: Here are a few suggestions to get you started [1]. As a result of the increasing need for reliable personal identification in computerized access control systems, biometrics has received more attention in recent years. Through the use of biometric identification, it is possible to identify or authenticate a person based on a physical characteristic

---

R. C. Poonia · P. Prabu
CHRIST (Deemed to be University), Bangalore, India
e-mail: prabu.p@christuniversity.in

D. Samanta (✉)
RIT Kosovo (A.U.K), Rochester Institute of Technology—RIT Global, Dr. Shpetim Rrobaj, Germia Campus, Prishtina 10000, Kosovo
e-mail: debabrata.samanta369@gmail.com

C. So-In et al. (eds.), *Information Systems for Intelligent Systems*, Smart Innovation, Systems and Technologies 324, https://doi.org/10.1007/978-981-19-7447-2_6

or personal characteristic. In order for a biometric identification system to be termed "automatically" capable of detecting a human characteristic or trait quickly, there must be little or no user involvement [2]. Since its introduction, biometric technology has been used by security and law enforcement organizations to identify and track individuals. Most importantly, biometric technology's ability to identify and secure a person is its most crucial characteristic. The behavioral and physical aspects of a person's biometrics may be classified into a number of different categories. A broad range of activities and patterns, such as signatures and typing, are used as behavioral biometrics in a variety of applications [3, 4]. Physical biometric systems, which employ the user's eye, finger, hand, voice, and face traits to validate their identify, are used to authenticate an individual's identity in many situations. Recognition Systems Inc. has created a biometric-based system for use in government.

The following is a general outline of how to go about it: Use a database of faces to recognize one or more individuals in a scenario shown in still or video images. Unsupervised face recognition programs may be used in a wide variety of situations, ranging from highly controlled environments to completely uncontrolled ones. In a controlled environment, individuals are photographed in various poses, including frontal and profile photos, against a constant background and in similarly positioned postures [5]. Mug shots are images of the person's face that are taken in public places. With the help of cropping software, it is possible to obtain a canonical face image from each mug shot. When making a canonical face image, all characteristics of the face, including its size and placement, are standardized to a certain extent, and the background is limited to a bare minimum [6, 7]. It is difficult to control the environment in which individuals do routine activities such as face recognition and other similar jobs.

6.2   Problem Definition

A prospective worldwide notification on the topic may be made in the manner described below: Recognition of many guys in a context is accomplished by using a database of recorded faces and a video or still picture of the scene. If the environment in which face recognition software functions is well-managed, it will provide a broad variety of options for its operation. In a restricted setting, profile and frontal photos with uniform backdrops and equivalent poses on the list of participants have been approved. This kind of face photographs are sometimes referred to as mug shots. The size and location of their facial skin have been standardized in accordance with the aforementioned criteria, and the backdrop area is theoretically vast [8]. They have used a canonical face image to achieve this. Because of the completely uncontrolled environment in which humans live, they are capable of performing generic facial recognition. Identifying faces in photos are essential for face recognition algorithms to function properly. It is one of the initial phases in the face detection process to determine the average size and position of the individuals who are present in the room. Maintaining control over the identification of the faces in the given photo may be difficult. When there are a number of different photos of the same face, as well as facial shapes that are similar, it might be difficult to identify the person [9, 10]. When attempting to recognize someone's face in a setting where the surroundings is uncertain, there are several difficulties to overcome. Even simple diseases may have dangerous oscillations, and facial expressions can vary over time as a result of these variations. In addition, it may be required to address the facial traits as they change over time (as a result of aging). Despite the fact that existing algorithms perform well in limited conditions, researchers are still having difficulty dealing with fluctuations in light and occlusions. It is just occlusion and illumination alterations that will be handled in this project; the other two primary challenges will be tackled in a separate project [11]. Due to the fact that only the most sophisticated procedures are being used, face recognition when this strategy is used, it is feasible to distinguish between at least two distinct kinds of face recognition systems:

- Detecting the presence of a guy is a difficult task. Often, just a single image of a person may be found. Real-time comprehension is not necessary in the majority of instances.
- Detection and tracking (multiple images per individual are frequently designed for real and training time recognition required).

Unlike previous cases, this is the first time that more than one face photo is included for each person. It is envisaged that preliminary face detection would be carried out [12]. The objective of this is to give you with the appropriate identification for your account (e.g., name tag).

6.3   Literature Review

In this work, a summary of numerous face recognition methods is provided. It includes a selection process for discovering the origins of the techniques and other important theories and concepts. Principal component analysis (PCA) and a face image are the most effective techniques for describing data, claim [13]. This process allows us to considerably reduce the image's size and dimensions. A recognition system utilizing a 7-State HMM in conjunction with SVD Coefficients is presented by [14]. As a result of ongoing training provided by the Olivetti Research Laboratory, a 99% success rate has been attained on half of all images in the database of interest (ORL). The YALE database has a success rate of 97.7%. Reference [15] achieved 95 recognitions. They used Pseudo-2D HMM technique with recognition time of 240 s per image. Reference [16] achieved 99.5% recognitions. They used DCT-HMM technique with recognition time of 3.5 s per image. Reference [17] achieved 100% recognitions. They used DHMM + Wavelet technique with recognition time of 0.3 s per image. Reference [18] achieved 96% recognitions. They used PDBNN technique with recognition time of 0.1 s per image. Reference [19] achieved 99% recognitions. They used DHMM + SVD technique with recognition time of 0.28 s per image.

The methods of extraction described by Edwards et al. that correspond to each of
these groupings are mentioned below. According to his findings, appearance-based
procedures are equally effective as motion-based procedures and perform better than
version-based ones. Compared to other types of systems, motion-based systems,
which use lighting normalized graphic sequences as the basis for their appearance,
require a more rigorous statistical approach and a larger emphasis on static images.
Reference [20] take into account choice criteria created in favor of this category
having estimation while estimating the vector parameters for any assessment graphic
in comparison to the majority of model squares estimation. The closest subspace to
the user is classified using this technique.

6.4 Proposed Model

In this work, the SVM classifier was used to classify ten randomly selected feature
subsets, with each subset being allocated a different random number. Following that,
a vote mechanism was employed to decide the final categorization of the participants.
It was decided to employ 180 support vectors for the purposes of this investigation.

6.4.1 Algorithm: Classification

Input: Training Instances
Intermediate Output: output from SVM classification for each feature subset
Output: Classification Results of the hybrid approach
Procedure Begin
• **Initialize** the weight wi for each data vector ti ε D.
• **Generate** a new data feature subset Di from D using random replacement
method.
Begin
• **For** each random feature subset Di do
Begin
• Apply SVM to each feature subset
• Generate O SVM, the classification output
End
• Depending on the results of the classification, adjust the weights of each data
vector in the training set. If an example was incorrectly classified, its weight will
be increased; otherwise, it will be dropped.
Until all of the input data vectors are correctly classified or until the iteration
limit is reached, repeat steps next through last one while generating new random
subsets.

• Use a majority voting mechanism to choose output O for the entire dataset by comparing the results of each Random feature subset's final outputs. Di of the initial set D acquired following Step end.

Return 0

End

They may be found in several forms, such as a mental posture, facial motion in a photograph, and an excerpt from another stance. It is anticipated that a face identification methodology, such as the Viola-Jones method, was used in this work to identify and crop pictures captured by monocular cameras that had been localized to their present position using a face identification technique [21, 22]. Instead, we use a conventional facial image that encompasses a variety of different facial orientations as well as the needs of different lighting conditions and facial emotions. Picture sets are assessed according to the degree of freedom of individual faces, with pre-defined spinning angles being used as a guideline. When everything seems to be the same, it is much simpler to compare and contrast [23].

Thus, an image is pre-processed using a graphic pyramid that is composed of a sequence of steps. Prior to feature extraction, SVM classification is conducted on the input pictures in order to decrease the chance of LBP features being particularly sensitive to localized noise or occlusions during the feature extraction process. A feature that can be extracted without noise is identified using the SVM classification [24]. The original image is then filtered repeatedly using the conventional variation and set to 0 or 1 when the feature is identified. As a consequence, this characteristic is assigned a value of 0. The extracted feature that has been tagged must be assessed, and the arbitrary woods technique is utilized to do this. Figure 6.1 shows model architecture combination of SVM + Random Forest.

In practice, a randomized tree was constructed, but the optimal parameters were kept in the nodes on the interior of the tree. The central node makes use of the ideal settings in order to maximize the utility of information. Every successive node gets the same optimization applied to it as the one before it. For each small to medium-sized perennial shrub, an approach is utilized that picks random training samples from the photo collection from the picture collection. It ensures consistent performance results despite the vast volumes of data while using less memory capacity for training than other approaches do. When the conclusion condition is satisfied, the practice comes to an end. A common observation in everyday radio frequency practice is that when the tree grows to a thickness more than the prescribed maximum thickness or when a significant number of samples remain in the current node, the tree is no longer developing. We'll present the results of experiments on a range of illnesses by putting the thickest trees and the smallest samples at the same node as the results of the experiments. In order to get things started, we've collected 100 images of 40 different individuals, each with ten photos of themselves. These individuals appear in a variety of stances, moods, and glasses across these 100 photographs. The rotation angle changes slightly between each individual's ten photos collected during a 10-day period. The bulk of the tests were developed in Python, and we utilized a range of evaluation techniques, including image evaluations, to determine their effectiveness.

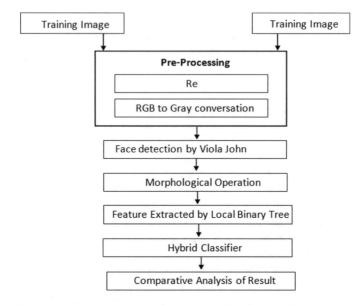

Fig. 6.1 Proposed model architecture combination of SVM + Random Forest

The database of 100 photographs is first organized into 3×3 matrices, and then a single image is chosen from among them. Following this, a photograph from the 100-image matrix was removed from consideration, and a new database with 399 photographs was established. Talk about all of the different algorithms that were used in this research. We used both the random forest technique and the support vector machine assessment process in our research to discover face recognition. Both techniques look for differences in characteristics and information from the existing dataset, and they both look for differences in characteristics and information from the existing dataset.

6.4.2 Algorithm: Proposed Model Architecture

Input: Tanning Image
Output: Classification of Machine learning Model
Procedure Begin
• The 100 image database is loaded into 3×3 matrices.
• Then randomly select to search for a picture. Then the image searched from the matrix of 100 images was omitted, and again a new database was created with 399images.
• 99 images are calculated and then subtracted from the images, and a related matrix is created.

• The Eigenvector of the Correlation Matrix is calculated. So suppose we took 20 pictures for which Eigenvectors are counted and the signature for the images is counted with different facial expression and size 205 × 274 px.

• Then the Eigenvector is calculated for searching the image, and the results for matching are obtained with the minimum Euclidean distance. The picture with the nearest distance is given as output.

• In this work, we evaluated the acceptance time for each photo and the.

• Overall recognition of 99 images.

End

6.5   Performance Evaluation

The proposed system is calculated on the basis of False Accept Rate (FAR), False Reject Rate (FRR), True Success Rate (TSR), and Error Rate (ERR). Table 6.1 shows Performance of the proposed model. Figure 6.2 represents Graphical representation for Image size versus FAR, FRR (%), Fig. 6.3 shows Graphical representation for Image size versus TSR (%). The image size varies from 10 – 400. The below Table 6.1 shows their performance:

The proposed model performance is also compared with the existing model. Their comparative analysis is listed below in Table 6.2.

According to the table, the proposed model shows a higher accuracy rate of 97.47% than previous findings.

Table 6.1 Performance of the proposed model

Image size	FAR (%)	FRR (%)	TSR (%)	ERR
400	2.51	2.56	97.44	0.85
200	2.51	2.56	97.44	0.85
100	5.03	5.13	94.87	0.85
50	5.03	5.13	94.87	0.85
40	10.05	10.26	89.74	0.85
35	7.54	7.69	92.31	0.85
30	2.51	2.56	97.44	0.85
25	12.56	12.82	87.18	0.85
20	15.08	15.38	84.62	0.85
15	10.05	10.26	89.74	0.85
10	15.08	15.38	84.62	0.85

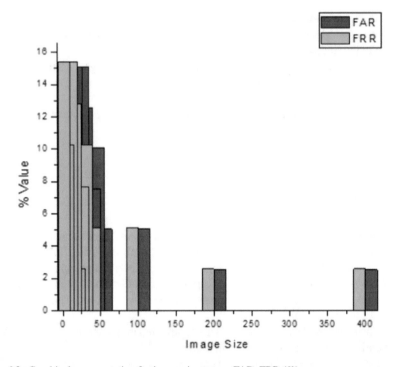

Fig. 6.2 Graphical representation for image size versus FAR, FRR (%)

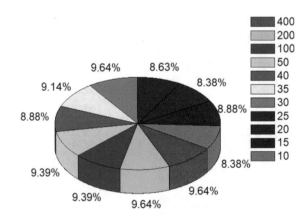

Fig. 6.3 Graphical representation for image size versus TSR (%)

6.6 Conclusion

The research system with accuracy is depends on FAR, FRR, TSR, and ERR when compared with the existing machine learning model. In the future author tends to

Table 6.2 Comparative analysis

Approach used	Mean recognition rate
SVM with linear and RBF kernel	96.05, 94.73, 96.05, 94.73%
LBP + SVM (face recognition system)	93%
RF + SVM (proposed framework)	97.47%
LBP (face description process)	94.60%
RF (histograms with oriented gradients)	92.60%

use the attribute extraction techniques, such as face alignment, which may allow to attain more accuracy for the larger database. By aligning the face, it is possible to extract facial alternatives. The approach that has been presented has the potential to be used in the future. In order to categories satellite-collected vector and raster remote sensing data as well as other geographic information systems such as Spot the Satellite, it is efficient and effective model for face recognition system.

References

1. Belhumeur, P.N., Hespanha, J.P., Kriegman, D.J.: Eigenfaces vs. Fisherfaces: recognition using class specific linear projection. IEEE Trans. Pattern Anal. Mach. Intell. **19**, 711–720 (1997)
2. Bianconi, F., Fernández, A.: On the occurrence probability of local binary patterns: a theoretical study. J. Mathe. Imag. Vis. **40**, 259–268 (2011)
3. Breiman, L.: Random forests. Mach. Learn. **45**, 5–32 (2001)
4. Brunelli, R., Poggio, T.: Face recognition: features versus templates. IEEE Trans. Pattern Anal. Mach. Intell. **15**, 1042–1052 (1993)
5. Chakraborty, D., Saha, S.K., Bhuiyan, M.A.: Face recognition using eigenvector and principle component analysis. Int. J. Comput. Appl. **50**, 42–49 (2012)
6. Chellappa, R., Wilson, C.L., Sirohey, S.: Human and machine recognition of faces: a survey. Proc. IEEE **83**, 705–741 (1995)
7. Chitaliya, N.G., Trivedi, A.L.: An efficient method for face feature extraction and recognition based on Contourlet transform and principal component analysis using neural network. Int. J. Comput. Appl. **6**, 28–34 (2010)
8. Geurts, P., Ernst, D., Wehenkel, L.: Extremely randomized trees. Mach. Learn. **63**, 3–42 (2006)
9. Geurts, P., Fillet, M., de Seny, D.D., Meuwis, M.A., Malaise, M., Merville, M.P., Wehenkel, L.: Proteomic mass spectra classification using decision tree based ensemble methods. Bioinformatics **21**, 3138–3145 (2005)
10. Grudin, M.A.: On internal representations in face recognition systems. Pattern Recogn. **33**, 1161–1177 (2000)
11. Guo, G., Li, S.Z., Chan, K.L.: Support vector machines for face recognition. Image Vis. Comput. **19**, 631–638 (2001)
12. Heisele, B., Ho, P., Wu, J., Poggio, T.: Face recognition: component-based versus global approaches. Comput. Vis. Understanding **91**, 6–21 (2003)
13. Ho, T.K.: A data complexity analysis of comparative advantages of decision forest constructors. Pattern Anal. Appl. **5**, 102–112 (2002)

14. Ho, T.K.: Random decision forests. In: Proceedings of the 3rd International Conference on Document Analysis and Recognition, held at Montreal during August 14–16, 1995, pp. 278–282
15. Howell, A.J., Buxton, H.: Invariance in radial basis function neural networks in human face classification. Neural Process. Lett. **2**, 26–30 (1995)
16. Lee, T.S.: Representation using 2D Gabor wavelets. IEEE Trans. Pattern Anal. Mach. Intell. **18**, 959–971 (1996)
17. Liu, K., Kehtarnavaz, N.: Real-time robust vision-based hand gesture recognition using stereos. J. Real-Time Process. **11**, 201–209 (2013)
18. Kaushik, S., Poonia, R.C., Khatri, S.K., Samanta, D., Chakraborty, P.: Transmit range adjustment using artificial intelligence for enhancement of location privacy and data security in service location protocol of VANET. Wireless Commun. Mobile Comput. **2022**, 13, Article ID 9642774 (2022). https://doi.org/10.1155/2022/9642774
19. Mahmoudi, F., Shanbehzadeh, J., Moghadam, A.M.E., Zadeh, S.H.: Retrieval based on shape similarity by edge orientation autocorrelogram. Pattern Recogn. **36**, 1725–1736 (2003)
20. Matas, J., Jonsson, K., Kittler, J.: Fast face localisation and verification. Image Vis. Comput. **17**, 575–581 (1999)
21. Moore, D.: Classification and regression trees. Cytometry **8**, 534–535 (1987)
22. Naseem, I., Togneri, R., Bennamoun, M.: Linear regression for face recognition. IEEE Trans. Pattern Anal. Mach. Intell. **32**, 2106–2112 (2010)
23. Nefian, A.V., Hayes, M.H.: Hidden Markow models for face recognition In: Proceedings of the 1998 IEEE International Conference on Acoustics, Speech and signal processing held at Seattle during May 15, 1998, pp. 2721–2724
24. Paul, L.C., Sumam, A.A.: Face recognition using principal component analysis method. Int. J. Adv. Res. Comput. Eng. Technol. (IJARCET) 1, 135–139

Chapter 7
Taxonomy for Classification of Cloud Service (Paas) Interoperability

Zameer Ahmed Adhoni⊙ and **N. Dayanand Lal**⊙

Abstract The decreasing cost of hardware has opened the doors to the development of various concepts like big data and huge storage spaces. The development of the cloud is one major development in this string of developments. The economic and other important contract features like the locking time gave rise to the usage of multi-clouds by various companies for the same or similar applications. The extended form of this scenario leads to operation on data on different clouds giving rise to interoperability. The Infrastructure as a service, Platform as a service, and Software as a service in the cloud force us to define taxonomy based on various views. Each view addresses its own customized problem. There is a need to generalize the taxonomy of interoperability. This paper presents a taxonomy for the classification of the cloud for Platform as a Service.

7.1 Introduction

7.1.1 Cloud Computing

The buzzword in the IT industry these days has been Cloud computing. The computing paradigm is not an unnecessary hype but a genuine emergence over a period of time. Cloud computing is a concept of vendors offering its customers a wide range of hardware, Software, network, computation, data, services, storage, and other allied services on a pay-as-you-use basis for agreed pricing, which could be used anytime, anywhere [1].

National Institute of Standards and Technology defines cloud computing as the cloud computing model of five essential characteristics: on-demand self-service,

Z. A. Adhoni (✉) · N. Dayanand Lal
Department of Computer Science and Engineering, Gitam School of Technology, Gitam University, Bangalore, India
e-mail: zadhoni@gitam.in

N. Dayanand Lal
e-mail: dnarayan@gitam.edu

broad network access, resource pooling, rapid elasticity, and measured service, along with three service models: Software as a Service (SaaS), Platform as a Service (PaaS), and Infrastructure as a Service (IaaS) [2].

Cloud computing is practical and is successfully implemented in projects that are storage-centric, computation-centric, platform oriented, bandwidth wanting, or web-based [3].

7.1.2 Cloud Interoperability

Major Providers like Microsoft, Amazon, Google have been actively providing cloud services and striving to provide efficient services in all three categories of service models [4]. These vendors have their Infrastructure, standards, methods to access the cloud, underneath hardware and Software. Which leads to hampering of building standardized standards, rules, and agreements. To access multiple clouds with different efficient services, it is necessary to have interoperability of cloud structure [5].

The IEEE Glossary of 2013 states that interoperability happened only by adequately implementing the standards. It states that interoperability is "the ability of a system or a product to work with other system or product without special effort on the part of the customer."

A broader definition of interoperability deals with the ability of the cloud and cloud computing to operate on different clouds as though they are the same. In general, from the computation point of view it should be possible for the user to exchange data between two or more clouds [6]. The definition goes beyond this to define interoperability as the ability to deal with abstractions supported by the clouds like environments/platforms, programs, run both locally and on cloud or hybrid, portability of data and services, workloads, management tools, server images between clouds and communicate to provide support for a single application [7].

The definition of interoperability lacks generalization and is true to a specific context. The main aim of all these definitions is to highlight the advantages of the cloud model and provide seamless computation capabilities across two or more clouds of different vendors or different cloud instances. This allows harvesting the best services provided by different vendors for the betterment of cloud computing [8].

7.1.3 Interoperability Issues

Applications that are built on one cloud should have the capacity to be tested and deployed on other clouds and should be operated on the other cloud. The major concern is managing the complexity of diversity, facilities of different clouds, portability, relocation of data & programs on the cloud platform [9]. It is seen that various clouds provide different facilities, which are most of the time well declared and known to both the client and vendor. Each vendor himself has a technology with bespoke APIs [10]. These APIs provided to the client help in the operation of the cloud. For example, Microsoft provides SQL Azure, and Google App Engine offers, among other things, for storing data into the cloud [11]. These are some of the major issues interoperability needs to address if operations between clouds need to happen in s smooth manner.

7.2 Review of Literature

7.2.1 Service Models

The service models are broadly accepted based on the classification of interoperability requirements and standardize cloud computing platforms [12]. The accepted models based on the heterogeneity faced during interoperations of different types of clouds are Infrastructure as a Service (IaaS), Platform as a Service (PaaS), and Software as a Service (SaaS) [3].

IaaS mainly deals with the Infrastructure of the cloud. This service mainly deals with providing minimum resources needed as resources to the client [13]. The resources mainly involved are processors of needed specifications, networking, storage, and operating system and applications. Amazon cloud and Google compute engine fall in this category [14, 15].

PaaS mainly deals with deploying on the cloud infrastructure, the client's application programs, needed services to this programs/Software, libraries, tools needed, etc. [16]. In brief all that is needed to ensure that the clients' Software performs to its best is to be provisioned here. Google APP Engine provides such a service [17].

SaaS mainly deals with the capability of the cloud to execute the clients' application program on the Infrastructure and Platform provided [18].

The pressing need for features like security, management, governance, portability, interoperability led to the thought of standardization of cloud computing. In the year 2010, Open Cloud Manifesto (OCM) was formed with the support of leading cloud vendors [19]. This group based its proceedings on the features of flexibility, speed, and agility in cloud computing.

Cloud computing interoperability form (CCIF) proposed to unify cloud APIs with a standard semantic interface called the Unified Cloud Interface (UCI) [20]. It also proposed the independence of the lower layers of Infrastructure. Orchestration layer and federation of clouds were proposed by this forum. This forum also suggested the architecture [21]. Major cloud vendors at that time, like Microsoft and Amazon, were unwilling for the architectural proposal. This unwillingness was more due to the commercial aspect of vendors, which led to the rejection of the proposal [22].

7.3 PaaS Level Interoperability

Cloud computing PaaS interoperability means movement of data and its related services from one platform to another, provided by different cloud vendors with different Infrastructure, without having to put in more effort by the client/customer [23]. Data movement would mean compatibility of data among different platforms and lower-level details the storage means and implementations. Services movement would mean one service working on a particular platform would have to work on a different platform provided by a different cloud service provider [19]. If both clouds same the same Platform it would only mean a movement of the service cloud from source to destination cloud [24]. However, if the platforms are different, then Packing, copying, instantiating, installing, deploying, customization, service type, its interactions with other processes, additional services needed, forking of child and existing forked children before portability starts, dependency on the operating system, are needed to enable the proper working of the service.

7.4 Taxonomy of PaaS Interoperability

PaaS is an instance where the customer is allowed to build and deploy applications and services using the environment, operating system, servers, networks, storage, other Infrastructure, programming language compiler/assembler, which is provided as a high-level software infrastructure by the vendor [22, 24].

PaaS is Domain-driven. Involves specifications, tools, environment, programming language, and other specifications needed to be well defined.

Infrastructure, though considered at the level of service as IaaS, it is seen that most of the time, it is impossible to have a system completely independent of infrastructure details. The taxonomy of PaaS can be pictorially represented as in Fig. 7.1. It is seen that PaaS is still not completely independent of other services of the cloud for various reasons. The figure hence includes Infrastructure as one of the interoperability categories.

7.5 PaaS Interoperability

The taxonomy of the PaaS interoperability is developed after considering all the essential features of cloud interoperability. The different levels defined and taxonomy-generated is based on the dependencies seen in the present cloud interoperability scenario. A brief explanation of the optimal levels is as follows:

The taxonomy is generated in the form of a hierarchical structure. The taxonomy starts with the PaaS interoperability, which mainly depends on the two major categories, namely, Domain and Infrastructure. The PaaS service is a domain-driven

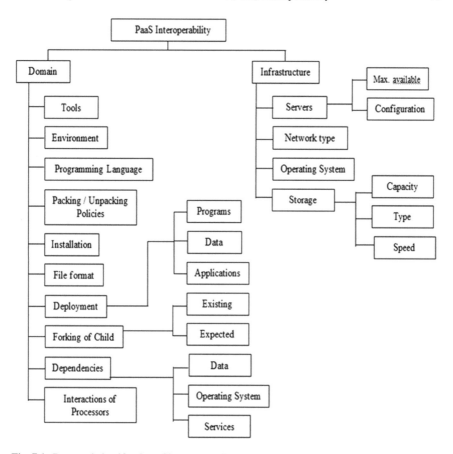

Fig. 7.1 Proposed classification of interoperability of PaaS taxonomy

service, and Domain plays an important role in the categorization. The second category is Infrastructure. Infrastructure is seen as a bleed into PaaS here. It is observed from various research and forums that Infrastructure is considered as a separate service. This holds well from the business point of view. However, when the scenario is considered from the interoperability point of view, the present situation does not allow complete bifurcation of PaaS from Infrastructure. Hence, PaaS has been broadly classified into brief Infrastructure as a part.

Infrastructure is further categorized into servers, network type, operating system, and memory. The maximum number of servers present in the existing system and its configuration will impact the application's work. The new cloud from where this application would run should have better configuration and should hold the same number of processors, if not more. Compatibility among operating systems is a major issue when interoperation is considered. However, it is seen there exist some APIs like the Tivoli Storage Manager API support cross-platform interoperability that helps operations between Linux and Windows systems, which needs to be provided by the vendor at the infrastructure level. Storage capacity, type or format, and speed also matter a lot in the successful execution of an application.

7.6 Conclusion

In this paper, an extensive survey of Platform as a service is surveyed, and the taxonomy for classification of the services are developed. The main reasons for the bleed of PaaS into IaaS are also justified based on the present scenario of the industry. It is necessary to understand the dependencies of these services if they are to be made independent in the near future. Hence a study of these services and the hindrance of these services for complete interoperability of the cloud is necessary. The taxonomy in this paper is capable of putting the cloud interoperability in order so as to help the developers decide on the level of interoperation they plan to achieve even before they attempt applications.

References

1. The Internet Engineering Task Force: http://www.ietf.org
2. Sheth, A., Ranabahu, A.: Semantic modeling for cloud computing, part 1. Internet Comput. IEEE **14**(3), 81–83 (2010)
3. Cloud Computing for Large-Scale Complex IT Systems: https://gow.epsrc.ukri.org/NGBOViewGrant.aspx?GrantRef=EP/H042644/1 (2010). Accessed Oct 2019
4. OVF Members List: http://www.dmtf.org/about/list
5. OpenStack: http://www.openstack.org/
6. Mourad, M.H., Nassehi, A., Schaefer, D., Newman, S.T.: Assessment of interoperability in cloud manufacturing. Robot. Comput. Integr. Manuf. **61**, 101832 (2020)
7. Tubishat, M., Ja'Afar, S., Alswaitti, M., Mirjalili, S., Idris, N., Ismail, M.A., Omar, M.S.: Dynamic Salp swarm algorithm for feature selection. Expert Syst. Appl. **164**, 113873 (2021)
8. The Xen Hypervisor: http://www.xen.org/
9. Lachmann, A., Clarke, D.J., Torre, D., Xie, Z., Ma'Ayan, A.: Interoperable RNA-Seq analysis in the cloud. Biochim. Biophys. Acta(BBA)-Bioenerg. 1863 (2020)
10. Zarko, I.P., Mueller, S., Płociennik, M., Rajtar, T.: The symbIoTe solution for semantic and syntactic interoperability of cloud-based IoT platforms. Paper presented at: Proceedings of the 2019 Global IoT Summit (GIoTS), Aarhus, Denmark, pp. 1–6 (2019)
11. Ramasamy, V., Pillai, S.T.: An effective HPSO-MGA optimization algorithm for dynamic re-source allocation in cloud environment. Clust. Comput. **23**, 1711–1724 (2020)
12. Petcu, D.: Portability and interoperability between clouds: challenges and case study. In: Abramowicz, W., Llorente, I.M., Surridge, M., Zisman, A., Vayssiere, J. (eds.) Towards a Service-Based Internet: Proceedings of 4th European Conference, ServiceWave 2011, Poznan, Poland, 26–28 October, 2011, pp. 62–74. Springer, Berlin, Heidelberg. https://doi.org/10.1007/978-3-642-24755-2_6
13. Libvirt: http://libvirt.org/
14. Amazon Web Services: http://aws.amazon.com
15. Microsoft Azure: http://www.microsoft.com/windowsazure
16. DMTF (Distributed Management Task Force): http://www.dmtf.org/

17. Dreibholz, T., Mazumdar, S., Zahid, F., Taherkordi, A.: Mobile edge as part of the multi-cloud ecosystem: a performance study. Paper presented at: Proceedings of the 2019 27th Euromicro International Conference on Parallel, Distributed and Network-Based Processing (PDP), Pavia, Italy, 2019, pp. 59–66
18. Shan, C., Heng, C., Xianjun, Z.: Inter-cloud operations via NGSON. IEEE Commun. Mag. **50**(1), 82–89 (2012)
19. Marković, M., Gostojíc, S.: A knowledge-based document assembly method to support semantic interoperability of enterprise information systems. Enterp. Inf. Syst. 1–20 (2020)
20. Habibi, M., Fazli, M., Movaghar, A.: Efficient distribution of requests in federated cloud computing en-vironments utilizing statistical multiplexing. Future Gener. Comput. Syst. **90**, 451–460 (2019)
21. Hilley, D.: Cloud computing: a taxonomy of platform and infrastructure-level offerings. Tech Rep GIT-CERCS-09-13, CERCS, Georgia Institute of Technology (2009)
22. Google App Engine: http://appengine.google.com/
23. Open Grid Forum: http://www.gridforum.org/
24. Google Compute Engine: http://cloud.google.com/products/computeengine.html

Chapter 8
Form Scanner & Decoder

Sharmila Sengupta, Harish Kumar, Anshal Prasad, Ninad Rane, and Nilay Tamane

Abstract In India, most of the people use a pen & paper for filling various application forms. Also, they are not comfortable with the English language. This project aims to automatically convert applications written in Hindi to English and therefore assists mainly the rural people who have the inhibition of first filling a form and that too in English. Technology may be all around us, but people are still congenial with the pen and paper. So, this project is based on the recognition of different handwritten characters written in Hindi language and converting them to English. It tries to develop a word recognition system to separate several Hindi words from handwritten forms using image segmentation techniques. Nowadays, all form reading processes are done digitally. This system will facilitate such processes in banking, agriculture, education, etc.

8.1 Introduction

Handwriting identification is a difficult sector in researching when discussing fields of character recognition and image processing. Since many years, quite a few researchers have worked on methods that make the processing time faster for identifying and extracting handwritten text whilst having a high accuracy, but all of

S. Sengupta · H. Kumar (✉) · A. Prasad · N. Rane · N. Tamane
Computer Engineering, Vivekanand Education Society's Institute of Technology,
Mumbai, India
e-mail: 2018.harish.kumar@ves.ac.in

S. Sengupta
e-mail: sharmila.sengupta@ves.ac.in

A. Prasad
e-mail: 2018.anshal.prasad@ves.ac.in

N. Rane
e-mail: 2018.ninad.rane@ves.ac.in

N. Tamane
e-mail: 2018.nilay.tamane@ves.ac.in

© The Author(s), under exclusive license to Springer Nature Singapore Pte Ltd. 2023
C. So-In et al. (eds.), *Information Systems for Intelligent Systems*, Smart Innovation,
Systems and Technologies 324, https://doi.org/10.1007/978-981-19-7447-2_8

them are available for non-Hindi languages. The difficulty arises because different people possess varied patterns of writing, and there is a broad range of characters present in this language consisting of 10 numerals and 36 consonants. The science created to recognize text in images of printed corpus is called optical character recognition (OCR). An OCR will not be applicable for handwritten images because of various challenges like different handwriting styles, effect of noise, blurring and distortions on the image. This can be tackled by locating the words in a segmentation-based multi-writer scenario. This paper proposes a system of hand-written word recognition and translation.

8.2 Literature Survey

The results in [1] are displayed based on seven public benchmarks which consist of standard text, non-standard text and long non-Latin text which improves the efficiency of the project. Also, this paper consists of capturing Latin text and translating it into English. The precision of the output is really low here. The research as mentioned in [2] uses CNN in more than one way, including—(i) Training the convolutional neural network model from the beginning in sequence way. (ii) Use of MobileNet: Transfer learning paradigm from pre-trained model to the Tamizhi database. (iii) Creating a model with convolutional neural networks and support vector machines. (iv) Support vector machines give the highest accuracy for Bangla handwriting identification. The research which is mentioned in [3] uses digitization of handwritten Devnagari text using CNN. This paper consists of a customer support system having texts in different languages. Comenia script, used in [4], is a modern handwritten font similar to block letters used at primary schools in the Czech Republic (Fig. 8.1).

Here, in this paper, a limited training set of handwritten letters is used to propose a new method to artificially create image samples which increases the accuracy of the system. The objective of this review paper [5] is to summarize several

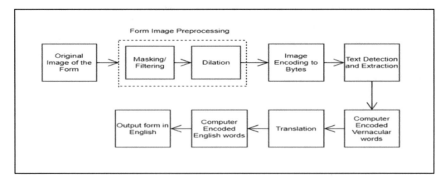

Fig. 8.1 System diagram

researches that have been conducted on character recognition of handwritten documents and provides a deep insight for our problem statement.

8.3   Methodology

The input folder area contains images which are assigned to the list; then, the image is rendered with a dynamic image. The word spotter finds the function of plain text links to handwritten text. Previous processing was performed on the inserted image where encryption was performed with the help of floor and ceiling hue saturation values. Post-encryption, this technique of encoding is performed when the conversion of that photo into a data character unit takes place.

The image is then filtered through processes of expansions and erosion followed by filtering the black text on the image. The binding boxes present around the documents are removed after finding their places, and afterwards, they are expanded and eroded for improvement. This photo is then encoded in a data character unit.

Later, only, the green ink text is available. To replace this handwritten text, the word spotter provides links next to the blue ink text. The next release is done on this, which is converted and then added to the new list. Now, the boxes have to be attached to the translated script. The first image is now horizontal, then with the help of links, delete the handwritten text and then change it with the words in the existing list.

A photograph of the handwritten input application is taken. Later, the required pre-processing is carried out.

In addition to this, procedures like dilation and erosion are performed. With the help of dilation, pixels are added to the parameters of the image. The total pixel quantity that is added or subtracted from objects within the image depends on two parameters—size and shape of the structural element used for processing images. At the time of the expansion of morphological and erosion functions, the shape of any given pixel in the outgoing image is arbitrated by the following factors—law on the next pixel and its neighbour. In the next section, image coding by bytes is done after which the handwriting appears.

The well-known text is then released for translation, and the resulting image is produced after recognizing input written script and its translation into the English language. Only, after completion of the English translation of the characters is done, the resulting images are produced in a part of the form in the handwritten form instead of the existing computer-generated English text.

- Text Recovery—uses CNN to find sentences and create compound boxes. With the use of multi-layer neural networks, potent text-finding modules can be trained. Convolutional neural networks can be trained with regional tagged images to boost the accuracy. CNN is of great help for processing low-quality features and high-quality content.

- Directions ID—separates direction for each combination box.
- ScriptID—finds text in compilation box. It allows multiple texts for each image, but for the first time, only, one script is taken per box.
- Text Visibility—this is the most important part of visual recognition where every part of the text is visible in the image. CNN detects alphabets and letters by finding differences amongst the characteristics.
- The neural network has a number of flexible layers that make up the output element, and the layers are fully integrated, followed by a soft layer of partition size.
- Structural Analysis—this determines the order study and separates topic from topic (Fig. 8.2).

Fig. 8.2 Flow diagram of OCR model

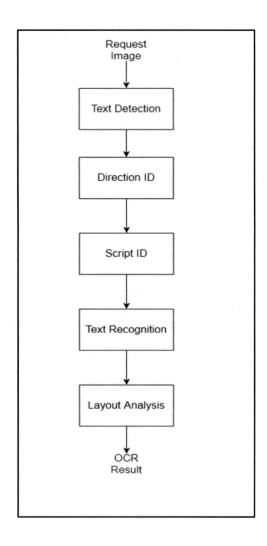

The OCR model was built with the help of Devanagari handwritten dataset which not only consists of letters but also handwritten digits. The dataset has a total of 92 thousand letter entries of which 78 thousand letters were of training set and 13 thousand letters were of testing set. The testing set includes all the entries based on alphabets from क to ज्ञ and numerals from ० to ९.

Firstly, the characters should be handwritten. These characters are then scanned and also cropped if needed.

Below are few character-based handwritten texts:

Character 'क':
See Fig. 8.3
Character 'ख':
See Fig. 8.4
Character 'ग':
See Fig. 8.5

Fig. 8.3 Devanagari handwritten dataset for character 'क'

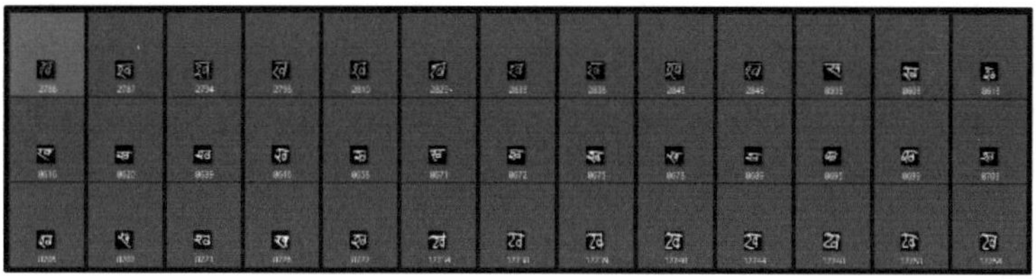

Fig. 8.4 Devanagari handwritten dataset for character 'ख'

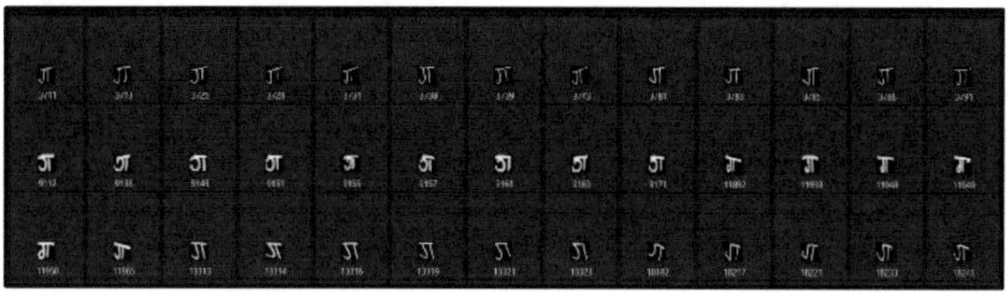

Fig. 8.5 Devanagari handwritten dataset for character 'ग'

The input to the CNN is two-dimensional character matrices derived after scanning the segmented line from the document. CNN layers are organized into three dimensions, width, length, and depth. Information about layers is as follows -

- Input Layout—this is a bath to capture the input and transfer it towards the following layers.
- Convolution Layout—this character removal function takes place here.

The convolution process involves smoothing the kernel over the input and creating a total of outputs. Many description maps are created by different convolution functions performed by kernels in the input. Layer depth can also be known as the number of feature maps.

- Fixed Line Unit—linear variability is present due to this. To speed up the learning process, most of the sub-zero values are replaced by zero. The results produced in the above layer are transmitted between the activation layer.
- Integration layer—reduces the area of all feature maps. This results in the reduction of calculation. A sliding window is also used that cuts through the feature map and converts them for obtaining values.
- FC layer—joins all the neurons of the earlier layer to every neuron present in the current layer. Layer layout is different from planning and retrieval problems. With the backtracking problem, the FCL is present before the output predictor. With this problem of separation, the following layer is a soft layer that helps to find opportunities for each class.

8.4   Results and Discussion

Input for the system is an image of a form which should be in either of jpg, png or bmp format. This image is uploaded as an input which should be provided with an image variable (Figs. 8.6 and 8.7).

Bounding Boxes: (Intermediate step)

In this step, the input image gets completely converted to black and white-based intensity values with enhanced quality which is also called thresholding. The list of coordinates calculated earlier is used during the process. Now, using the cv2 line method, the handwritten input sentences are marked. This black and white-enhanced colour representation of the original image is called the thresholded image, and then, the bounding boxes present around the sentences are added.

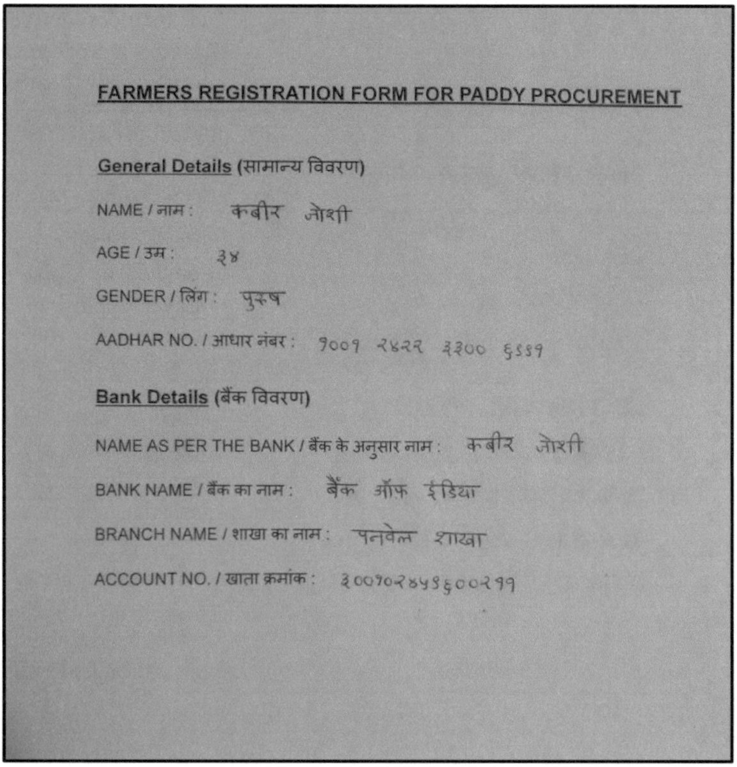

FARMERS REGISTRATION FORM FOR PADDY PROCUREMENT

General Details (सामान्य विवरण)

NAME / नाम : कबीर जोशी

AGE / उम्र : ३४

GENDER / लिंग : पुरुष

AADHAR NO. / आधार नंबर : १००१ २४२२ ३३०० ६९९१

Bank Details (बैंक विवरण)

NAME AS PER THE BANK / बैंक के अनुसार नाम : कबीर जोशी

BANK NAME / बैंक का नाम : बैंक ऑफ इंडिया

BRANCH NAME / शाखा का नाम : पनवेल शाखा

ACCOUNT NO. / खाता क्रमांक : ३००१०२४५९६००२११

Fig. 8.6 Input image

Masked image: (Intermediate step)

In this step, upper and lower HSV values are calculated which are helpful in masking areas apart from the bounding boxes (Fig. 8.8).

Dilated Image: (Intermediate step)

With the help of dilation, pixels are added to the parameters of the object. The total pixel quantity that is joined or subtracted from objects within the image depends on two parameters—size of the structural element used and shape of the structural element used for processing the image. At the time of the expansion of morphological and erosion functions, any pixel's shape in the outgoing image is arbitrated by the following factors—law on the corresponding pixel and its neighbour in the input image. Post-extension encoding is the process that is performed when conversion of that image into a data character unit takes place (Figs. 8.9 and 8.10).

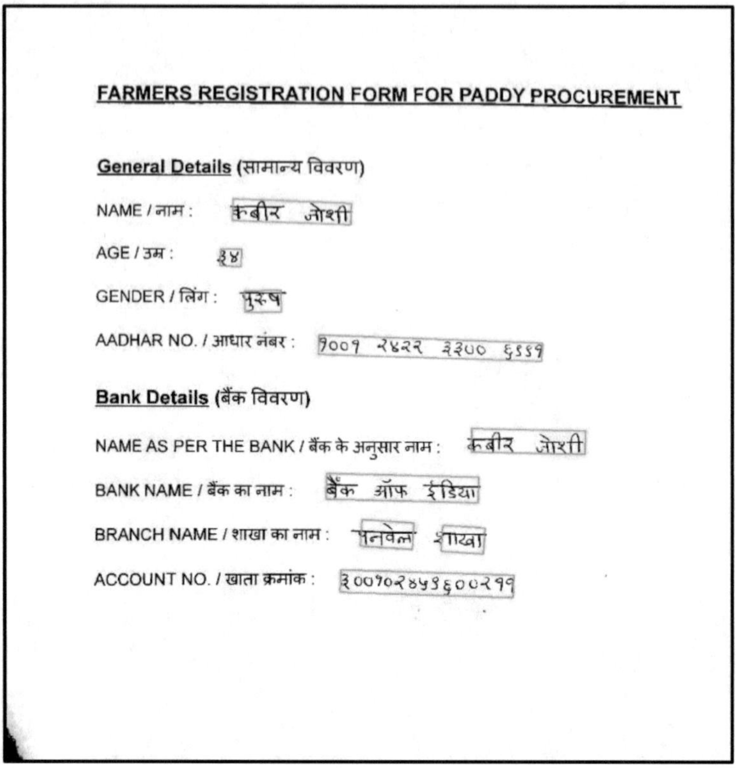

FARMERS REGISTRATION FORM FOR PADDY PROCUREMENT

General Details (सामान्य विवरण)

NAME / नाम : कबीर जोशी

AGE / उम्र : ३४

GENDER / लिंग : पुरुष

AADHAR NO. / आधार नंबर : ७००१ २४२२ ३३०० ६९९१

Bank Details (बैंक विवरण)

NAME AS PER THE BANK / बैंक के अनुसार नाम : कबीर जोशी

BANK NAME / बैंक का नाम : बैंक ऑफ इंडिया

BRANCH NAME / शाखा का नाम : पनवेल शाखा

ACCOUNT NO. / खाता क्रमांक : ३००७०२४५९६००२११

Fig. 8.7 Image after bounding boxes step

After following all the processes, the output image is generated where handwritten text in Hindi is converted to English text which is enclosed in the same areas of the border image. Converting handwriting from one language to another language is a task of utmost importance which has been done with the help of a Python package known as py-translate.

Experimental results

Initially, the system detects coordinates of Hindi handwritten text, and then, the process of translating it into equivalent English text is done, also overlaying the translated text back in the same area.

On testing various styles of Hindi handwriting, the system was found to be 60% accurate. The accuracy of the detection was found to be decreasing in case of increase in noise and distortion whereas in case of clean and proper handwriting, the Hindi handwritten text is detected with ease and further processes of translating the text and replacing it back on the same area yields better results.

Fig. 8.8 Image after masking process

8.5 Conclusion and Future Work

This software will turn out to be beneficial for people from countryside areas as it is a better technique to complete the forms without worrying about the language. They just need to put the image of the document which they have written in the vernacular language and then the software converts non-printed in the vernacular language to English, and also, the translated text is overlaid on the form, and the output document is shown and given back to the user.

The most important procedures in the functioning of our system are text identification, recognition and translation. This should reduce the dependence of people on other people and keep the process of document-filling simple and less troublesome.

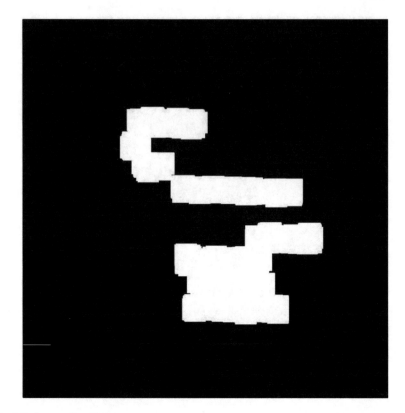

Fig. 8.9 Image after dilation process

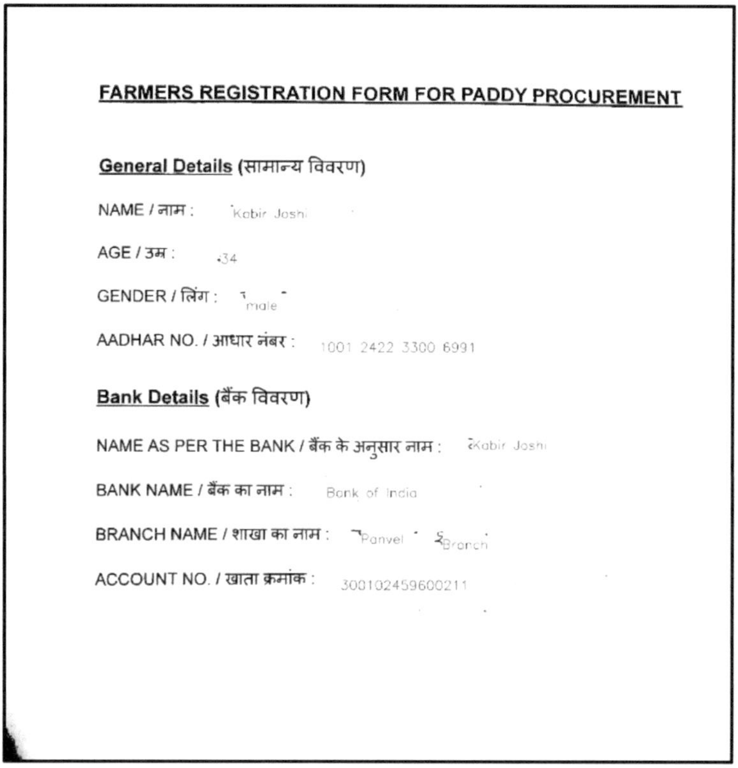

FARMERS REGISTRATION FORM FOR PADDY PROCUREMENT

General Details (सामान्य विवरण)

NAME / नाम :   Kabir Joshi

AGE / उम्र :   34

GENDER / लिंग :   male

AADHAR NO. / आधार नंबर :   1001 2422 3300 6991

Bank Details (बैंक विवरण)

NAME AS PER THE BANK / बैंक के अनुसार नाम :   Kabir Joshi

BANK NAME / बैंक का नाम :   Bank of India

BRANCH NAME / शाखा का नाम :   Panvel   Branch

ACCOUNT NO. / खाता क्रमांक :   300102459600211

Fig. 8.10 Final output image

References

1. Yu, D., et al.: Towards accurate scene text recognition with semantic reasoning networks. In: 2020 IEEE/CVF Conference on Computer Vision and Pattern Recognition (CVPR), 2020, pp. 12110–12119. https://doi.org/10.1109/CVPR42600.2020.01213
2. Ghosh, R., Vamshi, C., Kumar, P.: RNN based online handwritten word recognition in Devanagari and Bengali scripts using horizontal zoning. Pattern Recognit. J. **92**, 203–218 (2019)
3. Pande, S.D., Jadhav, P.P., Joshi, R., Sawant, A.D., Muddebihalkar, V., Rathod, S., Gurav, M. N. and Das, S.: Digitization of handwritten Devanagari text using CNN transfer learning—a better customer service support. Neurosci. Inf. **2**(3), 100016 (2022). ISSN 2772-5286
4. Memon, J., Sami, M., Khan, R.A., Uddin, M.: Handwritten Optical Character Recognition (OCR): A Comprehensive Systematic Literature Review (SLR). School of Computing, Quest International University Perak, Ipoh 30250, Malaysia , July 2020. IEEE Access
5. Rajnoha, M., Burget, R., Dutta, M.K.: Handwriting comenia script recognition with convolutional neural network. In: 2017 40th International Conference on Telecommunications and Signal Processing (TSP), 2017, pp. 775–779. https://doi.org/10.1109/TSP.2017.8076093

Chapter 9
Intrusion Detection Using Feed-Forward Neural Network

Anshumaan Mishra and Vigneshwaran Pandi

Abstract Arbitrarily, in a corporate environment, there might be a DOS attack where there is unusual traffic in the network environment. The early exploration around here and monetarily accessible Intrusion Detection Systems (IDS) are generally signature-based. The issue related to signature-based identification is that the updates are required which leads to an interruption in the IDS services. When zero-day DOS attacks occur, they contain new signatures which cannot be detected by current IDS, and thus it isn't appropriate for the constant monitoring of the network. This paper presents a study of interruption identification frameworks that are then used to survey and study them. Their qualities and shortcomings are additionally examined. The utilization of machine learning algorithms is perused over. We propose a neural network that can classify if the network traffic flowing is malicious or benign. We collect five features which are used in the classification. In the end, we collate a confusion matrix for a few thresholds and establish a relation between precision and recall to get better insights of the performance of the model.

9.1 Introduction

Cyber criminals create their own malicious traffic as one method for disrupting the organization's system(s). This recon technique can be extraordinarily significant to distinguish weak frameworks in the organization. A ton of research on IDS is based on the use of machine learning algorithms. Mrutyunjaya et al. [1] utilized Naïve Bayes approach on the KKD'99 dataset and can get an exceptionally high detection rate for their methodology yet their methodology has plenty of false positives. Kingsly et al. [2] utilized a framework clustering algorithm for abnormal network traffic identification. They have utilized a similar KKD'99 dataset and had the option to arrive at fair outcomes, however have low execution contrasted with different characteristics. Xianwei et al. [3] proposed an ensemble learning algorithm and compares it with numerous algorithms. Although their ensemble algorithm has the

A. Mishra · V. Pandi (✉)
SRM Institute of Science and Technology, Kattankulathur 603203, India
e-mail: vigneshwaranpandi1981@gmail.com

© The Author(s), under exclusive license to Springer Nature Singapore Pte Ltd. 2023   89
C. So-In et al. (eds.), *Information Systems for Intelligent Systems*, Smart Innovation,
Systems and Technologies 324, https://doi.org/10.1007/978-981-19-7447-2_9

highest accuracy feature selection process needs to be improved and their work does not describe which features helped them get the results. Tarek et al. [4] has used a pattern matching and protocol analysis (using decision trees) approach. Since the pattern matching approach works using signature detection, the disadvantage of the pattern matching approach is that there are many new signatures that are created for novel intrusion detection attacks this leads to the pattern matching approach being expensive. Chuanlong et al. [5] has used the RNN algorithm to create an IDS. This method has given them a higher accuracy but has taken a lot of time to detect and contain vanishing gradient which shows that their model may not be able to retrieve long-term memory. The issue with the signature-based strategy is that it isn't appropriate for continuous organization malicious discovery since it requires refreshing the dataset signature when another novel signature is introduced. Subsequently, more exploration is expected to recognize network traffic abnormalities and distinguish new kinds of abnormalities utilizing progressed machine learning classification techniques. With the traffic developing quickly, meeting constant abnormality discovery prerequisites is a significant test. Machine learning has become an ever-increasing number of well-known lately. This is because of the coming of numerous new PC innovations and the accessibility of more information. Machine learning advances have been around for quite a while; however, figuring out how to utilize them effectively and progressively is a recent problem. As referenced before, the utilization of machine learning techniques to identify interruptions has been contemplated by many. In any case, supposedly, the business interruption location instrument doesn't have these methods. As of now, existing methods are signature-based. Additionally, of the 4444 distinct methods endeavored by specialists, it isn't set up which technique is more appropriate for this application. Likewise, various techniques should be analyzed based on a typical benchmark dataset. Recognition accuracy is a significant attribute when fostering an interruption identification framework. The framework ought to perform fitting discovery undertakings with a high location pace of malevolent action yet a low number of false positives during typical computer use. Machine learning is a novel technique that has been used by researchers to classify denial of service attacks. There are many supervised machine learning algorithms used by researchers such as SVM, artificial neural network, Random Forest, etc. Supervised machine learning depends on historic data. On the other, unsupervised learning does not require historic data, and its models can be trained without any dataset. But in most cases, supervised machine learning algorithms perform better.

This survey paper is structured as follows: Sect. 9.2 discusses the fundamental knowledge on network attacks. Section 9.3 provides information on the algorithms present to aid detection and Sect. 9.4 provides data on how an algorithm's performance is measured. The second last Sect. 9.5 provides relevant literature.

9.2   Preliminary Discussion

The objective of network security is to prevent damage to resources present in the network, prevent downtime of services provided by these resources, and ensure data integrity and confidentiality. We discuss the types of attacks in this section.

9.2.1   Denial of Service

This attack is performed to disrupt the services of a system. For instance, a web application is hit with a DOS attack when too many users try to log in and the absence of a load balancer causes the Web site to load and process data slower. If it is an ecommerce Web site, users may not be able to see the products, payments will not be completed successfully, loss of data, etc.

9.2.2   Probe

Before attacking the network, an attacker may do a little investigation about the devices present in the network. Probing attacks are very normal methods of gathering information about the kinds and quantities of machines associated with a network, and a host can be attacked to decide the sorts of software introduced and additionally applications utilized. A probe attack is viewed as the first phase in a genuine attack to think twice about the attack vectors present for hosts in the network. Albeit no particular damage is brought about by this phase, they are viewed as genuine dangerous abnormalities to organizations since they may acquire valuable data for dispatching a coordinated attack.

9.2.3   User to Root

The purpose of this attack is to gain confidential user information which would help control or exploit significant organization assets. Utilizing a social engineering approach or sniffing credentials, the attacker can get to a typical client record and afterward exploit some weakness to acquire the advantage of being an administrator.

9.2.4 Remote to User

This type of attack is performed in-game superuser access in a device connected to the same network as the attacker. These types of attacks are called R2L. The attackers use the trial and error method to find out the passwords, this can be done using brute-forcing or using automated scripts, etc. One of the sophisticated ways includes an attacker using a network monitoring tool to capture the password before attacking the system.

9.2.5 Botnets

These kinds of attacks are very traditional but still occur even today. Online PCs, particularly those with a high-transmission capacity association, have turned into a helpful objective for attackers. Attackers can deal with these PCs employing immediate exploitation. The most common attacks suggest sending files containing a malicious payload that exploits a vulnerable PC, for instance, an unpatched eternal blue vulnerability in Windows 7. For the most part, these attacks are led through automated software which helps them select their targets with ease. The necessity for dispatching direct attacks is that publicly accessible services on the designated PCs contain software vulnerabilities.

9.2.6 Dataset Used

These kinds of attacks are very traditional but still occur even today. Online PCs, particularly those with a high-transmission capacity association, have turned into a helpful objective for attackers. Attackers can deal with these PCs employing immediate exploitation. The most common attacks suggest sending files containing a malicious payload that exploits a vulnerable PC, for instance, an unpatched eternal blue vulnerability in Windows 7. For the most part, these attacks are led through automated software which helps them select their targets with ease. The necessity for dispatching direct attacks is that publicly accessible services on the designated PCs contain software vulnerabilities.

9.3 Algorithms Presented

In this survey, a myriad of algorithms had been used to detect anomalies. In the literature presented, most of them are supervised models used for the prediction of malicious attacks. Since the classification of attacks performed by most of the authors in this survey is multiclass algorithms that perform better in multiclass classification have provided higher accuracies. We describe the various techniques used by authors in this survey.

9.3.1   Supervised Learning Algorithms

Supervised algorithms map an input to output and learn about the function that helps it in the mapping process. To learn this, function the supervised algorithm needs historic data. In this scenario that historic data is the network datasets (KDD 99, NSL-KDD, KDD Test+, etc.), SVM uses a hyperplane to separate data samples of one class from another. SVMs can perform well even with limited scope training sets. In any case, SVMs are susceptible to noise near the hyperplane. The dimensions of the hyperplane depend upon the number of features, if there are two features, there is only a single hyperplane. SVM kernel is used to provide complex data transformations which help in distinguishing data samples with different labels. KNN has also been used in intrusion detection. It uses an imaginary boundary line to classify data. In the event that the vast number of a sample's neighbors fit in the same class, the sample has a high probability of having its place in that class. The parameter k incredibly impacts the functioning of KNN models. The more modest k is, the higher the risk of overfitting is high. Conversely, the bigger k is, the lesser the chance of overfitting. The Naïve Bayes calculation uses conditional probability from Bayes Theorem. Naïve Bayes classifier determines the conditional probability for each sample for different classes. When the attribute independence hypothesis is satisfied, the ideal result is reached by Naïve Bayes. The conditional probability formula is calculated as shown in Eq. (9.1).

$$P(Y = ck) = \Pi_{i=1}^{n} P(Y = ck) \qquad (9.1)$$

The decision tree algorithm is mainly used for classification. The tree-like struc-ture, which makes it easy to understand and automatic removal of inappropriate and redundant features makes it a top choice for classification problems. The learning system requires feature selection, generation of a tree, and tree pruning. During the training of the model, the tree generates child nodes from root nodes after selecting relevant features. Progressive algorithms, for instance, the Random Forest and the extreme gradient boosting (XGBoost), contain many decision trees stacked or clus-tered together. Clustering depends on similarity among data, grouping profoundly comparable data into one cluster and assembling less-comparable data into various other clusters. Different from classification, clustering is a kind of unsupervised learning. Past knowledge about labeled data is required for these types of algo-rithms; however, external knowledge is required. Therefore, the requirements for the dataset are moderately less. K-means is an example of a clustering algorithm, where the number of clusters is denoted by K and the mean of attributes is denoted by means. K-means clustering algorithm involves distance as a similarity measure criterion. *K*-means uses a centroid-based algorithm, which has a centroid for every cluster. The main motive of this algorithm is to reduce the sum of the distances of the data samples and their corresponding cluster. Data samples are split into *K* number of clusters. Less distance between two data points increases the chances of them being in the same class. Each classifier can be distinguished based on its advantages and limitations. One method is to add classifiers together to form a stronger classifier. Hybrid classifiers consist of different stages each having its classifier model. From past research, it is evident that ensemble and hybrid classifiers were better as compared to singular classifiers so more and more research has been done using these classifiers. For better performance, choosing the specific classifier is very important.

The decision tree algorithm is mainly used for classification. The tree-like structure, which makes it easy to understand and automatic removal of inappropriate and redundant features makes it a top choice for classification problems. The learning system requires feature selection, generation of a tree, and tree pruning. During the training of the model, the tree generates child nodes from root nodes after selecting relevant features. Progressive algorithms, for instance, the Random Forest and the extreme gradient boosting (XGBoost), contain many decision trees stacked or clustered together. Clustering depends on similarity among data, grouping profoundly comparable data into one cluster and assembling less-comparable data into various other clusters. Different from classification, clustering is a kind of unsupervised learning. Past knowledge about labeled data is required for these types of algorithms however external knowledge is required. Therefore, the requirements for the dataset are moderately less. *K*-means is an example of a clustering algorithm, where the number of clusters is denoted by *K* and the mean of attributes is denoted by means. *K*-means clustering algorithm involves distance as a similarity measure criterion. *K*-means uses a centroid-based algorithm, which has a centroid for every cluster. The main motive of this algorithm is to reduce the sum of the distances of the data samples and their corresponding cluster. Data samples are split into *K* number of clusters. Less distance between two data points increases the chances of them being in the same class. Each classifier can be distinguished based on its advantages and limitations. One method is to add classifiers together to form a stronger classifier. Hybrid classifiers consist of different stages each having its classifier model. From past research, it is evident that ensemble and hybrid classifiers were better as compared to singular classifiers so more and more research has been done using these classifiers. For better performance choosing the specific classifier is very important.

9.3.2 Deep Learning Algorithms

From 2015 to now, more research on deep learning-based IDSs has been underway. There is no need for feature engineering. The dataset provides an ample number of data samples that can be used by the deep learning models to learn the features. As a result, deep learning approaches can be used from start to finish. When dealing with massive datasets, deep learning models have advantages as compared to shallow models. A typical deep learning model has three layers an input, hidden, and output layer. The input layer has neurons equal to the features of the labeled data. During the training system, algorithms utilize unknown elements in the input distribution to extricate features, bunch protests, and find valuable data patterns. Optimization strategy hyperparameter selection and network architecture are important factors while determining the performance of the neural network. The activation function is calculated using the values of the hyperparameters, it is shown by formula (9.2).

$$a^i = g(x^i\ W_X + b_x) \tag{9.2}$$

In the formula above, there is a weight W_X that is associated with the input matrix x^i and b_x is the bias.

9.3.3 Sequence Models

Sequence models usually are employed to find patterns and learn from those patterns through features they learn. One of the widely used sequence models is recurrent neural network (RNN). Each neuron in the recurrent neural network is called a unit which takes into consideration current input and the information from the previous input while making a decision. The attributes of sequential data are contextual, analyzing disconnected data from the sequence makes no sense. To obtain relevant information, each unit in an RNN gets not only the current state yet additionally past states. This trademark makes RNNs often experience the ill effects of exploding or vanishing gradients. Actually, standard RNNs manage only limited length sequences. To tackle the long-term dependence issue, many RNN variants have been proposed, like long transient memory (LSTM). LSTM(s) solve the problem of exploding/ vanishing gradients. They retain more information which helps in better prediction; however, they are not utilized as much as RNN for intrusion detection.

9.3.3.1 Hyperparameters

In this section, we discuss various parameters used for setting the configuration for the neural network there are many types of parameters that are discussed further.

Number of Neurons

In every deep learning model, there are neurons present indie input–output and hidden layers according to the dataset, and the arrangement of neurons must be taken into consideration. For instance, the data samples in the dataset will be converted into a matrix consisting of binary numbers so the number of input neurons should be equal to the features of the dataset if it is a labeled dataset. If there is a multiclass classification, then the output neurons will be based on the number of classes.

Learning Rate

It is the adjustment value for the weights provided which the input so to make the deep neural network converge. If the learning rate is high, the model executes faster; however, the convergence chance is low. It is the converse for a small learning rate.

Optimizer

They are a preset configuration of methods that would help neural networks in learning with ease and reducing loss function. Although there are many optimizers used, for classification purposes, this survey will cover a few of them. Gradient descent is a first-order optimization algorithm is used by backpropagation neural, also in classification and regression problems. It reduces the loss function by altering the weights and the main benefit of this optimizer is that it's easy to implement; however, it requires a large Ben a larger dataset is provided for training. Stochastic gradient descent overcomes some of the disadvantages of the gradient descent optimizer in stochastic gradient descent; the derivative is taken one at a time, which requires less memory when loading larger datasets. Min-Batch Gradient Descent is a modification of the original gradient descent optimizer and for a vendor standard gradient descent and the stochastic gradient descent optimizers. One of its advantages is that it updates the model parameters after dividing the dataset into batches. It also doesn't require a high amount of memory. However, this is not useful if better meters such as the learning rate needs to be constant.

9.4 Metrics

Metrics are required to measure the performance of the machine learning algorithms. Best performing algorithms can be observed through the outcome of these metrics. To broadly measure the intrusion identification result, various metrics are applied concurrently in IDS research. Accuracy is defined as the quantity of correctly classified samples to the sum of all samples. Accuracy is best used when the dataset is balanced that is the number of classes has an equal number of data samples; however, if there is an imbalance in the dataset accuracy may not prove the best metric imager performance. It is measured using the formula Eq. (9.3)

$$\frac{TP + TN}{TP + FP + FN + TN} \tag{9.3}$$

The required parameter for each of the metrics presented in this survey is as follows.

True Positive (TP): The number of samples correctly identified as malicious traffic.

True Negative (TN): The number of samples correctly identified as legitimate traffic.

False Positive (FP): The number of samples wrongly identified as malicious traffic.

False Negative (FN): The number of samples wrongly identified as legitimate traffic.
  Precision is demarcated as the number of correct predictions made and the accuracy of the minority class. When false positives need to be reduced precision is a good metric to measure. It is represented in Eq. (9.3)

$$\frac{TP}{TP + FP} \tag{9.4}$$

Recall helps in providing a value to the missed positive predictions. It should be used when the objective is to reduce false negatives. It is calculated using Eq. (9.4).

$$\frac{TP}{TP + FN} \tag{9.5}$$

F-measure contains the properties of both precision and recall. It is measured using the formula in Eq. (9.5). It can be used in the case of an imbalanced dataset used to train models.

$$\frac{2 * P * R}{P + R} \tag{9.6}$$

The false negative rate helps quantify samples that were falsely identified as legitimate traffic but turned out to be malicious traffic. In other terms, it measures the missed incorrect predictions. It is measured using Eq. (9.6).

$$\frac{FN}{TP + FN} \tag{9.7}$$

The false-positive rate is best explained as the proportion of false positive samples or samples that were incorrectly identified as benign traffic. In attack detection, the FPR is calculated by (9.7).

9.5   Literature Review

Kehe et al. [6] have used a CNN model to create an IDS. Their approach performs well as compared to outdated intrusion detection procedures. But the detection accuracy is too low, and the time required to detect needs to be reduced, Mohammed et al.

[7] have made a custom algorithm that is based on particle swarm optimization. While their fast learning network can provide higher accuracies when the number of hidden layers increases, they have not considered the accuracy of the minority class which is low due to a smaller number of samples of that class. Chou et al. [8] have used an ensemble technique on KDD99 Dataset and they have received good results, but their detection of normal and malicious activities is not up to the signature for R2L attacks, even after using combination techniques. Liang et al. [9] have utilized IoT to recognize DOS attacks and their work depends on a multi-specialist system, utilizing blockchain and profound learning in which every one of them enjoys their benefits. Nonetheless, the system can't identify the uncommon sorts of attacks in their current circumstance. Makious et al. [10] endeavor to track down the effect of component determination on the precision of their classifier Notwithstanding, their own did not depend on certifiable attacks and simply works in unreasonable attacks. Bhattacharjee et al. [11] have utilized the NSL-KDD dataset to prepare fuzzy algorithms like *K*-Means and C-implies and do a correlation of their presentation. They discovered their assault detection is exceptionally less. Dutkevych et al. [12] gave an irregularity-based answer for forestall convention-based attacks and intrusion into ongoing systems that break down multidimensional information traffic. Be that as it may, there are some change regions to reduce the accuracy of IPS. Zhengbing et al. [13] has reported a lightweight intrusion detection system for constant, productive, and compelling detection of intrusions. In this paper, conduct profiles and data mining procedures are naturally kept up with to recognize facilitated attacks.

9.6 Inference from Existing System

From the existing system, we can conclude that most of the methods use data mining techniques for intrusion detection and have achieved a very high accuracy or any other performance metric. Massive datasets suggest NSL-KDD have been used to train machine learning and deep learning models in most cases, accuracy is above 85%. Apart from data mining techniques such as mouse dynamics performed by authors in [14] have also shown a good performance in terms of detection but in all the methodologies presented in this paper all of them fail to identify zero-day intrusion detection techniques which can be used to create a more safer network.

9.7 Methodology

We have used a dataset containing 41 features and a binary label. We have plotted a comparison chart to find out the importance of each feature and we then select top 5 features. For the neural network, we have developed a feed-forward neural network with two hidden layers and one output layer shown in Fig. 9.2. Through trial and error, we discovered that a total of 16 neurons would be appropriate to get maximum performance. Before training the network, we perform data cleaning by removing null values, dropping columns with unique values for each sample. Next, we scale the values of the selected features between 0 and 1, this provides, better learning for the neural network. The feature importance graph, which is calculated using a Random Forest model.

We have chosen the top five features for our work, namely: Source bytes: Size of the network packet sent from the source. Destination Packet: size of the network packet sent from the destination to source. Flag: status of the connection between the machines/operating systems. Same service rate: percentage of connections to the same service. Destination host same service rate: percentage of connections between same ports running the same service.

9.7.1 Our Neural Network Architecture

In Fig. 9.1, our architecture for the neural network is presented with the green circles showing the input neurons, blue ones showing hidden layer neurons which are 16 in our case. The output layer contains a single neuron as this is a binary classification for which single neurons are required.

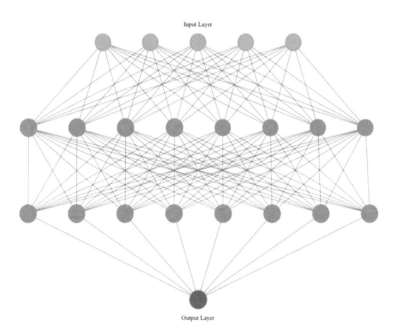

Fig. 9.1 Neural network architecture

9.8   Results

We discuss the results of our neural network output here. Since the output of the neural network is a probability, we measure the performance of the neural network at specific probabilities. Table 9.2 gives the precision, recall, and f-measure at different stages. Table 9.1 contains the true positive, true negative, false positive, and false negative for threshold starting from 0.6 to 0.9. Table 9.1 is important for determining the relationship false positive and the threshold. As the threshold increases, the wrongly identified samples decrease. However, the wrongly identified benign network (false negatives) samples have been increasing as the threshold has increased. With 0.9 being the highest threshold, we can see the false positive and true positive values to be 0. This is due to least number of wrongly identified.

As the probability threshold increases, we see a decrease in the precision and f-measure for the normal traffic flowing. However, there is an increase in the recall value for normal and a decrease for anomaly recall. We create a precision–recall curve to explain the above table in a better way.

In Fig. 9.2, the curve tells us different values of precision and recall. Apart from providing a relation between the two values for different thresholds, it also tells us about the performance of the neural network, where high precision relates to a low false positive rate, and high recall relates to a low false negative rate. A system with

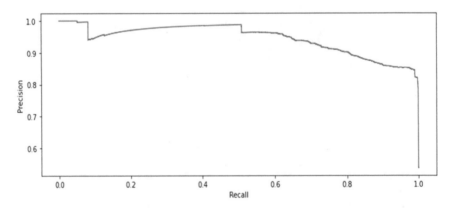

Fig. 9.2   Precision–recall curve for all thresholds

Table 9.1   Confusion matrix

Threshold	True negative	False positive	False negative	True positive
0.6	6814	1431	87	9302
0.7	6890	1355	161	9228
0.8	7192	1053	1339	8050
0.9	8245	0	9389	0

Table 9.2 Precision, recall, and f-measure at different stages

Probability threshold	Class type	Metrics		
		Precision	Recall	F-measure
0.7	Normal	0.89	0.81	0.85
	Anomaly	0.85	0.91	0.88
0.8	Normal	0.84	0.83	0.84
	Anomaly	0.85	0.86	0.86
0.9	Normal	0.75	0.87	0.81
	Anomaly	0.87	0.74	0.80

high recall but low precision returns many results, but most of its predicted labels are incorrect when compared to the training labels. A system with high precision but low recall is just the opposite, returning very few results, but most of its predicted labels are correct when compared to the training labels.

9.9   Conclusion

Our survey paper has presented a lot of exploration work dependent on identifying Denial of Service attacks. Attack classifications are introduced to accentuate the requirement for components to distinguish attacks. Our examination presents the most established and most recent in intrusion detection innovation and gives an itemized classification of various detection approaches utilizing various models. IDS abilities are probably going to be significant elements of organization network architecture (switches, spans, switches, and so forth) and working frameworks. IDS systems(s) predicting attacks without being updated continuously is wanted as most of the past work done depends on various information mining procedures; however, their detection rate for new attacks is very low.

9.9.1   Comparison with Existing Systems

We presented our approach that can distinguish most recent attacks or intrusions in the organization into a binary classification. Our approach has used five features which we found to be enormously contributing to the detection of anomalous traffic. We compare our precision, recall, and f-measure values obtained at 0.7 threshold with the work of others recent research as shown in Fig. 9.3.

   While our accuracy, our work performs a binary classification and does not provide the name type of network attack occurring, it can only sense malicious intent in the network and thus for future work, a classifier which can accurately pinpoint a type of attack is desired. Thus, future work regarding accurate detection of the type of attack

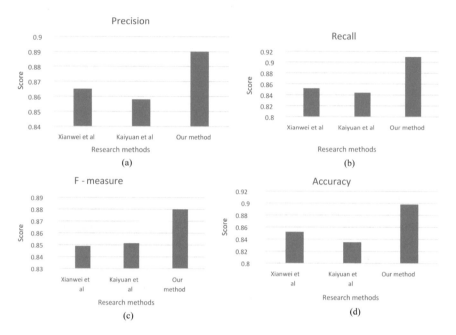

Fig. 9.3 Comparing precision (**a**), recall (**b**), *f*-measure (**c**) and accuracy (**d**)

is required. Binary classifiers do identify malicious intent but a proper classification of these attacks is required.

References

1. Meng, W., Lu, Y., Qin, J.: A dynamic mlp-based ddos attack detection method using feature selection and feedback. Comput. Secur. **88** (2020)
2. Brown, D.J., Suckow, B., Wang, T.: A Survey of Intrusion Detection Systems, vol. 146. Department of Computer Science, University of California, San Diego
3. Bakshi, A., Yogesh, B.: Securing cloud from ddos attacks using intrusion detection system in virtual machine. In: Second International Conference on Communication Software and Networks, vol. 123, pp. 260–264 (2010)
4. Lo, C.C., Huang, C.C., Ku, J.: Cooperative intrusion detection system framework for cloud computing networks. In: First IEEE International Conference on Ubi-Media Computing, vol. 23, pp. 280–284 (2008)
5. Dutkevyach, T., Piskozub, A., Tymoshyk, N.: Real-time intrusion prevention and anomaly analyze system for corporate networks. In: 4th IEEE Workshop on Intelligent Data Acquisition and Advanced Computing Systems: Technology and Applications, vol. 136, pp. 599–602 (2007)
6. Zhengbing, H., Jun, S., Shirochin, V.P.: An intelligent lightweight intrusion detection system with forensic technique. In: 4th IEEE Workshop on Intelligent Data Acquisition and Advanced Computing Systems: Technology and Applications, vol. 36, pp. 647–651 (2007)
7. Han, H., Lu, X.L., Ren, L.Y.: Using data mining to discover signatures in network- based intrusion detection. In: Proceedings of the First International Conference on Machine Learning and Cybernetics, Beijing, vol. 1, pp. 1–12 (2002)

8. Zhengbing, H., Zhitang, L., Jumgi, W.: A novel intrusion detection system (nids) based on signature search of datamining. In: WKDD First International Workshop on Knowledge discovery and Data Ming, vol. 28, pp. 10–16 (2008)

9. Mehra, M., Saxena, S., Sankaranarayanan, S., Tom, R.J., Veeramanikandan, M.: IOT based hydroponics system using deep neural networks, vol. 155 (2018)

10. Zhengbing, H., Jun, S., Shirochin, V.P.: An intelligent lightweight intrusion detection system with forensics technique. In: 2007 4th IEEE Workshop on Intelligent Data Acquisition and Advanced Computing Systems: Technology and Applications, September 2007, vol. 11, pp. 647–651 (2007)

11. Cannady, J.: Artificial neural networks for misuse detection. In: Proceedings of the 1998 National Information Systems Security Conference (NISSC'98) pp. 443–456 (1998)

12. Grediaga, Á., Ibarra, F., García, F., Ledesma,. B., Brotóns, F.: Application of neural networks in network control and information security. In: International Symposium on Neural Networks, May 2006, vol. 3973, pp. 208–213. Springer, Berlin, Heidelberg (2006)

13. Vieira, K., Schuler, A., Westphall, C.: Intrusion detection techniques in grid and cloud computing environment. In: Proceeding of the IEEE IT Professional Magazine (2012)

14. Roschke, S., Feng, C., Meinel, C.: An extensible and virtualization compatible ids management architecture. In: Fifth International Conference on Information Assurance and Security, vol. 2, pp. 130–134 (2009)

Chapter 10
Design and Development of Smart Waste Management Automatic Bin

Vipin Bondre, Kunal Tajne, Sampada Ghode, Rithik Gondralwar, Prajakta Satpute, and Sanket Ramteke

Abstract A dustbin is a garbage receptacle constructed of metal, plastic, or any other hard-to-store waste material that is used for temporarily keeping trash. Smart containers help to provide. The smart bin is suitable for high-traffic areas such as campuses, theme parks, airports, train stations, and shopping malls. They help to keep the environment clean by storing in a variety of renewable and non-renewable materials. Smart dustbin, as its name suggests, functions intelligently or can be described as an automatic dustbin. It works by using a servo motor to open the trashcan automatically when you walk in front of it. As a result, some sensors are in use to detect the object in front of the trash can. This undertaking the smart waste dustbin system is an ingenious device that will aid in the clean-up of our towns. Although dustbins are supplied in many metropolitan areas for people to use, they are not properly maintained, which is a major factor in environmental pollution and is degrading our environment day by day, resulting in serious unfavorable impacts for mankind. The garbage used to be collected manually in the conventional system versus automated system were presented through this proposed work. As a result, time that has been consumed in manual conventional system is comparatively more. This proposed system will allow them to save time and effort in a more efficient manner. Nowadays, automation is the most desired characteristic. Smart dustbins are the most appropriate solution for this. It will aid in the development of a green and smart city. To do so, we'll need to create an autonomous smart dustbin that can first recognize the garbage bin's current state. They will be able to empty the container right away. It ultimately aids in maintaining cleanliness in society, reducing the spread of diseases caused by trash.

V. Bondre (✉) · K. Tajne · S. Ghode · R. Gondralwar · P. Satpute · S. Ramteke
Yeshwantrao Chavan College of Engineering, Nagpur 441110, India
e-mail: vipin.bondre@gmail.com

C. So-In et al. (eds.), *Information Systems for Intelligent Systems*, Smart Innovation,
Systems and Technologies 324, https://doi.org/10.1007/978-981-19-7447-2_10

10.1   Introduction

Population increases and proportionally total garbage in urban areas. It's the perfect size for little spaces. The movement sensor on this bin is unique in that it opens the lid mechanically when it detects movement around it. This is a fantastic feature that allows you to throw trash without having to touch the bin. As you get further away from the bin, the lid closes on its own. Using IoT and sensor-based circuitry, we suggest a smart trash can that operates automatically to help alleviate this problem [1]. Ordinary trash cans must be opened by pressing your foot against the lever and then thrown away. Also, a person must keep track of when it is full in order to empty it and prevent it from overflowing. Here, we suggest a Smart Dustbin, Swachh Bharat With IoT that performs everything on its own. Our device comprises a sensor that detects human clap signal, and when a motion is detected, it opens automatically without anyone having to press the lever. When receiving the signal, the sensor will automatically open and close the hatch.

A level measuring ultrasonic sensor is also included in the dustbin, which continuously measures the amount of waste in the bin and automatically identifies when it is about to fill up. As a result, we have a fully automated smart dustbin with garbage level, monitoring and lid opener.

10.2   Literature Survey

The suggested scheme detects the status of the smart bin and evaluates whether it is full or empty in able to cater the garbage pickup schedule. As a result, informs the authorities, reducing the cost and risk. Saving time Waste management in real time using a smart dustbin system the dustbin is at a certain level [1]. It will inform the relevant authority about just the state of each dustbin. As a consequence, the waste management vehicle can only be used when it is extremely necessary. The proposed system consists of two dustbins, with Dustbin B not being used until Dustbin A is filled up. When Dustbin A is full, Dustbin B can only be used, and Rubbish bin A will not reopen until the garbage in the bin A is cleared. Visible wavelength sensors are installed at the front of the bins so when someone passes in front of it, it automatically opens and closes using a servo motor. Furthermore, an ultrasonic sensor is mounted within the bins to check the level of automatic bin whether it is full. When the trashcan A or B is full, a signal or status is sent to the appropriate agency via the GSM system [2]. Due to a surge in waste, rubbish bins in several cities are overflowing at various public locations. It generates environmental contamination and unpleasant odors, allowing deadly illnesses and human illness to spread. In today's world, most megacities are undergoing transformations and will most likely be reformed as smart cities. The author proposed designing a Waste Management

System for Smart Cities based on IoT to avoid an unsanitary scenario created by ineffective rubbish collecting methods. Multiple waste bins will be positioned across the city or campus in this proposed system; these automatic trash bins will be fitted with an inexpensive embedded device that the level of disposal in each bin including its exact area [3]. The trash collection system currently in use includes garbage truck rounds on a daily or weekly basis, which not only do not touch each part of the town but are also remains of public sectors. An author of this study [4] recommends a cost-effective IoT-based system for the administration to use available resources to address the situation properly the massive amounts of garbage collected each day, as well as a better option for citizens' garbage evacuation discomfort. This is achieved through an arrangement of smart bins which monitors and analyzes data acquired in order to produce predictive garbage truck routes using cloud-based technologies utilizing algorithms checked to see whether the trash one was filled or not. The Internet of Things was used to build an automatic trash segregation system in this project. In particular, we developed a system of automatic bin which includes sensors that can intelligently separate the waste and provide a waste pickup depends on its observation report. To conduct automatic trash classification using image recognition, machine learning was applied. The created models were used to classify garbage effectively after training more than thousands biodegradable and non-biodegradable trash samples [5]. Waste disposal has recently become a major concern around the world. Garbage is created in large quantities and dumped in environmentally dangerous ways. To resolve this concern, an Internet of Things-based smart bin is a nice decision that will assist to save environment. The author in its proposed system [6] presented an automatic waste segregator as a simple, rapid solution for a waste segregation system that can be provided to processing without any delay. The conventional way of it takes too long to manually monitor trash in trashcans that requires a major human effort, time, and money, all of which can be avoided with history's advancements. The goal of this article is to utilize IoT's powerful tools to completely automate the garbage monitoring process using ultrasonic sensors and Node MCU, as well as to provide an optimal garbage collection route [7]. The article suggested a real-time garbage collection and management system based on smart bins in this document in which waste is first sorted and collected in dedicated bins that are monitored in real time on the cloud. We receive an email notification as an alert if the garbage bins were full, which is forwarded to an authorized person. If a fire is detected, a message notice is also sent [8]. The author developed a cloud-based system to manage the solid waste management process, as well as a mobile application for waste collection drivers and the city council to monitor and control this operation [9].

The survey based on above research paper indicates that there is always scope of its future upgradation and development.

10.3 The Ease of Use

We begin by taking a standard dustbin and measuring its dimensions. Then, at the center of the front side of the dustbin, we make a round hole. We drill a hole, the size of a Passive Infrared Sensor (PIR), in the dustbin and install the sensor in the center of the hole.

We use a copper PCB board (8cmx6cm) to connect four LEDs (shown in Fig. 10.3) that display the amount of waste in the trash can. These four LEDs are White, Green, Blue, and Red, and they come in four different colors. Where white denotes the lowest level of waste or no garbage, green denotes a low garbage level, blue denotes a medium garbage level, and red denotes a full dustbin.

Then we place an ultrasonic sensor under the lid to sense the trash in the trashcan. The ultrasonic sensor is aimed downwards and toward the dustbin's end.

Now we need to make room for the servo motor on the backside of the trashcan. The hole is drilled on the right side of the lever and vertically in the dustbin's middle.

We now attach a thread to the servo motor's knob, which will be used to pull the dustbin lid open. The other end of the thread is attached to the dustbin's inner lever, which is pulled by a servo motor to open the lid. 12.

The Arduino UNO is then placed on the right side of the trashcan inside a plastic container, along with a PCB Board for +ve and −ve supply.

10.4 Working

Let's start programming with Arduino. Arduino Uno is a microcontroller board with ATmega328 processor. Due to its simplicity, Arduino is a great open source prototyping tool for enthusiasts as well as professionals. The Arduino Uno includes four digital I/O pins, six analog inputs, a 16 MHz crystal oscillator, a USB connection, a power connector, an ICSP header, and a reset button. It includes everything you need to get started with the microcontroller, including a USB cable connecting it to your computer and an AC–DC adapter or battery to power it. The proposed design and its integration schematic is shown in Fig. 10.1.

The Arduino Uno differs from previous Arduino boards in that it does not have a USB-to-serial FTDI controller chip. Instead, it uses the ATmega 8U2 microprocessor, which is used as a USB-to-serial converter. The Italian word "uno" means one. The Arduino Uno and version 1.0 will continue to be the standard version of Arduino going forward Uno is the latest in the line of Arduino USB boards, as well as a popular platform the most common, its illustration is shown in Fig. 10.2.

After you've set up the smart dustbin and made all of the necessary connections, submit the code to Arduino and power the device with 5 V. When the system is turned ON, Arduino keeps an eye out for anything that comes close to the PIR Sensor.

Passive (PIR) sensors are infrared rays that measure the thermal energy of the environment using a pair of pyro electric sensors as shown in Fig. 10.3. These two

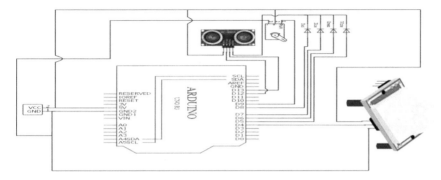

Fig. 10.1 Design and integration of smart dustbin

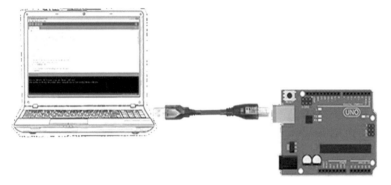

Fig. 10.2 Illustration of code transfer from Laptop to The Arduino open-sourced IDE drivers can be downloaded for free

sensors are placed side by side and the sensors (the signal difference between them) It is activated when it changes, for example, if a person enters the room. It can be set to turn OFF the LED light activation alarm or warn the authorities. The set of lenses in the sensor housing focuses IR radiation on each of the two charcoal sensors.

These lenses increase the detection area of the device. Despite the complex physics of lens design and sensor circuits, these products are easy to use. The sensor only needs power and ground to generate a sufficiently strong discrete output for the microcontroller. Adding a sensitivity potentiometer and adjusting the amount of time the PIR is active after power ON is two things. This is a common change. Sensor parameters can also be changed as follows:

1. Staying on for a set period of time after detecting movement.
2. Pulsing ON and OFF in a "non-retriggering" manner.

Servo motors are electromechanical devices that generate torque and velocity by combining current and voltage. A servo motor is a component of a closed loop system

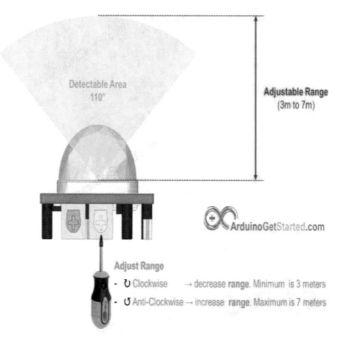

Fig. 10.3 PIR sensor range and motion

that produces torque and velocity in response to servo controller inputs and is closed by a feedback mechanism as shown in Fig. 10.4.

The motor, the feedback circuit, and, most significantly, the motor driver are all included in a servo motor. There is only one power line, one ground line, and one control pin that must be used.

The following are the instructions for connecting a servo motor to an Arduino:

Fig. 10.4 Servo motor and its three pins

1. A three-pin female connector is found on the servo motor. Frequently, the earth is the darkest, if not completely black. Connect this to the Arduino's GND pin.
2. Connect the Arduino's power cable to 5 V (by all standards, it should be red).
3. Connect the remaining line from the servo connector to an Arduino digital pin. After receiving the signal, the Servo Motor pulls the knob downwards, stretching the thread and opening the dustbin lid. According to the time delay established in the potentiometer, the dustbin remains open. When the timer expires, the PIR sensor provides an output to Arduino, which instructs the servo motor to return to its original position, reducing the tension in the string and closing the dustbin lid.

The dustbin's other component is rubbish level detection, which is performed with an ultrasonic sensor. This sensor measures the depth of trash in the can and transmits the data to the Arduino.

In ultrasonic sensing, a transducer sends and receives ultrasonic pulses to transmit information about the vicinity of an object. Different types of echoes are produced when high frequency sound waves bounce off objects as shown in Fig. 10.5.

Ultrasonic sensors work by emitting sound waves louder than humans can hear. "The sensor's transducer acts like a microphone that receives and transmits ultrasonic waves". Many people, including you, use a single transducer to transmit pulses and receive echoes. "The sensor measures the time between sending and receiving an ultrasonic pulse to estimate the distance to the target".

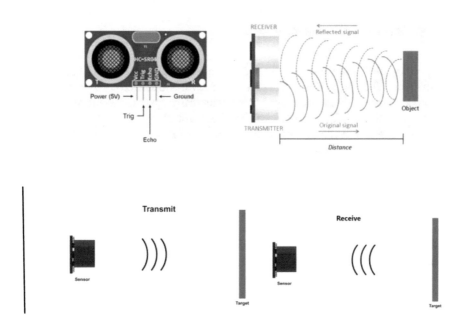

Fig. 10.5 Ultrasonic sensor working illustration

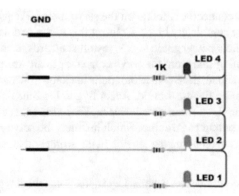

Fig. 10.6 LED Panel circuit to show level of garbage white (Distance1 >= 15) && (Distance1 <= 20): It is the lowest garbage level. Green (Distance1 >= 10) && (Distance1 <= 15): It is the low garbage level. Blue (Distance1 >= 05) && (Distance1 <= 10): It is the medium garbage level. Red (Distance1 >= 00) && (Distance1 <= 05): It is the highest garbage level/dustbin is full

When the ultrasonic sensor detects the distance and sends the information to Arduino, the Arduino uses its code to switch ON the LED light. Four LEDs indicate the amount of trash in the dustbin. The colors and their traditional connotations are as follows shown in Fig. 10.6.

10.5 Result and Conclusion

Depending on the garbage level, a smart dustbin might provide an LED notification to the user. When the rubbish reaches a certain level, the user can empty the dustbin. It also responds to human presence by automatically opening the lid, allowing the user to toss the rubbish without having to deal with the dustbin. Our results show the comparison between the conventional bin and automatic bin. This study of our proposed design and development is carried on five days span of time. Our results show the outstanding results I terms of efficiency as shown in Figs. 10.7 and 10.8.

Above graph show the basic comparative efficiency performance difference between our automated system and conventional system. Things were used to build an IoT-based smart trashcan were Uno, Ultrasonic, PIR, and a servo motor Although an ultrasonic sensor can detect the distance between waste materials and the sensor using a range of materials such as vegetables, paper boxes, paper, plastic, poly foam, glass, and a mixture, it cannot detect the distance between waste materials and the sensor. There is a reading error of less than 2–3 cm in the actual value.

Because there is sufficient area to loss waste items, the result suggests that 5 cm or less is the best distance to transmit red LED signal to user. The majority of those who were interested in this concept believed that the product had future commercial

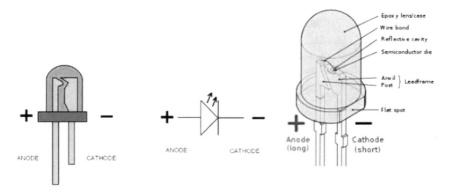

Fig. 10.7 LED interior and information about its part

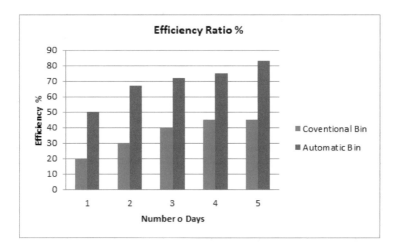

Fig. 10.8 Efficiency ratio

potential, but that the design needed to be improved in terms of efficiency and effectiveness. 5 out of 8 people are very happy with the system's features. Overall, user was pleased with the product's functionality. However, a small percentage of users felt that the product was too expensive.

The waste management is also carried on our regular conventional bin and our developed automated bin. As shown in Fig. 10.9. Here from the our five day survey, we again found that automated bin performance is again better than the existing conventional bin performance is able to achieved only up to 43% of waste management, whereas the automated bin has performed up to 88%. The five out of eight people are very happy with the system's features. Overall, user was pleased with the product's functionality. However, a small percentage of users felt that the product was too expensive.

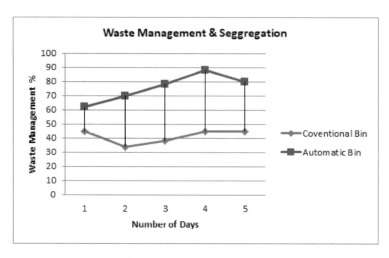

Fig. 10.9 Waste management and segregation

Acknowledgements We thank our project guide Dr. Vipin D. Bondre, for his thorough guidance throughout the project. We are deeply grateful and appreciative to them for their skilled, sincere, and invaluable advice and support, which has been of enormous use to us.

We would like to thank, Head, Department of Electronics and Telecommunications, for his unwavering support in ensuring the successful completion of our project.

We would like to thank, our college's Principal, for providing us with all of the essential resources and infrastructure, without which we would not have been able to complete our project successfully.

We'd also want to thank, our Project Coordinator, for his constant direction, which allowed the project to take shape.

We thank to our technical assistant, for providing us with the essential technological assistance. Last but not least, we'd want to express our gratitude to the entire faculty and non-teaching staff members who volunteered to help us despite their hectic schedules.

References

1. Murugaanandam, S., Ganapathy, V., Balaji, R.: Efficient IOT based smart bin for clean environment. In: International Conference on Communication and Signal Processing, April 3–5, 2018, India (2018)
2. Sai Rohit, G., Bharat Chandra, M., Saha, S., Das, D.: Smart dual dustbin model for waste management in smart cities. In: 2018 3rd International Conference for Convergence in Technology (I2CT) The Gateway Hotel, XION Complex, Wakad Road, Pune, India, Apr 06–08 (2018)
3. Rambhia, V., Valera, A.: Rahul Punjabi smart dustbins automatic segregation &efficient solid waste management using IoT solutions for smart cities. Int. J. Eng. Res. Technol. (ijert) **8**(12) (2019)
4. Lokuliyana, S., Jayakody, A., Dabarera, G.S.B., Ranaweera, R.K.R., Perera, P.G.D.M., Panangala, P.A.D.V.R.: Location based garbage management system with IoT for smart city. In: The 13th international conference on computer science & education (ICCSE 2018) August 8–11, 2018. Colombo, Sri Lanka (2018)

5. Pamintuan, M., Mantiquilla, S.M., Reyes, H., Samonte, M.J.: iBIN: An intelligent trash bin for automatic waste segregation and monitoring system. In: 2019 IEEE 11th International Conference on Humanoid, Nanotechnology, Information Technology, Communication and Control, Environment, and Management (HNICEM)
6. Anusree, K., Michael, M., Asok, V.P., Sreya, R.S., Sandra, P.V.: Iot based waste management and segregation. Int. J. Innov. Sci. Res. Technol. **6**(6) (2021). ISSN No: 2456-2165
7. Medehal, A., Annaluru, A., Bandyopadhyay, S., Chandra, T.S.: Automated smart garbage monitoring system with optimal route generation for collection. In: 2020 IEEE international smart cities conference (ISC2), 28 Sept–1 Oct 2020
8. Jain, R., Halder, O., Sharma, P., Jain, A., Elamaran, E.: Development f smart garbage bins for automated segregation of waste with realtime monitoring using Iot. Int. J. Eng. Adv. Technol. (IJEAT), **8**(6S) (2019). ISSN: 2249–8958
9. Chaudhari, S.S., Bhole, V.Y.: Solid waste collection as a service using IoT solution for smart cities. In: 2018 International conference on smart city and emerging technology (ICSCET), IEEE (2018)

Chapter 11
A Digital Socio-Technical Innovation to Bridge Rural–Urban Education Divide: A Social Entrepreneurial Perspective

Somprakash Bandyopadhyay, Arina Bardhan, Priyadarshini Dey, and Jayanta Basak

Abstract The paper addresses the problem of extant rural–urban divide regarding access to quality education and training and describes a sustainable social entrepreneurial model using social technologies (*Social technology* is an umbrella term used to capture a wide variety of terminologies depicting internet-enabled communications, platforms and tools, which has the potential to establish *collaborative connectivity* among billions of individuals over the globe) to bridge this divide. In the context of West Bengal, India, we describe our social entrepreneurial initiative named *NexConnect* which is a digital teaching–learning platform that connects learners from all socio-economic background and geographical locations with knowledgeable senior citizens. Unfortunately, in today's world, the educated senior citizens in India, who possess knowledge from their acquired experiences, are no longer considered to be a part of the mainstream socio-economic activities of the nation. NexConnect uses the *dormant* knowledge resource of those educated elderly or retired teachers by connecting them online with marginalized children and young learners. The paper is a descriptive understanding of *NexConnect* as an example of Digital Socio-Technical Innovation of a social entrepreneurial initiative in solving the problem of access to quality education and to include marginalized communities into the mainstream.

11.1 Introduction

Social entrepreneurship, usually defined as "entrepreneurial activity with an embedded social purpose" [1], has become a significant economic phenomenon at a global scale [9]. While commercial entrepreneurs in the business sector identify

S. Bandyopadhyay (✉)
Indian Institute of Management Calcutta, Joka, Kolkata 700104, India
e-mail: somprakash@iimcal.ac.in

A. Bardhan · P. Dey · J. Basak
NexConnect Ventures Pvt Ltd, Kolkata 700068, India

C. So-In et al. (eds.), *Information Systems for Intelligent Systems*, Smart Innovation, Systems and Technologies 324, https://doi.org/10.1007/978-981-19-7447-2_11

untapped market potentials and mobilize resources to enter into those markets to satisfy customers and generate profit out of it, social entrepreneurs use the same profit orientation but to solve social problems for the betterment of underprivileged communities. A social entrepreneur can be for-profit or it can re-invest the profit to grow the business. Gregory Dees [5] describes social entrepreneurs as change agents, by: "adopting a mission to create and sustain social value (not just private value), recognizing and relentlessly pursuing new opportunities to serve that mission, engaging in a process of continuous innovation, adaptation, and learning". Thus, a social enterprise is an "organization that trades, not for private gain, but to generate positive social and environmental externalities" [15]. From these definitions, we can derive two important characteristics of social enterprise: a business model that incorporates some form of commercial activity to generate revenue; and a mission to satisfy one or more major social needs [8, 9, 11, 12].

The disruptive effects of Internet-driven social technologies to transform business practices are also positively influencing the social entrepreneurial practices that have the potential to do good for society and to transform local economies [2, 3]. In this context, the paper describes our initiative toward a Digital Socio-Technical Innovation and its implementation through our social entrepreneurial initiative named "NexConnect" (https://www.nexconnect.co.in), which is a digital teaching–learning platform that connects learners from all socio-economic background and geographical locations with knowledgeable senior citizens online. Educated senior citizens of India, who possess knowledge from their acquired experiences, are no longer considered to be a part of the mainstream socio-economic activities of the nation [4]. NexConnect is a digital platform that connects those educated elderly or retired teachers with children and young learners online. This model uses social technologies to digitally connect communities in rural areas with various external agencies including educated senior citizens as teachers from urban space that assures access to quality education and information.

However, the introduction of socio-technological innovations in society to solve social problems requires "a deep transition from older system to a new technology-mediated system" [7]. In the present context, we are trying to make a transition from a face-to-face, localized physical classroom environment to a digital platform-driven and personalized virtual classroom environment. These socio-technical transitions focus on technology adoption through observing changes in user practices and socio-cultural-political-economic context [10]. According to Bandyopadhyay et al. [3], "The concept of socio-technical transition stresses the interdependence of technological, social, cultural, and political dimensions, as well as the mutual adjustment of these dimensions. It puts digital platform at the core of the framework, connected to other socially enabled digital platforms and service. All actors are connected to this platform to develop a cooperative connectivity". However, this digital or virtual interactions may also be integrated with physical interactions between different entities. In fact, some form of blending of digital and physical is needed to support teaching–learning process [6].

By describing our social entrepreneurial journey of NexConnect, we explain the challenges encountered while doing physical and virtual coordination among different agents through a digital platform. Here, the social entrepreneur plays an additional role of a *transition intermediary* to help in the process of technology adoption. Such intermediation is a mandatory prerequisite in the context of underprivileged rural community, who are digitally naïve. The rural–urban education divide can only be mitigated, if the digital platform can enable effective virtual collaborations between different participating entities.

We are focusing on three participating entities such as adult learners and children in rural communities, knowledgeable elderly community who are the providers of knowledge and local agents who are acting as enablers of the rural community. On one hand, remote rural schools are shrouded with problems such as lack of quality teachers, teaching learning materials, and poor learning environment. On the other hand, senior citizens, traditionally considered as the knowledge providers and guardians of the society, are now relegated to the background and are no longer engaged in any productive activities. Through our model, we wish to connect these two disadvantaged communities and include them into the mainstream by using our digital teaching–learning platform. We require the third community of local agents to act as a bridge between the rural community and our online learning platform. We aim to sustain both the community of senior citizens and local community agents both socially and economically by proposing a business model.

11.2   NexConnect: Bridging Rural–Urban Education Divides

As mentioned earlier, NexConnect is an online teaching–learning platform, based on a cooperative relationship between various agents such as teachers, local micro-entrepreneurs/NGO and students who forms a network to achieve the following goals:

- *Bridging rural–urban education divide*: India is currently facing a huge rural–urban education divide, where, on one hand, urban schools are focusing more on high-cost privatization of learning and disseminating quality education to the urban students; and, on the other hand, rural schools are shrouded with problems such as non-availability of good schools, good teaching learning materials and quality teachers [6]. In this context, NexConnect builds a virtual live classes using social technologies where experienced and senior urban teachers get connected with all kinds of learners online using structured and audio-visual content.
- *Up skilling young adults and disseminate holistic education*: Experienced or retired teachers have acquired years of experiences that can be shared to the rural students and adult learners not only in forms of academic tutoring but also in forms of up-skilling or disseminating value education to all. There is a large section of rural population who discontinue academics after completing class X. NexConnect encourages vocational training in subjects such as tailoring, spoken

English, basic computing skills, etc., to adult learners, which would help them avail various non-mainstream activities or opportunities in building their career.

NexConnect have architected and built a digital platform that uses Internet to create virtual classrooms (using in-bulit online video conferencing system), connecting expert teachers. The day-to-day teaching–learning process is assisted by on-site classroom coordinator appointed at each rural classroom (e.g., para-teachers or local micro-entrepreneurs). The role of an on-site classroom coordinator would be to schedule classes, manage the class physically and coordinate the teaching process including assignment corrections. The students are located in remote classrooms (termed as NexConnect Internet School), which are equipped with computer, large TV or projector, audio–video system and high-speed Internet connectivity. NexConnect's digital platform uses a *blended learning framework*, with blending of three components: (i) online synchronous teaching imparted by remote teachers using virtual classrooms; (ii) online audio-visual contents in regional languages available online or made by teachers; and (iii) offline classroom coordinator, who manages the classroom physically. The platform is also equipped with a learning analytics app for easy assessment of students' performance [3].

WhatsApp communication is an integral part of this digital learning platform. WhatsApp is used by the teachers to send study materials, assignments, audio-videos clippings, etc., to learners. Students also have the opportunity to get connected with their online teachers for clarification of doubts using their parents' smartphones. The classroom coordinators interact with the teachers and students and also upload the answer sheets for the online teachers to correct. WhatsApp thus help to create virtual learning communities between teachers, students and classroom coordinators, enabling personalized interactions.

NexConnect follows a hybrid model, combining components of both profitability and non-profitability. This model, termed as the hybrid revenue model, brings in elements of non-profit entity and a business (for-profit) entity together to solve the social problem of bridging rural–urban education divide. The for-profit entity is able to earn revenue to support the non-profit entity. The hybrid model helps organizations aiming to achieve high social impact by offering them cross subsidy options that strike a balance between customer acquisition and quality access of services to benefactors [13, 14]. NexConnect creates a revenue model that successfully subsidizes rural education by earning maximum revenue from urban students:

- *Direct teaching*: Senior and experienced teachers teach both academic and non-academic subjects to students in urban areas. These students belong from upper socio-economic strata of society. These students pay standard hourly fees to the online learning platform.
- *Franchise mode of teaching or NexConnect Internet Schools*: The social entrepreneur (NexConnect) identifies some individuals (local agents) who have a study center or are willing to open study centers to host NexConnect online classes. These individuals must have the infrastructure of computer/tablet/laptop and Internet connectivity. They will have to assimilate students from the locality who needs assistance with studies. The local individuals will have to create class

of students ranging from 10 to 20. These students will pay a nominal hourly fee to the local agent. If the local agent is part of a NGO, the learners can participate free of cost.

11.3   An Example from the Field

NexConnect has been involved in academic, competency, skill based and vocational training programs to rural children and adult learners online across six districts of West Bengal and one district of Jharkhand. NexConnect has impacted about 600 students and have trained and up-skilled 200 artisans (primarily women) in rural areas within a span of two years. NexConnect generally works in partnership with several NGOs, semi-governmental bodies and governmental bodies, who has helped in creating rural entrepreneurs and establish study centers in remote rural locales. An example is given below:

Chandanpiri Ramkrishna Ashram, Namkhana, West Bengal: NexConnect has conducted more than 200 sessions in a school for underprivileged children in Chandanpiri, Namkhana, West Bengal, India (Fig. 11.1). The intervention aimed at providing the students with quality teachers and quality teaching learning material through its online digital platform. Students were taught Spoken English and Mathematics. Competency-based curriculum was provided to the teacher along with relevant videos which were shared with the students. In English, the students of class VII were taught detailed Grammar along with their text following West Bengal Board Syllabus. With rigorous efforts from both sides, students could successfully read and understand sentences in English toward the end of six months of intervention. For Mathematics, there were two remote expert teachers teaching for a period of three months each. Moreover, their capacity of grasping and understanding problems on numeracy was also very slow. However, the students were obedient and hardworking in nature. The use of graphical representations and animations to explain numerical problems made it very easy for the students to understand the concepts.

Initially when the classes started, it occurred on an irregular basis owing to problems in audio, video and Internet connectivity. After struggling for about a month, the Internet connection became stable. The teachers who were teaching them also started feeling enthusiastic about the program. Gradually with the progress in these online sessions, a marked improvement could be observed in the teacher–student online interaction level, in spite of virtual presence of the teachers. The students became very spontaneous in responding back to the online teacher. They also felt at ease to discuss their problems with the teacher. This in turn automatically had a positive effect on their learning outcomes. It further enhanced their understanding of a particular topic. The caring nature of the teacher helped to develop a strong teacher–student relationship. It also brought a sense of discipline among students. The spoken English classes remarkably improved their communicative skills and helped them to overcome their fear of speaking in English. They also learned to introduce themselves in different contexts.

Fig. 11.1 NexConnect internet school at *Namkhana*, West Bengal

During initial field visits to this center, a pilot-study was conducted to evaluate the academic standards of the students. A basic ASER survey was conducted on the students along with grade level competency tests in the subjects Mathematics and English. The pre-tests were conducted on a group of ten students from class 7, aged 12–13 years. Though they study in class VII, the students failed to write a single sentence on their own in English. However, on probing and through the results, it was revealed that they are failing to cope with the academic structure owing to poor learning conditions and lack of academic support from parents. Trial live sessions started and continued for 6 months and a total of 144 online sessions were conducted with them.

Similar to pre-study, post-studies in forms of continuous assessments were taken in both the subjects. There were two main assessments conducted along with weekly assessments. The results of the two assessments depicted a steady growth among the children. In English, it was observed that regular practice helped students retain the new vocabulary and spellings. The graph (Fig. 11.2) depicts the scores obtained by the students in the pretest and the subsequent post-tests in English. The figure shows gradual improvements among students. The first post-test included assessments on basic grammar, common vocabularies, reading short passages and answering basic questions from texts. The first post study was conducted after the first three months of intervention. The post study on Mathematics too shows gradual improvement in scores among the students, although the improvement in English is much more than that in Mathematics (Fig. 11.2).

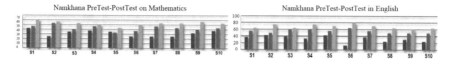

Fig. 11.2 A Longitudinal evaluation of students' performance (10 students across *x*-axis, scores across *y*-axis (out of 100 marks))

The second post-test has shown noticeable improvements among students. Once the teacher shifted to colorful visual aids, showing calculations in white boards and relating mathematical problems with real-life problems, students got more interested and started solving sums more intuitively. This resulted in higher scores in the final post study.

Despite these positive outcomes, absenteeism continued to be one of the major problems in the area. Because they had to travel huge distance to take these online classes, often it was not possible for them to come on time. Moreover, due to unfavorable weather conditions and frequent power-cut, it became difficult to retain the regularity of the classes. Nonetheless the intervention created a strong learning interest among students and expanded their learning achievements to a considerable extent.

11.4   Implementation Challenges

The scaling-up process of NexConnect's digital teaching–learning platform in different rural locations has to cope up with several implementation challenges that needs to be addressed in order to sustain this venture. They are as follows:

- *Student Drop-outs due to economic factors*: The Internet Schools those are chargeable can retain a smaller number of students. On the other hand, the free Internet Schools run by NGOs have less drop-outs of students.
- *Students Drop-outs due to social obligations*: Female learners, both young and adults, have household works, which make them irregular in attending the Internet Schools. The rural adult males also find it difficult to attend classes because of their day-long work obligations.
- *Apprehensive rural parents*: Usually, it is difficult for the rural parents to understand the benefits of learning through digital platform. So, their children are often demotivated to participate in online classes.
- *Lack of family support*: The female adult learners in rural areas often do not get family support to participate in vocational or skill development trainings, since they fail to appreciate the advantages of such trainings in creating newer livelihood opportunities.

- *Lack of support from local private tutors or local teachers*: Local private tutors and teachers feel threatened by the increasing popularity of the online classes conducted in the Internet schools. They often demotivate parents and students to participate in virtual classes conducted by NexConnect.
- *Infrastructure related issues*: Power supply disruptions, weak Internet connectivity, poor road condition and lack of availability of proper transport to NexConnect Internet Schools are some of the important parameters that hinders consistent participation in NexConnect Internet Schools.

To mitigate these challenges, NexConnect, in collaboration with local NGOs and local civic agencies, has initiated the following additional activities:

- *Conducting awareness workshops*: These periodic workshops mainly for adult learners and parents illustrate the importance of virtual classroom to get the benefit of involving expert teachers from remote locations, which, in turn, will improve the quality of learning and training.
- *Conducting digital literacy training and motivational programs for the classroom coordinators.*
- *Involving local private tutors and local teachers as classroom coordinators* to make them directly involved in the day-to-day learning activities within NexConnect Internet Schools.

11.5   Conclusion

The potentials of social business using social technologies can be used to address issues of disparity in quality education between rural and urban areas. NexConnect as a hybrid business model aims toward bridging rural–urban education divide and improve quality of learning among students in remote rural communities by cross subsidizing revenue from urban to rural sector. Thus, to sustain such a concerning issue in domain of education, a social entrepreneur needs to create networks of students, teachers, local agents, volunteers, governmental and non-governmental bodies. This points out the necessity to use components of collaboration and cooperation in empowering underserved communities using Internet-driven digital platforms. Both adult and young learners can receive quality content from the cyber-resource and interact with good quality teachers online. The paper thus aims at highlighting a relevant social business model specifically aimed at mobilizing the *dormant* knowledge capital in contributing effectively and positively toward affordable and easily accessible education for all.

References

1. Austin, J., Stevenson, H., Wei-Skillern, J. (2006). Social and Commercial Entrepreneurship: Same, Different, or Both? Entrepreneurship Theory Pract. **30**(1)
2. Bandyopadhyay, S., Bhattacharyya, S., Basak, J.: Social Knowledge Management for Rural Empowerment: Bridging the Knowledge Divide Using Social Technologies. Taylor and Francis Group, Routledge (2021a). ISBN: 978-0-367-70708-8
3. Bandyopadhyay, S., Bardhan, A., Dey, P., Bhattacharyya, S.: Bridging the Education Divide using Social Technologies: Explorations in Rural India. Springer (2021b). ISBN: 978-981-336-738
4. Bardhan, A., Bandyopadhyay, S.: Productive ageing: Insights from an action research dealing with senior citizens' engagement with an E-learning platform to educate underprivileged children. Indian J. Gerontol. **33**(2) (2019)
5. Dees, J.G.: Enterprising non-profits. Harvard Business Review, pp. 55–67 (1998)
6. Dey, P., Bandyopadhyay, S.: Blended learning to improve quality of primary education among underprivileged school children in India. Education and Information Technologies. Springer (2018)
7. Kivimaa, P., Boon, W., Hyysalo, S., Klerkx, L.: Towards a typology of intermediaries in sustainability transitions: A systematic review and a research agenda. Res. Policy Elsevier **48**(4), 1062–1075 (2019)
8. Laville, J.-L., Nyssens, M.: The social enterprise: Towards a theoretical socio-economic approach. In: Borzaga, C., Defourny, J. (eds.) The Emergence of Social Enterprise, pp. 312–332. Routledge, London (2001)
9. Mair, J., Marti, I.: Social entrepreneurship research: A source of explanation, prediction, and delight. J. World Bus. **41**(1), 36–44 (2006)
10. Markard, J., Raven, R., Truffer, B.: Sustainability transitions: An emerging field of research and its prospects. Res. Policy Elsevier **41**, 955–967 (2012)
11. Peattie, K., Morley, A.: Eight paradoxes of the social enterprise research agenda. Soc. Enterprise J. 91–107 (2008)
12. Peredo, A.M., McLean, M.: Social entrepreneurship: a critical review of the concept. J. World Bus. 56–65 (2006)
13. Porter, M.E., Kramer, M.R.: The big idea: Creating shared value. Harv. Bus. Rev. **89**(1/2), 62–77 (2011)
14. Ramírez, R.: Value co-production: Intellectual origins and implications for practice and research. Strateg. Manag. J. **20**, 49–65 (1999)
15. Santos, F.M.: A Positive Theory of Social Entrepreneurship. Working Paper 2009/23/EFE/ISIC, INSEAD Business School, Fontainebleau (2009)

Chapter 12
Robust Plant Leaves Diseases Classification Using EfficientNet and Residual Block

Vinh Dinh Nguyen, Ngoc Phuong Ngo, Quynh Ngoc Le, and Narayan C. Debnath

Abstract Vietnamese agriculture aims to develop sustainable intensification with the help of modern technology. Therefore, it is necessary to provide an intelligent system-based artificial intelligence to help farmers to analyze and diagnose the plant leaves disease to increase the plant health and productivity. Various systems have been proposed to detect and classify plant leaf diseases, such as convolutional neural networks (CNN), support vector machines (SVM), decision trees, KNN, and random forests. However, the performance of these existing systems is still not satisfactory for real applications. Therefore, we proposed a novel approach to accurately detect and classify plant leaf disease by using the EfficientNet and Residual block. The proposed system obtained better results than existing methods, such as CNN, SVM, KNN, random forest, and decision tree. The proposed system was able to detect and classify the blueberry healthy with a detection rate of 97.24%, the blueberry disease with the detection rate of 94.24%, the apple black rot with a detection rate of 95.14%, the tomato bacterial spot with a detection rate of 98.24%, and the potato late blight with the detection rate of 98.87%.

V. D. Nguyen (✉) · N. C. Debnath
School of Computing and Information Technology, Eastern International University, Binh Duong, Vietnam
e-mail: vinh.nguyen@eiu.edu.vn

N. C. Debnath
e-mail: narayan.debnath@eiu.edu.vn

N. P. Ngo · Q. N. Le
Department of Plant Physiology-Biochemistry, College of Agriculture, Can Tho University, Can Tho City, Vietnam
e-mail: npngoc@ctu.edu.vn

Q. N. Le
e-mail: ngocquynh@ctu.edu.vn

© The Author(s), under exclusive license to Springer Nature Singapore Pte Ltd. 2023
C. So-In et al. (eds.), *Information Systems for Intelligent Systems*, Smart Innovation, Systems and Technologies 324, https://doi.org/10.1007/978-981-19-7447-2_12

12.1 Introduction

Nowadays, Vietnamese agriculture aims to develop sustainable intensification with the help of modern technology. Therefore, it is necessary to provide an intelligent system-based artificial intelligence to help farmers to analyze and diagnose the plant leaves disease to increase the plant health and productivity. Various systems have been proposed to detect and classify plant leaf diseases, such as deep convolutional neural network (CNN) [1], support vector machine (SVM) [2], K-nearest neighbor [3], decision tree [4], and random forest [5]. However, the performance of these existing systems is still not satisfactory for real applications in agriculture. Therefore, we proposed a novel approach to accurately detect and classify plant leaf disease by using the EfficientNet [6] and Residual block [7]. The proposed system was developed to detect and classify five plant leaf diseases including blueberry disease, apple black rot, tomato bacterial spot, blueberry healthy, and potato late blight by using the Kaggle Plant Diseases Dataset [8] and our custom data in Vietnam. The proposed system obtained better results than existing methods, such as CNN, SVM, KNN, random forest, and decision tree. The proposed system was able to detect and classify the blueberry healthy with a detection rate of 97.24%, the blueberry disease with the detection rate of 94.24%, the apple black rot with a detection rate of 95.14%, the tomato bacterial spot with a detection rate of 98.24%, and the potato late blight with the detection rate of 98.87%.

The remainder of this paper is organized as follows. Section 12.2 presents an overview of the existing plant leaf disease detection systems. Section 12.3 introduces a novel approach for detecting and classifying plant leaf diseases. Section 12.4 presents the system configuration and experimental results of the proposed systems. Section 12.5 concludes the paper and discusses limitations and future work.

12.2 Related Work

Root, kernel, stem, and leaf of plants are often used to identify plant disease by plant pathologists. In this research, we are interested in detecting and classifying plant leaf diseases. Various approaches have been introduced to solve the problems of plant leaf disease detection involving CNN [1], support vector machine (SVM) [2], K-nearest neighbor [3], decision tree [4], and random forest [5]. Ferentinos et al. proposed an efficient approach to accurately detect plant disease by using CNN-based architecture [9]. Lakshmanarao et al. also investigated CNN to detect 15 types of diseases by using the PlantViallge dataset from Kaggle [10]. Lakshmanarao et al. divided the PlantVil-lage into two datasets: 400 images for training, and 100 images for testing. Lakshmanarao's system achieves the accuracy of 90% for classifying tomatoes healthy, the accuracy of 90% for tomatoes bacterial spot, the accuracy of 85% for tomatoes late light, and the accuracy of 80% for tomatoes Septoria spot, the accuracy of 85% for tomatoes yellow curved. Sardogan et al. found that the CNN can combine with Learning Vector Quantization (LVQ) algorithm to increase the performance of identifying tomato leaf disease [11]. Recently, Förster et al. proposed a Generative Adversarial Network (GAN) for barley plants [12]. Ganesan et al. used fuzzy-based method and CIELuv color space to detect diseases of plant leaves [13]. In this research, we aim to develop a novel deep learning-based method with fuses the benefits of EfficientNet [6] and Residual block [7].

12.3   The Proposed Method

The proposed system consists of the following steps (Fig. 12.1): First, the EfficientNet [6] is used to learn and extract robust and stable 1280 features. Second, we proposed a new deep convolutional neural network architecture to classify the plant leaf diseases, such as blueberry disease, apple black rot, tomato bacterial spot, blueberry healthy, and potato late blight by investigating the residual block [7]. To improve the performance of the classification systems, various approaches have been proposed to learn and extract the robust feature from the input data involving hand-design features, such as local binary patterns (LBP) [14], histogram of gradient (HOG), local density encoding (LDE) [15] following by using support vector machine (SVM), K-nearest neighbor (KNN), decision tree or random forest. Another approach is to use a learning-based feature, such as ConvNet [1]. The learning-based methods proved that their accuracy is higher than hand-design features in various experiments in the real world.

Recently, EfficientNet [6] has been applied to develop various applications (object detection and classification) because of its benefits to improving the performance of the existing systems by considering the compound scaling method with depth, width, and resolution scaling. Without loss of generalization, the deep convolutional neural network (as in Ref. [6]) can be constructed as a list of layers as follows:

$$\text{DeepConvNet} = L_k \otimes L_{k-1} \ldots \otimes L_1\{X1\}$$

Recognizing that the *depth (d), resolution (r) and width (w)* should changes based on the target application, Tan et al. [6] introduced a method to estimate *d, r, w* as follows:

$$d = \alpha^\Phi, \ w = \beta^\Phi, \ r = \gamma^\Phi$$

where α, β, and γ are hyperparameters. Φ is a coefficient that controls how many more resources are available. Therefore, this research aims to use EfficientNet [6] to create a stable features by considering *d, r, and w* as shown in Fig. 12.1.

From the stable feature obtained by using the base-net (EfficientNet0), we proposed a deep neural network architecture as in Fig. 12.2 to classify the plant leaf disease. In this research, we also proposed an approach to integrating the residual

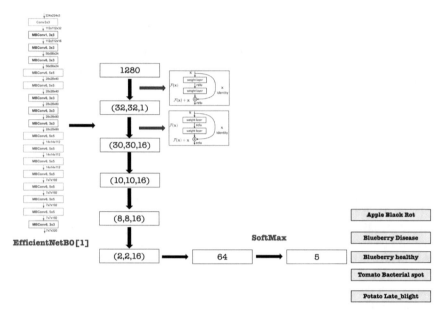

Fig. 12.1 Workflow of the proposed approach for classifying the plant leaf diseases

block technique [7] to increase the performance of the proposed method as described in Fig. 12.2.

12.4 Experimental Results

12.4.1 Dataset

To evaluate the performance of the proposed system, we use Kaggle Plant Diseases Dataset [8]. The Kaggle Plant Diseases provides 87,000 RGB images of healthy and diseased leaves with 38 different classes. This research performs the experiments by using four types of classes from the Kaggle Plant Diseases Dataset. We conducted experiments with blueberry healthy (1817 images), apple black rot (1988 images), tomato bacterial spot (1703 images), and potato late blight (1940 images) as shown in Figs. 12.3a–d. In addition, we also set up and captured our dataset with the blueberry disease as shown in Fig. 12.3e. This dataset was captured at Can Tho University, Vietnam.

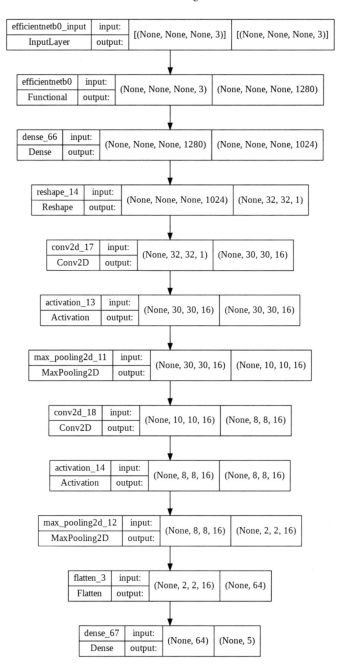

Fig. 12.2 Proposed deep neural network without the residual block

Fig. 12.3 Images from the Kaggle Plant Diseases Dataset [8]: **a** samples images of blueberry healthy, **b** sample images of apple black rot, **c** tomato bacterial spot, and **d** potato late blight. Images from our dataset were captured at Can Tho, Vietnam: **e** this dataset contains images of blueberry disease

12.4.2 System Configuration, Comparison Methods and Evaluation Metrics

The proposed system was trained on a PC with Intel Core i9 CPU @ 3.7 GHz, 64 GB, and9 Nvidia GTX 3080 10 GB. The proposed system was evaluated using the detection rate (DR, number of correct detection/total number of grouth truth) and false alarm rate (FA, number of incorrect detection/total number of grouth truth). The detected result is considered as correct if its IoU with the grouth truth is greater than or equal to 0.5 and classifies the correct class. The detected result is considered incorrect if it returns the wrong class. To evaluate the performance of our system to related works, we set up an experiment with existing methods, such as deep convolutional neural network (CNN) [1], support vector machine (SVM) [2], K-nearest neighbor [3], decision tree [4], and random forest [5].

12.4.3 Results and Discussion

For the blueberry healthy from the Kaggle dataset, we conducted experiments by using 1272 images for training and 545 images for testing as given in Table 12.1. The proposed system obtained the best result with DR of 97.24% and FA of 1.2% because of the benefits of stable feature from the compound scaling of the EfficientNet and the Residual block, while the DR and FA of the CNN [1] is 95.45% and 1.51%, respectively. The DR and FA of the SVM [2] are 93.23% and 2.33%, respectively. The DR and FA of the KNN [3] are 72.15% and 4.65%, respectively. The DR and FA of the decision tree [4] are 60.21% and 5.76%, respectively. The DR and FA of the random forest [5] are 86.12% and 3.45%, respectively. The KNN obtained the worst performance because of its imagination to operate with the large dataset and depends on the distance-based algorithm. For the blueberry disease from our dataset, we conducted experiments by using 350 images for training and 150 images for testing as given in Table 12.2. The proposed system obtained the best result with DR 94.24% and FA 1.56% because of the benefits of stable features from the compound scaling of the EfficientNet and the Residual block. The performance of the proposed system is degraded because of the difficulty of the testing dataset. The DR and FA of the CNN [1] are 92.34% and 1.76%, respectively. The DR and FA of the SVM [2] are 90.45% and 3.34%, respectively. The DR and FA of the KNN [3] are 68.18% and 5.15%, respectively. The DR and FA of the decision tree [4] are 57.12% and 5.78%, respectively. The DR and FA of the random forest [5] are 84.35% and 4.23%, respectively. The proposed system obtained better results than the CNN because we considered the balancing of depth, width, and resolution scaling.

For the apple black rot from the Kaggle dataset, we conducted experiments by using 1272 images for training and 545 images for testing as given in Table 12.3. The proposed system obtained the best result with DR of 95.14% and FA of 1.23% because of the benefits of stable feature from the compound scaling of the EfficientNet and the Residual block, while the DR and FA of the CNN [1] is 94.15% and 4.23%, respectively. The DR and FA of the SVM [2] are 85.95% and 2.33%, respectively. The DR and FA of the KNN [3] are 65.43% and 5.34%, respectively. The DR and FA of the decision tree [4] are 55.23% and 6.05%, respectively. The DR and FA of the random forest [5] are 83.33% and 3.24%, respectively. For the tomato bacterial spot from the Kaggle dataset, we conducted experiments by using of 1192 images for training and 511 images for testing. The proposed system obtained the best result

Table 12.1 Experimental results on the blueberry healthy dataset

Method	Train samples	Test samples	DR (%)	FA (%)
CNN1	1272	545	95.45	1.51
SVM [2]	1272	545	93.23	2.33
KNN [3]	1272	545	72.15	4.65
Decision tree [4]	1272	545	60.21	5.76
Random forest [5]	1272	545	86.12	3.45
The proposed	**1272**	**545**	**97.24**	**1.20**

Table 12.2 Experimental results on our custom dataset with blueberry disease

Method	Train samples	Test Samples	DR (%)	FA (%)
CNN [1]	350	150	92.34	1.76
SVM [2]	350	150	90.45	3.34
KNN [3]	350	150	68.18	5.15
Decision tree [4]	350	150	57.12	5.78
Random forest [5]	350	150	84.35	4.23
The proposed	**350**	**150**	**94.24**	**1.56**

Table 12.3 Experimental results on the apple black rot dataset

Method	Train samples	Test samples	DR (%)	FA (%)
CNN [1]	1392	596	94.15	1.89
SVM [2]	1392	596	85.95	4.23
KNN [3]	1392	596	65.43	5.34
Decision tree [4]	1392	596	55.23	6.05
Random forest [5]	1392	596	83.33	3.24
The proposed	**1392**	**596**	**95.14**	**1.23**

with DR of 98.24% and FA of 1.18% because of the benefits of stable feature from the compound scaling of the EfficientNet and the Residual block, while the DR and FA of the CNN [1] is 96.11% and 1.45%, respectively. The DR and FA of the SVM [2] are 87.94% and 3.25%, respectively. For the potato late blight from the Kaggle dataset, we conducted experiments by using 1358 images for training and 582 images for testing. The proposed system obtained the best result with DR of 98.87% and FA of 1.35% because of the benefits of stable feature from the compound scaling of the EfficientNet and the Residual block, while the DR and FA of the CNN [1] is 88.91% and 2.65%, respectively. The DR and FA of the SVM [2] are 70.34% and 3.21%, respectively.

12.5 Conclusions

This research proposes a novel approach for classifying plant leaf disease by using the benefit of EfficientNet and Residual block. The proposed system was able to detect and classify the blueberry healthy with a detection rate of 97.24%, the blueberry disease with the detection rate of 94.24%, the apple black rot with a detection rate of 95.14%, the tomato bacterial spot with a detection rate of 98.24%, and the potato late blight with the detection rate of 98.87%. The proposed system obtained better results than existing methods, such as CNN, SVM, KNN, random forest, and decision tree. However, the proposed system still has several limitations: (1) the performance of the proposed system was degraded when the input image is affected by noise. (2). The processing time is still not satisfied for real-time application. For future work, we aim to improve the performance of the proposed system by integrating the benefits of the unsupervised method, such as GAN. Moreover, we also plan to integrate this system into a mobile device to help the farmer to know the plant disease, quickly, without needing help from agriculture specialists.

Acknowledgements This project was funded by the Ministry of Education and Training, Vietnam, with grant number: B2021-TCT-10. The authors are grateful to the Department of Software Engineering, School of Computing and Information Technology, Eastern International University (EIU), Vietnam.

References

1. Rahul, S., Amar, S., Kavita, Jhanjhi, N.Z.: Plant disease diagnosis and image classification using deep learning. Comput. Mater. Continua **71**(2), 2125–2140 (2022)
2. Ertekin, S., Bottou, L., Giles, C.L.: Nonconvex online support vector machines. IEEE Trans. Pattern Anal. Mach. Intell. **33**(2), 368–381 (2011). https://doi.org/10.1109/TPAMI.2010.109
3. Huang, J., Wei, Y., Yi, J., Liu, M.: An improved kNN based on class contribution and feature weighting. In: 2018 10th International Conference on Measuring Technology and Mechatronics Automation (ICMTMA), pp. 313–316 (2018). https://doi.org/10.1109/ICMTMA.2018.00083
4. Yang, F.-J. (2019) An extended idea about decision trees. In: 2019 International Conference on Computational Science and Computational Intelligence (CSCI), pp. 349–354 (2019).https://doi.org/10.1109/CSCI49370.2019.00068
5. Liu, J., Lv, F., Di, P.: Identification of sunflower leaf diseases based on random forest algorithm. In: 2019 International Conference on Intelligent Computing, Automation and Systems (ICICAS), pp. 459–463 (2019). https://doi.org/10.1109/ICICAS48597.2019.00102
6. Tan, M., Le, Q.: EfficientNet: Rethinking model scaling for convolutional neural networks. In: Proceedings of the 36th International Conference on Machine Learning, in Proceedings of Machine Learning Research, vol. 97, pp. 6105–6114 (2019). https://proceedings.mlr.press/v97/tan19a.html
7. He, K., Zhang, X., Ren, S., Sun, J.: Deep residual learning for image recognition. In: 2016 IEEE conference on computer vision and pattern recognition (CVPR), pp. 770–778 (2016). https://doi.org/10.1109/CVPR.2016.90
8. Kaggle Plant Diseases Dataset, https://www.kaggle.com/datasets/vipoooool/new-plant-diseases-dataset. Accessed June, 2022
9. Ferentinos, K.P.: Deep learning models for plant disease detection and diagnosis. Comput. Electron. Agric. **145**, 311–318 (2018)
10. Lakshmanarao, A., Babu, M.R., Kiran, T.S.R.: Plant disease prediction and classification using deep learning ConvNets. In: 2021 International Conference on Artificial Intelligence and Machine Vision (AIMV), pp. 1–6 (2021). https://doi.org/10.1109/AIMV53313.2021.9670918
11. Sardogan, M., Tuncer, A., Ozen, Y.: Plant leaf disease detection and classification based on CNN with LVQ algorithm. In: 2018 3rd International conference on computer science and engineering (UBMK), pp. 382–385 (2018). https://doi.org/10.1109/UBMK.2018.8566635
12. Förster, A., Behley, J., Behmann, J., Roscher, R.: Hyperspectral plant disease forecasting using generative adversarial networks. In: IGARSS 2019—2019 IEEE International Geoscience and Remote Sensing Symposium, pp. 1793–1796 (2019). https://doi.org/10.1109/IGARSS.2019.8898749
13. Ganesan, P., Sajiv, G., Leo, L.M.: CIELuv color space for identification and segmentation of disease affected plant leaves using fuzzy based approach. In: 2017 Third International Conference on Science Technology Engineering & Management (ICONSTEM), pp. 889–894 (2017). https://doi.org/10.1109/ICONSTEM.2017.8261330
14. Ojala, T., Pietikainen, M., Maenpaa, T.: Multiresolution gray-scale and rotation invariant texture classification with local binary patterns. IEEE Trans. Pattern Anal. Mach. Intell. **24**(7), 971–987 (2002). https://doi.org/10.1109/TPAMI.2002.1017623.(2002)
15. Nguyen, V.D., Nguyen, D.D., Lee, S.J., Jeon, J.W.: Local density encoding for robust stereo matching. In: IEEE Transactions on Circuits and Systems for Video Technology, vol. 24, no. 12, pp. 2049–2062 (2014). https://doi.org/10.1109/TCSVT.2014.2334053

Chapter 13
Breast Cancer Classification Model Using Principal Component Analysis and Deep Neural Network

M. Sindhuja⬤, **S. Poonkuzhali**⬤, **and P. Vigneshwaran**⬤

Abstract Life-threatening cancer is prevalent over the globe. According to statistics, most people are diagnosed with cancer in the later stages, even though cancer can be prevented and cured in early stages. The goal of this research is to diagnose a breast cancer is either benign or malignant, as well as to forecast the likelihood of a cancer recurrence even after a course of therapy has been completed. Despite the fact that many machine learning algorithms have resulted in strong predictions, the accuracy in the early stages of categorization is not up to the expected level. Deep learning (DL), a higher degree of machine learning, can forecast breast cancer types and recurrences. Classifiers were built using a deep neural network (DNN) that used Principal Component Analysis (PCA) to choose features. Different machine learning techniques are compared with the proposed system's accuracy. Early-stage breast cancer prediction is more accurate with the DNN-based method. The clinical management system will benefit from the proposed system since it will aid in identification of cancer at an early stage and the subsequent provision of appropriate therapy.

13.1 Introduction

Uncontrolled cell division and tissue destruction are hallmarks of the cancer condition. Any part of the body might be affected by cancer, which can begin anywhere

M. Sindhuja (✉)
Department of Artificial Intelligence and Machine Learning, Rajalakshmi Engineering College, Thandalam, Chennai, Tamil Nadu 602105, India
e-mail: sindhujavignesh06@gmail.com

S. Poonkuzhali
Department of Computer Science and Engineering, Rajalakshmi Engineering College, Thandalam, Chennai, Tamil Nadu 602105, India
e-mail: poonkuzhali.s@rajalakshmi.edu.in

P. Vigneshwaran
Department of Networking and Communications, SRM Institute of Science and Technology, Kattankulathur, Chengalpattu, Tamil Nadu 603203, India

© The Author(s), under exclusive license to Springer Nature Singapore Pte Ltd. 2023
C. So-In et al. (eds.), *Information Systems for Intelligent Systems*, Smart Innovation, Systems and Technologies 324, https://doi.org/10.1007/978-981-19-7447-2_13

and spread throughout the body. As per the WHO's International Agency for Cancer Research (IARC), 8.2 million people died from cancer in 2012, and 27 million more will by 2030.The most common malignancies in women are those of the breast, cervix, colorectum, ovary, oral and lip; Oral cavity, lung, lip, stomach, colorectum and throat are the most prevalent malignancies in men. Women are more likely than males to be diagnosed with BC, which is the second-most common malignancy in women (excluding skin cancer). Treatment options for breast cancer include both local and systemic approaches. Surgical procedures and radiation therapy are examples of local treatments; chemotherapy and hormone therapy are local ones. To get the best outcomes, these two treatments are often combined. Stomach, lip, oral, colorectum, lung, and throat are the most prevalent malignancies in men. In the first five years following breast cancer therapy, the majority of recurrences occur. Recurrences of breast cancer can occur locally, regionally, or as distant metastases. The lymph nodes, bones, liver, lungs, and brain are some of the most common locations of recurrence outside the breast. Because of this, predicting the likelihood of breast cancer recurrence is critical. Breast cancer detection technology has progressed in recent years, with new methods of imaging the breast, new ways of identifying dangerous and benign tumor cells, and novel detection methodologies. Age, diet, marital status, disease stage, and treatment all play a role in a woman's risk for breast cancer.

There are varying degrees of correlation and uncertainty when it comes to the stage of the breast cancer. This section, therefore, focuses on predicting a patient's susceptibility to breast cancer. Deep learning was used to investigate this issue, employing CNNs in the MATLAB environment to forecast the intensity range of breast cancer. In recent research for nuclei segmentation, diverse algorithms were compared and tested on a dataset of 500 photos, and the accuracy percentage ranged from 96% to 100%. A new study on the diagnosis of BC used cytological pictures analysis of needle biopsies to distinguish between benign and malignant images. The study claims a 98% performance accuracy on 737 photos using four different classifiers trained with a 25-dimensional feature vector. When using nuclei segmentation and support vector machine and neural network, the study provided an accurate diagnosis technique for breast cancer. As a result of this investigation, the accuracy ranged from 76% to 94%. It's not uncommon for a traditional feature extraction strategy to result in a solution that's unique to the situation at hand, as it requires significant time and effort, as well as extensive expertise in a related field. The CNN architecture has been utilized to handle high-resolution texture images.

Computers may "learn" from prior examples and find hard-to-diagnose patterns from huge, noisy, or complex datasets using machine learning, a subset of artificial intelligence. These qualities are ideal for medical applications that rely on complicated proteomic and genomic studies. Support vector machines, Bayesian belief networks, and artificial neural networks are commonly utilized in cancer diagnosis and detection. Recently, machine learning has been used to predict cancer outcomes. The survey revealed numerous high-performing algorithms for analyzing datasets properties. Instead of task-specific algorithms, deep learning is part of a larger family of machine learning methods. We use deep learning to implement neural networks.

13.2 Related Works

This section includes the related papers that are relevant to the proposed work. Some of the literatures are discussed in below.

Naive Bayes (NB) and k-nearest neighbor (KNN) classifiers have been proposed by Amrane et al. [1] for the breast cancer classification. Breast cancer classification techniques are provided in [2–4]. Donated to the University of California, Irvine, is the Breast Cancer Dataset (BCD) (UCI). There are eleven qualities, nine of which are examined in detail, and the final feature provides a binary value that is used to decide whether a tumor is benign or malignant (benign tumour-2 and malignant tumor-4). A total of 699 cases are included in the dataset. We were unable to expand our dataset beyond 683 samples because of the presence of 16 observations with blank BCDs. Cross-validation is used to compare the suggested classifiers' accuracy between the two new implementations. KNN scored good accuracy (97.51%) and the decreased error rate compared to NB classifiers (96.19%). Various machine learning techniques, including support vector machine (SVM), decision tree (C4.5), Naive Bayes, and k-Nearest Neighbors (KNN), were compared using the Wisconsin Breast Cancer (original) datasets. The primary goal of this article was to evaluate the accuracy, precision, sensitivity, and specificity of each algorithm in terms of efficiency and efficacy. SVM scored the increased accuracy (97.13%) and the decreased error rate in experiments. Two different convolutional neural networks (CNNs) were used by Kumar et al [5], one for detecting individual nuclei even in densely populated areas, and the other for classifying them. The first CNN uses the input HE image to forecast the distance transform of the underlying (but unknown) multi-nuclear map. Cases and controls are categorized in the second CNN's analysis of nuclear center patches. The chance of recurrence for a patient can be calculated by voting on patches generated from images of the patient. For a sample of 30 recurrent cases and 30 non-recurrent controls, the proposed method yielded an AUC of 0.81, after training on an independent set of 80 case-control pairs. It's possible that if this technique is confirmed to be correct, it could aid in the selection of various treatment alternatives, such as active surveillance or radical prostatectomy or radiation and hormone therapy, if further research is done. As a result, it can also be used to predict treatment outcomes in other tumors.

Using convolutional neural networks and hematoxylin-and-eosin (H&E) stained breast tissues, Bejnordi et al. [6] suggested a system for classifying H&E stained breast specimens for the diagnosis of breast cancer patients. An evaluation of 646 breast tissue biopsies utilizing the suggested approach achieved an area under the ROC of 0.92, revealing the diagnostic value of hitherto overlooked tumor-associated stroma. Participants' risk factors for cancer and non-cancerous diseases were examined by Atashi et al. [7]. Risk factors were grouped into three priority levels, then fuzzified and the subtractive clustering approach was used to input them in the same order. In order to train and test the new model, 70% and 30% of the dataset was randomly partitioned into two halves. Following the training, the system was put to the test using data from the Wisconsin Clinic and the real Clinic, with promising results.

The variables were given the necessary fuzzy functions, and the model was then trained using the combined dataset. First, the model was tested on 30% of the dataset, and then on the real data from a real Clinic (BCRC). The model's precision for the above phases was 81 and 84.5%, respectively, with a sensitivity of 85.1 and a specificity of 74.5%.

Studies on BCC and M1 have been done extensively; while mammograms can miss about 15% of breast cancer cases, alternative methods use the genome or phenotypes to categorize the illness [8, 9]. SDC, Linear Discriminant Analysis (LDA), and Fuzzy C Means Clustering are some of the approaches used to classify breast cancer. One of the most commonly used machine learning methods is the k nearest neighbors algorithm [10, 11]. A similarity measure must be used before a new element can be classified [12]. KNN can be used to assess the rate of false positives in cancer classification [13, 14].

Biological, chemical, and physiologic features are commonly predicted using naive Bayesian classifiers. Combining NBC with other classifiers, like as decision trees, can help determine prognosis or classification models. The WBC database was used to examine the accuracy of various breast cancer diagnosis classification algorithms [15]. The optimized learning vector approach performed at 96.7%, the large LVQ method obtained 97.13%, and the SVM for cancer diagnosis has the greatest accuracy of 97.13% [16]. Ashrita Kannan et al. [17] proposed random forest model to predict whether a Type 2 diabetic patient is vulnerable to cancer or not.

13.3 Methodology

At an early stage, the system tries to differentiate between benign and malignant breast cancers, and then use textual data to forecast a return of the disease. Principal component analysis (PCA) was used to select the subset of characteristics from the supplied collection of features for breast cancer categorization. As a data classifier, deep neural network (DNN) uses metrics taken directly from tumor images to categorize breast cancer tumors. Using this information, doctors will be able to determine the cancer's risk level so that they can provide the appropriate treatment.

Because of its great accuracy, deep learning has become widely used in medicine in the last few years. Because of this, a variety of techniques have been developed, including CNN, DNN, RNN, Auto-encoder-based methods, and sparse coding techniques. As compared to current methodologies, convolutional neural network-based approaches exhibited substantial gains. To categorize breast cancer tumors based on measures received directly from the tumors, deep neural network (DNN) is being used as a data classifier.

As part of this strategy, a DNN model is employed in classification, and PCAs are used in feature selection. The proposed method's phases are outlined in the following steps and illustrated in Fig. 13.1.

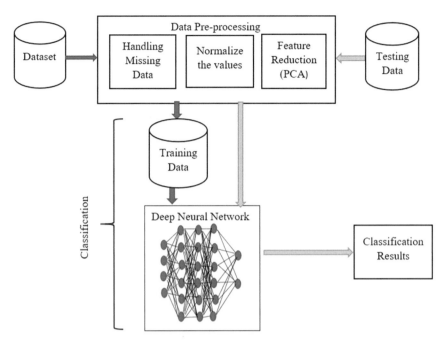

Fig. 13.1 Architecture of the proposed system

1. Remove all occurrences with missing values from the WBC dataset.
2. PCA is used to identify the features having good impact on the dataset.
3. The eigen values and cumulative values are calculated and the best feature is extracted and other features are eliminated through PCA.
4. Use a deep neural network to classify the dataset.

13.3.1 Dataset Description

To conduct the analysis, the Wisconsin Breast Cancer (WBC) Dataset from the UCI Repository is used, which contains 699 cases with 9 feature attributes. Because it has two categories, benign and malignant, this dataset can be classified as binary. Essentially, benign tumors don't spread to other sections of the body, but they are still tumors. The term "malignant" refers to cancerous cells that have metastasized to other organs or tissues in the body. Table 13.1 provides a list of each of the nine characteristics.

Table 13.1 Attribute information for WBC dataset

S. No	Attribute	Domain
1	Clump thickness	1–10
2	Uniformaity of cell size	1–10
3	Uniformaity of cell shape	1–10
4	Marginal adhesion	1–10
5	Single epithelial cell size	1–10
6	Bare nuclei	1–10
7	Bland chromatin	1–10
8	Normal nuclei	1–10
9	Mitoses	1–10

13.3.2 Preprocessing

Two extremely important terms which will improve the ability our model to learn efficiently are quality of data and usefulness of data. To achieve the quality of data, preprocessing is necessary to standardize the data with feature reduction.

Sixteen out of the 699 cases have no data. Prior to feature selection, all 16 instances of that dataset were deleted. In order to increase performance and precision, this system selects the best features from among the feature variables. The numbers have been reduced in size to make them more relatable to the average person.

13.3.3 Feature Selection

Feature selection is an essential part of any machine learning model. It removes any uncertainty from the data, making it easier to use. Training a model is easier and takes less time when the data is less. In this way, over-fitting data is avoided. Choosing the optimal subset of features from the complete feature set can enhance accuracy. Using wrapper methods, filters, and embedding methods to select characteristics is one such example.

Extraction and selection of features were done using PCA. It was possible to get rid of the original dataset by using PCA, or Principal Components Analysis (PCA). The loading of a variable determines its influence on a component. Relative dimensions of objects were often provided by the first PC. The most important information will be stored on the first few PCs.

To lower the dataset's dimensionality, researchers used three simple rules of thumb:

- Tests for Scree
- Cumulative Variance
- Rules established by Kaiser Guttman.

Tests for Scree

Retaining just the PCs known to as Scree Plot for testing is a regular occurrence. For a long time, the eigenvalue plotter is used. The "elbow" in the curve is important to look for when the curve flattens down, using a scree plot. That which begins with "elbow" is how many PCs a person should have.

Cumulative Variance

The eigenvalues or the percentage of variance explained by each PC determines the number of PCs [9]. It is possible to calculate the kth-longest segment's predicted length by randomly dividing the first p segments in half. And if the PC's proportion of variance explained is larger than, the PCs are kept. The permissible range of percentages is 80–100%.

Rules established by Kaiser Guttman

As a result of the work of Henry Kaiser and Louis Gutman on a method for selecting components fewer than the number required for perfect reconstruction, the approach was universally regarded as accurate. This rule is commonly used in common factor analysis and PCA. The number of factors m is calculated by counting the number of eigenvalues greater than 1. As a result, computers with an eigenvalue larger than 1 must be kept.

13.3.4 Classification

An 80/20 training-test split is used in the proposed method. Randomly-split datasets are used to create the datasets. After partitioning, a dataset is used to train the classifier. There are eight input nodes, six hidden layers, and one output node in this deep neural network classifier. As a result of its complexity, this network is computationally intensive but provides promising results after training.

13.3.5 Deep Neural Network (DNN)

There is no difference in structure between deep neural networks and conventional artificial neural networks. The models and models' complex hierarchies are made easier with its assistance. When the weights of each node are changed and the network is back-propagated, it contains 'n' hidden layers that process data from the preceding layer, known as the input level. Number of inputs can be mapped to a single input node. It is common for DNNs to include more nodes than the input layer to speed up the process of learning. The output layer's output nodes can each be customized independently. An input-output layer bias, learning rate (initial weights), hidden layer count, and nodes per layer, and an epoch-ending stop condition are all examples of these parameters. In order to avoid nullified network outcomes, this model's bias value is 1. It is also possible to alter the default learning rate, which is set at 0.15, to achieve alternative outcomes. The weight of a node can be updated after each epoch based on the error rate measured during back propagation. If have a large number of inputs and data, will have a large number of hidden layers. When the number of epochs has been reached or the learning model's desired outcome has been achieved, the network comes to an end. A model that has more layers and nodes to train demands more time and resources.

13.3.6 Algorithmic Steps for PCA-DNN

- Step 1. The Wisconsin Breast Cancer dataset is preprocessed to remove the missing values from the instances.
- Step 2. Extract the features using PCA from the dataset.

  - Step 2.1: Calculate the eigen values and cumulative values
  - Step 2.2: Select the Principal Components
  - Step 2.3: Reduce the data dimensions.

- Step 3. The ranked attributes are extracted and eliminates other features.
- Step 4. Apply DNN to classify the dataset.

  - Step 4.1: Define input layers with input nodes.
  - Step 4.2: Prepare the data for training by establishing the hidden layers.
  - Step 4.3: For each node, set the learning rate to (1) and the bias value to (0.15) to alter the network's weights.
  - Step 4.4: Activation Function: Rectified linear unit (ReLU): $f(x) = \max(0, x)$
  - Step 4.5: Specify the number of epochs you wish an output node's value to travel back over when back propagation is desired.
  - Step 4.6: The network should be trained using the specified set of training data before being used.
  - Step 4.7: To discover the model's classification rate, feed the test data to the trained network.
  - Step 4.8: Train the network until all epochs have been finished.
  - Step 4.9 Analyze the model's precision by determining its accuracy with respect to evaluation measures.

- Step 5. Test data should be used to verify the model.

13.4   Results and Discussion

There are 699 cases in the WBC dataset, with 9 variables. The binary classification models are possible since the class label values are only binary, 0 for benign and 1 for malignant, in this dataset. When it comes to cancer, the dataset provides two values for benign and four for malignant. Convert the class label's 2 and 4 values to zero and one. 16 of the 699 cases are of no use. To cut down on the number of mistakes in the system, the missing occurrences are removed. In the end, 683 occurrences select the features. The WBC dataset is depicted in Table 13.1.

The WBC dataset is first subjected to PCA. The three methods for determining how many PCs should be held before being classified are used to determine the PCs. The number of PCs that should be preserved based on the Scree Test is the number of PCs that remain stable and unchanging. According to the results shown in Figure 13.2, it is reasonable to keep six PCs.

As a result, the number of PCs is determined depending on their eigenvalues or the proportion of variance explained by the particular PC. The cumulative variance of the eigenvalues is shown in Fig 13.3. PCs that meet the 90% requirement for keeping are shown by a blue solid line and a dotted line in the graph. It suggest the first five PCs should be kept.

Table 13.2 summarized the dataset's eigenvalues and total variances. Almost 90.61% of the information is found in the top five PCs.

Finally, the third rule, the KG-Rule, recommended selecting the PC with the greatest fraction of eigenvalues. According to Table 13.2, the eigenvalues of the first eight PCs are all bigger than 1, which means that only eight PCs need to be preserved.

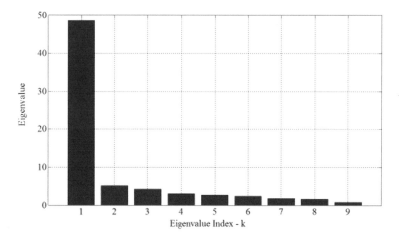

Fig. 13.2 Level of stabilization at k = 6

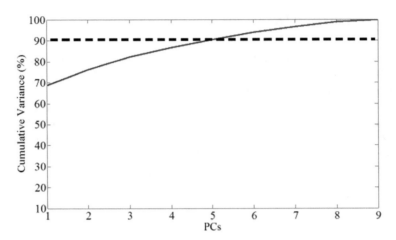

Fig. 13.3 Cumulative variance of eigen values

Table 13.2 Eigen values of PCs

S No	Eigen Value	Cumulative
1	48.54	68.9
2	5.17	76.2
3	4.28	82.3
4	3.1	86.7
5	2.73	90.6
6	2.44	94
7	1.77	96.6
8	1.59	98.8
9	0.8	100

Moreover, three inputs will be used and evaluated as feature vectors for the DNN classifier, namely Dataset 1 (6 PCs using Scree Test), Dataset 2 (5 PCs using Cumulative Variance) and Dataset 3 (8 PCs using KG rule). Table 13.3 gives the classification results for both training and testing stages.

Table 13.3 Classification results

	Scree test	Cumulative variance	KG rule
No. of PCs	6	5	8
Training accuracy	98.4	97.3	97.3
Testing accuracy	95.1	96.6	94.9

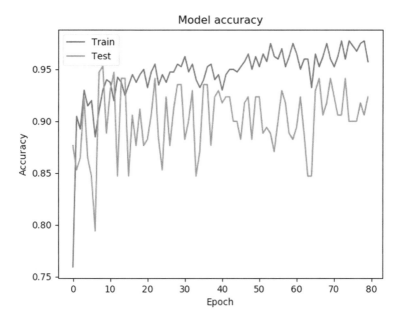

Fig. 13.4 Classification accuracy

Table 13.4 Performance comparison of proposed system with existing systems

Classification model	Accuracy	Model evaluation
NB	71.67	tenfold
EM-PCA-CART	93.2	tenfold
SVM	95	70–30
SVM	78	tenfold
DNN(Proposed)	**98.4**	**80–20**

Figure 13.4 shows the training and testing accuracy for the dataset 1 which comprises 6 attributes selected through Scree test. The proposed system achieved better accuracy of 98.4% for 80–20% train-test split.

Table 13.4 gives the performance comparison of the proposed system with existing system. This model behaves as expected and provides a promising result of 98.4%.

13.5   Conclusion and Future Work

Many people today are struggling with modern-day ailments. Breast cancer is the most common and fatal diseases rising globally. Death rates will rise due to lack of knowledge and late diagnosis of diseases. Computer-aided diagnosis will be the best way for everyone to diagnose accurately. The CAD system will not replace expert doctors, but it will help them make better decisions by evaluating patient information. Practitioners can make mistakes owing to inexperience or inadequate report analysis. So it will be a better medicinal cure. If the model used to train the system is clear, the system can only make accurate decisions. This system outperforms earlier models and just requires minor tweaking. The algorithm takes a long time to train because it uses a deep neural network. This system will run faster on GPU-based platforms than on commercial hardware. As a result, the user's data will be tested and processed using a more accurate computational model.

PCA and deep neural networks were utilized to classify data in the proposed system. Its DNN-PCA model accuracy is 98.4%. In this system, input may be carried and parsed across many layers, each of which contains numerous neurons. Because of back-propagation, node weights and network fine-tuning node values in each layer gradually lower the system's error rate. Due to the network's complexity, training time will inevitably increase. In the future, researchers can use particle swarm optimization or genetic algorithms to select features that will improve the overall model's accuracy. Deep learning methods demand high-end computational resources for training and testing in order to be implemented on local devices. Computing power can be increased by using cloud-based virtual machines or parallel processing. It reduces training time and makes the system computationally cheap.

References

1. Amrane, M., Oukid, S., Gagaoua, I., Ensari̇, T.: Breast cancer classification using machine learning. In: International Conference on Electric Electronics, Computer Science, Biomedical Engineerings' Meeting (EBBT), 18th & 19th April 2018, Istanbul, Turkey. https://doi.org/10.1109/EBBT.2018.8391453
2. Asria, H., Mousannif, H., Al Moatassime, H., Noeld, T.: Using machine learning algorithms for breast cancer risk prediction and diagnosis. In: 6th International Symposium on Frontiers in Ambient and Mobile Systems (FAMS 2016), Procedia Computer Science, vol. 83, pp. 1064–1069 (2016)
3. Alarabeyyat, A., Alhanahnah, M.: Breast cancer detection using K-nearest neighbor machine learning algorithm. In: 9th International conference on IEEE, v.i.e.E.(DeSE), pp. 35–39 (2016)
4. Akay, M.F., Support vector machines combined with feature selection for breast cancer diagnosis. Expert Syst. Appl. **36**(2), 3240–3247 (2009)
5. Kumar, N., Verma, R., Arora, A., Kumar, A., Gupta, S., Sethi, A., Gann, P.H.: Convolutional neural networks for prostate cancer recurrence prediction. In: Proceedings of SPIE, vol. 10140, Medical Imaging 2017: Digital Pathology, 101400H (1st March 2017). https://doi.org/10.1117/12.2255774
6. Bejnordi, B.E., Lin, J., Glass, B., Mullooly, M., Gier-ach, G.L., Sherman, M.E., Karssemeijer, N., Van Der Laak, J., Beck, A.H.: Deep learning-based assessment of tumor-associated stroma for diagnosing breast cancer in his-topathology images. In: IEEE 14th International Symposium on Biomedical Imaging, ISBI 2017, Apr 18th to Apr 21st 2017, Melbourne, Australia, pp. 929–932
7. Atashi, A., Nazeri, N., Abbasi, E., Dorri, S., Alijani_Z, M.: Breast Cancer Risk Assessment Using adaptive neuro-fuzzy inference system (ANFIS) and Subtractive Clustering Algorithm. Multi. Cancer Invest. **1**(2), pp. 20–26 (2017)

8. Bhatia, N.: Survey of nearest neighbor techniques. Int. J. Comput. Sci. Inf. Secur. **8**(2) (2010)
9. Francillon, A., Rohatgi, P.: Smart card research and advanced applications. In: 12th International conference CARDIS 2013. Springer International Publishing, Berlin, Germany
10. Prabhakar, S.K., Rajaguru, H., Maglaveras, N., Chouvarda, I., de Carvalho, P.: Performance analysis of breast cancer classification with softmax discriminant classifier and linear discriminant analysis. In: Precision Medicine Powered by pHealth and Connected Health. IFMBE Proceedings, vol. 66. Springer, Singapore (2018)
11. Snchez, J.S., Mollineda, R.A., Sotoca, J.M.: An analysis of how training data complexity affects the nearest neighbor classifiers. Pattern Anal. Appl. **10**(3), 189–201 (2007)
12. Baldi, P., Brunak, S.R.B., Baldi, P.: Bioinformatics: The Machine Learning Approach (2001)
13. Raniszewski, M.: Sequential reduction algorithm for nearest neighbor rule. Comput. Vis. Graph. (2010)
14. Bhuvaneswariaa, P., Therese, B.: Detection of cancer in lung with K-NN classification using genetic algorithm. Proc. Mater. Sci. **10**, 433–440 (2015)
15. Zhou, Z., Jiang, Y., Yang, Y., Chen, S.F.: Lung cancer cell identification based on artificial neural network ensembles artificial intelligence. Med. Elsevier **24**, 25–36 (2002)
16. Pradesh, A.: A.o.F.S.w.C.B.C.D. Indian J. Comput. Sci. Eng. **2**(5), 756–763 (2011)
17. Kannan, A., Vigneshwaran, P., Sindhuja, R., Gopikanjali, D.: Classification of cancer for type 2 diabetes using machine learning. In: ICT Systems and Sustainability, Springer Advances in Intelligent Systems and Computing, vol. 1, 1077, pp. 133–141 (2020)

Chapter 14
A Novel Hierarchical Face Recognition Method Based on the Geometrical Face Features and Convolutional Neural Network with a New Layer Arrangement

Soroosh Parsai and **Majid Ahmadi**

Abstract This paper represents the face recognition system utilizing a hierarchical arrangement for feature extraction and classification. The proposed method uses the active appearance graph model (AMM) as the first feature extractor and CNN as the second extractor. Also, for each stage, a classifier is considered. The paper investigates the results of two different types of classifiers, SVM and Softmax. The support vector machine shows superior function and outcomes. For the first stage, the AMM extracts six points from the face, then calculates three axillary points based on those to later measure and create five axes or distances on the face as the feature map for each image. It is worth mentioning that the axillary axes are divided by half to eliminate the effect of different facial orientations by the algorithm. In the second stage, for the CNN to give desired results, various forms of data augmentation, such as horizontal flip, shift, scaling, and rotation, are implemented into the algorithm. The color FERET database is used in this research for evaluation. The AMM divides 1208 classes of the FERET into 64 new classes. The biggest new class has 42 members. Applying the algorithm to the FERET database, we compare the accuracy of the system to the other existing methods. Our method shows a better accuracy of 96.35% with Softmax and 97.68% with SVM. Also, compared to the other techniques, the number of required data augmentation is reduced drastically, which translates to lower computational complexity and faster process.

14.1 Introduction

Biometric measurement systems have become an inseparable part of our life. Recent improvements in technology, alongside the changes in the requirements of the security systems, have forced us to implement those parameters into almost every security

S. Parsai (✉) · M. Ahmadi
University of Windsor, Windsor, ON N9E4T7, Canada
e-mail: parsai@uwindsor.ca

M. Ahmadi
e-mail: ahmadi@uwindsor.ca

© The Author(s), under exclusive license to Springer Nature Singapore Pte Ltd. 2023
C. So-In et al. (eds.), *Information Systems for Intelligent Systems*, Smart Innovation, Systems and Technologies 324, https://doi.org/10.1007/978-981-19-7447-2_14

system that we use today. Fingerprint, IRIS scan, voice recognition, and face recognition are mainly the techniques we employ. There are two main applications for these techniques: verification and identification. In verification, we are dealing with a system based on the information from an independent individual. Whenever a new input is put through the system, the algorithm determines if the input belongs to the group of data that exists in the system. This eventually will lead to accepting the input as an instance of the in question-individual or rejecting it. On the other hand, identification is to set the system on multiple individuals, whom we refer to as classes, and the purpose of the system is to identify the class of each input based on the relations between the input and the information from the database of classes. The proposed method in this chapter is to be employed for identification, and the biometric measure that we consider is face recognition.

We can confidently say that face recognition is the only technique among biometric measures that we use daily, and of course, we do it naturally. This recognition is mainly done by distinguishing people based on the unique combination of their facial features, albeit unconsciously. Hence, it makes sense to base the design of an automated face recognition system on the features that can be extracted from our source. However, these features could be completely different from those we use in our daily interactions with others. For the feature extraction, two different techniques exist among the current methods in use. Geometrical facial features [1] such as shape, location, and distances between the face parts such as the nose, eyes, eyebrows, and lips. This approach eliminates the effect of irrelevant information, such as background, different coloring methods, and mainly the data unrelated to face geometry. However, this method makes the system very sensitive to the uncontrolled variation of the image, such as face rotation, illumination, and sometimes facial expressions. The other approach, on the other hand, uses the global elements of the whole image, regardless of their relevance to the face. It is needless to say that the downside of this holistic method is its sensitivity to the unrelated information of the image. Active appearance graph models (AAM) [2] and convolutional neural networks (CNN) [3] are examples of the first and second methods, respectively. Also, a combination of some methods exists as the hierarchical technique.

CNN has been proven to be one of the most efficient techniques in image processing [4]. Because of the nature of the convolution operation and the existence of different types of filters, CNNs give desirable results in most feature extraction and classification, especially in image processing [5]. However, they are not free of shortcomings. One of the obstacles in employing CNNs for face recognition is that these networks need a relatively large number of samples per person to show acceptable performance, and evidently, the high number of samples is not the case in most artificial and real-life situations. One of the ways to overcome this problem is to use data augmentation. The main goal of data augmentation is to create enough samples from existing data to satisfy the system's prerequisites.

The other disadvantage of these systems is their computational complexity [6]. This issue is more critical in the instance with a relatively high number of individuals in the database, which is more likely to happen in real-life situations. The introduction of the GPU to the process and utilizing the computational power of new Graphical Processing Units helped to some extent [7].

The hierarchical approaches try to utilize various techniques simultaneously, to address multiple obstacles and deficiencies of the system [8]. In these techniques, usually, several methods are employed for different stages of a face recognition system in a hierarchical arrangement [9, 10]. This approach helps to improve the accuracy by adding multiple layers to the feature extraction. However, in most cases, because the hierarchy is implemented in the structure of the feature extractor rather than the structure of the entire system, the impact is only on the system's accuracy.

The rest of the paper is organized as follows: Sect. 14.2 briefly explains the proposed method. The structure of the proposed algorithm is discussed in detail in Sect. 14.3. Section 14.4 presents the experimental setup, and Sect. 14.5 is dedicated to the results of the experiments. Finally, in Sect. 14.6, the conclusion is represented.

14.2   The Proposed Hierarchical Face Recognition Method

Dealing with large databases with a large number of individuals is one of the obstacles of face recognition systems. This issue is more noticeable when a new face is needed to be added to the database. The other weakness of conventional face recognition systems is that the system's accuracy is affected by the shortcoming of the single extractor and classifier. On the other hand, the proposed method uses a simple and fast feature extractor based on the geometrical features of the face in the first stage to create new classes of individuals, however, with more members from the initial individual classes in each new class. In the process of recognition, the first stage shortlists the probable classes that the input image might belong to.

Meanwhile, the CNN in the second stage is employed to search through the short-listed classes determined by the first stage to precisely establish the class of the input image. The advantage of this method is that the CNN, which is the most resource-demanding part of the system, works on a comparatively smaller database. Also, reducing the number of individuals decreases the number of data augmentations that are required to enhance the individuals to samples ratio.

14.2.1   Geometrical Feature Extraction [11]

Face landmarking, or the localizing of the face parts, is the main task in geometrical feature extraction. Figure 14.1 shows some of the landmark features of a human face.

AMM is one of the most effective techniques for extracting geometrical features [12]. Points and locations extracted by the algorithm can be used to form the feature axes, which are the distances between certain locations of the face image.

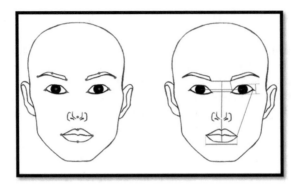

Fig. 14.1 Left: Face landmarks, Right: Feature axes

14.2.2 Convolutional Neural Network

CNNs are feed-forward types of neural networks that have been used especially in image processing and face recognition. They are made up of different layers. The most important part of these networks is the kernel or filter.

The dot product of the convolution is between the pixels of the input image and the kernel. The layer that contains the convolution operation is called the convolution layer. Other layers, such as the Pooling layer, ReLU layer, and fully connected layer, can be found in the structure of a CNN. Figure 14.2 shows a general overview of a CNN.

Convolution Layer. The main process of convolution happens in this layer. Because of that, we can consider this layer to be the carrier of the main computational load of the neural network. The main idea here is to extract hidden features of the face image. This idea is achieved by dividing the input image into smaller sections and convolving those parts with the kernel. The convolution operation preserves the spatial relations of the pixels in that small section. A set of learnable neurons are responsible for creating the convolution and constructing the activation or feature map. Note that the

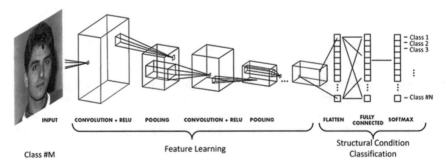

Fig. 14.2 Overview of the convolutional neural network

Fig. 14.3 Block diagram of the CNN

output of each convolution layer could be the input of another convolution or another layer.

Pooling Layer. A dimensionality reduction step with the nature of pooling is applied to the input to reduce the system's complexity. In the pooling layer, the target is to reduce the dimension of the data without losing any important parts. The procedure starts with dividing the input into smaller non-overlapping blocks. Then a process of a nonlinear down-sampling is applied to each region. The most famous nonlinear operations in the pooling layer are max pooling and average pooling. This layer specifically helps with faster convergence, more suitable generalization, and robustness against distortion, and this layer's usual placement is between convolution layers.

ReLU Layer. Rectifying happens in the nonlinear ReLU layer. The layer operates on each pixel and replaces all the negative values of the feature map with zero. To better understand the operation of the ReLU function, we can look at the mathematical expression of the layer as $f(i) = \max(0, i)$ in which i is the input value representing a pixel of the input image.

Fully Connected Layer. The term FCL is basically a property of the network rather than being a separate layer. It refers to the fact that every kernel in the layer is connected to the kernel of the next layer. The layers that we mentioned before are the interpretation of the complex features of the input. FCL enables the system to put those complex features into use. We consider the FCL to be the last pooling layer that forwards the extracted features into our classifier.

The main steps of the CNN as a face recognition system are shown in the block diagram in Fig. 14.3.

14.3   The Proposed Algorithm

The block diagram of the proposed method is illustrated in Fig. 14.4. The technique mainly consists of two major sections; Geometrical and CNN feature extraction and classification, and a few other subsections, as shown in Fig. 14.4.

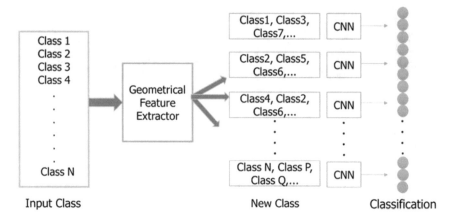

Fig. 14.4 Block diagram of the proposed algorithm

14.3.1 AMM

In the geometrical feature extractor, the algorithm employs AMM to extract the geometrical features of the face based on the location of some of the face parts. Six main points and three axillary points are considered to calculate five axes or distances as a feature map for each input. The tags of those five distances are WA, EA, WEA, LA, and ELA, as shown in Fig. 14.5. Each distance creates two axes, the left and the right side of the distance. **This is mainly done to prevent the effect of variation in face orientation.**

In the next stage, the algorithm feeds the feature map into an SVM classifier. The SVM ranks the dependency of the input to the classes. Specifically, the first ten most probable classes are assigned to each input. This process leads to the creation of new classes containing multiple members of the original input classes.

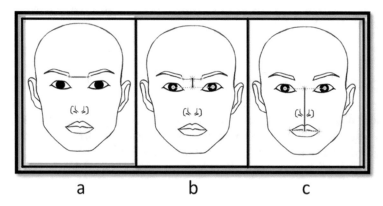

Fig. 14.5 a WA, **b** EA and WEA, and **c** LA and ELA

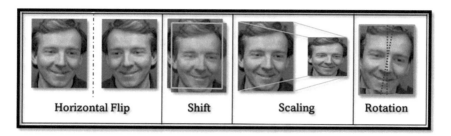

| Horizontal Flip | Shift | Scaling | Rotation |

Fig. 14.6 Data augmentation

14.3.2 Data Augmentation

Before putting the data into the CNN for training, a process of augmentation is done on the classes of the database. The augmentation step helps to overcome the negative effect of the low number of input samples. Some of the techniques we used for data augmentation are horizontal flip, shift, scaling, and rotation. These methods are shown in Fig. 14.6.

14.3.3 CNN

The convolutional neural network stage employs a new proposed arrangement of layers. Multiple layers of convolution, ReLU, and pooling exist in the arrangement. Other than those layers, we proposed the addition of normalization layers to the arrangement. Figure 14.7 shows the proposed arrangement of layers of the feature extraction section of the CNN, and Table 14.1 displays the parameters of those layers.

SVM. Support vector machines (SVM) are characterized as maximum margin classifiers since they reduce classification error while also increasing geometric margin [13]. An SVM creates a separating hyperplane in the feature space that maximizes the margin between the datasets. To determine the margin, two parallel hyperplanes are built, one on either side of the separating one. These hyperplanes are pushed up against the two datasets, allowing the hyperplane with the farthest distance to the adjoining support vectors of both classes to achieve a decent separation, as shown in Fig. 14.8. The greater the margin or distance between these parallel hyperplanes, the more likely unknown samples will be accurately identified [14].

Softmax. Softmax measures the probability of input belonging to a class. As a classifier, it enables the algorithm to predict the probability of the label in a multi-label database. Figure 14.9 illustrates the Softmax operation and its location in a conventional CNN.

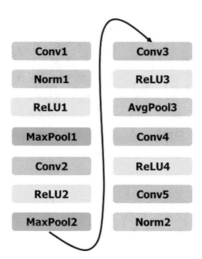

Fig. 14.7 Proposed arrangement of layers

Table 14.1 Parameter of the CNN's layers

Type	Patch size	Stride
Conv1	5×5	2
Conv2	3×3	2
Conv3	3×3	1
Conv4	3×3	2
Conv5	3×3	1
Pool1	3×3	1
Pool2	3×3	1
Pool3	8×8	–

14.3.4 Database

The color FERET [15, 16] database is a face recognition dataset. It comprises 11,338 color photos at a resolution of 512×768 pixels taken in a semi-controlled setting with 1208 individuals in multiple distinct stances. The database contains different poses for each individual ranging from frontal to half-face images. Also, different facial expressions and lighting conditions exist for each individual. In our experiment, we only used frontal images of each person, and the images were resized to 64×64 whenever it was fed to CNN. A few samples of the dataset are shown in Fig. 14.10.

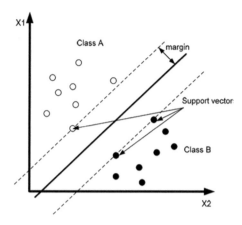

Fig. 14.8 Maximum-margin hyperplane for an SVM classifier. Samples on the margin are called the support vectors

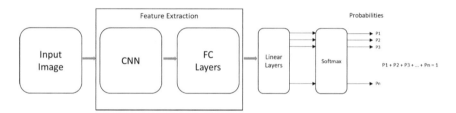

Fig. 14.9 Location and relation of the Softmax to the CNN

14.4 Experimental Setup

In the experiment, all the classes of the color FERET database are used. AMM is the algorithm used to extract the geometrical features, and the classifier in this stage is SVM. The combination of AMM and SVM creates 64 new classes out of 1208 original classes in the database. The average number of individuals in new classes is around 21 persons. The total number of individuals the algorithm deals with after AAM is 1384 groups. The biggest new class in terms of the number of individuals has 42 members, and the smallest has 5.

Before putting the data into the CNN, the samples are resized to 64 × 64. The data augmentation is carried out in the new classes to increase the number of samples to 10,000 in each new class. The recognition results are gathered for both SVM and Softmax as the classifiers of the convolutional neural network.

Half of the data are used for training, and the remaining half for the test. If CNN does not provide an output, i.e., does not merge, a return to the previous stage will happen, and the algorithm will consider the next tag in rank as the new class for AMM output.

Fig. 14.10 Samples of the color FERET database

14.5   Experimental Results

The results of running the algorithm on all 1208 classes of the color FERET database (50% training 50% test) and a comparison between the proposed method and other existing techniques are given in Table 14.2. Table 14.3 illustrates some of the parameters of the test and results for different geometrical feature extraction methods, and Table 14.4 gives the results of different parameters such as input size and results for CNN. As we can see, the highest recognition rate of around 97% happens when the size of the input image is 64×64 and also, using the SVM classifier increases the recognition rate compared to the Softmax (97.68% over 96.35%).

14.6   Conclusion

This research proposes a new system for face recognition applications. The new method has a hierarchical structure and consists of two stages of feature extraction and classification. The first stage extracts the geometrical features of the face image by employing the AMM technique and uses SVM as the classifier. The output of the first classifier forms the new classes of the data from the original input classes.

Table 14.2 Recognition rate on the color FERET database and comparison of the results

Method	%Accuracy
SVM-Linear [17]	81
Eigenfaces and DoG filter [18]	84
SVM-Polynomial [17]	87
SVM-RBF [17]	91
HMM [17]	90
AMM + CNN + Softmax (Proposed)	96.35
AMM + CNN + SVM (Proposed)	97.68

Table 14.3 The parameters and outputs of the different geometrical feature extractors

Method	Image size	Training time per sample	Classes (Highest)	Number of returns	Number of sub-groups
Tzimiro [2]	Original	1.012	1208	89	64
Cootes [19]	Original	0.88	1208	145	115
FACS [20]	Original	0.98	1208	107	71

Table 14.4 Parameters and output results of different sizes of input and classifiers in CNN

Method	Image size	Time per epoch Sec (Avg)	#Classes (Highest)	%Accuracy	Incorrect prediction
Proposed + SVM	16×16	14	42	71.85	340
Proposed + SVM	32×32	19	42	91.14	107
Proposed + SVM	64×64	33	42	97.68	28
Proposed + Softmax	16×16	18	42	69.86	364
Proposed + Softmax	32×32	24	42	90.39	116
Proposed + Softmax	64×64	41	42	96.35	44

We propose a new layer arrangement for the CNN and add two normalization layers to the structure. The data augmentation is also employed to compensate for the low number of input samples. The second stage utilizes the proposed CNN to extract the global features of the samples. We compared the recognition rate for two different classifiers, Softmax and SVM, as the second classifier. This paper uses the color FERET database for evaluation purposes. The results show a better performance of the proposed method by providing a higher recognition rate.

References

1. Marcolin, F., Vezzetti, E.: Novel descriptors for geometrical 3D face analysis. Multimedia Tools Appl. **76**(12), 13805–13834 (2017)
2. Tzimiropoulos, G., Pantic, M.: Fast algorithms for fitting active appearance models to unconstrained images. Int. J. Comput. Vis. **122**(1), 17–33 (2017)
3. Coşkun, M., et al.: Face recognition based on convolutional neural network. In: 2017 International Conference on Modern Electrical and Energy Systems (MEES). IEEE (2017)
4. Yan, K., et al.: Face recognition based on convolution neural network. In: 2017 36th Chinese Control Conference (CCC). IEEE (2017)
5. Albawi, S., Mohammed, T.A., Al-Zawi, S.: Understanding of a convolutional neural network. In: 2017 International Conference on Engineering and Technology (ICET). IEEE (2017)
6. Alzubaidi, L., et al.: Review of deep learning: Concepts, CNN architectures, challenges, applications, future directions. J. Big Data **8**(1), 1–74 (2021)
7. Sati, V., et al.: Face detection and recognition, face emotion recognition through NVIDIA Jetson Nano. In: International Symposium on Ambient Intelligence. Springer, Cham (2020)
8. Wang, M., Deng, W.: Deep face recognition: A survey. Neurocomputing **429**, 215–244 (2021)
9. Li, H., Hua, G.: Hierarchical-pep model for real-world face recognition. In: Proceedings of the IEEE Conference on Computer Vision and Pattern Recognition (2015)
10. Gao, G., et al.: Hierarchical deep CNN feature set-based representation learning for robust cross-resolution face recognition. In: IEEE Transactions on Circuits and Systems for Video Technology (2020)
11. Amarapur, B., Patil, N.: The facial features extraction for face recognition based on geometrical approach. In: 2006 Canadian conference on electrical and computer engineering. IEEE (2006)
12. Tzimiropoulos, G., et al.: Generic active appearance models revisited. In: Asian Conference on Computer Vision. Springer, Berlin, Heidelberg (2012)
13. Suthaharan, S.:Support vector machine. In: Machine Learning Models and Algorithms for Big Data Classification. Springer, Boston, MA, pp. 207–235 (2016)
14. Kecman, V.: Learning and Soft Computing: Support Vector Machines, Neural Networks, and Fuzzy Logic Models. MIT press (2001)
15. Phillips, P.J., et al.: The FERET database and evaluation procedure for face-recognition algorithms. Image Vis. Comput. **16**(5), 295–306 (1998)
16. Phillips, P.J., et al.: The FERET evaluation methodology for face-recognition algorithms. IEEE Trans. Pattern Anal. Mach. Intell. **22**(10), 1090–1104 (2000)
17. Nabatchian, A.: Human Face Recognition (2011)
18. Kortli, Y., et al.: Face recognition systems: A survey. Sensors **20**(2), 342 (2020)
19. Cootes, T.F., Edwards, G.J., Taylor, C.J.: Active appearance models. In: European Conference on Computer Vision. Springer, Berlin, Heidelberg (1998)
20. Lien, J.J., et al.: Automated facial expression recognition based on FACS action units. In: Proceedings Third IEEE International Conference on Automatic Face and Gesture Recognition. IEEE (1998)

Chapter 15
Contextual Academic Achievement Analysis Affected by COVID-19 Pandemic of Higher Education Learners in Thailand Using Machine Learning Techniques

Kanakarn Phanniphong⑩**, Wongpanya S. Nuankaew**⑩**, Direk Teeraputhon**⑩**, and Pratya Nuankaew**⑩

Abstract This research aims to present the context and impact that the Thai education system has experienced from the COVID-19 pandemic in Thailand. It consists of three research objectives: (1) to study the context of the impact on academic achievement from the COVID-19 pandemic in higher education, (2) to develop a model for clustering the academic achievement of students in higher education during the COVID-19 pandemic in Thailand, and (3) to compare the academic achievement of students in higher education during the COVID-19 pandemic in Thailand. The research data were 43,230 transactions (1961 students) from four educational programs at the Faculty of Business Administration and Information Technology, Rajamangala University of Technology Tawan-ok, the results showed that the context of the impact on the education system among tertiary learners has decreased in the number of graduates during the COVID-19 pandemic. However, students graduating during the COVID-19 pandemic in Thailand had higher levels of academic achievement than those in normal circumstances. The findings reflect those learners who achieved academic achievement during the COVID-19 pandemic were more persevering and tolerant than those in the traditional system.

K. Phanniphong
Rajamangala University of Technology Tawan-Ok, Bangkok 10400, Thailand

W. S. Nuankaew
Rajabhat Maha Sarakham University, Maha Sarakham 44000, Thailand

D. Teeraputhon · P. Nuankaew (✉)
University of Phayao, Phayao 56000, Thailand
e-mail: pratya.nu@up.ac.th

© The Author(s), under exclusive license to Springer Nature Singapore Pte Ltd. 2023
C. So-In et al. (eds.), *Information Systems for Intelligent Systems*, Smart Innovation, Systems and Technologies 324, https://doi.org/10.1007/978-981-19-7447-2_15

15.1 Introduction

The recent COVID-19 pandemic in Thailand has had a huge impact on students and the education system. Students and educational institutions are affected in different ways [1, 2]. Such as students need to adapt and invest in technology for online learning. Educational institutions searched for solutions to enable teaching and learning for learners. The cause was due to the city lockdown during the severe COVID-19 pandemic. Therefore, the study of satisfaction and attitudes toward changing learning styles is gaining more and more attention in Thailand. It includes internationally seeking solutions to address the issue of education management during the COVID-19 pandemic [1, 3–7]. Moreover, the researches on learner satisfaction and attitudes affects a wide range and at all levels of education, including secondary education [1], higher education [4, 6, 7], and the general public [3, 5]. At the same time, organizations and educational institutions have adapted to the impact of the COVID-19 pandemic. Discussion and review of literature on the issue of distance learning and the use of e-learning as the primary educational tool were given importance. It is the current and future trends that educational institutions are driving toward their goals [8–11].

However, it was only a small group of researchers who studied the context of the academic achievement of learners affected by the COVID-19 pandemic. It therefore inspires researchers to study and research the issue. For this reason, this research aims to present the context and impact of the Thai education system during the COVID-19 pandemic by comparing the learning achievements of learners over two time periods: normal situation and during the COVID-19 pandemic. There are three major goals of this research. The first goal is to study the context of the impact on academic achievement from the COVID-19 pandemic in higher education in Thailand. The second goal is to develop a model for clustering the academic achievement of students in higher education during the COVID-19 pandemic in higher education in Thailand. The last goal is to compare the academic achievement of students in higher education during the COVID-19 pandemic in Thailand over two periods. Research tools are divided into two main tools: basic statistical analysis tools and clustering model development tools. The basic statistical analysis tools were used to describe and detail the context of the collected data. It contains the frequency, mean, standard deviation, percentage, and interpretation. The clustering model development tools include k-means, k-medoids, and k-determination.

The research hypothesis designed for this study consists of two hypotheses.

H1: Machine learning tools and data mining techniques can effectively and appropriately develop a model for clustering learner's academic achievement.

H2: An effective learner's academic achievement clustering model be able to design and plan an appropriate learning promotion plan for learners in all situations.

Based on the research objectives and research hypothesis that has been established, the researchers strongly believe that this research will be of great benefit to learners and educational institutions in Thailand.

15.2 Materials and Methods

The materials and methods of the research consist of four elements. The first element is to determine the population and research sampling. The second element is to perform data collection. The third element is to design research tools for analyzing data. Finally, the fourth element is to design research tools for cluster modeling and to determine model performance.

15.2.1 Population and Sample Size

This research applied quantitative research techniques to achieve the research objectives. The research population was the students who enrolled and achieved academic achievements in the academic year 2015–2022 from the Faculty of Business Administration and Information Technology at the Rajamangala University of Technology Tawan-ok: Chakrabongse Bhuvanarth Campus, Bangkok, Thailand.

There are two categories of research sample definitions. The first category is a sample group representing students who were not affected by the COVID-19 pandemic in Thailand. It was the students who had received graduation in the academic year 2018 to the academic year 2020, which was not at the time of the COVID-19 pandemic in Thailand. The secondary category is a sample group representing students who were affected by the COVID-19 pandemic in Thailand. It was the students who had received graduation in the academic year 2021 and the academic year 2022, which was at the time of the COVID-19 pandemic in Thailand.

15.2.2 Data Collection

Before starting the data collection process, the researchers attended a 5-h training course in the Basic Human Subject Protection Course in the ThaiMOOC.org system. During this period, the researchers took a pretest, a posttest, participated in four modules and received a certificate of study. In addition, the researchers presented the research project to the researchers' university for approval of the research project through the Human Research Committee. All data collected by researchers was kept secret and destroyed once the research was complete. The research data were collected and summarized in Table 15.1.

Table 15.1 provides an overview of the data collected. It contains student achievement data for four educational programs: Bachelor of Accountancy Program, Bachelor of Business Administration Program in Information System, Bachelor of Business Administration Program in Management, and Bachelor of Business Administration Program in Marketing from the Faculty of Business Administration and

Table 15.1 Summary of data collection

Program/academic year	Number of students	Number of transaction	Academic achievement results				
			Max	Min	Mean	Mode	Median
Bachelor of accountancy program							
2015–2018	159	2105	3.80	2.00	2.81	2.90	2.77
2016–2019	251	5355	3.86	2.00	2.74	2.69	2.72
2017–2020	121	4688	3.90	2.00	2.79	2.70	2.80
2018–2021	51	1296	3.93	2.01	2.78	2.86	2.75
2019–2022	45	958	3.84	2.00	2.82	2.24	2.79
Total/Average	627	14,402	3.93	2.00	2.78	2.00	2.76
Bachelor of business administration program in information system							
2015–2018	73	932	3.76	2.00	2.62	2.07	2.48
2016–2019	114	2236	3.90	2.00	2.80	2.50	2.74
2017–2020	75	2000	3.77	2.00	2.66	2.57	2.57
2018–2021	43	1152	3.66	2.01	2.64	2.32	2.50
2019–2022	37	789	3.81	2.01	2.73	2.11	2.62
Total/Average	342	7109	3.90	2.00	2.70	2.50	2.60
Bachelor of business administration program in management							
2015–2018	135	1928	3.80	2.02	2.77	2.54	2.68
2016–2019	199	4708	3.92	2.00	2.82	2.63	2.76
2017–2020	111	4001	3.80	2.02	2.77	2.67	2.74
2018–2021	59	1508	3.63	2.06	2.77	2.21	2.68
2019–2022	43	906	3.65	2.31	2.92	3.13	2.85
Total/Average	547	13,051	3.92	2.00	2.80	2.67	2.74
Bachelor of business administration program in marketing							
2015–2018	108	1972	3.82	2.00	2.76	3.22	2.72
2016–2019	141	3722	3.96	2.05	2.77	2.35	2.72
2017–2020	68	2337	3.89	2.05	2.91	2.50	2.88
2018–2021	11	289	2.93	2.29	2.58	2.29	2.54
2019–2022	17	348	3.47	2.20	2.99	2.88	3.11
Total/Average	345	8668	3.96	2.00	2.80	2.89	2.76
Overall, the data collected							
2015–2018	475	6937	3.82	2.00	2.76	2.07	2.69
2016–2019	705	16,021	3.96	2.00	2.78	2.64	2.74
2017–2020	375	13,026	3.90	2.00	2.78	2.67	2.76
2018–2021	164	4245	3.93	2.01	2.73	2.50	2.66
2019–2022	142	3001	3.84	2.00	2.85	2.80	2.83
Total/Average	1861	43,230	3.96	2.00	2.77	2.50	2.73

Information Technology, Rajamangala University of Technology Tawan-ok, Thailand. There were 1,861 students' academic achievements during the academic year 2015–2022.

The most obvious observation was that students had a decrease in graduation across all educational programs. In the academic year 2019–2022, there were 142 graduates, representing 7.63%. It was the lowest in five years during which Thailand was hit by the COVID-19 pandemic.

15.2.3 Research Tools

This section presents the procedures and methods of selecting tools to develop an appropriate achievement clustering model with machine learning tools and data mining techniques.

Based on the nature of the data collected in the research as well as the goals of developing a model for clustering the learning achievement of learners during the COVID-19 pandemic. The ideal machine learning tools for this research were only unsupervised learning techniques with the following objectives and procedures. It was an analysis to find significant achievement clusters based on the impact of the COVID-19 pandemic situation. The data were categorized by educational program and classified according to the period unaffected by COVID-19 (academic year 2019–2020) and the period affected by COVID-19 (academic year 2021–2022). The machine learning tools in this section include k-means clustering, k-medoids clustering, and k-value analysis with k-determination.

15.2.4 Model Evaluation and Model Deployment

The purpose of the evaluation is to serve as a guide for selecting effective and acceptable models. The evaluation in this section is based on the elbow principle. It is presented in determining the k-value known as k-determination. This principle is used to determine a suitable cluster. This principle is used to determine the appropriate cluster to select the appropriate k-value further.

The purpose of the deployment is to take advantage of the selected model. Utilization can occur in several ways, for example, defining an organization's strategy, application development, and research extension.

For this research, the Rajamangala University of Technology Tawan-ok: Chakrabongse Bhuvanarth Campus would like to know the impact of the COVID-19 pandemic in Thailand on the academic achievement of students at the institution. Therefore, the utilization of this research is to know the context of the impact of the COVID-19 pandemic. It is also used as a strategy for predicting students' chances of delaying or failing to complete their studies.

15.3 Research Results

15.3.1 Context of Students' Academic Achievement

Although the global COVID-19 pandemic started around the end of 2019, Thailand was clearly affected in the period 2021–2022. The context of academic achievement affected by COVID-19 in Thailand is therefore limited to the scope of the academic year 2021–2022. The data was compiled for analysis, and contextual comparison of the academic achievement was achieved for five years, comprising graduates in the academic year 2018 to the academic year 2022.

A summary of the context for student achievement in the research is given in Table 15.1. It can be summarized as the following points categorized according to the educational program. Graduates in the last two years have fallen in the same direction across all educational programs. It appears that the COVID-19 pandemic has dramatically impacted the dimensions of the Thai education system. This part of the clear evidence is the context studied in this research.

15.3.2 Students' Academic Achievement Clustering Model

An appropriate student clustering uses GPA and duration affected and unaffected by the COVID-19 pandemic as crucial attributes of the clustering criteria.

Appropriate Learners Clustering with K-Means Clustering Techniques

The results analyzed and reported in this section were classified according to the four educational programs, as shown in Tables 15.2 and 15.3.

Table 15.2 found that the k-value corresponded to k-determination and the elbow technique for all educational programs during the non-affected COVID-19 pandemic in Thailand; the k-value was 5. It matches the appropriate k-value of the results calculated for the duration of all educational programs impacted by the COVID-19 pandemic in Thailand; the k-value was 5. Additionally, the distribution of student grade point average (GPA) and the number of members in each cluster period is shown in Table 15.3.

In the next section, it presents a comprehensive analysis of all data divided into two sections. Table 15.4 found that the k-value corresponded to k-determination and the elbow technique for the aggregate data collected during Thailand's non-affected and affected COVID-19 pandemic; the k-value was 5.

The distribution of student grade point average (GPA) for all collected data and the number of members in each of the two cluster periods are shown in Table 15.5.

Tables 15.2, 15.3, 15.4 and 15.5 analyzed the appropriate clustering for each educational program. An exciting and noteworthy conclusion was that all analyzes yielded a consistent number of clusters. It is a k-value equal to 5. The comparison in Table 15.5 reveals that the GPA during the pandemic of COVID-19 is higher than the

Table 15.2 K-means clustering analysis for four educational programs

K-Value	ACD	Distance	K-Value	ACD	Distance
K-means clustering analysis for bachelor of accountancy program					
Non-affected			Affected		
2	0.06332	0.00000	2	0.06834	0.00000
3	0.03011	0.03322	3	0.02673	0.04161
4	0.01602	0.01409	4	0.01605	0.01068
5*	0.01173*	0.00429*	5*	0.00947*	0.00658*
6	0.00784	0.00389	6	0.00696	0.00251
7	0.00566	0.00218	7	0.00574	0.00122
8	0.00446	0.00120	8	0.00472	0.00102
9	0.00399	0.00047	9	0.00400	0.00071
10	0.00289	0.00110	10	0.00246	0.00154
K-means clustering analysis for B.B.A. in information system					
Non-affected			Affected		
2	0.06506	0.00000	2	0.06111	0.00000
3	0.02988	0.03518	3	0.02920	0.03191
4	0.01685	0.01303	4	0.01528	0.01392
5*	0.01112*	0.00574*	5*	0.00989*	0.00539*
6	0.00783	0.00328	6	0.00584	0.00405
7	0.00588	0.00195	7	0.00492	0.00092
8	0.00432	0.00157	8	0.00338	0.00153
9	0.00401	0.00030	9	0.00303	0.00035
10	0.00371	0.00030	10	0.00243	0.00060
K-means clustering analysis for B.B.A. in management					
Non-affected			Affected		
2	0.06432	0.00000	2	0.05028	0.00000
3	0.02911	0.03522	3	0.02365	0.02663
4	0.01635	0.01276	4	0.01064	0.01301
5*	0.01137*	0.00498*	5*	0.00660*	0.00404*
6	0.00737	0.00399	6	0.00494	0.00167
7	0.00581	0.00156	7	0.00356	0.00138
8	0.00407	0.00174	8	0.00219	0.00137
9	0.00312	0.00095	9	0.00171	0.00048
10	0.00271	0.00041	10	0.00155	0.00016

Table 15.2 (continued)

K-Value	ACD	Distance	K-Value	ACD	Distance
K-means clustering analysis for B.B.A. in marketing					
Non-Affected			Affected		
2	0.06258	0.00000	2	0.02931	0.00000
3	0.02876	0.03382	3	0.01474	0.01457
4	0.01640	0.01236	4	0.00935	0.00539
5*	0.01039*	0.00601*	5*	0.00393*	0.00543*
6	0.00795	0.00244	6	0.00287	0.00106
7	0.00543	0.00252	7	0.00170	0.00117
8	0.00426	0.00117	8	0.00123	0.00047
9	0.00343	0.00083	9	0.00101	0.00022
10	0.00275	0.00068	10	0.00061	0.00040

* Optimal number of clusters

period not affected by the COVID-19 pandemic in Thailand. To make this research valuable and clear for further comparative study of the impact of the COVID-19 pandemic in Thailand on its effect on the higher education system, the researchers used the k-medoids technique to analyze clustering to compare the results of this section.

Appropriate Learners Clustering with K-Medoids Clustering Techniques

The results analyzed and reported in this section were classified according to the four educational programs as follows:

Table 15.6 found that the k-value corresponded to k-determination and the elbow technique for all educational programs during the non-affected COVID-19 pandemic in Thailand; the k-value were 5 and 6. It matches the appropriate k-value of the results calculated for the duration of all educational programs impacted by the COVID-19 pandemic in Thailand; the k-value were 5 and 6. Additionally, the distribution of student grade point average (GPA) and the number of members in each cluster period is shown in Table 15.7.

In this section, it presents a comprehensive analysis of all data divided into two sections. It was found that the optimal cluster of two time periods comprising the non-affected period and the affected period of the COVID-19 pandemic in Thailand were the same. Table 15.8 is presented using the k-medoids clustering technique and the results calculated from Average within Centroid Distance (ACD) to represent the change in k-values. It was found that the k-value corresponded to k-Determination and the elbow technique for the aggregate data collected during the non-affected COVID-19 pandemic in Thailand, the k-value was 5. It matches the appropriate k-value of the results calculated impacted by the COVID-19 pandemic in Thailand, the k-value was 5.

Table 15.3 The GPA and membership distribution for each educational program

Cluster	Centroid	Items	Cluster	Centroid	Items
K-means clustering analysis for bachelor of accountancy program					
Non-affected			Affected		
cluster_0	2.187	138 (25.99%)	cluster_0	2.240	22 (22.92%)
cluster_1	2.607	141 (26.55%)	cluster_1	2.563	20 (20.83%)
cluster_2	2.974	129 (24.29%)	cluster_2	2.847	28 (29.17%)
cluster_3	3.307	87 (16.38%)	cluster_3	3.246	17 (17.71%)
cluster_4	3.656	36 (6.78%)	cluster_4	3.691	9 (9.38%)
Total:		531 (100.00%)	Total:		96 (100.00%)
K-means clustering analysis for B.B.A. in information system					
Non-affected			Affected		
cluster_0	2.172	72 (27.48%)	cluster_0	2.205	27 (33.75%)
cluster_1	2.515	77 (29.39%)	cluster_1	2.546	22 (27.50%)
cluster_2	2.888	51 (19.47%)	cluster_2	2.894	14 (17.50%)
cluster_3	3.246	35 (13.36%)	cluster_3	3.273	10 (12.50%)
cluster_4	3.641	27 (10.31%)	cluster_4	3.660	7 (8.75%)
Total:		262 (100.00%)	Total:		80 (100.00%)
K-means clustering analysis for B.B.A. in management					
Non-affected			Affected		
cluster_0	2.238	104 (23.37%)	cluster_0	2.252	15 (14.71%)
cluster_1	2.609	133 (29.89%)	cluster_1	2.524	20 (19.61%)
cluster_2	2.939	106 (23.82%)	cluster_2	2.762	27 (26.47%)
cluster_3	3.268	57 (12.81%)	cluster_3	3.065	22 (21.57%)
cluster_4	3.645	45 (10.11%)	cluster_4	3.503	18 (17.65%)
Total:		445 (100.00%)	Total:		102 (100.00%)
K-means clustering analysis for B.B.A. in marketing					
Non-affected			Affected		
cluster_0	2.189	54 (17.03%)	cluster_0	2.262	4 (14.29%)
cluster_1	2.495	78 (24.61%)	cluster_1	2.531	8 (28.57%)
cluster_2	2.813	81 (25.55%)	cluster_2	2.872	5 (17.86%)
cluster_3	3.153	63 (19.87%)	cluster_3	3.174	9 (32.14%)
cluster_4	3.589	41 (12.93%)	cluster_4	3.445	2 (7.14%)
Total:		317 (100.00%)	Total:		28 (100.00%)

Table 15.4 K-Means clustering analysis for the aggregate data collected

K-Value	ACD	Distance	K-Value	ACD	Distance
Non-affected			Affected		
2	0.06471	0.00000	2	0.06060	0.00000
3	0.02989	0.03482	3	0.02767	0.03294
4	0.01667	0.01323	4	0.01494	0.01273
5*	0.01167*	0.00499*	5*	0.00946*	0.00548*
6	0.00771	0.00396	6	0.00685	0.00261
7	0.00572	0.00199	7	0.00492	0.00193
8	0.00456	0.00116	8	0.00392	0.00101
9	0.00373	0.00083	9	0.00381	0.00010
10	0.00272	0.00101	10	0.00311	0.00070

* Optimal number of clusters

Table 15.5 K-means clustering analysis for the aggregate data collected

Cluster	Centroid	Items	Cluster	Centroid	Items
Non-affected			Affected		
cluster_0	2.207	389 (25.02%)	cluster_0	2.242	73 (23.86%)
cluster_1	2.589	433 (27.85%)	cluster_1	2.591	85 (27.78%)
cluster_2	2.936	357 (22.96%)	cluster_2	2.942	85 (27.78%)
cluster_3	3.267	231 (14.86%)	cluster_3	3.322	38 (12.42%)
cluster_4	3.640	145 (9.32%)	cluster_4	3.650	25 (8.17%)
Total:		1555 (100.00%)	Total:		306 (100.00%)

The distribution of student grade point average (GPA) for all collected data and the number of members in each of the two cluster periods are shown in Table 15.9.

From Tables 15.2, 15.3, 15.4, 15.5, 15.6, 15.7, 15.8 and 15.9, it shows that using the two techniques for clustering analysis, there was no difference in the learning achievements of the two time periods. It was found that using two techniques consisting of K-means and k-medoids, clustered 5 were identical. A further conclusion is that although graduates declined during the COVID-19 pandemic, learners had higher levels of academic achievement. It reflects those graduates during the COVID-19 pandemic are more diligent than regular learners.

Table 15.6 K-medoids clustering analysis for four educational programs

K-Value	ACD	Distance	K-Value	ACD	Distance
K-medoids clustering analysis for bachelor of accountancy program					
Non-Affected			Affected		
2	0.09961	0.00000	2	0.10547	0.00000
3	0.05830	0.04131	3	0.03156	0.07391
4	0.03081	0.02749	4	0.02151	0.01005
5*	0.01573*	0.01508*	5*	0.01712*	0.00439*
6	0.01116	0.00456	6	0.01368	0.00344
7	0.00872	0.00245	7	0.00723	0.00645
8	0.00927	0.00055	8	0.00916	0.00192
9	0.00564	0.00363	9	0.00550	0.00365
10	0.00541	0.00023	10	0.00426	0.00125
K-medoids clustering analysis for B.B.A. in information system					
Non-Affected			Affected		
2	0.11606	0.00000	2	0.14068	0.00000
3	0.06355	0.05251	3	0.06998	0.07070
4	0.03427	0.02927	4	0.04488	0.02510
5	0.01517	0.01911	5	0.01999	0.02489
6*	0.00872	0.00645	6*	0.01168*	0.00831*
7	0.00852	0.00019	7	0.01143	0.00025
8	0.00837	0.00015	8	0.00809	0.00334
9	0.00683	0.00154	9	0.00729	0.00080
10	0.00510	0.00173	10	0.00390	0.00339
K-medoids clustering analysis for B.B.A. in management					
Non-Affected			Affected		
2	0.17028	0.00000	2	0.07196	0.00000
3	0.05141	0.11888	3	0.03603	0.03593
4	0.02305	0.02836	4	0.02262	0.01341
5	0.01484	0.00821	5	0.00971	0.01291
6*	0.01187*	0.00298*	6*	0.00747*	0.00224*
7	0.00822	0.00364	7	0.00595	0.00152
8	0.00651	0.00171	8	0.00545	0.00050
9	0.00557	0.00093	9	0.00508	0.00037
10	0.00515	0.00042	10	0.00358	0.00150

(continued)

Table 15.6 (continued)

K-Value	ACD	Distance	K-Value	ACD	Distance
K-medoids clustering analysis for B.B.A. in marketing					
Non-affected			Affected		
2	0.10024	0.00000	2	0.10088	0.00000
3	0.03395	0.06629	3	0.03897	0.06191
4	0.02047	0.01347	4	0.01662	0.02234
5*	0.01241*	0.00806*	5*	0.00950*	0.00713*
6	0.00869	0.00372	6	0.00448	0.00501
7	0.00848	0.00021	7	0.00381	0.00067
8	0.00790	0.00058	8	0.00192	0.00189
9	0.00522	0.00267	9	0.00180	0.00012
10	0.00485	0.00038	10	0.00165	0.00015

15.4   Conclusion

This research aims to study the context and impact on the Thai education system during the COVID-19 pandemic. It found that learners were affected by the COVID-19 pandemic in different ways. The most significant finding is that the proportion of college students graduating has declined during the COVID-19 pandemic. It harkens back to the research goal of exploring the context and impact of the COVID-19 pandemic in Thailand on the education system. This research reflects three key objectives: (1) to study the context of the impact on academic achievement from the COVID-19 pandemic in higher education, (2) to develop a model for clustering the academic achievement of students in higher education during the COVID-19 pandemic in Thailand, and (3) to compare the academic achievement of students in higher education during the COVID-19 pandemic in Thailand. The research data is the academic achievement results of 1861 students (43,230 transactions) from the Faculty of Business Administration and Information Technology at the Rajamangala University of Technology Tawan-ok: Chakrabongse Bhuvanarth Campus, Thailand. It was found that there was no difference in the clustering of successful learners. However, it has been found that learners who are in the situation and impact of the COVID-19 pandemic in Thailand have more effort and patience than normal learners. Tables 15.2, 15.3, 15.4, 15.5, 15.6, 15.7, 15.8 and 15.9 show that there is no difference in the appropriate clusters. The optimal cluster count for both periods affected and non-affected by the COVID-19 pandemic is a value of K equal to 5. A curious and questionable question in the future is why learners who succeeded in learning during the COVID-19 pandemic had better achievement than normal learners. In this regard, the researchers plan to expand the research results to develop a model for predicting the achievement of learners affected by the COVID-19 pandemic in Thailand.

Table 15.7 K-medoids clustering analysis for bachelor of accountancy program

Cluster	Centroid	Items	Cluster	Centroid	Items
K-medoids clustering analysis for bachelor of accountancy program					
Non-affected			Affected		
cluster_0	2.130	116 (21.85%)	cluster_0	2.450	41 (42.71%)
cluster_1	2.510	139 (26.18%)	cluster_1	2.890	27 (28.13%)
cluster_2	2.930	121 (22.79%)	cluster_2	3.080	11 (11.46%)
cluster_3	3.170	70 (13.18%)	cluster_3	3.430	11 (11.46%)
cluster_4	3.380	85 (16.01%)	cluster_4	3.770	6 (6.25%)
Total:		531 (100.00%)	Total:		96 (100.00%)
K-medoids clustering analysis for B.B.A. in information system					
Non-affected			Affected		
cluster_0	2.120	52(19.85%)	cluster_0	2.260	27 (33.75%)
cluster_1	2.350	45 (17.18%)	cluster_1	2.520	17 (21.25%)
cluster_2	2.570	60 (22.90%)	cluster_2	2.680	10 (12.50%)
cluster_3	2.980	54 (20.61%)	cluster_3	3.020	15 (18.75%)
cluster_4	3.400	34 (12.98%)	cluster_4	3.560	8 (10.00%)
cluster_5	3.750	17 (6.49%)	cluster_5	3.810	3 (3.75%)
Total:		262 (100.00%)	Total:		80 (100.00%)
K-medoids clustering analysis for B.B.A. in management					
Non-affected			Affected		
cluster_0	2.120	55 (12.36%)	cluster_0	2.320	20 (19.61%)
cluster_1	2.430	92 (20.67%)	cluster_1	2.580	21 (20.59%)
cluster_2	2.680	107 (24.04%)	cluster_2	2.800	19 (18.63%)
cluster_3	2.980	76 (17.08%)	cluster_3	2.920	13 (12.75%)
cluster_4	3.150	64 (14.38%)	cluster_4	3.200	12 (11.76%)
cluster_5	3.680	51 (11.46%)	cluster_5	3.470	17 (16.67%)
Total:		445 (100.00%)	Total:		102 (100.00%)
K-medoids clustering analysis for B.B.A. in marketing					
Non-affected			Affected		
cluster_0	2.330	90 (28.39%)	cluster_0	2.430	10 (35.71%)
cluster_1	2.630	69 (21.77%)	cluster_1	2.660	2 (7.14%)
cluster_2	2.880	65 (20.50%)	cluster_2	2.880	5 (17.86%)
cluster_3	3.200	58 (18.30%)	cluster_3	3.130	9 (32.14%)
cluster_4	3.660	35 (11.04%)	cluster_4	3.470	2 (7.14%)
Total:		317 (100.00%)	Total:		28 (100.00%)

Table 15.8 K-medoids clustering analysis for the aggregate data collected

K-Value	ACD	Distance	K-Value	ACD	Distance
Non-affected			Affected		
2	0.07215	0.00000	2	0.06322	0.00000
3	0.04918	0.02297	3	0.04842	0.01480
4	0.01985	0.02933	4	0.02406	0.02436
5*	0.01413*	0.00571*	5*	0.01352*	0.01054*
6	0.01097	0.00316	6	0.00927	0.00425
7	0.01047	0.00050	7	0.00766	0.00162
8	0.00671	0.00376	8	0.00618	0.00148
9	0.00654	0.00018	9	0.00672	0.00054
10	0.00451	0.00203	10	0.00389	0.00282

* Optimal number of clusters

Table 15.9 K-medoids clustering analysis for the aggregate data collected

Cluster	Centroid	Items	Cluster	Centroid	Items
Non-affected			Affected		
cluster_0	2.120	303 (19.49%)	cluster_0	2.260	66 (21.57%)
cluster_1	2.510	470 (30.23%)	cluster_1	2.450	74 (24.18%)
cluster_2	2.930	450 (28.94%)	cluster_2	2.890	76 (24.84%)
cluster_3	3.400	218 (14.02%)	cluster_3	3.130	44 (14.38%)
Cluster_4	3.660	114 (7.33%)	Cluster_4	3.470	46 (15.03%)
Total:		1555 (100.00%)	Total:		306 (100.00%)

The limitation of the research is the inability to predict future impacts on the education system. What the education system can do is prepare for unexpected situations and prepare learners for lifelong learning.

Acknowledgements This research project was supported by the Thailand Science Research and Innovation Fund and the University of Phayao (Grant No. FF65-UoE006). In addition, this research was supported by many advisors, academics, researchers, students, and academic staffs. The authors would like to thank all of them for their support and collaboration in making this research possible.

Conflict of Interest The authors declare no conflict of interest.

References

1. Kornpitack, P., Sawmong, S.: Empirical analysis of factors influencing student satisfaction with online learning systems during the COVID-19 pandemic in Thailand. Heliyon **8**, e09183 (2022). https://doi.org/10.1016/j.heliyon.2022.e09183

2. Pathak, S., Laikram, S.: Cooperative education during Covid-19 pandemic: enhancing legal rights and professional development of interns in Thailand. Int. J. Disaster Resilience Built Environ. **13**, 133–149 (2022). https://doi.org/10.1108/IJDRBE-08-2021-0098

3. Srisathan, W.A., Naruetharadhol, P.: A COVID-19 disruption: The great acceleration of digitally planned and transformed behaviors in Thailand. Technol. Soc. **68**, 101912 (2022). https://doi.org/10.1016/j.techsoc.2022.101912

4. Noori, A.Q.: The impact of COVID-19 pandemic on students' learning in higher education in Afghanistan. Heliyon **7**, e08113 (2021). https://doi.org/10.1016/j.heliyon.2021.e08113

5. Sheldon, P., Antony, M.G., Charoensap-Kelly, P., Morgan, S., Weldon, L.: Media and interpersonal channels uses and preferences during the COVID-19 pandemic: the case of the United States, Thailand, and Croatia. Heliyon **7**, e07555 (2021). https://doi.org/10.1016/j.heliyon.2021.e07555

6. Raccanello, D., Balbontín-Alvarado, R., da Silva Bezerra, D., Burro, R., Cheraghi, M., Dobrowolska, B., Fagbamigbe, A.F., Faris, M.E., França, T., González-Fernández, B., Hall, R., Inasius, F., Kar, S.K., Keržič, D., Lazányi, K., Lazăr, F., Machin-Mastromatteo, J.D., Marôco, J., Marques, B.P., Mejía-Rodríguez, O., Méndez Prado, S.M., Mishra, A., Mollica, C., Navarro Jiménez, S.G., Obadić, A., Mamun-ur-Rashid, M., Ravšelj, D., Vorkapić, S.T., Tomaževič, N., Uleanya, C., Umek, L., Vicentini, G., Yorulmaz, Ö., Zamfir, A.-M., Aristovnik, A.: Higher education students' achievement emotions and their antecedents in E-learning amid COVID-19 pandemic: A multi-country survey. Learn. Instr. 101629 (2022). https://doi.org/10.1016/j.learninstruc.2022.101629

7. Mengistie, T.A.: Higher education students' learning in COVID-19 pandemic period: The ethiopian context. Res. Glob. **3**, 100059 (2021). https://doi.org/10.1016/j.resglo.2021.100059

8. Fauzi, M.A.: E-learning in higher education institutions during COVID-19 pandemic: Current and future trends through bibliometric analysis. Heliyon **e09433** (2022). https://doi.org/10.1016/j.heliyon.2022.e09433

9. Metchik, A., Boyd, S., Kons, Z., Vilchez, V., Villano, A.M., Lazar, J.F., Anand, R.J., Jackson, P., Stern, J.: How we do it: Implementing a virtual, multi-institutional collaborative education model for the COVID-19 pandemic and beyond. J. Surg. Educ. **78**, 1041–1045 (2021). https://doi.org/10.1016/j.jsurg.2020.12.012

10. Khan, Md.S.H., Abdou, B.O.: Flipped classroom: How higher education institutions (HEIs) of Bangladesh could move forward during COVID-19 pandemic. Soc. Sci. Humanit. Open **4**, 100187 (2021). https://doi.org/10.1016/j.ssaho.2021.100187

11. Pal, K.B., Basnet, B.B., Pant, R.R., Bishwakarma, K., Kafle, K., Dhami, N., Sharma, M.L., Thapa, L.B., Bhattarai, B., Bhatta, Y.R.: Education system of Nepal: impacts and future perspectives of COVID-19 pandemic. Heliyon **7**, e08014 (2021). https://doi.org/10.1016/j.heliyon.2021.e08014

Chapter 16
Medical Records Sharing System Based on Blockchain: A Case Study in Vietnam

An Cong Tran, Long Phi Lam, and Hai Thanh Nguyen

Abstract In recent years, blockchain-related technologies and applications have gradually emerged. Blockchain technology is essentially a decentralized database maintained by the collective, and it is now widely applied in various fields. At the same time, with the development of medical technology, medical information is becoming increasingly important in monitoring the patient's medical examination and treatment process. Medical information must be the most private information about a person. However, due to issues such as operation errors within the network or a hacking attack by a malicious person, there have been major leaks of sensitive personal information. Therefore, it is an issue worth studying to ensure patients' privacy and protect these medical materials. On the other hand, the patient's Electronic Medical Record (EMR) cannot be searched across the hospital under the current medical system. As a result, repeated examinations can occur when the patient attends the hospital for treatment, resulting in a waste of medical resources. This study has deployed Blockchain to store and share Electronic Medical Records in a transparent and non-repudiation manner. In addition, we evaluated and tested the proposed system with various test cases and examined the performance with various numbers of transactions. Moreover, we have deployed Symmetric and Asymmetric key cryptography to exchange and share patients' Electronic Medical Records. The proposed system is expected to provide an interesting solution for exchanging medical data between medical facilities, especially for hospital transfers in Vietnam.

16.1 Introduction

In the digital age, digitizing paper documents into electronic ones is necessary, especially in the medical field. The digitization of traditional medical records (in papers) into electronic medical records helps hospitals easily solve the problems of storing,

A. C. Tran · L. P. Lam · H. T. Nguyen (✉)
Can Tho University, Can Tho, Vietnam
e-mail: nthai.cit@ctu.edu.vn

A. C. Tran
e-mail: tcan@ctu.edu.vn

C. So-In et al. (eds.), *Information Systems for Intelligent Systems*, Smart Innovation, Systems and Technologies 324, https://doi.org/10.1007/978-981-19-7447-2_16

retrieving, sharing, and searching for medical records. Moreover, electronic medical records also have a long storage time, even forever. If there is a good backup and backup plan, compared to the original paper, medical records can be damaged according to the law. On the other hand, in the current health system, the patient's medical records cannot be inter-hospital searchable. For example, when a patient is initially diagnosed and treated at hospital A, then at hospital B, hospital B must do the same examination again, wasting medical resources. Furthermore, when a patient presents to a doctor, the physician does not know the patient's medical history concerning other past hospital visits. Therefore, it increases medical risks and reduces the accuracy of diagnosis. Therefore, accessing inter-hospital medical records is a very important goal. The top issue is privacy and security when all important information is on the computer. Unfortunately, high-tech criminals always find security holes in systems and exploit this vulnerability to steal user information. The same goes for electronic medical records. As a result, only authorized people can view medical records. Blockchain can solve this problem, a technology that increases transparency and security for electronic medical records, helps hospitals manage and share medical records better, and helps patients feel secure more about the security of personal information.

In Vietnam, sharing electronic medical records is necessary. Especially patients who need to go to a hospital transfer. According to policies of Vietnam [1], the medical examination and treatment facility where the patient is transferred must provide information so that the doctors can know about the scope of benefits and the payment level of medical examination and treatment expenses covered by health insurance for patients. Electronic medical records help patients not need to carry documents such as medical records or test results of previous visits every time they go to the doctor. In addition, the patients do not need to redo the paraclinical services because their information is saved and shared between the hospitals. In addition, patients can also review their medical records or family members' medical records, thereby proactively planning health care and reasonable disease prevention. Furthermore, electronic medical records have helped transmit patient data between departments, clinics, or hospitals quickly, accurately, uniformly, and synchronously. Therefore, it can help avoid duplicate clinical indications, save patients' costs, and improve medical facilities' quality of diagnosis and treatment.

In this study, we propose an electronic medical record management system in a case study in Vietnam to help medical facilities manage more effectively. Patients do not need to carry cumbersome papers at the next medical examination-no need to re-examine and Avoid duplicate clinical indications when transferring. Instead, doctors can make accurate and effective diagnoses by sharing medical records between departments/treatment departments/hospitals and reviewing the medical records online.

16.2   Related work

In recent years, there have been many medical organizations around the world that have piloted blockchain technology in numerous studies.

As stated in [2], Some medical organizations have deployed Blockchain to store their data. For example, the American medical organization Delaware conducted a pilot project on Blockchain technology, which focuses on the authorization process for healthcare providers and health insurance companies UnitedHealth Group and Humana [3] deployed two large health insurance companies are conducting a pilot project of Blockchain decentralization feature into the current confusing and inaccurate information block of users. It can be the optimal method to help insurance companies avoid the error of user information. Change Healthcare [4] launched the first Blockchain network model in the medical field, Intelligent Healthcare Network. This network allows 550 transactions per second to improve transparency and efficiency in healthcare, control revenue, and payment status, and avoid organizational losses. Scientists in numerous studies also attract the research on Blockchain. The study in [5] revealed benefits from electronic health records and provided insights and experiences to prevent electronic health records system implementations fail.

The authors in [6] provided an overview on analyzing the latest literature on Blockchain and healthcare services relating to Electronic Health Records using Blockchain techniques. Another study in [7] evaluated the current landscape, design selection, and imitated issues and future research on blockchain-based personal health records. The work [8] concerned some privacy aspects of electronic health records and proposed a privacy-protecting information system to control the disclosure of personal data to third parties. The research in [9] indicated the health organizations' privacy and security concerns in healthcare settings and discussed some solutions. In [10], the authors also stated that the sharing of Electronic Health Records with medical practitioners could improve the accuracy of the diagnosis. However, they are also concerned about the privacy and security preservation of patients' records. They deployed a private and consortium blockchain using the consensus, data structures, and mechanisms. The work in [11] focused on deploying blockchain technology for Electronic Health Records storage and leveraging secure storage of electronic records by defining granular access rules.

The study in [12] introduced OmniPHR to integrate PHRs, providing data for patients' and healthcare providers' use, and they evaluated the feasibility of OmniPHR in elasticity/scalability capabilities. They expected that patients could maintain their health history with their devices. In [13], the authors proved their proposed method which can be necessary for patients, providers, and third parties. In addition, the study was to illustrate how the framework can maintain privacy and security. In [14], the authors examined different barriers to Health Information Systems and Electronic Medical Records Implementation to propose some suggestions on beneficial actions and options. In our study, we investigate a case study in Vietnam and propose an information system deploying Blockchain to exchange electronic health records transparently for hospital administrative and professional procedures.

16.3   Methods

The application focuses on two main parts: sharing medical records between clinics/treatment departments/hospitals and managing detailed inpatient/outpatient medical records and paraclinical indications. Safety requirements, the system proposes to meet the tasks related to updating the database must be confirmed by the user. In addition, data must be periodically backed up and stored securely in case the system is suddenly stopped due to power failure or equipment damage. In addition, sensitive information such as passwords needs to be encrypted when stored. Functions are required to authenticate with tokens to decentralize users. Medical record information must be encrypted when stored and can only be viewed and manipulated by authorized user groups and share transactions. Medical record information must be encrypted.

The system's general architecture includes 03 main components, including Python Web Service, Node.js Web Service interacting with a blockchain network, and a React web client interacting with two Web Services and users, as shown in Fig. 16.1.

16.3.1   Database Based on Blockchain

In our architecture, Node.js Web Service and Hyperledger Fabric network provide APIs written in Express for blockchain-based electronic medical record management and sharing. First, it will receive requests from the web client from the JWT token decoded by the server. Node.js then looks up the user's secret keys against the token

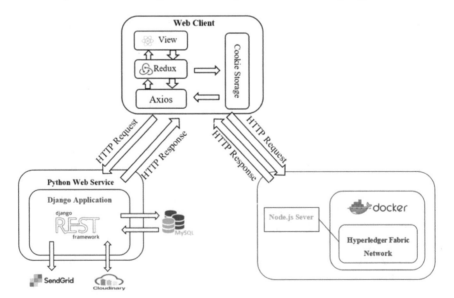

Fig. 16.1 The overall architecture of the proposed method for electronic medical record management system with blockchain.

to connect to the Channel corresponding to the organization. Next, if the connection is established successfully, the server can call the chaincode corresponding to the HTTP request and return the result via the HTTP Response. Finally, a fabric network will be built and launched using Docker with main components: Organization, CA, Channel, Peer, Chaincode, and Ledger.

After receiving the return data from the Web service, the Dispatcher sends the data and the action to the Reducer. Here, the Reducer uses the action and data from the Dispatcher and combines it with the old state to create a new state. Then, the Reducer notifies the view to update the interface.

The tables colored blue in the header as presented in Fig. 16.2 include data related to electronic medical records stored on the Blockchain because medical record infor-

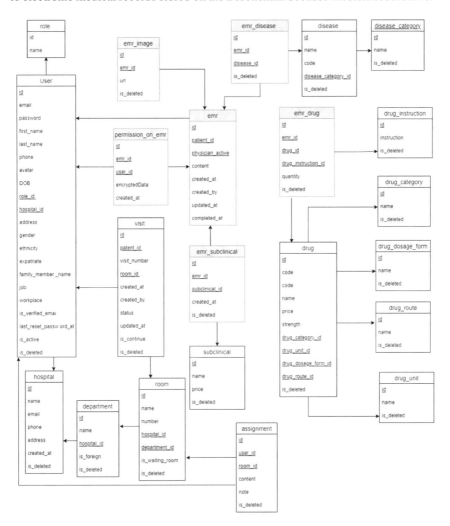

Fig. 16.2 Database scheme for electronic medical records management system.

mation requires high security, transparency, and longevity. Tables not directly related to the medical record information can be processed by the Python Web Service and stored in the MySQL database with various tables.

16.3.2 Python Web Service

Python Web Service provides APIs written in the Django [15] REST framework according to RESTful architecture to manage basic objects of the system: drugs, diseases, users, medical examination information, clinics, treatment departments, clinical services ready, and personal records. The attached data is the drug details in JSON format. First, DRF will check whether the URL is valid or not. If valid, it will switch to view. Here, JWT authentication performs request authentication. If valid, it will switch to Serializer, where Serializer converts data from JSON to Python. Next, the Model creates a new row in the database in the medicine table and returns the newly created data line to the Serializer in Python. Next, the Serializer will convert the result to JSON and return it to the view. Finally, the view will return the corresponding result for the request as an HTTP response.

16.3.3 The Process of Encoding an Electronic Medical Record

As illustrated in Fig. 16.3, When creating a medical record for patient John (1), Doctor Nguyen Van A can enter a medical record (plaintext M). Then, a randomly generated symmetric key (SK) based on the AES-256 cryptosystem with a key length

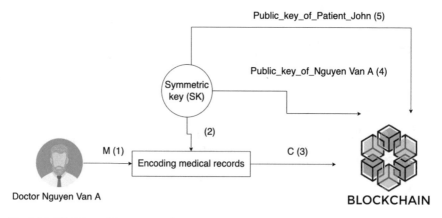

Fig. 16.3 Workflow of the process of creating a medical record.

of 256 bits can be generated in (2). Doctor Nguyen Van A can use this symmetric key (SK) to encrypt patient data and Save encrypted medical record information (C) to the Blockchain (3). Doctor Nguyen Van A can also use his own "public_key_Nguyen Van A" to encrypt the symmetric key (SK) and save it in the decentralized sharing of medical records on the Blockchain. At the same time, Doctor Nguyen Van A also uses the "public_key_John" of patient John to encrypt the symmetric key data (SK) and save it in the decentralized sharing of medical records on the Blockchain to share medical records later (5).

16.3.4   The Process of Sharing an Electronic Medical Record

As shown in Fig. 16.4, Doctor Nguyen Van A needs to share the medical record when a patient wants to transfer hospital/clinic/department. Let us say Nguyen Van A needs to share this medical record with Doctor Alice. At (1), Doctor Nguyen Van A gets the authorization table information on the medical record from the Blockchain. In (2), Blockchain returns encrypted data (C_EMR_1) containing the symmetric key (SK). Next, in (3), Doctor Nguyen Van A uses his own "private_key_Nguyen Van A" to decrypt. The decryption process is successful. First, Nguyen Van A obtains the symmetric key (SK) (4). Then in (5), Doctor Nguyen Van A uses Nguyen Thi B's "public_key_Nguyen Thi B" to encrypt the symmetric key data (SK) and save it to the decentralized shared medical record table on the Blockchain (C_PERMISSION_EMR_1). On Nguyen Thi B's side, she can get the authorization table information on the medical record from the Blockchain (6). The Blockchain returns encrypted data (C_PERMISSION_EMR_2) containing the symmetric key (SK) (7). Then, she used her own "private_key_Nguyen Thi B" to decrypt (8). The decryption is successful. Nguyen Thi B obtains the symmetric key (SK) (9). Then,

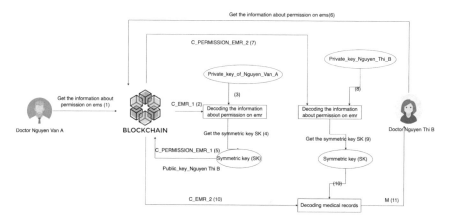

Fig. 16.4 Workflow of the process of sharing a medical record.

Alice uses the symmetric key (SK) to decrypt the medical record information from the Blockchain (10). The decryption process is successful. Then, Nguyen Thi B reads the medical record of John (M) (11).

16.4 Experiments

This section presents possible scenarios with test cases to detect application errors for necessary maintenance, determine whether the system meets the set needs, and check whether the interface is easy to use. In addition, we evaluated the performance with various transaction rates. The test cases and experiments were deployed on a machine equipped with a FUJITSU CPU Core(TM) i5-4005U, with a RAM of 2×4 GB DDR3L 1333 MHz, an SSD of 128 GB, installed a Ubuntu 16.04 operating system.

In order to evaluate the performance, we created new electronic medical records, granted a new authorization on medical records, and shared medical records when moving clinic/treatment department/hospital. In the experiment, 1200 transactions of medical records, granting new permissions on medical records, and sharing medical records were generated for the evaluation-the number of transactions per second ranges from 15 to 40. The experiments have deployed on a machine equipped with a CPU Core i5-4005U, with a RAM of 2×4 GB DDR3L 1333 MHz, an SSD of 128 GB, installed a Ubuntu 16.04 operating system.

Table 16.1 presents the performance of the proposed method, where *TPS* is the number of transactions Per Second in Caliper's send frequency option. *Success* denotes the number of successful transactions. Fail is the number of failed transactions, *Send Rate* is the caliper rate required to transact, *Latency(Max, Min, Avg)* denotes time from beginning to complete, *Throughput* shows the average number of transactions processed per second. As shown in the table, the number of transactions per second gradually increased from 15 to 35, the sending rate and throughput gradually increased from 15.1 to 35.2, and the minimum latency and average latency remained constant at 0.01 s and 0.02 s, respectively. In addition, the maximum latency

Table 16.1 Performance evaluation of the proposed blockchain network

TPS	Success	Fail	Send Rate (TPS)	Max Latency (s)	Min Latency (s)	Avg Latency (s)	Throughput (TPS)
15	1200	0	15.1	0.07	0.01	0.02	15.1
20	1200	0	20.1	0.07	0.01	0.02	20.1
25	1200	0	25.1	0.13	0.01	0.02	25.1
30	1200	0	30.2	0.11	0.01	0.02	30.2
35	1200	0	35.2	0.11	0.01	0.02	35.2

reaches the maximum threshold of 0.13 s when the number of transactions per second is 25. The maximum latency reaches the minimum threshold of 0.07 s, while the number of transactions per second is 15–20.

16.5 Conclusion

This study introduced a system that allows sharing of electronic medical records between clinics, treatment departments, and hospitals based on Blockchain. We also presented numerous test cases to evaluate the proposed method's performance regarding privacy concerns and transactions to proceed. As a result, electronic medical records can be transferred and exchanged via medical facilities, hospitals, and other services.

In the future, we expect to enhance the efficiency of the system. We can develop a mobile app with the same features as the web app and evaluate the efficiency of distributed systems.

References

1. Ministry of Public Health (Vietnam). 14/2014/TT-BYT: Circular of vietnamese ministry of public health on the transfer between medical examination and treatment facilities. https://thuvienphapluat.vn/van-ban/The-thao-Y-te/Thong-tu-14-2014-TT-BYT-viec-chuyen-tuyen-giua-co-so-kham-chua-benh-228285.aspx
2. Zhao, W.: Medical society of delaware tests blockchain to improve healthcare access. https://www.coindesk.com/markets/2017/08/10/medical-society-of-delaware-tests-blockchain-to-improve-healthcare-access/. section: Markets
3. ViVE.: How UnitedHealth, humana are thinking about 'techquity' | fierce healthcare. https://www.fiercehealthcare.com/payers/vive-2022-how-unitedhealth-humana-are-thinking-about-techquity (2022)
4. Change healthcare case study — hyperledger foundation. https://www.hyperledger.org/learn/publications/changehealthcare-case-study
5. Smelcer, J.B., Miller-Jacobs, H., Kantrovich, L.: Usability of electronic medical records. J. Usability Stud. **4**(2), 70–84 (Feb 2009)
6. Mayer, A.H., da Costa, C.A., da Rosa Righi, R.: Electronic health records in a blockchain: a systematic review. Health Inform. J. **26**(2), 1273–1288 (Sept 2019). https://doi.org/10.1177/1460458219866350
7. Fang, H.S.A., Tan, T.H., Tan, Y.F.C., Tan, C.J.M.: Blockchain personal health records: systematic review. J. Med. Internet Res. **23**(4), e25094 (April 2021). https://doi.org/10.2196/25094
8. Haas, S., Wohlgemuth, S., Echizen, I., Sonehara, N., Müller, G.: Aspects of privacy for electronic health records. Int. J. Med. Inf. **80**(2), e26–e31 (Feb 2011). https://doi.org/10.1016/j.ijmedinf.2010.10.001
9. Keshta, I., Odeh, A.: Security and privacy of electronic health records: concerns and challenges. Egypt. Inform. J. **22**(2), 177–183 (July 2021). https://doi.org/10.1016/j.eij.2020.07.003
10. Shamshad, S., Minahil, Mahmood, K., Kumari, S., Chen, C.M.: A secure blockchain-based e-health records storage and sharing scheme. J. Inf. Sec. Appl. **55**, 102590 (Dec 2020). https://doi.org/10.1016/j.jisa.2020.102590

11. Shahnaz, A., Qamar, U., Khalid, A.: Using blockchain for electronic health records. IEEE Access **7**, 147782–147795 (2019)
12. Roehrs, A., da Costa, C.A., da Rosa Righi, R.: OmniPHR: a distributed architecture model to integrate personal health records. J. Biomed. Inform. **71**, 70–81 (July 2017). https://doi.org/10.1016/j.jbi.2017.05.012
13. Vora, J., Nayyar, A., Tanwar, S., Tyagi, S., Kumar, N., Obaidat, M.S., Rodrigues, J.J.P.C.: Bheem: a blockchain-based framework for securing electronic health records. In: 2018 IEEE Globecom Workshops (GC Wkshps). pp. 1–6 (2018)
14. Khalifa, M.: Barriers to health information systems and electronic medical records implementation. a field study of Saudi Arabian hospitals. Procedia Comput. Sci. **21**, 335–342 (2013). https://doi.org/10.1016/j.procs.2013.09.044
15. Django Software Foundation: Django. https://djangoproject.com

Chapter 17
An Efficient Detection and Classification of Sarcastic by Using CNN Model

Summia Parveen, S. Saradha, and N. Krishnaraj

Abstract There have been a lot of new ways to figure out how people feel about things on Twitter and other microblogging platforms in the last few years. Using both types of semantics, experiments are carried out to see how they perform in three popular tasks of sentiment investigation on Twitter: sentiment analysis at the level of entities, sentiment analysis at the level of tweets, and context-sensitive sentiment lexicon adaptation. These three tasks include as follows: Find and categorize sarcastic texts were the objectives of this module. After the initial sentiment classification, texts that were initially classified as positive will be classified as moreover positive sarcastic or true positive. Texts classified as negative by the primary sentiment classification module can be confidential in two ways. Negative sarcasm and true negativity are two different types of negativity. In this research, a CNN model integrates the implicit as well as explicit representations of short text for sarcastic classification and is abbreviated as CNN-SC model. The proposed classification model incorporates three main processes, namely feature engineering, FS, and classification. Using data from Amazon and Twitter, it was tested to see if it worked. Detailed information on the dataset, experiments, and findings is presented in the section on experimental setup and results.

17.1 Introduction

SA is an important key feature in decision-making processes. It is obtained from classifying the sentiments from client's online opinion. SA is capable of accessing

S. Parveen (✉) · S. Saradha
Department of Computer Science and Engineering, Sri Eshwar College of Engineering, Coimbatore, Tamil Nadu 641202, India
e-mail: summiaparveen.h@sece.ac.in

S. Saradha
e-mail: saradha.s@sece.ac.in

N. Krishnaraj
Department of Networking and Communications, School of Computing, SRM Institute of Science and Technology, Kattankulathur, Tamil Nadu 603203, India

© The Author(s), under exclusive license to Springer Nature Singapore Pte Ltd. 2023
C. So-In et al. (eds.), *Information Systems for Intelligent Systems*, Smart Innovation, Systems and Technologies 324, https://doi.org/10.1007/978-981-19-7447-2_17

a person's specific entity. This is widely applied in different fields like marketing, education, e-commerce, etc.. SA aims in examining the Internet suggestions as well as to determine the sources of sentiment. Various resources presented for sentimental data are online, Web reviews, social database, and so on. Any user could share the opinions by means of comments, posts, and reviews. These reviews are utilized by business people and researchers in order to implement application programming interfaces (APIs). It motivates the clients for providing their own ideas and suggestions regarding a product which could be a simple way of collecting information and used for research work. There are 3 models of API on Twitter: REST API, Search API as well as Streaming API. First category is often applied for gathering latest news and profile data. Search API activates researchers and developers to examine specific information obtained from Twitter. Streaming API is employed for collecting Twitter data simultaneously. A new application is developed using three interconnected APIs. Hence, SA is considered to be a significant reason for massive information existing through online.

17.2 Literature Review

17.2.1 Sarcasm Sentiment Analysis

SD processes are rapidly developed domains of NLP by exploring ideas from word, expression as well as sentence level classifier [1] to file [2]. Many investigations have been carried out in discovery methods to analyze the emotions through optimal accuracy in written text and analyzing sarcasm, humors with irony inside social media information. Sarcastic emotion finding is classification into 3 types dependent on text features utilized to classifier, that is, lexical, pragmatic, and hyperbolic.

17.2.2 Lexical Feature-Based Classification

Text assets, namely unigram, bigram, n-grams, and so on, come under the lexical features of a text. Authors utilized these features for identification of irony. Wang et al. [3] established this model in the initial period, while they detected that lexical features participated in an essential function in identifying irony with sarcasm in text. In their following work, Riloff et al. [4] utilized these lexical features together through syntactic features for identifying sarcastic tweets. Barbieri et al. utilized a design dependent model (higher frequency words with content words), while punctuation dependent techniques are created using a weighted k-nearest neighbor (KNN). Muresan et al. [5] examined that bigram dependent features generate optimal outcomes in identifying irony in tweets and Amazon product reviews. Boia et al. [6]

discovered numerous lexical features (gained from LWIC [7], and WordNet involves [8]) the process of identifying sarcasm.

17.2.3 Pragmatic Feature-Based Classification

Pragmatic feature-based classification utilizes representative as well as symbolic text in tweets regularly because of the restrictions in communication distance. This representative symbolic text is known as pragmatic features (namely smilies, emoticons, replies, @user, and so on.). They have used major dominant features for recognizing sarcasm in tweets as some author has utilized this feature in their effort for detecting sarcasm. Pragmatic feature is a major key attributes utilized in Ref. [4] for detecting sarcasm in words. Reyes et al. [9] utilized pragmatic features such as emoticons with special punctuations for identifying sarcasm from newspaper text information. Boia et al. [6] searched this feature by a few more attributes like smileys and extended a sarcasm recognition scheme utilizing the pragmatic features of Twitter information. Joshi et al. [10] further utilized the pragmatic features in political tweets for forecasting that party will win the election. Likewise, Bamman and Smith [11] utilized emotional with behavioral features on users' as well as past tweets for identifying irony statement.

17.3 The Proposed Cnn-Based Sarcasm Classification Model

The proposed classification model incorporates three main processes, namely feature engineering, FS, and classification. The entire architecture is signified in Fig. 17.1. The proposed CNN model integrates the implicit as well as explicit representations of short text for sarcastic classification.

17.3.1 Overview of CNN

CNN is a type of a multiple layer which is employed in FC feedforward NN widely employed for extracting the local features in an automated way. Although several kinds of CNN exist, the structure of standard CNN is developed using a convolutional layer, a pooling layer, and a FC layer which is comes under multiple NN. Several types of layers act diverse roles. For instance, the convolutional layer is employed for detecting the personal concatenation of features from input information by local association of weights.

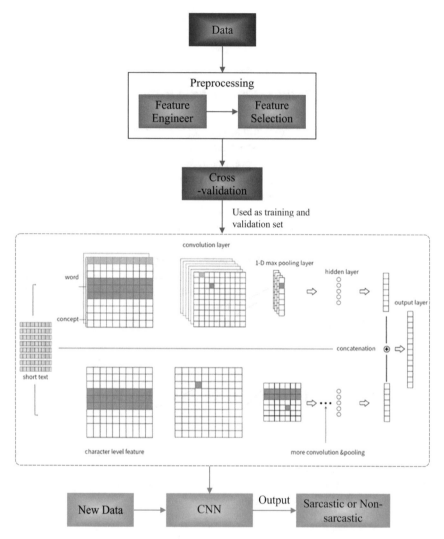

Fig. 17.1 Overall process of the proposed CNN model

Next, the pooling layer is applied for merging identical features into single network parameters and achieves translation invariant characteristics. The FC layer is helpful in converting the input to a vector and achieves various processes. On comparing with the full-connected network, the CNN involves less number of parameters using local connections, distributes weights, and pooling functions.

Softmax regression

It is frequently included at the right-handed side of the FC layer in final stage of classification. Softmax regression technique will determine possible input sample which comes under the class of all labels. The softmax function can be explained by

$$q(\chi_j^l) = \frac{e^{x_j^{l-1}}}{\sum_{k=1}^{J} e^{x_k^{l-1}}} \tag{17.1}$$

17.3.2 Model Design

The proposed model deals with a concept of CNN, applying 2 sub-networks in order to filter the word concepts and character features. The short text words are conceptualized by the use of knowledge base, and the words are made in a short text. Next, the model is explained and showed the way of learning the features from inducing data from every word, model as well as unique characters.

17.3.3 Short Text Conceptualization

Initially, the first process begins with the conceptualization of short text method. This is attained by available details like DB-pedia, Yago, Freebase, and Probase [12]. Here, Probase is applied because of it comprises maximum analysis of wordly data over other methods. It has the probabilistic data for calculating the short text, like classification, popularity, and typicality. While managing maximum number of relatives in Probase, a list of models and relevancy in data could be derived.

At this point, the concept vector can be represented as $\mathcal{C}p = \{\langle cp_1, wd_1 \rangle, \langle cp_2, wd_2 \rangle, \cdots, \langle cp_k, wd_k \rangle\}$ is introduced, where cp_z denotes the method of knowledge base, $and\,wd_z$ indicates the weight of data connected by cp_z. It calculates the concept vector of provided data utilizing a new knowledge exhaustive manner. In Probase, the number k is assigned to 10 concepts; therefore, first ten results would be derived. For instance, a short text of "CNOOC signed a PSC by ROC", and the resultant concept would be attained as {<client, 0.9>, <channel, 0.6>, <mythological creature, 0.6>, <international famous enterprise, 0.2>, <chinese oil main, 0.2>}.

17.3.4 Overall Architecture of the Model

Once the conceptualization results are obtained, knowledge base concepts and short text should be incorporated. To embed word and concept, predefined word embedding is used and maintains it as stationary. However, since there is no pre-trained embedding system, the embedding of characters should be changed in the training

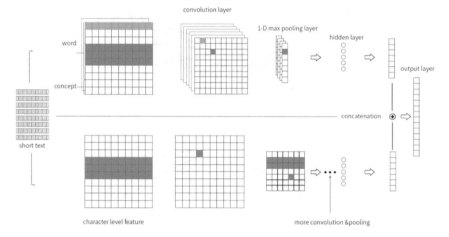

Fig. 17.2 Network architecture

period. This is an efficient way of making the embedding characters an optimal one. They are upper and lower sub-networks. The first one is helpful in embedding words and concepts present in short-text, whereas lower sub-network is applied for embedding characters or alphabets. These models are assumed to be CNN, where the characteristics could be learned in lower and higher levels correspondingly. The upper unit comprises of 7 layers: 1 input layer, 2 *conv* layers, 2 pooling layers, and 2 hidden layers.

17.4 Results and Discussion

17.4.1 Results Analysis on Amazon Dataset

Here, Table 17.1 depicts the confusion matrix achieved by various classification models on the employed Amazon database. A table value represents the projected CNN technique which provides an optimal classification by dividing 71 instances as non-sarcastic and 83 instances as sarcastic. Then, PSO-SVM method classifies 69 instances as non-sarcastic and 89 instances as sarcastic. Followed by, the DT classifier undergoes classification of 61 instances as non-sarcastic and 62 instances as sarcastic. Afterward, the NB classifier model provides effective classification process by accomplishing 71 instances as non-sarcastic and 66 instances as sarcastic ones. Consequently, AdaBoost classifier classifies 64 instances as non-sarcastic and 60 instances as sarcastic. Subsequently, the NB classifier efficiently classifies by obtaining 66 instances as non-sarcastic and 76 instances as sarcastic ones. According to the table, it is noticeable that PSO-SVM method consumes maximum number of optimally classified instances on the applied Amazon dataset.

Table 17.1 Confusion matrix of Amazon dataset using various classifiers

Experts	CNN		PSO-SVM		DT		NB		AdaBoost		SVM	
	NS	S	NS	S	NS	S	NS	S	NS	S	NS	S
NS	71	4	69	6	61	14	71	4	64	11	66	9
S	6	83	9	80	27	62	23	66	29	60	13	76

Table 17.2 Classifier results analysis on Amazon dataset

Classifiers	Precision	Recall	F1-score
CNN	0.95	0.92	0.93
PSO-SVM	0.92	0.88	0.90
DT	0.76	0.75	0.75
NB	0.86	0.84	0.83
AdaBoost	0.77	0.76	0.76
SVM	0.87	0.87	0.87

Table 17.2 and Figs. 17.3, 17.4 and 17.5 illustrate the outcome obtained by different classifiers on Amazon dataset with precision, recall, and F1-score, respectively. By estimating the results in terms of precision, it produces ineffective results whereas the DT with the precision value of 0.76.

Additionally, a little higher precision value of 0.77 is achieved by AdaBoost classification model. On the other hand, NB model implies reasonable classification with precision value of 0.86. Then, slightly higher precision value of SVM model is provided with the value of 0.87. Simultaneously, the projected PSO-SVM technique displayed superior classification by attaining the maximum precision value of 0.92.

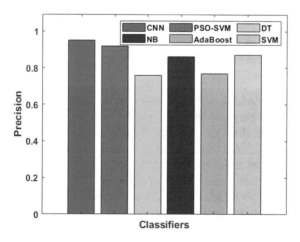

Fig. 17.3 Precision analysis of various methods on Amazon dataset

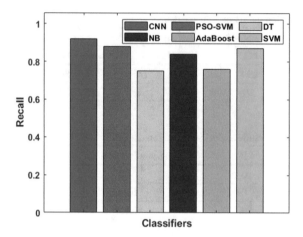

Fig. 17.4 Recall analysis of various methods on Amazon dataset

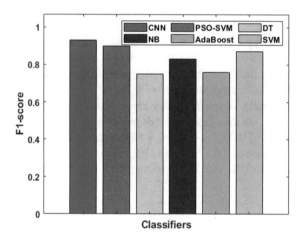

Fig. 17.5 F1-Score of various methods on Amazon dataset

Though PSO-SVM model gives best precision value, it is not reputed as CNN model which attained a highest precision value of 0.95.

By calculating the simulation outcome in terms of recall, it offers worst results which are by DT with the recall value of 0.75. In addition, AdaBoost classifier obtains a slight higher recall value of 0.76. Followed by, the NB model shows a manageable classification with the recall value of 0.84. Next, better recall value of 0.87 is derived from SVM approach. The PSO-SVM method provides optimal classification by attaining maximum recall value of 0.88. Even though the PSO-SVM model produced maximum recall value, it does not signify the performance of CNN technology. As an interesting factor, the CNN model attained a highest recall value of 0.95.

Table 17.3 Confusion matrix of Twitter dataset using various classifiers

Experts	CNN		PSO-SVM		DT		NB		AdaBoost		SVM	
	NS	S	NS	S	NS	S	NS	S	NS	S	NS	S
NS	1827	218	1604	441	1585	460	1100	945	1346	699	1192	853
S	257	1698	299	1656	611	1344	555	1400	504	1451	448	1507

On the basis of results obtained from F1-score, the inefficient outcome acquired from DT with F1-score value of 0.75. Then, a little rise in F1-score value of 0.76 is achieved by AdaBoost classifier. Followed by, the NB model reaches appreciable classification with F1-score value of 0.83. On the other side, a better F1-score value is achieved from SVM model with the estimate of 0.87. Then, PSO-SVM approach resembles the best classification by accomplishing a good F1-score value of 0.90. Therefore, it is monitored that superior classification is offered by the proposed CNN model by obtaining higher F1-score value of 0.93.

17.4.2 Results Analysis on Twitter Dataset

Table 17.3 illustrates the confusion matrix obtained by diverse classifying models based on the Twitter dataset. The table value indicates that CNN model gives superlative classification by classifying 1827 instances as non-sarcastic and 1698 instances as sarcastic. Followed by, PSO-SVM model undergoes classification of 1604 instances as non-sarcastic and 1656 instances as sarcastic. Secondly, the DT method classifies 1585 instances as non-sarcastic and 1344 instances as sarcastic ones. Afterward, the NB classification process offers efficient simulation outcome by achieving 1100 instances as non-sarcastic and 1451 instances as sarcastic.

Finally, AdaBoost classifier computes the classification with 1346 instances as non-sarcastic and 1451 instances as sarcastic. Next, the SVM classifier model offers better outcome by attaining 1192 instances as non-sarcastic and 1507 instances as sarcastic. According to the table values, it denotes that CNN model reaches great number of efficient classified instance from the Twitter dataset.

Table 17.4 depict the outcome attained by diverse classification models with respect to precision, recall, and F1-score. On estimation of results in terms of precision, it displays the ineffective results from NB classifier with the precision value of 0.65. Similarly, a slightly higher precision value of 0.68 is retrieved by SVM model. Besides, the DT model shows reasonable classification with the precision value of 0.70. Next to that, even slightly higher precision value of AdaBoost model is offered with the value of 0.71. Followed by, PSO-SVM model showed improved classification by attaining the maximum precision value of 0.78. Hence, the presented CNN model offered superior classification with the highest precision value of 0.89.

By evaluating the results with respective to recall, it gives inefficient outcome from NB with the recall value of 0.65. Additionally, SVM classifier derives a slight

Table 17.4 Performance evaluation of proposed method with various classifiers on Twitter dataset

Classifiers	Precision	Recall	F1-score
CNN	0.89	0.87	0.88
PSO-SVM	0.78	0.84	0.81
DT	0.70	0.70	0.70
NB	0.65	0.65	0.65
AdaBoost	0.71	0.71	0.71
SVM	0.68	0.67	0.67

higher recall value of 0.67. Besides, the DT model offers a managing classification with the recall value of 0.70. Next, a better recall value of 0.71 is obtained from AdaBoost model. Simultaneously, the proposed PSO-SVM method proves to be an optimal classification by attaining the greater recall value of 0.84. As an interesting factor, the CNN model showed standard computation by accomplishing maximum recall value of 0.87.

According to the results extracted from F1-score, the worst outcome acquired from NB with F1-score value of 0.65. Then, a slightly increase in F1-score value of 0.67 is attained by SVM classification process. Next, the DT model reaches considerable classification with the F1-score value of 0.70. Besides, a better F1-score value is achieved from AdaBoost model with the measure of 0.71. Afterward, the proposed PSO-SVM technique denotes better classification by obtaining good F1-score value of 0.81. Interestingly, the CNN model offered optimal results over the compared methods by attaining the maximum F1-score value of 0.88.

17.5   Comparative Analysis

Table 17.5 and Fig. 17.6 represent the result of different classifiers based on the average classifier results. The table values denote least classifier results which are achieved by the DT model with the average precision of 0.73, recall of 0.72, and F1-score of 0.72. Next, the slightly better classifier results are derived by AdaBoost technique with the average precision of 0.74, recall of 0.73, and F1-score of 0.73. Afterward, the NB model controls well with the average precision of 0.76, recall of 0.74, and F1-score of 0.74.

Consecutively, the SVM method produces reasonable results with the average precision of 0.78, recall of 0.77, and F1-score of 0.77. Followed by, PBLGA model attempts to control the average precision of 0.84, recall of 0.81, and F1-score of 0.82.

At the same time, the IWS method shows a competitive result with the average precision of 0.83, recall of 0.91, and F1-score of 0.87. Simultaneously, PSO-SVM achieves better classification with the maximum average precision of 0.85, recall of 0.86, and F1-score of 0.86. Though PSO-SVM offered better results, it fails to perform as CNN model which attains the maximum precision of 0.89, recall of 0.87,

Table 17.5 Performance evaluation of proposed method with various classifiers on average

Classifiers	Precision	Recall	F1-score
CNN	0.92	0.89	0.90
PSO-SVM	0.85	0.86	0.86
DT	0.73	0.72	0.72
NB	0.76	0.74	0.74
AdaBoost	0.74	0.73	0.73
SVM	0.78	0.77	0.77
PBLGA	0.84	0.81	0.82
IWS	0.83	0.91	0.87

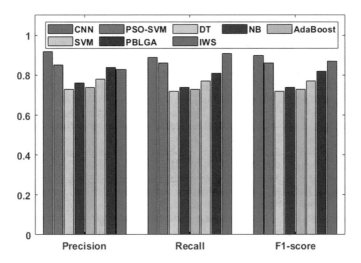

Fig. 17.6 Comparative analysis of diverse models in terms of different measures

and F1-score of 0.88. Based on the above tables and figures, it is proved that the presented CNN model performs optimally when compared with other measures.

17.6   Conclusion

This research study has presented a CNN-based sarcasm detection and classification model to classify the tweets or reviews to sarcastic or non-sarcastic ones. The proposed classification model incorporates three main processes, namely feature engineering, FS, and classification. The presented CNN-based model is tested using two kinds of dataset from Amazon and Twitter. The experimental results stated the optimal classification results are offered by the CNN model with the highest average precision of 0.92, recall of 0.89, and F1-score of 0.90.

References

1. Attardo, S.: Irony as relevant inappropriateness. In: Irony in Language and Thought. Psychology Press, New York, NY, USA, pp. 135–174 (2007)
2. Ghosh, D., Guo, W., Muresan, S.: Sarcastic or not: Word embeddings to predict the literal or sarcastic meaning of words. In: Proceedings of EMNLP, pp. 1003–1012 (2015)
3. Wang, Z., Wu, Z., Wang, R., Ren, Y.: Twitter sarcasm detection exploiting a context-based model. In: Proceedings of Web Information Systems Engineering (WISE), pp. 77–91 (2015)
4. Riloff, E., Qadir, A., Surve, P., De Silva, L., Gilbert, N., Huang, R.: Sarcasm as contrast between a positive sentiment and negative situation. In: Proceedings of the Conference on Empirical Methods in Natural Language Processing, pp. 704–714 (2013)
5. Muresan, S., Gonzalez-Ibanez, R., Ghosh, D., Wacholder, N.: Identification of nonliteral language in social media: A case study on sarcasm. J. Assoc. Inf. Sci. Technol. (2016)
6. Boia, M., Faltings, B., Musat, C.-C., Pu, P.: A: Is worth a thousand words: How people attach sentiment to emoticons and words in tweets. In: Proceedings of International Conference on Social Computing, pp. 345–350 (2013)
7. Fersini, E., Pozzi, F.A., Messina, E.: Detecting irony and sarcasm in microblogs: The role of expressive signals and ensemble classifiers. In: Proceedings of IEEE Data Science and Advanced Analytics (DSAA), pp. 1–8 (2015)
8. Bharti, S.K., Babu, K.S., Jena, S.K.: Parsing-based sarcasm sentiment recognition in Twitter data. In: Proceedngs of IEEE/ACM ASONAM, pp. 1373–1380 (2015)
9. Reyes, A., Rosso, P., Veale, T.: A multidimensional approach for detecting irony in Twitter. Lang. Resour. Eval. **47**(1), 239–268 (2013)
10. Joshi, A., Sharma, V., Bhattacharyya, P.: Harnessing context incongruity for sarcasm detection. In: Proceedings of 53rd Annual Meeting Association for Computational Linguistics, International Joint Conference Natural Language Processing (ACL-IJCNLP), vol. 2, pp. 757–762 (2015)
11. Bamman, D., Smith, N.A.: Contextualized sarcasm detection on Twitter. In: Proceedings of AAAI International Conference Web Social Media (ICWSM), pp. 574–577 (2015)
12. Wu W, Li H, Wang H, Zhu KQ.: Probase: A probabilistic taxonomy for text understanding. In SIGMOD, pp. 481–492 (2012)

Chapter 18
Internet of Things Security and Privacy Policy: Indonesia Landscape

Sidik Prabowo, Maman Abdurohman, and Hilal Hudan Nuha

Abstract IoT development is growing rapidly in hardware and software technology; this presents challenges, especially for the government as a regulator to follow suit. However, there are many gaps in IoT policy and regulation, especially in terms of security and privacy in Indonesia. In this paper, we have identified gaps in existing conditions in Indonesia compared with the USA, Singapore, and UEA as major countries that have advanced first in IoT security policy. This paper presents a gap related to the availability of guidelines, regulations, and laws that regulate IoT security in Indonesia and proposes a stakeholder map that needs to cooperate in producing regulations and policies that comply with the conditions of government and industry in Indonesia.

18.1 Introduction

Internet of things (IoT) is one of the technology trends that has entered the "trough of disillusionment" phase in 2020 and enter the "slope of enlightenment" phase, which means that this technology is successful and accepted by consumers. IoT can survive this phase and appears on the Gartner Impact Radar 2022 [7], which shows that IoT is specifically for IoT platforms that have a "very high" impact and will continue to grow for the next 3 to 6 years. There are 12.3 billion active endpoints IoT devices worldwide by 2021. Meanwhile, in 2025, it is predicted that it will reach 27 billion device [9]. In Indonesia itself, it is predicted that the IoT market share will grow

Supported by Telkom University.

S. Prabowo (✉) · M. Abdurohman · H. H. Nuha
School of Computing, Telkom University, Bandung 40257, Indonesia
e-mail: pakwowo@telkomuniversity.ac.id

M. Abdurohman
e-mail: abdurohman@telkomuniversity.ac.id

H. H. Nuha
e-mail: hilalnuha@telkomuniversity.ac.id

201

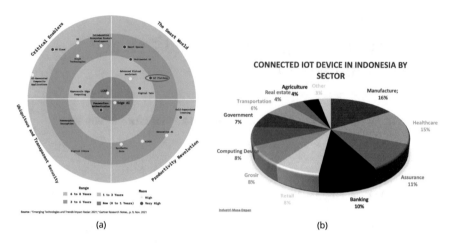

Source: "Emerging Technologies and Trends Impact Radar: 2021," Gartner Research Notes, p. 9, Nov. 2021

(a)   (b)

Fig. 18.1 **a** Gartner impact radar 2022, **b** connected IoT device prediction in 2022 by sector in Indonesia

rapidly, indicating that by 2022, the IoT industry in Indonesia will reach a value of 444 trillion rupiahs with more than 400 million connected devices distributed in various sectors as shown in Fig. 18.1b. It can be seen that the manufacturing and healthcare sectors are predicted to have a fairly large level of adoption of IoT technology compared to other industries in Indonesia. This is believed to be partly due to the trend of industrial digitization in Indonesia and support from the government through "making Indonesia 4.0" as a vision of industrial transformation in Indonesia.

Our research shows that there is a very large gap in the availability of policies and regulations needed for the growth of IoT in Indonesia. To the best of our knowledge, there has never been a publication that focus on this issue especially in Indonesia.

- We present a gap analysis related to IoT security regarding regulatory and policy aspects based on literature and best practices compared to the existing Indonesian condition.
- We proposed the formation of a stakeholder map of parties who must cooperate in formulating policies related to IoT security policy and regulation in Indonesia.

The reminder of this paper has the following systematic: Chap. 2 describes study related to IoT security and policy best practices around the world. Chapter 3 describes the components that make up the IoT security framework, based on existing best practices. Chapter 4 presented an overview of the current conditions related to the IoT security framework and policies that exist in Indonesia, based on the components described in the previous chapter. Chapter 5 Conclusion of this study.

18.2   IoT Security and Privacy Policy

The magnitude of the potential for developing IoT services in various sectors through-out the world still cannot be matched with protection related to the information security involved; this is, of course, a very important issue to pay attention to. Fur-thermore, the results of studies related to solutions that have been developed are described, including the development of the IoT security framework, which is sup-ported both in the form of laws and new guidelines. In addition to the general purpose security framework, there is also an IoT security framework issued and used by the governments of countries globally to provide guidance and direction for development related to IoT in their country.

18.2.1   United States of America (USA)

The United States (USA) is a country that has complete regulations related to cyber-security, one of which is related to IoT. Based on The Internet of things (IoT), Cyber-security Improvement Act of 2020 (Public Law (PL) 116 207) as the legal basis for the National Institute of Standard and Technology (NIST) as a non-regulatory agency that acts as an extension of the U.S Department of Commerce to issue guidelines, standards, and derivative rules related to the application of IoT, especially in the application in the realm of government. No significant difficulties were found in the deriving from the cybersecurity standards that NIST had previously held. This is sup-ported by NIST's standards, which are technology/device-neutral and easy to apply in the IoT context. Until this paper was written, NIST had issued many publications related to IoT, which contain various guidelines and rules that must be applied by stakeholders related to IoT in the USA. Because it already has comprehensive regu-lations and guidelines related to cybersecurity before, reducing it to comply with IoT is not a difficult thing for NIST to do. The IoT security framework compiled by NIST explicitly categorizes documents and standards issued for each relevant stakeholder (Fig. 18.2).

18.2.2   Singapore

Singapore as Indonesia's closest neighbor, 2013 has formed an independent state body that functions as an IoT Technical Committee, which has a role in formulating and providing recommendations regarding the implementation of IoT by all relevant stakeholders in Singapore. One of the results of this committee is incorrectly issuing ITSC TR 64: "Guidelines for IoT Security for Smart Nation" [6] from 2018 as a foothold related to IoT regulations in Singapore. As a derivative of this document, the Singapore government issued the "Guidelines: Internet of things (IoT) Cybersecurity guide" [2] in 2020. This guide is used by IoT developers, IoT providers, and IoT

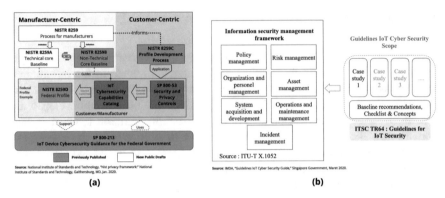

Fig. 18.2 **a** IoT security framework NIST architecture, **b** IoT cybersecurity guidelines Singapore scope

users in Singapore. Developing this guideline refers to the ITU-T X.1052 Information Security Management framework and ITSC TR64: 2018 "Guidelines for IoT security for smart nations," previously issued. This guide recommends four principles in IoT development: secure by defaults, rigor in defense, accountability, and resiliency. This guide does not regulate the implementation of data privacy in the IoT domain; this is because privacy in Singapore is held in the Personal Data Protection Act (PDPA) [13].

18.2.3 United Arab Emirates (UAE)

The United Arab Emirates (UAE) has issued several specific guidelines and regulations regarding IoT implementation. Starting in 2018 where the UAE government issued an IoT Regulatory Policy, then continued in 2019 with IoT regulatory Procedures. The guidance document is an offshoot of the UAE Telecommunication Law (Federal Law Decree 3 of 2003). Both regulations were developed with the General Data Protection Regulation (GDPR) as one of the references. The purpose of developing this framework is as a guide and regulation that must be complied with by relevant stakeholders to support the IoT ecosystem in the UAE, namely telecommunications service providers, IoT service providers, IoT users (individual, Business, and Government) supported by applicable law (Federal Law decree 3 of 2003), applying this framework is mandatory for all stakeholders. Meanwhile, if found, parties who do not meet the rules contained therein will be subject to sanctions [11]. With regulations accompanied by laws, the application of IoT in the UAE becomes more focused and monitored and provides separate guarantees for consumers. In addition to Singapore and the UAE, other countries such as the United Kingdom (UK), South Africa, Japan, India, New South Wales, South Korea, the Republic of China, and Brazil are some countries that have issued policies and derivative regulations related to IoT in their countries.

18.3 IoT Security Framework Component

From the results of the previous literature review, there is a general description regarding the application of regulations related to IoT in several countries grouped into three components, namely.

18.3.1 Code of Practice/Guidelines

A code of practice/guidelines is a written guide issued by an institution or association that functions as a guideline or standard by related stakeholders, one of which is to support the implementation of Good Corporate Governance (GCG). This document contains more detailed guidelines but still maintains the principle of complying with applicable regulations or laws. A simple example of this is the one applied by the United Kingdom (U.K) in a document entitled "Code of Practice for Consumer IoT Security" [1]. From Table 18.1, it can be seen that there is a 13 component that relevant stakeholders must obey in the IoT industry in the U.K. The code of practice was issued in 2018 by U.K Department for Digital, Culture, Media and Sport, and reviewed every two years to adjust to the latest technology trend and regulation. Code of practice/guidelines acts as a technical guide and can adapt quickly to current technological developments. In its application, this code of practice needs to be supported by rules and regulations that are stronger in nature and have legal consequences in their application. This regulation is usually regulated regarding the obligations that must be fulfilled by each relevant stakeholder and the values that must be borne when not implementing these obligations. This regulation will assist the adoption process of the existing code of practice. Continuing the case study in the U.K., after launching the code of practice document in 2018, the U.K government subsequently issued The Product Security and Telecommunications (PSTI) Bill in November 2021. This regulation provides legal support for the consumer device industry, including IoT devices.

18.3.2 Reference Standard

In the preparation of practice/guidelines, it will be very helpful for the process if there are references or standards to be followed. The standards used have received recognition from professional institutions in their respective fields. For example, related to data protection in the world, several references and institutions issue them, including:

- General Data Protection Regulation (GDPR)—EU [8]
- Lei Geral de Proteção de Dados (LGPD)—Brasil [5]

Table 18.1 U.K. code of practice for consumer IoT security

Component	Applied to			
	1	2	3	4
No default password	V			
Implement a vulnerability disclosure policy	V	V	V	
Keep software updated	V	V		
Securely store credentials and security-sensitive data	V	V	V	
Communicate securely	V	V	V	
Minimize exposed attack surface	V	V		
Ensure software integrity	V			
Ensure that personal data is protected	V	V	V	V
Make system resilient to outages	V	V		
Monitor system telemetry data		V		
Make it easy for consumers to delete personal data	V	V	V	
Make installation and maintenance of device easy	V	V	V	
Validate input data	V	V	V	

- Act on Protection on Personal Information (APPI)—Japan [3]
- Personal Data Protection Bill (PDPB)—India [12]
- Notifiable Data Breach (NDB)—Australia [4]
- California Consumer Privacy Act (CCPA)—California [10]

The standards related to data protection can then be used as a reference in compiling IoT codes of practice or guidelines for areas related to data or information generated by IoT devices. The next chapter describes the security conditions and policies related to IoT in Indonesia seen from the constituent components.

18.4 Indonesia IoT Security Landscape

As stated in the previous chapter, the following is an overview of the components that make up the IoT security framework in Indonesia today.

18.4.1 Code of Practice

From the ministries and institutions involved in the IT sector in Indonesia until this publication was written, no one has issued any code of practice or guidelines for implementing IoT in Indonesia. So that in its performance, relevant stakeholders tend to look for regulations and policies from outside that do not violate existing laws in Indonesia.

18.4.2 Regulation

In Indonesia, cybersecurity policies began to be initiated in 2007 when the Minister of Communications and Information Technology (KOMINFO) issued regulation No. 26/PER/M.Kominfo/5/2007 concerning Securing the Utilization of Internet Protocol-Based Telecommunication Networks, followed by regulation No. 16/PER/M KOMINFO/10/2010 and revised again by Ministerial Regulation No. 29/PER/M.KOMINFO/12/2010. Until now, there are several regulations related to cybersecurity that can be used as a reference even though they are not specifically regulated related to IoT, including the following:

- Law (UU) Number 19 of 2016 concerning Information and Electronic Transactions (UU ITE)
- Minister of Communication and Informatics Regulation No. 20 of 2016 concerning Personal Data Protection
- Government Regulation number 71 of 2019 for the Implementation of Electronic Systems and Transactions (PSTE)
- Presidential Regulation Number 39 of 2019 concerning One Data Indonesian
- Regulation of the Minister of Communications and Information Technology Number 5 of 2020 concerning: Electronic System Operators in Private Scope
- Regulation of the National Cyber and Crypto Agency (BSSN) Number 8 of 2020 concerning Security Systems in the Implementation of Electronic Systems
- Regulation of the National Cyber and Crypto Agency Number 11 of 2020 concerning the Technical Competency Dictionary in the Cybersecurity and Encryption Sector
- Regulation of the National Cyber and Crypto Agency Number 4 of 2021 concerning Guidelines for Information Security Management of Electronic-Based Government Systems and Technical Standards and Procedures for Security of Electronic-Based Government Systems

However, among the related regulations above, there are no specific regulations that regulate IoT technology both from the producer's side and the protection of data and information security from consumers themselves. This is, of course, a fairly large gap related to the security of IoT consumer data and information in Indonesia.

18.4.3   Reference Standard

Until now, there is no reference standard used for IoT in Indonesia. The existing standard related to information technology security in Indonesia currently available is SNI ISO/IEC 27001:2013 on Information Security Management Systems. However, this document is very broad to be applied to IoT technology.

18.4.4   Recommendation

From the results of this study, we designed the architecture of relevant stakeholders in Indonesia that can formulate regulations and guidelines for IoT in Indonesia. Based on the literature on several countries that have previously issued guidelines and rules related to IoT security policies, it is proposed to involve all parties, ranging from users/consumers, industry, and regulators to academics, to be able to provide comprehensive input (Fig. 18.3).

- Regulator: KOMINFO, as the ministry authorized to regulate various matters related to information and communication technology in Indonesia, can cooperate with BSSN as an institution that is competent and has the authority about information security assurance can be the main pillar in preparing regulations and

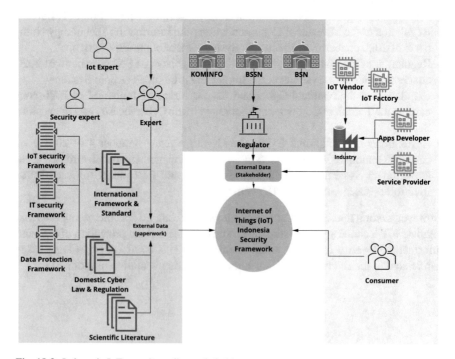

Fig. 18.3 Indonesia IoT security policy stakeholder map

codes of practice related to IoT security in Indonesia. BSN can then standardize the formulations produced by the two institutions to ensure their application by all parties related to the IoT industry in Indonesia.

- Industry: Many industries are related to IoT services in Indonesia itself.

  - IoT Device Manufacturers/Vendors such as sonoff, Bardi, Xirca, Xiaomi, Schneider, as well as other IoT product producing companies operating in Indonesia
  - Service Providers: In addition to cellular providers such as XL and 3, which provide data and communication services, in Indonesia, there is PT Telkom which currently has deployed LoRaWAN networks in several regions in Indonesia, and PT Telkomsel, which has also deployed NB-IoT networks, can be included in giving opinions or useful input

    Mobile/Application Developers: One of the parties who plays a role in the development of IoT is a mobile developer who plays a role in developing IoT applications according to user needs. Many companies are engaged in this field in Indonesia, ranging from home to enterprise scale. To get input from the point of view of IoT application developers, the contribution of this party is also very necessary in the preparation of the IoT security framework that is to Indonesian conditions.
  - IoT Platform Providers: There are many IoT platforms currently circulating in the Indonesian market, including Antares and Iotera, a domestic product that can also be involved in formulating policies related to IoT security in Indonesia from the point of view of the IoT platform provider.

These parties also need to accommodate their inputs and suggestions so that the produced regulations and guidelines can later be applied easily by these actors.

- Researcher/Expert: To provide comprehensive input and literature from currently available sources, researchers and experts need input, especially in IoT and information security.
- User/Consumer: An entity that is no less important is the user of IoT services, both personal and business users. The user's concerns regarding the expected security will be obtained.

From experience shown by countries that have previously issued regulations related to IoT security, all these stakeholders have an important role in producing comprehensive rules and guidelines.

18.5 Conclusion

There are huge gaps related to IoT security and privacy policy in Indonesia. This publication documents these gaps and provides recommendations that can be used as options to close these gaps immediately. A comprehensive strategy must be carried out in an organized manner by the relevant ministries and institutions to close these gaps. This needs to be addressed directly to increase security and convenience for both users and service providers related to IoT in Indonesia.

References

1. Department for Digital, Culture, Media & Sport: Code of Practice for Consumer IoT Security. United Kingdom Government (Oct 2018)
2. IMDA: Guidelines IoT Cyber Security Guide. Singapore Government (2020)
3. Iwase, H.: Japan overview of the act on the protection of personal information. Eur. Data Prot. Law Rev. 5(1), 92–98 (2019)
4. Leonard, P.: The new Australian notifiable data breach scheme. SSRN Electron. J. (2018)
5. Machado, D.D.: Lei geral de proteção de dados pessoais—LGPD (2021)
6. National Institute of Standards and Technology: NIST Privacy Framework: Tech. rep. National Institute of Standards and Technology, Gaithersburg, MD (Jan 2020)
7. Nguyen, T., Reynolds, M., Kandaswamy, R., et al.: Emerging technologies and trends impact radar. In: Gartner Research Notes, p. 9 (2021)
8. Parliament E. of the European Union C.: The EU General Data Protection Regulation (GDPR). European Union (Apr 2016)
9. Praveen, P.: IoT analytics research 2022 report. Tech. rep, IoT Analytics (May (2022)
10. Stallings, W.: Handling of personal information and deidentified, aggregated, and pseudonymized information under the California consumer privacy act. IEEE Secur. Priv. 18(1), 61–64 (2020)
11. White, M., Mennie, P., Chudzynski, R.: Regulating the internet of things in the UAE. Tech. rep. PWC (2019)
12. Yadav, D.A., Yadav, G.: Data protection in India in reference to personal data protection bill 2019 and IT act 2000. Int. Adv. Res. J. Sci. Eng. Technol. 8(8) (2021)
13. Yip, M.: Personal data protection act 2012: understanding the consent obligation. Personal Data Prot. Digest 2017, 266 (2017)

Chapter 19
Fluctuating Small Data Imputation with Lagrange Interpolation Based

Ikke Dian Oktaviani, Maman Abdurohman, and Bayu Erfianto

Abstract The data imputation process, the subject of this research, is a solution needed in the current era of big data. In addition to big data, small data often experiences similar problems such as data loss, especially on time series-based data. Several things can cause data loss, one of which is the data communication problem caused by the network on the Internet of Things (IoT) system. Losing data in data conditions that are quite volatile is also quite a challenge. Not all methods can produce good accuracy in dealing with fluctuating data cases. The method used to impute small volatile data in this study is Lagrange interpolation. Several methods used for the data imputation process consist of Lagrange interpolation of degree one (linear), degree two (quadratic), and degree three (cubic). Lagrange cubic interpolation generally provides better accuracy with data loss percentages of 10%, 30%, and 50% compared to several other types of polynomial interpolation indicated by the MSE value. Indirectly, the results of this research can predict that if the application of Lagrange interpolation on fluctuating data has good accuracy, then Lagrange interpolation can have the same accuracy as non-fluctuating data.

19.1 Introduction

Data imputation is a method to recover lost data in a series of data so that the dataset can be intact [1]. Data imputation can affect data analysis, especially on time series data which emphasizes the completeness of the data. Incomplete data is a separate problem that often appears in several systems, one of which is an IoT-based system.

I. D. Oktaviani (✉) · M. Abdurohman · B. Erfianto
School of Computing, Telkom University, Bandung 40257, Indonesia
e-mail: idoktaviani@student.telkomuniversity.ac.id

M. Abdurohman
e-mail: abdurohman@telkomuniversity.ac.id

B. Erfianto
e-mail: erfianto@telkomuniversity.ac.id

© The Author(s), under exclusive license to Springer Nature Singapore Pte Ltd. 2023
C. So-In et al. (eds.), *Information Systems for Intelligent Systems*, Smart Innovation, Systems and Technologies 324, https://doi.org/10.1007/978-981-19-7447-2_19

Several methods can overcome the problem of data loss. The method applied must be adapted to the data conditions. The solutions offered range from statistical methods [2, 3] to artificial intelligence (AI)-based methods [4–6]. In choosing the right data imputation method, it is necessary to consider several components such as the amount of data, data trends, data stationarity, and others.

The application of statistical methods in performing data imputation can be found in many studies that have been carried out. The search for the mean is an example of the application of simple statistical methods often used. Another fastest method still often used is removing the missing data itself [7]. That method can affect the process of data analysis to be carried out. In addition to these methods, searching for the best value in recovering lost data can use AI methods [8].

The output parameters of the research objectives are also the basis for choosing the data imputation method. Several parameters, such as accuracy and cost, cannot be coupled in some cases. For example, applying the mean or linear interpolation will be more cost-effective than the AI method. However, in terms of accuracy, the AI method can give much better results than conventional methods [1]. That example is also supported by the amount of data used.

In this study, solving the problem of missing data with a small amount of data (small data) uses conventional methods. This small amount of data is often found in IoT-based systems with live data and systems. Thus, the contributions of this research are as follows:

1. Application of the Lagrange interpolation method for data imputation on fluctuating small data.
2. Improved accuracy of Lagrange interpolation by finding the best data retrieval position.

This paper has several sections starting from this section, which deals with the research's background, objectives, and contributions—followed by Sect. 19.2, which briefly describes the previous research on the data imputation process. Next is Sect. 19.3, which discusses the data imputation process, and Sect. 19.4, which explains the study's results. This paper is closed by Sect. 19.5, which is the conclusion of this study.

19.2   Related Work

The data imputation process has been carried out by many researchers, including Ref. [7], who apply the data imputation process to small data using the mean method and replace missing data with zero. The results of the imputation process are then continued with data classification using particle swarm optimization and Naïve Bayes. This imputation method can increase the accuracy to 4.33% in the dataset. In addition, there are Ref. [9] who apply predictive mean matching (PMM) to impute data with small sample sizes. Although Ref. [9] cites the statement Ref. [10] that PMM is not suitable to be applied to small data, this is a challenge. Based on the experiments conducted, PMM gave good results for imputing small data with various scenarios such as sample size, percentage of missing data, and other scenarios. Another imputation research on small data is Ref. [11], which uses the dynamic Bayesian network (DBN) method with support vector regression (SVR) on time series data. The results of this study indicate that DBN in feature selection does not provide a significant reduction in the error rate model. Several other methods for data imputation are described in a survey paper [12]. The paper explains that finding the mean value is a suitable method to overcome the problem of missing data in small data.

19.3 Data Imputation Process

Data imputation is done using a series of data that has gone through a random data removal process. Completing the missing data is by applying the Lagrange interpolation formula to Eq. (19.1) based on polynomials. Through Eq. (19.1), Lagrange interpolation is not based on a certain amount of data.

$$F(x) = y = \frac{(x - x_1)(x - x_2) \ldots (x - x_n)}{(x_0 - x_1)(x_0 - x_2) \ldots (x_0 - x_n)} y_0 + \frac{(x - x_0)(x - x_2) \ldots (x - x_n)}{(x_1 - x_0)(x_1 - x_2) \ldots (x_1 - x_n)} y_1$$
$$+ \ldots + \frac{(x - x_0)(x - x_1) \ldots (x - x_{n-1})}{(x_n - x_0)(x_n - x_1) \ldots (x_n - x_{n-1})} y_n \qquad (19.1)$$

If the above equation undergoes a simplification form, it will become Eq. (19.2) and Eq. (19.3)

$$F(x) = y = \sum_{i=0}^{n} P_i(x) y(x_i) \qquad (19.2)$$

where

$$P_i(x) = \prod_{j=0}^{n} \frac{x - x_j}{x_i - x_j} \qquad (19.3)$$

and $j \neq i$ [13].

The selection of the position of the reference data is the most important thing to get better accuracy, considering that interpolation uses a comparison of the amount of data that is not the same on polynomials with more than one degree (non-linear).

Table 19.1 shows an example of the data used in this study. The greater the percentage of missing data, the fewer reference points that will be used in the calculation process. This increasing percentage should impact the accuracy of the method. Lagrange interpolation processes the data so that the Nan value can be replaced with the interpolation result, then it will be compared with the actual value. The time

Table 19.1 Sample data with missing data percentage

Time (x)	Air_temp (y)	10%	30%	50%
00:00:49	64.76	64.76	64.76	64.76
00:01:49	63.86	Nan	Nan	Nan
00:02:49	64.22	64.22	64.22	64.22
00:03:49	64.4	64.4	Nan	Nan
00:04:49	64.4	64.4	64.4	Nan
00:05:49	63.5	63.5	63.5	63.5
00:06:49	62.78	62.78	62.78	Nan
00:07:49	62.42	62.42	62.42	62.42
00:08:49	62.24	62.24	62.24	62.24
00:09:49	62.24	62.24	Nan	Nan

data becomes the independent variable, the x variable, and the air temperature data becomes the dependent data, which is then called the y variable. Each variable is then processed by Eq. (19.1).

19.4   Result and Discussion

By applying the Lagrange interpolation method, the Lagrange interpolation has more accurate results when compared to data imputation by finding the average value and its relatives, namely linear, quadratic, and cubic polynomial interpolations. This result is related to selecting data positions used as reference data. Table 19.2 summarizes the best results from applying Lagrange interpolation compared to polynomial interpolation for data imputation using the metric mean square error (MSE).

Table 19.2 MSE value of imputation data method

Method	Missing value percentage (%)	Polynomial	Lagrange
Linear interpolation	10	0.083	0.083
	30	0.034	0.034
	50	0.062	0.061
Quadratic interpolation	10	0.051	0.051
	30	0.033	0.033
	50	0.043	0.043
Cubic interpolation	10	0.123	0.041
	30	0.069	0.051
	50	0.144	0.039

Table 19.3 MSE value of Lagrange interpolation

Comparison of total data	Missing value percentage (%)	Linear		Quadratic		Cubic	
		A	B	A	B	A	B
1 : n	10	0.083	–	0.051	–	0.041	–
	30	0.034	–	0.033	0.031	0.051	0.054
	50	0.061	0.061	0.043	0.049	0.039	0.056
2 : (n − 1)	10	–	–	0.148	–	0.146	–
	30	–	–	0.058	0.06	0.053	0.052
	50	–	–	0.039	0.056	0.341	0.065

All polynomial degrees in the Lagrange interpolation in this study underwent a selection of the position of the reference data. In this study, the derivation of the method is symbolized by the number 1, which represents the comparison of the composition of data collection of 1 : n applies to linear, quadratic, and cubic Lagrange, and number 2 for composition 2 : $(n − 1)$ applies to quadratic and cubic Lagrange where $n = 1, 2, 3$. Then, there is the symbol A which means taking the value as it is, and the symbol B which represents that the data is taken in time sequence by first processing the Nan data between the points that want to find the value using linear interpolation.

From Table 19.3, the selection of the composition of the reference data can affect the accuracy value, which is then proven by the MSE value obtained. Although the percentage of data loss is quite large, namely >10%, the MSE shown in Table 19.3 does not experience a significant change in either the type A or type B method. What is unique in this study is that the interpolation only depends on the values in a data series. When the reference data is out of range (extrapolation), this can increase the error value. This case is handled by taking the value before or after, which is known to impute the missing value at the beginning or the end of the data. That also applies if the interpolate process's reference value is outside the interpolated data range.

Based on that result, the best method for Lagrange interpolating data imputation in 10% missing value is cubic with the composition of data reference 1 : n, and its MSE value was 0.041. The best method for the 30% missing value is quadratic with the composition of data reference 1 : n, and its MSE value was 0.031. Last, the best method for 50% missing value was quadratic with data reference composition 2 : $(n − 1)$ and cubic with data reference composition 1 : n. Both have the same MSE value of 0.039 and type A.

19.5   Conclusion

Data imputation is an important thing that is done early before a series of data is finally processed. The condition of incomplete data impacts the data analysis process, especially if the data used is important. By imputing data using Lagrange interpolation, fluctuating data can be imputed with good results (see Fig. 19.1). The process of random data removal consists of three types, namely the percentage of 10%, 30%, and 50%. That percentage can affect the level of accuracy of the data imputation process. In addition, the selection of reference data points can also have an effect. However, with the increase in the amount of missing data, it does not give a significant change in the error rate, as evidenced by the MSE value. That proves that Lagrange interpolation is a good enough interpolation to solve cases of data imputation on fluctuating small data compared to several other interpolation methods. Figure 19.1 shows that quadratic 2A was a good method for 10% missing value, but quadratic 1A was a better method for 30% missing value. 50% missing data was an important problem, and cubic 1A can impute the missing data well than other methods. In general, Lagrange cubic interpolation type 1A can solve the problem quite well for every percentage of missing data.

With the development of the interpolation method, it is hoped that trials can be carried out on the same data using other types of interpolation. This hope is due to the possibility that other interpolation methods can give better accuracy results. The development of the data imputation method with interpolation needs to be explored more deeply, considering that one of these conventional methods is sufficient to provide accurate results with relatively lower computational time than other current methods. In addition, the performance of the interpolation method, which is not affected by the amount of data, is one of the advantages of this method so that it is suitable for solving problems with small data.

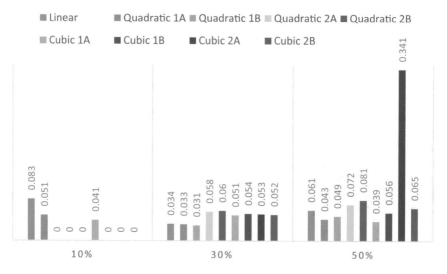

Fig. 19.1 MSE of Lagrange interpolation imputation method

References

1. Jerez, J.M., Molina, I., García-Laencina, P.J., Alba, E., Ribelles, N., Martín, M., Franco, L.: Missing data imputation using statistical and machine learning methods in a real breast cancer problem. Artif. Intell. Med. **50**, 105–115 (2010). https://doi.org/10.1016/J.ARTMED.2010.05.002
2. Quinteros, M.E., Lu, S., Blazquez, C., Cárdenas-R, J.P., Ossa, X., Delgado-Saborit, J.M., Harrison, R.M., Ruiz-Rudolph, P.: Use of data imputation tools to reconstruct incomplete air quality datasets: A case-study in Temuco, Chile. Atmos. Environ. **200**, 40–49 (2019). https://doi.org/10.1016/J.ATMOSENV.2018.11.053
3. Hughes, R.A., Heron, J., Sterne, J.A.C., Tilling, K.: Accounting for missing data in statistical analyses: multiple imputation is not always the answer. Int. J. Epidemiol. **48**, 1294–1304 (2019). https://doi.org/10.1093/IJE/DYZ032
4. Pereira, R.C., Santos, M.S., Rodrigues, P.P., Abreu, P.H.: Reviewing autoencoders for missing data imputation: Technical trends, applications and outcomes. J. Artif. Intell .Res. **69**, 1255–1285 (2020). https://doi.org/10.1613/JAIR.1.12312
5. GAIN: Missing Data Imputation using Generative Adversarial Nets. https://proceedings.mlr.press/v80/yoon18a.html
6. Duan, Y., Lv, Y., Liu, Y.L., Wang, F.Y.: An efficient realization of deep learning for traffic data imputation. Transp. Res. Part C Emerg. Technol. **72**, 168–181 (2016). https://doi.org/10.1016/J.TRC.2016.09.015
7. Misdram, M., Noersasongko, E., Syukur, A., Faculty, P., Muljono, M., Agus Santoso, H., Ignatius Moses Setiadi, D.R.: Analysis of imputation methods of small and unbalanced datasets in classifications using naïve bayes and particle swarm optimization. In: Proceedings—2020 International Seminar on Application for Technology of Information and Communication: IT Challenges for Sustainability, Scalability, and Security in the Age of Digital Disruption, iSemantic 2020, pp. 115–119 (2020). https://doi.org/10.1109/ISEMANTIC50169.2020.9234225
8. Lin, W.C., Tsai, C.F.: Missing value imputation: a review and analysis of the literature (2006–2017). Artif. Intell. Rev. **53**, 1487–1509 (2020). https://doi.org/10.1007/S10462-019-09709-4
9. Kleinke, K.: Multiple Imputation by Predictive Mean Matching When Sample Size Is Small. https://doi.org/10.1027/1614-2241/a000141
10. van Buuren, S.: Flexible Imputation of Missing Data (2012). https://doi.org/10.1201/B11826
11. Susanti, S.P., Azizah, F.N.: Imputation of missing value using dynamic Bayesian network for multivariate time series data. In: Proceedings of 2017 International Conference on Data and Software Engineering, ICoDSE 2017. 2018-January, pp. 1–5 (2018). https://doi.org/10.1109/ICODSE.2017.8285864
12. Osman, M.S., Abu-Mahfouz, A.M., Page, P.R.: A survey on data imputation techniques: water distribution system as a use case. IEEE Access **6**, 63279–63291 (2018). https://doi.org/10.1109/ACCESS.2018.2877269
13. Manembu, P., Kewo, A., Welang, B.: Missing data solution of electricity consumption based on Lagrange Interpolation case study: IntelligEnSia data monitoring. In: Proceedings—5th International Conference on Electrical Engineering and Informatics: Bridging the Knowledge between Academic, Industry, and Community, ICEEI 2015, pp. 511–516 (2015). https://doi.org/10.1109/ICEEI.2015.7352554

Chapter 20
Shuffle Split-Edited Nearest Neighbor: A Novel Intelligent Control Model Compression for Smart Lighting in Edge Computing Environment

Aji Gautama Putrada, Maman Abdurohman, Doan Perdana, and Hilal Hudan Nuha

Abstract One of the IoT case studies is to apply intelligent control to smart lighting using k-nearest neighbor (KNN). However, migrating computing from the cloud to the edge becomes more challenging due to the limited capacity of the microcontroller's flash memory and the large size of the KNN model. This paper proposes shuffle split-edited nearest neighbor (SSENN), a novel method to compress intelligent control models for smart lighting while maintaining its performance. The method adopts the high compression ratio (CR) of random under-sampling (RU) and the high accuracy of edited nearest neighbor (ENN). The methodology trains basic KNN and observes its performance and characteristics as the number of datasets increases. Then we synthesize the SSENN on the KNN and further evaluate and benchmark the model with RU and ENN. The test results show that KNN + SSENN has higher Average $\pm$ Std. Dev. accuracy than KNN + RU, which is 0.915 ± 0.01. In addition, the CR of KNN + SSENN is higher than KNN + ENN, which is 2.60. With a model size of 14.2 kB, the KNN + SSENN model can fit in the microcontroller's flash memory.

Keywords K-nearest neighbor · Edited-nearest neighbor · Model compression · Smart lighting · Edge computing · Arduino Uno

A. G. Putrada (✉) · D. Perdana
Advanced and Creative Networks Research Center, Telkom University, Bandung 40257, Indonesia
e-mail: ajigautama@student.telkomuniversity.ac.id

D. Perdana
e-mail: doanperdana@telkomuniversity.ac.id

M. Abdurohman · H. H. Nuha
School of Computing, Telkom University, Bandung 40257, Indonesia
e-mail: abdurohman@telkomuniversity.ac.id

H. H. Nuha
e-mail: hilalnuha@telkomuniversity.ac.id

© The Author(s), under exclusive license to Springer Nature Singapore Pte Ltd. 2023
C. So-In et al. (eds.), *Information Systems for Intelligent Systems*, Smart Innovation,
Systems and Technologies 324, https://doi.org/10.1007/978-981-19-7447-2_20

20.1 Introduction

The k-nearest neighbor (KNN) machine learning method suffers space complexity problem as the model size becomes bigger when the dataset increases [7, 11]. This property becomes a challenge when implementing the edge computing concept on, in our case study, intelligent control for smart lighting [12, 15]. Whereas microcontrollers such as Arduino Uno have limited flash memory capacity, only 32 kB [13, 14].

Several other studies have used model compression methods to smaller the KNN model. Mukahar et al. [9] proved that the application of edited nearest neighbor (ENN) to prototype selection could reduce the KNN model while increasing its Accuracy. However, several other studies show that ENN suffers a limited compression rate (CR) [20]. On the other hand, random under-sampling (RU) can have a CR up to 2.00; however, limited Accuracy [8]. There is a research opportunity for a novel model compression method with optimum CR while maintaining the classifier's prediction performance.

Our research proposes a novel method called shuffle split-edited nearest neighbor (SSENN) that takes inspiration from RU and ENN. We aim to prove that SSENN can have an optimum CR while maintaining the classifier's prediction performance. We evaluate our proposed method by benchmarking it with RU and ENN.

To the best of our knowledge, there has never been a study that applied a compression model on smart lighting intelligent control in edge computing concept. The contributions made by our research include the following:

1. A novel smart lighting intelligent control solution by proposing the concept of edge computing
2. A novel compression method that can compress a KNN model while maintaining its performance
3. A novel way to embed the KNN model in Arduino Uno

The remainder of this paper has the following systematic: Chap. 20.2 describes studies related to the use and performance of state-of-the-art under-sampling methods. Chapter 20.3 details the methods used and other concepts. Chapter 20.4 reports the test results and benchmarks our results with state-of-the-art research. Finally, Chap. 20.5 highlights the important findings of this study.

20.2 Related Works

Besides ENN and RU, several other methods are applicable for model compression in related studies. Zheng et al. [21] uses the clustered centroid (CC) method and has a CR value up to 0.11. Tyagi et al. [16] used condensed neighboring neighbors (CNN) to correct skewed data and can increase the imbalance ratio from 1.44 to 2.55. Wijanarto et al. [18] used repeated ENN (RENN) to improve the performance of a sentiment analysis case and get the highest accuracy compared to other methods,

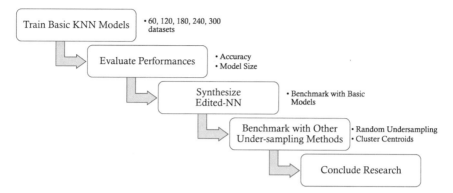

Fig. 20.1 The proposed research methodology

which is 0.98. Alfaiz et al. [2] used all KNN (AllKNN) to detect credit card fraud and got better accuracy than other methods, which is 0.99. Verdikha et al. [17] used the instance hardness threshold (IHT) in the case of hate speech classification and got an index balanced accuracy (IBA) improvement value of 1.44. Alamsyah et al. [1] used the nearmiss (NM) method in disease classification and obtained an accuracy of 0.76. There is a research opportunity to compare the performance of ENN with other methods mentioned in the case of intelligent control on smart lighting.

20.3   Methodology

We have briefly explained the research methodology in Sect. 20.1. Figure 20.1 shows the complete methodology of this research.

20.3.1   Edge Computing Characteristics

In edge computing, the computing process that was originally in the cloud migrates to the microcontroller. In order to migrate, it is important to know the characteristics of the microcontroller [4]. Table 20.1 contains the comparison of important edge computing and cloud computing characteristics.

20.3.2   KNN

In making predictions, KNN measures the distance from a test data to its train data, collects k train data with the closest distance, then groups the test data into the largest

Table 20.1 Environment comparisons of cloud servers and edge microcontrollers

Item	Cloud	Edge
Typical device	Server systems	Arduino Uno
Typical CPU	Intel Xeon	ATMega328P
Market price	US$6,881.25	US$6.88
Memory type	RAM	Flash Memory
Memory size	256 GB	32 kB

class from the k train data [3]. In measuring distance, several methods can be used, including Euclidean distance [10]. The following is the formula for calculating each distance of a data train to test data based on Euclidean distance:

$$Distance = \sqrt{\sum_{i=1}^{n}(x_i - y_i)^2} \tag{20.1}$$

where n is the number of features, x is the train data, and y is the test data [6].

20.3.3   SSENN

Our SSENN is a modification of ENN and is inspired by RU. The RU provides satisfying model size on imbalanced data, however suffers from the random sampling process. On the other hand, ENN increases the performance by removing unneeded training data. Here we split the data to the optimum size and then run ENN to provide satisfying prediction performance. The shuffle process is necessary to obtain unbiased dataset before splitting the training data. The Algorithm 20.1 describes the process, where OS is the original training data, n is the desired split ratio, OS_s is the shuffled training data, OS_{ss} is the split shuffled training data, and ES is the edited training data [21]. Previously before running the algorithm is an empiric experiment to determine the optimum n that provides the CR result most approximate to the CR of RU.

20.3.4   Benchmark Model Compression Methods

RU and ENN are used to benchmark the proposed SSENN performance. RU is one of the most well-known and simple under-sampling methods. The basic concept of under-sampling is to reduce the amount of data in the majority class to overcome the problem of data imbalance, which also reduces the number of datasets. RU reduces the number of majority class data by randomly discarding class members with a

---

Algorithm 20.1: SSENN algorithm

Data: OS, n
Result: ES
1 $OS_s = RandomShuffle(OS)$;
2 $OS_{ss} = Split(OS_s, n)$;
3 $ES \leftarrow OS_{ss}$;
4 **for** $x_i \in OS$ **do**
5     $class = KNN(x_i)$;
6   **if** $class$ is False **then**
7 | Discard x_i from ES;
8   **end**
9 **end**

---

certain random state value [5]. ENN was introduced by Dennis L. Wilson in 1972 [19]. ENN reduces training data by eliminating misclassified data using the KNN method, resulting a more efficient dataset also an increased prediction performance.

20.3.5 Testing Parameters

The intelligent control model performance measurement uses Accuracy. The Accuracy formula is as follows:

$$\text{Accuracy} = \frac{\text{TP} + \text{TN}}{\text{TP} + \text{TN} + \text{FP} + \text{FN}} \tag{20.2}$$

where TP is the true positive value, TN is the true negative value, FP is the false positive value, and FN is the false negative value.

The model compression method performance measurement uses CR. Where the CR formula is as follows:

$$\text{CR} = \frac{\text{Original Model Size}}{\text{Compressed Model Size}}. \tag{20.3}$$

The higher the value of CR, the better the performance of the under-sampling method.

20.4 Results and Discussion

20.4.1 Results

The first step is to observe the characteristics of the KNN model to changes in the number of datasets. A 10-time repetition in every training is to examine the characteristics of the biased model. Figure 20.2a shows the change in accuracy with

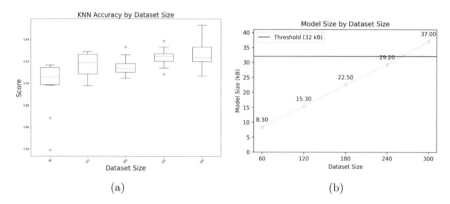

Fig. 20.2 The relationship between KNN training size and: **a** accuracy, **b** model size

increasing the number of datasets. There is an increasing trend in which the 300 training datasets have the highest average ± std. dev. 0.928 ± 0.01. Figure 20.2b is a line plot that shows the growth of the model size against the number of datasets used for training. There is an escalation in model size, where in the model with 300 datasets, the model size is 37 kB. This size has exceeded the threshold of 32 kB, which is the size of the flash memory of the Arduino Uno. The two results show the urgency of a method that can decrease the model size while maintaining the prediction performance.

Next is the SSENN proof of concept. Figure 20.3a is a bar plot that shows the accuracy of KNN + SSENN. The Average ± Std. Dev. Accuracy of KNN + SSENN is 0.923 ± 0.01. This value is higher than KNN + RU, which has an Average ± Std. Dev. Accuracy worth 0.889 ± 0.02. KNN + SSENN and KNN + ENN has no significant difference in Average ± Std. Dev. Accuracy. Figure 20.3b is a bar chart showing the measurement results of KNN + ENN's model size. KNN + SSENN has a model size of 14.2 kB. The model size is smaller than the threshold value and benchmark methods, where KNN + RU has a model size of 14.4 kB and KNN + ENN has a model size of 34.4 kB and is larger than the threshold size, indicating that it does not fit in the flash memory of an Arduino Uno. Thus KNN + SSENN, KNN + RU, and KNN + ENN have CR values 2.60, 2.57, and 1.08 respectively.

20.4.2 Discussion

As mentioned in paper [20], ENN has satisfying Accuracy however poor CR, whereas in [8] RU has promising CR however poor Accuracy. In this research, our proposed SSENN proves to have better Accuracy than RU and higher CR than ENN. Our contribution is a novel compression method that can compress a KNN model while maintaining its performance.

We compare model compression methods proposed by state-of-the-art (SOTA) papers with our proposed KNN + SSENN method. Table 20.2 shows the comparison.

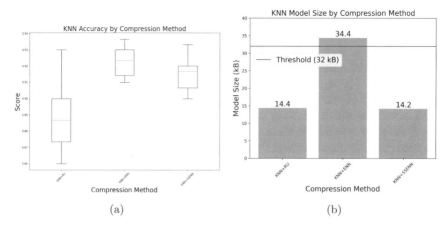

Fig. 20.3 The relationship between the model compression methods with: **a** accuracy, **b** model size

Table 20.2 Comparison with model compression methods used in related works

Cite	Method	Accuracy[a]	CR	Compared to SSENN	
				Accuracy	CR
Proposed method	SSENN	0.915 ± 0.01	2.60	–	–
[21]	CC	0.892 ± 0.01	2.57	Lower	Lower
[16]	CNN	0.866 ± 0.04	3.36	Lower	Higher
[18]	RENN	0.914 ± 0.01	1.09	Lower	Lower
[2]	AllKNN	0.920 ± 0.01	1.08	Higher	Lower
[17]	IHT	0.887 ± 0.02	1.65	Lower	Lower
[1]	NM	0.770 ± 0.05	2.57	Lower	Lower

[a] Average ± Std. Dev. of parameter

SSENN has lower Accuracy compared to AllKNN [2]. Then SSENN has lower CR compared to CNN [16]. In addition, SSENN are better in both terms compared to CC [21], RENN [18], IHT [17], and NM [1]. However, none of the methods best our SSENN in both Accuracy and CR, which further proves that our proposed method is a solution that optimizes model size and performance, hence suitable for the characteristics of an edge computing environment.

20.5   Conclusion

We propose a novel method called shuffle split-edited nearest neighbor (SSENN) for intelligent control models on smart lighting in an edge computing environment. Our aim is to compress the intelligent model while maintaining its performance. The novel method is based on random under-sampling (RU) and edited nearest neighbor (ENN) methods, where we also use them to benchmark our proposed method. The test results show that KNN + SSENN has higher Average $\pm$ Std. Dev. Accuracy compared to RU, which is 0.915 ± 0.01. In addition, the compression ratio (CR) of KNN + SSENN is higher than of KNN + ENN, that is 2.60. With a model size of 14.2 kB, the KNN + SSENN can compress a KNN model to fit in an Arduino Uno's flash memory while maintaining its performance. For future work, the direction of our research is to embed this compressed model in edge computing solutions and measure its performance, especially latency.

References

1. Alamsyah, A.R.B., Anisa, S.R., Belinda, N.S., Setiawan, A.: Smote and nearmiss methods for disease classification with unbalanced data: Case study: IFLS 5. In: Proceedings of The International Conference on Data Science and Official Statistics, vol. 2021, pp. 305–314 (2021)
2. Alfaiz, N.S., Fati, S.M.: Enhanced credit card fraud detection model using machine learning. Electronics **11**(4), 662 (2022)
3. Aulia, M.S., Abdurrahman, M., Putrada, A.G.: Pendeteksian kadar glukosa dalam darah pada gejala diabetes tipe 1 menggunakan algoritma k-nearest neighbor dengan metode nafas. SMARTICS J. **5**(1), 14–21 (2019)
4. De Vita, F., Nocera, G., Bruneo, D., Tomaselli, V., Giacalone, D., Das, S.K.: Porting deep neural networks on the edge via dynamic k-means compression: a case study of plant disease detection. Perv. Mob. Comput. **75**, 101437 (2021)
5. Elhassan, T., Aljurf, M.: Classification of imbalance data using tomek link (t-link) combined with random under-sampling (rus) as a data reduction method. Glob. J. Technol. Optim. S **1** (2016)
6. Fakhruddin, R.I., Abdurohman, M., Putrada, A.G.: Improving PIR sensor network-based activity recognition with PCA and KNN. In: 2021 International Conference on Intelligent Cybernetics Technology & Applications (ICICyTA), pp. 138–143. IEEE (2021)
7. Ghassani, F., Abdurohman, M., Putrada, A.G.: Prediction of smartphone charging using k-nearest neighbor machine learning. In: 2018 Third International Conference on Informatics and Computing (ICIC), pp. 1–4. IEEE (2018)
8. Goyal, S.: Handling class-imbalance with KNN (neighbourhood) under-sampling for software defect prediction. Artif. Intell. Rev. **55**(3), 2023–2064 (2022)
9. Mukahar, N., Rosdi, B.A.: Performance comparison of prototype selection based on edition search for nearest neighbor classification. In: Proceedings of the 2018 7th International Conference on Software and Computer Applications, pp. 143–146 (2018)
10. Nando, P., Putrada, A.G., Abdurohman, M.: Increasing the precision of noise source detection system using KNN method. In: Kinetik: Game Technology, Information System, Computer Network, Computing, Electronics, and Control, pp. 157–168 (2019)
11. Putrada, A.G., Abdurohman, M., Perdana, D., Nuha, H.H.: Machine learning methods in smart lighting towards achieving user comfort: a survey. IEEE Access (2022)
12. Putrada, A.G., Ramadhan, N.G., Makky, M.: An Evaluation of Activity Recognition with Hierarchical Hidden Markov Model and Other Methods for Smart Lighting in Office Buildings. ICIC International (2022)
13. Rahman, A.M., Hossain, M.R., Mehdi, M.Q., Nirob, E.A., Uddin, J.: An automated zebra crossing using Arduino-Uno. In: 2018 International Conference on Computer, Communication, Chemical, Material and Electronic Engineering (IC4ME2), pp. 1–4. IEEE (2018)

14. Shang, F., Lai, J., Chen, J., Xia, W., Liu, H.: A model compression based framework for electrical equipment intelligent inspection on edge computing environment. In: 2021 IEEE 6th International Conference on Cloud Computing and Big Data Analytics (ICCCBDA), pp. 406–410. IEEE (2021)
15. Tiruvayipati, S., Yellasiri, R.: Practicability of embarrassingly parallel computations for enormous miniature workloads over massive underutilized IoT. In: 2019 IEEE International WIE Conference on Electrical and Computer Engineering (WIECON-ECE), pp. 1–4. IEEE (2019)
16. Tyagi, S., Mittal, S., Aggrawal, N.: Neighbours online (NOL): an approach to balance skewed datasets. In: International Conference on Information Management & Machine Intelligence, pp. 387–392. Springer (2019)
17. Verdikha, N.A., Adji, T.B., Permanasari, A.E.: Study of undersampling method: Instance hardness threshold with various estimators for hate speech classification. IJITEE (Int. J. Inf. Technol. Electr. Eng.) 2(2), 39–44 (2018)
18. Wijanarto, W., Brilianti, S.P.: Peningkatan performa analisis sentimen dengan resampling dan hyperparameter pada ulasan aplikasi bni mobile. Jurnal Eksplora Informatika 9(2), 140–153 (2020)
19. Wilson, D.L.: Asymptotic properties of nearest neighbor rules using edited data. IEEE Trans. Syst. Man Cybern. 3, 408–421 (1972)
20. Zhai, J., Qi, J., Zhang, S.: An instance selection algorithm for fuzzy k-nearest neighbor. J. Intell. Fuzzy Syst. 40(1), 521–533 (2021)
21. Zheng, H., Sherazi, S.W.A., Lee, J.Y.: A stacking ensemble prediction model for the occurrences of major adverse cardiovascular events in patients with acute coronary syndrome on imbalanced data. IEEE Access 9, 113692–113704 (2021)

Chapter 21
Intelligent Remote Online Proctoring in Learning Management Systems

Muhammad Arief Nugroho, Maman Abdurohman, Sidik Prabowo, Iis Kurnia Nurhayati, and Achmad Rizal

Abstract An essential component of the learning phase is the administration of assessments using online testing platforms. It is the responsibility of the platforms used for online education to guarantee that every student successfully completes the evaluation procedure without cheating. Because of the widespread COVID-19 epidemic, all educational institutions are required to alter the process by which they administer online examinations and to take measures to reduce the amount of cheating that takes place during online examinations. In this paper, we create a proctoring system that is integrated with a learning management system. It has the following features: user verification, browser lockdown, face counter, and automatic cheating counter. The goal of this proposed system is to solve the problem that was presented. On the client side, there is no need for any extra software or programs to support any of these functions. System experiment is done under actual exam situations in order to gauge the dependability of the proctoring system.

M. A. Nugroho · M. Abdurohman · S. Prabowo (✉)
School of Computing, Telkom University, Bandung 40257, Indonesia
e-mail: pakwowo@telkomuniversity.ac.id

M. A. Nugroho
e-mail: arif.nugroho@telkomuniversity.ac.id

M. Abdurohman
e-mail: abdurohman@telkomuniversity.ac.id

I. K. Nurhayati
School of Communication and Bussiness, Telkom University, Bandung 40257, Indonesia
e-mail: iiskurnia@telkomuniversity.ac.id

A. Rizal
School of Electrical Engineering, Telkom University, Bandung 40257, Indonesia
e-mail: achmadrizal@telkomuniversity.ac.id

C. So-In et al. (eds.), *Information Systems for Intelligent Systems*, Smart Innovation, Systems and Technologies 324, https://doi.org/10.1007/978-981-19-7447-2_21

21.1 Introduction

E-learning is gaining popularity fast owing to its capacity to supply academic materials and its accessibility to students who lack access to a campus or are otherwise
restricted, especially during the COVID-19 pandemic for the last two years [1, 2].
A student can use their own computer or mobile device with an Internet connection
to participate in learning activities and get help from the lecturer.

Assessment is a critical component of every educational process, and online learning programs are no different [3]. Cheating is always possible during an online assessment. As a result, it is critical to avoid and identify its occurrence [4]. On a traditional
exam held in a proctored class environment, the exam participant is proctored by a
human proctor who observes the classes throughout the exam. However, on an online
exam, a reliable and cheap proctoring capability is required to provide the education
provider's guarantee of the student's learning result.

The online proctoring system will monitor student's activity during online assessment by getting access to their cameras, displays, or video recording devices in order
to guarantee that they comply to the exam standards [5]. Typically, the online proctor
integrated with an artificial intelligence system that observes students' movements
and surroundings in order to assess if suspected cheating activities should be flagged.
Online proctoring tools/plugins are often accessed via for-profit firms that contract
with educational institutions to provide real time, online proctoring services from
any location with an Internet connection. Students utilize these online proctoring services to schedule examinations and then communicate with their assigned proctor.
Recently, commercial solutions as well as academic articles are available to solve
this issue. Some of them just validate the identity [6–8], while others watch in real
time and record the exam [9].

In response to previous research findings, we developed a proctoring system that
integrates as a plugin into e-learning management systems and utilizes machine
learning and artificial intelligence for authentication and verification of users, real-
time people counting during examinations, browser lockdown, and continuous exam
monitoring via image capture. By integrating it with the browser, we incorporated
cheating detection into our suggested solution. In our proposed method, cheating is
detected when the user tries to go to another tab or window, or when the picture taken
by the camera shows two or more people.

21.2 Related Works

Technology's learning and teaching benefits expand. This has boosted online education and teaching. In the proper hands, online learning may help universities spread
their curriculum to a global audiencecite10. As colleges expand online education,
concerns have grown about how to maintain academic integrity. Distance or flexibility between students and instructors in an online learning setting may make it

harder to guarantee online assessment results [10]. References [11–14] also argued that online assessment's biggest issue is academic honesty. Online students are more likely to acquire responses from others than face-to-face classmates. Online examinations are thus harder to maintain. Online exam cheating may be prevented by proctoring the exam offline (in person), developing cheat-resistant questions (using subjective metrics instead of objective measures), and lowering the weight of exam results in the final course grade. Proctored experiments remain a common way to measure student learning, but their delivery methods are expanding, from online examinations to other online testing platforms. This has raised academic and non-academic questions about students' exam conduct and the design and administration of online examinations. Cheating and fraud are included.

Choosing the correct online exam proctoring technology might be tough. Reference [15] offers three characteristics that may impact the choosing of an online exam proctoring service. Cost, security, and instructor/student tech comfort.

21.2.1 Cost

Several of the first solutions have been offered on the market as commercial solutions. Reference [16] describes the online proctoring solutions into these groups :

Fully Live Online Proctoring A live proctor watches exam students on webcam. Students use live proctoring for online examinations. When students book an appointment, a proctor from one of two online proctoring centers will communicate with them via webcam. Students attach their screens to the proctor.

Reviewed Proctoring Proctoring records student exam activity. There's no automatic exam monitoring. Examiners must review each student's tape to see whether they cheated. Each student's PC camera and microphone capture exam video [17]. Both live and reviewed proctoring restrict student identity during examinations.

Automated Proctoring Computers monitor students throughout examinations to detect cheating. Students must produce a photo/ID card to authenticate themselves before taking the exam. During examinations, cameras and microphones will monitor student activities. When a student cheats, the system uses AI and computer vision to detect it.

21.2.2 Security

Online exam security is an ongoing issue [18, 19]. This affects the institution's reputation and reduces the credential's worth to employers [20]. Security-enhanced online assessment tools decrease cheating and academic dishonesty [21]. These systems prevent and detect cheating in various ways. Exam-takers are given a username and password and must verify their identity and agree not to commit any crimes. Proctoring systems provide efficiency, convenience, and a high-tech appearance [22, 23]. Despite these qualities, experts are concerned about the scores' authenticity and security. Exam-takers breached non-proctoring conduct agreements by asking assistance, accessing the Internet, and copying the exam knowledge, according to studies [24]. Unsupervised score distributions were higher than proctored [25]. Impersonation by applicants is another security risk [26]. Users may misuse Proctoring Software, which lets us take examinations at home. Users may give their exam certificates to anyone [27]. Authenticating users before letting them take the exam is crucial.

21.2.3 Student Comforts Levels

There is a dearth of understanding about online proctored examinations and exam anxiety. Exam-takers must present a picture ID via webcam for online proctoring. The proctor advises each exam-taker to use his camera. The proctor checks for notes, phones, notepads, and books. In most circumstances, the exam-taker must move a camera around the room or maintain his head motionless to complete online proctoring verification. Examinees' stress levels may increase, affecting their performance. It prohibits the exam-taker from moving and compels his eyes to stay on the display [28]. Most examinations required students to install Safe Exam Browser. Certain exams need SEB. Examinees will have to put up an application to take the exam, which will be challenging. The examinee will feel nervous. The educator must manually configure SEB in moodle for the exam to run on SEB.

21.3 System Overview

We designed a proctoring model system with the following characteristics: browser lockdown, face identification and verification, automatic cheat detector, and real-time people counter. These features were inspired by the benefits offered by the proctoring system that was integrated into the e-learning platform. All of these capabilities are rolled into a single module or plugin, and then that module or plugin is immediately included into the e-learning platform. The current system process flow is shown in Fig. 21.1.

21.3.1 Profile Picture Integration

This module synchronizes academic system with e-learning platform data. The data included names, emails, courses, usernames, and passwords, as well as profile images. This picture data is used for user verification during the exam. The

Fig. 21.1 System overview

feature that permits students to change their profile images has been disabled for security reason. This module requires an API to integrate both systems' data. This API was only called once, at the beginning of the semester.

21.3.2   User Verification

Before taking an online exam, three rounds of user verification are necessary, including authenticating a participant's face with their profile picture, Internet connection speed, and share screen. If any of the verification stages aren't completed, students can't take the exam. Below are sub-module sections:

Face Verification Facial identification and comparison are two main components on the face verification process. Face detection system determines whether a photo contains a face. Face detection systems establish the existence, position, size, and (potentially) orientation of faces in still images and videos. This method detects faces despite gender, age, or facial hair.

Bandwidth Validation Bandwidth verification determines exam-takers' Internet speeds. The exam requires 2 mbps minimum bandwidth. This bandwidth requirement ensures a smooth exam run. This bandwidth verification method might be required or optional (when the bandwidth is less than 2 mbps, a warning is given to the user).

Screen Share Validation The screen activity that occurs during the exam will be captured by the screen share validation. Every student must validate. Along with the results of face verification performed during the exam, the results of the screen share validation will also be saved in the user proctoring log. During the exam, the proctoring system will take a screenshot of the current screen that shows exam questions every time a question is moved or every 30 s, whichever comes first.

21.3.3   Browser Lockdown

Students are unable to switch screens during exams because of the browser lockdown feature. Students often move their attention to other windows while the exam is in progress. Browser lockdown may prevent this. The browser lockdown module uses a web browser and requires no other software. On the e-learning platform, custom scripts have been implemented in order to monitor window changes while the students are taking their exams. The cheating counter will increase by one each time the user changes the window since the browser lock will notice the occurrence and add a counter for each infraction.

21.3.4 Real-Time Face Counter Detection

Face counter detection is used to identify two student incidents:

- During the time period of the exam, a student face is not identified for a period of five seconds.
- The proctoring system detects multiple persons on the camera at the same time.

The process of face counting is identical to that of face verification, with the addition of these two prerequisites. If the system determines that any of the aforementioned requirements have been met, the violation counter will automatically advance by one.

21.3.5 Cheating Counter

Cheating counter counts exam breaches. The amount of infractions may be adjusted dynamically by the instructor. If they encounter any of the following situations, there will be a rise in the frequency of violations: Examinees switch windows during exam; no faces were seen throughout the exam in 5 s; more than one individual was discovered during the exam.

21.4 Results and Discussion

21.4.1 System Experiment

The purpose of this experiment is to verify that the specialized plugin that is connected with proctoring is capable of functioning effectively on the Moodle platform, which is used for online education. The scenario of the experiment is played out in accordance with the order in which the students are taking the examination, namely as follows:

User Validation Students are needed to check their bandwidth, face, and screenshare throughout the initial step of the process. Students will not be permitted to take the examination if the three validation steps are not completed successfully.

(a)   (b)

Fig. 21.2 **a** User validation. **b** No person detected

Figure 21.2a illustrates the student verification procedure. Students must check bandwidth to identify the access speed of the exam device, faces matched to the profile photo in the LMS, and screenshare to capture all screen actions throughout the exam. The green line in the face verification shows a successful face verification.

After completing the user authentication procedure, the LMS will display a browser lockdown page where students may work on exam questions. The system will identify violations using the following criteria on the page:

- No persons were identified in the webcam: When the proctoring system in the monitoring camera finds that there is no face present, the number of violations that have been accumulated is immediately raised by one. If it has been more than 5 s since the last time the system recognized a face, this event will be triggered. Detailed illustration shows in Fig. 21.2b.
- Discovered two or more persons in the webcam: When two or more faces are recognized on the camera, the system will immediately identify the violations that have occurred. After the incident has been recorded by the proctoring system, the proctoring application programming interface (API) will perform a series of computations to calculate the total number of participants visible in the picture. The violation counter will automatically increase if there are more than two persons found in the frame region.
- The window change is performed by the student: Fig. 21.3b illustrates the detection of window changes that happened throughout the examination. Each window movement will be recognized automatically, increasing the violation counter by one.

21.4.2 User Verification Results

This user verification exam is comprised of two steps of user verification, the first of which took place before the exam and the second of which took place during the

Fig. 21.3 **a** More than one person detected. **b** Screen switch or apps detected

exam. Student data originates from students who took the exam, which was taken by 1220 students, with an average of each student doing face verification 201 times during the exam, and with an accuracy level of facial recognition and detection of violations based on facial conditions that were not identified, and there were more than two persons totaling 91.04 %

21.5 Conclusion

In this work, we present an improved plugin that can be embedded into an e-learning platform. This plugin allows identity verification of online students as well as continuous exam monitoring during examinations is being taken by utilizing a face recognition system. Browser lockdown and cheating counter provide secure examinations environment. Academic dishonesty among students during exam has dropped. It is possible to enhance future study by including audio recording and voice recognition modules that are combined with speech-to-text analysis in order to examine the acoustic circumstances surrounding the exam room during the exam. In addition, there is room for improvement with regard to the examination of screen captures. At this point in time, the study is solely concerned with user verification and validation; however, it does not go into detail on the analysis of the findings of the screen capture. In order to identify instances of cheating within screen grabs, further procedures using AI will be required to examine the outcomes of the screen captures.

References

1. Hamdan, M., Jaidin, J.H., Fithriyah, M., Anshari, M.: E-learning in time of covid-19 pandemic: challenges & experiences. In: 2020 Sixth International Conference on e-Learning (econf), pp. 12–16 (2020). https://doi.org/10.1109/econf51404.2020.9385507
2. Wang, X., Chen, W., Qiu, H., Eldurssi, A., Xie, F., Shen, J.: A survey on the e-learning platforms used during covid-19. In: 2020 11th IEEE Annual Information Technology, Electronics and Mobile Communication Conference (IEMCON), pp. 0808–0814 (2020). https://doi.org/10.1109/IEMCON51383.2020.9284840

3. Andersen, K., Thorsteinsson, S.E., Thorbergsson, H., Gudmundsson, K.S.: Adapting engineering examinations from paper to online. In: 2020 IEEE Global Engineering Education Conference (EDUCON), pp. 1891–1895 (2020). https://doi.org/10.1109/EDUCON45650.2020.9125273
4. Noorbehbahani, F., Mohammadi, A., Aminazadeh, M.: A systematic review of research on cheating in online exams from 2010–2021. Educ. Inf. Technol. (2022). https://doi.org/10.1007/s10639-022-10927-7
5. Atoum, Y., Chen, L., Liu, A.X., Hsu, S.D.H., Liu, X.: Automated online exam proctoring. IEEE Trans. Multimedia **19**(7), 1609–1624 (2017). https://doi.org/10.1109/TMM.2017.2656064
6. AV, S.K., Rathi, M.: Keystroke dynamics: a behavioral biometric model for user authentication in online exams. In: Research Anthology on Developing Effective Online Learning Courses, pp. 1137–1161. IGI Global (2021)
7. Khlifi, Y., El-Sabagh, H.A.: A novel authentication scheme for e-assessments based on student behavior over e-learning platform. Int. J. Emerg. Technol. Learn. **12**(4) (2017)
8. Zhu, X., Cao, C.: Secure online examination with biometric authentication and blockchain-based framework. Math. Prob. Eng. **2021** (2021)
9. Li, H., Xu, M., Wang, Y., Wei, H., Qu, H.: A visual analytics approach to facilitate the proctoring of online exams. In: Proceedings of the 2021 CHI Conference on Human Factors in Computing Systems, pp. 1–17 (2021)
10. Holden, O.L., Norris, M.E., Kuhlmeier, V.A.: Academic integrity in online assessment: a research review. In: Frontiers in Education, p. 258. Frontiers (2021)
11. Gamage, K.A., Silva, E.K.D., Gunawardhana, N.: Online delivery and assessment during covid-19: safeguarding academic integrity. Educ. Sci. **10**(11), 301 (2020)
12. Kharbat, F.F., Abu Daabes, A.S.: E-proctored exams during the covid-19 pandemic: a close understanding. Educ. Inf. Technol. **26**(6), 6589–6605 (2021)
13. Lee, J.W.: Impact of proctoring environments on student performance: online vs offline proctored exams. J. Asian Finan. Econ. Bus. **7**(8), 653–660 (2020)
14. Nguyen, J.G., Keuseman, K.J., Humston, J.J.: Minimize online cheating for online assessments during covid-19 pandemic. J. Chem. Educ. **97**(9), 3429–3435 (2020)
15. Brown, V.: Evaluating technology to prevent academic integrity violations in online environments. Online J. Distance Learn. Admin. **21**(1) (2018)
16. Labayen, M., Vea, R., Flórez, J., Aginako, N., Sierra, B.: Online student authentication and proctoring system based on multimodal biometrics technology. IEEE Access **9**, 72398–72411 (2021)
17. Cote, M., Jean, F., Albu, A.B., Capson, D.: Video summarization for remote invigilation of online exams. In: 2016 IEEE Winter Conference on Applications of Computer Vision (WACV), pp. 1–9. IEEE (2016)
18. Butler-Henderson, K., Crawford, J.: A systematic review of online examinations: a pedagogical innovation for scalable authentication and integrity. Comput. Educ. **159**, 104024 (2020)
19. Dadashzadeh, M.: The online examination dilemma: to proctor or not to proctor?. J. Instruct. Pedagogies **25** (2021)
20. Carrell, S.E., Malmstrom, F.V., West, J.E.: Peer effects in academic cheating. J. Human Resour. **43**(1), 173–207 (2008)
21. Slusky, L.: Cybersecurity of online proctoring systems. J. Int. Technol. Inf. Manage. **29**(1), 56–83 (2020)
22. Arthur, W., Glaze, R.M., Villado, A.J., Taylor, J.E.: The magnitude and extent of cheating and response distortion effects on unproctored internet-based tests of cognitive ability and personality. Int. J. Select. Assess. **18**(1), 1–16 (2010)
23. Gibby, R.E., Ispas, D., McCloy, R.A., Biga, A.: Moving beyond the challenges to make unproctored internet testing a reality. Ind. Organ. Psychol. **2**(1), 64–68 (2009)
24. Bloemers, W., Oud, A., Dam, K.V.: Cheating on unproctored internet intelligence tests: strategies and effects. Pers. Assess. Decis. **2**(1), 3 (2016)
25. Steger, D., Schroeders, U., Gnambs, T.: A meta-analysis of test scores in proctored and unproctored ability assessments. Euro. J. Psychol. Assess. **36**(1), 174 (2020)

26. Hylton, K., Levy, Y., Dringus, L.P.: Utilizing webcam-based proctoring to deter misconduct in online exams. Comput. Educ. **92**, 53–63 (2016)
27. Ghizlane, M., Hicham, B., Reda, F.H.: A new model of automatic and continuous online exam monitoring. In: 2019 International Conference on Systems of Collaboration Big Data, Internet of Things & Security (SysCoBIoTS), pp. 1–5. IEEE (2019)
28. Turani, A.A., Alkhateeb, J.H., Alsewari, A.A.: Students online exam proctoring: a case study using 360 degree security cameras. In: 2020 Emerging Technology in Computing, Communication and Electronics (ETCCE), pp. 1–5. IEEE (2020)

Chapter 22
Analysis of Crop Yield Prediction Using Random Forest Regression Model

N. Prasath[ID], **J. Sreemathy**[ID], **N. Krishnaraj**[ID], and **P. Vigneshwaran**[ID]

Abstract The agriculture sector is one of the most important application areas in India. It involves multiple decision-making situations of varying complexity according to the numerous factors influencing them. Use of modern technology can be leveraged to provide insights to the problem and find efficient solution. The necessary approach for obtaining optimal solution in this problem utilizes data analytics. Environmental readings, soil quality, and economic viability have made it relevant for the agricultural industry to use such information and make crucial decisions. The proposed work explores the use of regression analysis on agricultural data in predicting crop yield. It is based on major classification methods which show considerable success.

22.1 Introduction

Agriculture contributes significantly to the country's GDP. With the continuing and rapid expansion of human population, understanding the crop yield of the country can prove beneficial in ensuring food security and reducing the impacts of climate change. To help the farmer in maximizing the profit, crop yield can be very useful.

In a country like India where huge population depend on farming as only source of earning, crop yield prediction can be very useful. Sometimes due to insufficient rain or due to many other natural factors, the crop production goes down. This is very serious problem in the country. So, with the analysis of past data available, we can suggest some patterns or what changes they can have or what are the necessary factors which needs to be taken care of.

N. Prasath (✉) · N. Krishnaraj · P. Vigneshwaran
Department of Networking and Communications, SRM Institute of Science and Technology, Tamil Nadu, Kattankulathur, Chengalpattu 603 203, India
e-mail: prasath283@gmail.com

J. Sreemathy
Department of Computer Science and Engineering, Sri Eshwar College of Engineering, Tamil Nadu, Kinathukadavu, Coimbatore 641 202, India

© The Author(s), under exclusive license to Springer Nature Singapore Pte Ltd. 2023
C. So-In et al. (eds.), *Information Systems for Intelligent Systems*, Smart Innovation, Systems and Technologies 324, https://doi.org/10.1007/978-981-19-7447-2_22

The suitability for a particular crop to be grown in a certain region is based on the geographical properties of the region as well as the nutrient requirements of the crop. Since a majority of them are seasonal, different cultivation plan is required to ensure maximum output. Most researches are focussed on specific set of features and analyze the effect of production with respect to the chosen set of parameters. Climate is the feature which is used commonly in such research studies.

Natural disasters cause huge loss to crop production. The power of computing can be leveraged to reduce the risk involved with agricultural decisions. Also crop produce prediction can be done better. Yield prediction can be done by analyzing various types of information gathered from vast sources such as agricultural statistics, soil data, and meteorological data.

Data analysis is a process of handling large data in raw formats and transforming them into useful information by finding interesting patterns in the data. Data is first collected from various sources and cleansed, which involves removal of improper or missing data. Then, these various datasets are integrated together. Next step is to analyze the data by use of different algorithms to find patterns and correlations in the data.

22.2 Related Works

Existing approaches in crop yield prediction utilize classification and clustering techniques mainly. Classification is done when data is to be categorized into groups, where a set of data is available which is pre classified. So, training data is available which is used to classify the unknown data. When there is no knowledge of data to be categorized or when no training dataset is available, then clustering is performed which divides data into different clusters. An overview of the research works undertaken in this domain in the past years is presented chronologically.

In 2009, a paper by George Rub analyzed the comparison between techniques which included regression as well as support vector to figure out the best model amongst them. The attributes consist of yield, electric conductivity of soil, amount of fertilizer used, REIP data about plants' state of nutrition. Different models are evaluated on the same dataset. Support vector regression performed best on almost every dataset and also takes the least computation time [1]. Research was conducted about the effect on soyabean production due to different climatic and environmental factors. Decision tree algorithm was used in this study for making clusters and segregates the different parameters. The datasets used in the study were from agricultural department of Bhopal district in Madhya Pradesh. Decision tree consists of structure similar to a flow chart tree in which each child nodes indicate a condition applied on an attribute, a branch indicates a possible result of applied rule, and each leaf node indicates a class distribution. Beginning node of the tree is called as root node. Decision tree method used in this study is Interactive Dichotomize 3 which depends on information based on two assumptions. In this study, a decision tree is created for soybean crop in which relative humidity is the root node in the tree or deciding parameter for bulk production of the crop. The decision tree formed seems to suggest that there exists a relation between climatic condition and crop productivity of soybean, and this observation was confirmed from the accuracy of the model and Bayesian classification [2].

The monthly average temperature data for a selected region was used to check the impact on crop production. [3]. The area of research was 104,328 100 ha rectangular grids in Western Australia Datasets obtained according to the chosen field of study. All algorithms of classification have been checked, including Gaussian processes, decision stump, additive regression, and Lazy LWL [3]. All of these algorithms are based on regression. The algorithm which gave best results was Gaussian process. The use of the data mining classification feature indicates that the overall monthly temperature and wheat yield were strongly correlated [3]. Yield prediction model for a chosen crop is studied which uses adaptable clustering method on crop dataset for predicting yield. Bee hive cluster method was used to analyze agricultural data which helps in making decisions related to crop production. Various data mining algorithms are utilized to analyze a vast dataset of attributes. The dataset used is acquired from crop surveys of multiple regions in India. This research used data obtained for three crops, namely rice, sugarcane, and paddy, in order to find patterns and relations various crop properties. The CRY algorithm has improved performance in comparison with clustering and regression tree algorithm. It performs about 12% better as compared to cluster and regression tree algorithm [4]. This paper describes a general methodology for crop yield prediction. Feature selection forms a vital stage during pre-processing production information from the available knowledge base data. This algorithm operates on pre-processed data and aids in clustering and classifying crop types based on crop yield.

Notable authors, Ramesh and Vishnu Vardhan [5], in 2015, used multiple linear regression (statistical model) for predicting yield and they analyzed the production of rice based on the monsoon rainfall. Hierarchical clustering is done on a specific region's yearly average annual rainfall, and average production is mapped to each cluster. Multiple linear regression is applied on these cluster production values. A test scenario uses the average rainfall data of a particular year and map it onto the existing clusters. Once cluster is identified, the production can be estimated in that year for any region. With one variable, average yearly rainfall in the specific region, we can expect tentative rice production in the coming years. [5]. Multiple linear regression is used, with the predictors like sowing area, yield, year, rainfall, soil parameters, and the predict and being the year's rainfall. Three used datasets obtained by the department of state. First data collection for rainfall data was provided by the Indian Meteorological Department between 1951 and 2011. Basic information such as rice production, sowing area, yield per hectare, year, and soil parameters was obtained from the Andhra Pradesh Department of Agriculture. The last one concerns the Indian Department of Statistics where data is gathered for the crop yields of a given area and for one year. The output prediction method takes into account two variables: rainfall for 58 years and rice production, i.e. yield per hector. Using the equation of multiple linear regression, an analysis is made. The net result obtained is more than 90% accurate [5].

In the research by Veenadhari Suraparaju, the authors attempted to create a Website to analyze the effect of meteorological conditions on crop yields in a few districts of Madhya Pradesh. Decision tree algorithm was used to build the prediction model. The accuracy of the prediction was above 75% in all the crops [6]. Research was done to see the effects of different natural factors like temperature, humidity, pH of soil, etc., to forecast the production of major crops and propose various planting crops which is best suited for that region in Bangladesh. The source of dataset used in the survey was Bangladesh Agricultural Research Institute. The dataset is pre-processed, and the important factors which are required in the study are separated. This step is data pre-processing. The whole process of analyzing was divided into two steps: clustering and classification. First step was clustering in which data is analyzed to verify whether there is any strong correlation between the different factors and yield of crop. K-means clustering is applied here. Implementation of k-means cluster was done with help of RapidMiner Studio. Second step was classification or regression models which gives the results for the yield prediction. Linear regression, KNN, and neural net are applied here. The accuracy is between 90 and 95%. All the algorithms used perform prediction with varying accuracy [7].

Predicting the agricultural yield is done using the algorithms K-nearest neighbour and Naive Bayes. Both of the algorithms are tested on the soil dataset collected from a laboratory in Madhya Pradesh. The accuracy of each method is obtained by testing it against the datasets. The training dataset serves as the foundation for both algorithms, and their performance is observed according to the prediction performed based on the testing dataset. The datasets used in this study provide nutrient readings from various locations in Jabalpur. The dataset shows the presence of nutrients in the soil. These readings of the nutrients can be divided into various categories which can be used to analyze its effects on the production of crops [8]. Training data is used in developing model on which Naive Bayes and K-nearest neighbour are applied. As part of testing this technique, a new dataset consisting of nutrients is considered as test dataset, which has different readings of various locations in Jabalpur. Predictions are made using this info, with the training dataset's category serving as the label. Soil classification is done and divided to low, medium, and high categories. Utilizing data mining techniques prediction is made about the crop yield using chosen dataset.

The research by Niketa Gandhi et al. [9] aims to applying improvement in traditional statistical methods by efficient usage of data mining techniques with agricultural dataset in order to predict rice crop yield during Kharif season in semi-arid climatic zone in India. Data collected from ninety-seven districts that fall under the semi-arid climatic zone of India, for five years from 1998 to 2002. Different parameters under scrutiny were minimum, maximum, and average temperature, soil features, rainfall readings, production area and yield. The soil data is taken from National Bureau of Soil Survey and Land Use Planning Nagpur in Maharashtra [8].

Three districts from Andhra Pradesh, one district from Delhi, eight districts from Gujarat, nine districts from Haryana, fifteen from Karnataka, fourteen from Madhya Pradesh, seven from Maharashtra, five from Punjab, eighteen from Rajasthan, seven districts from Tamil Nadu, and ten districts from Uttar Pradesh were selected. Based on the data availability from the publicly accessible government records of India, five years data from 1998 to 2002 of various parameters were considered for the present research. Classification algorithms J48, LADTree, IBk, and LWL were executed in Weka using standard parameters. The performance of these four classification algorithms was compared in the present study. These four algorithms were compared on basis of different evaluation parameters. The results showed that J48 and LADTree achieved the best results in all criteria, whilst IBk had the lowest accuracy and specificity, LWL with lowest sensitivity [9]. The research paper by Priya, P., U. Muthaiah, and M. Balamurugan was all about predicting the yield of the crop on basis of the agricultural data from Tamil Nadu by implementing random forest algorithm. Rainfall, temperature, and production are the agricultural parameters used here. The predicted value of rainfall came as 250, temperature as 36, and predicted yield as 120,423.9 [10].

Research was done in which supervised machine learning method was used to predict crop yields based on the agricultural data from the Telangana state in India. Here, three methods, namely K-nearest neighbour, support vector machine, and linear square support vector machine, were used to train and build the model. The average accuracy of KNN method came to be around 60%, SVM method to be about 80%, and LS-SVM method to be approximately 90% [11]. The work done by R. Medar, V. S. Rajpurohit, and S. Shweta was all about crop yield prediction using different machine learning techniques. Here, two methods are used—Naive Bayes method and K-nearest neighbour method. The accuracy of Naive Bayes was 91.11%, and K-nearest neighbour method was 75.55% [12].

The scope of the project undertaken by Sangeeta and Shruthi G is to determine the crop yield of an area using dataset with some features which are important or related to crop production such as temperature, moisture, rainfall, and production of the crop in the previous years. To predict crop yield, regression models have been used like random forest, polynomial regression, decision tree, etc. [13]. Metrics like accuracy and precision is calculated for the proposed model. Amongst all the three-algorithm random forest gives the better yield prediction as compared to other algorithms with an accuracy of above 90%. Decision tree regression has accuracy of about 70%. The accuracy of all models varied significantly with change in split ratio of the dataset into train and test data.

Research work was done about finding what are the suitable or best condition for millet crop to give best yield. Random forest algorithm was used for analyzing and predicting crop yield. The accuracy of the model was 99.74% [14]. This paper by Vogiety, Abhigna was about building an UI for crop yield prediction. Random forest classifier was used here for crop yield prediction. The accuracy of the model has been around 75% [15].

22.3   Methodology

Random forest is an algorithm that is well-known and powerfully supervised machine learning algorithm and can carry out both regression and classification tasks. During preparation, the algorithm creates a set of decision trees and outputs the mode of the groups (classification) or mean prediction (regression) of the individual trees. The more trees in a forest, the more accurate the prediction. Random decision forests correct the problem that decision trees have of overfitting their testing dataset. In this analysis, features such as rain, production, and temperature are used to build a random forest, a cluster of decision-making trees via training dataset. The resulting forest random tree can be used to correctly predict crop yields on the basis of the input attributes [16].

Random forest is a regression model in which the dataset is divided into different subsets in which decision tree algorithm is applied, and at the end, the average is taken to improve the predictive accuracy of that dataset. The process of crop prediction system is explained here as shown in the Fig. 22.1 in steps initially data collection, i.e. collecting data by means of dataset, and then data was cleaned by pre-processing, and then, data transformation was done as per our research requirement, and then, the data was analyzed with the help of regression algorithm, and finally, we obtain the expected result.

Dataset has been taken from the government site (www.data.gov.in). In this dataset, several features have been included which highly influence crop yield.

The dataset contains more than 50,000 tuples. It comprises of total of 31 states and 537 districts. Apart from temperature and rainfall data, features such as area, crop, and season (kharif, rabi, summer) were added. Important features were appended to the dataset so that we can get better accuracy from our models and to minimize the mean absolute error. The features we added are potential of hydrogen (pH), conductivity, nitrogen, and electrical conductivity. Our dataset was divided into two parts: testing dataset and training dataset. Our final dataset consists of 9 important features, and these are temperature, rainfall, area, season, crop, production, pH, nitrogen, electrical conductivity.

The dataset description is shown as Fig. 22.2, where its witnessed the features that are considered as described above paragraph.

As shown in the Fig. 22.3, random forest algorithm is explained in individual steps as shown below:

1. Random K subsets are selected from the training dataset.
2. The decision tree is applied to the selected subset.

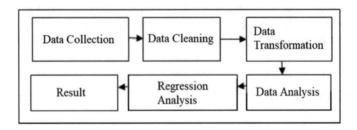

Fig. 22.1 Crop prediction system

Area	Production	Rainfall	Season	Temperature	Crop	pH	Nitrogen(kg/ha)	ElectricalConductivity(ds/m)
7800	3200	30.4003333	Kharif	28.007	Moong(Green Gram)	6.5	497	4.1
39922	75572	111.901	Kharif	27.23233333	Maize	5.6	473	3.9
44656	49099	3.3965	Rabi	20.277	Wheat	7.3	366	4.9
6540	3945	30.9325	Rabi	24.2415	Wheat	5.3	417	3.8
2911	2062	189.208333	Kharif	27.45633333	Maize	6.3	267	3.5
7	2	21.60975	Rabi	23.57	Moong(Green Gram)	6.3	435	3.4
18	21	206.31975	Summer	28.76025	Groundnut	5.8	370	2
39157	1878596	186.938333	Kharif	27.94066667	Sugarcane	5.1	312	5.3
10562	16812	81.5933333	Kharif	26.181	Jowar	7.1	334	3.8
179	2174	68.50175	Rabi	24.13975	Potato	6.8	365	2.9

Fig. 22.2 Description of dataset

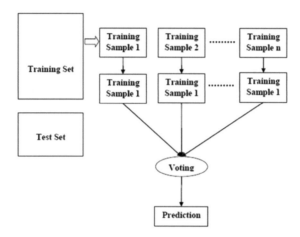

Fig. 22.3 Overview of random forest algorithm

3. Then, prediction is made for each decision tree.
4. Steps 2 and 3 are repeated for K subsets.
5. Average prediction of the subset will be the final prediction value.

After raw data is collected from different sources, pre-processing is done. Different methodologies are used to check for correlation amongst the features and the distribution of the data. We have used Python libraries Seaborn and matplotlib for making visualizations. Since we have our target variable (production) which has continuous values so we have applied different regression techniques and tested them using metrics such as R^2.

R^2 is a mathematical metric that indicates how similar the data is to the fitted regression line. It is also known as the coefficient of determination or the coefficient of multiple determination for multiple regression. R^2 describes the degree to which variation in the second variable explains. R^2 values vary from 0 to 1 and are usually indicated as percentages between 0 and 100%.

Considering a dataset has n values $y_1, \ldots, y_n$ (known as y_i), each associated with a fitted or predicted value $f_1, \ldots, f_n$ (known as f_i). Then, the residual values are defined as

$$e_i = \overline{y}_i - f_i \qquad (22.1)$$

$\overline{y}$ is mean of observed data

$$\overline{y} = \frac{1}{n} \sum_{i=1}^{n} y_i \qquad (22.2)$$

The variability of the dataset can be measured with two sums of squares formulas. The total sum of squares:

$$SS_{tot} = \sum_{i} (y_i - \overline{y})^2 \qquad (22.3)$$

The sum of squares of residuals:

$$SS_{res} = \sum_{i} (y_i - f_i)^2 = \sum_{i} e_i^2 \qquad (22.4)$$

The definition of the coefficient of determination is

$$R^2 = 1 - \frac{SS_{res}}{SS_{tot}} \qquad (22.5)$$

In the best-case scenario, the predicted values closely equal the observed values, what leads to $SS_{res} = 0$ and $R^2 = 1$. A base structure, that regularly predicts $\overline{y}$, will have $R^2 = 0$.

For our research, we developed a model for predicting yields of selected crops in certain parts of India. Predictions were calculated according to the input features using random forest regression model. Its performance is compared with other regression models like decision tree regression and gradient boosting regression.

22.4 Results and Discussion

The findings indicate that a precise crop yield prediction can be achieved by random forest regression. Random forest algorithm attains highest R^2 value as 0.89 which is slightly better than the other algorithms compared. The project has high success rate in predicting the production of a certain region when the given features (rainfall, temperature, season, crop, nitrogen, pH, and electrical conductivity) were fed to the model.

It is appropriate for agricultural planning to predict crop yields. Our model will be helpful for the farmers as it will predict the production as an output for the various parameters that will be input by the user.

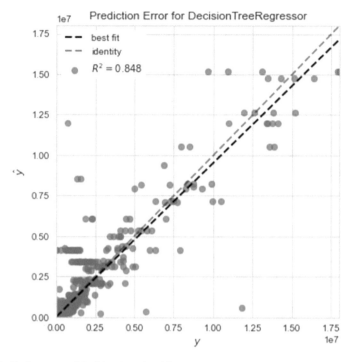

Fig. 22.4 Performance of decision tree algorithm

As shown in the Fig. 22.4, its clear that 84.8% of error is being eliminated by using decision tree regression model-based implementation of crop yield prediction.

As shown in the Fig. 22.5, its clear that 88% of error is being eliminated by using gradient boosting algorithm model-based implementation of crop yield prediction which is far greater than decision tree regression model and its witnessed that there is an increased variation of about 4% error free results than compared to the decision tree regression technique.

The findings as shown in the Fig. 22.6 indicate that a precise crop yield prediction can be achieved by random forest regression. Random forest algorithm attains highest R^2 value as 89% which is better than the other algorithms compared. Overall, its clearly shows that random forest algorithm outperforms well amongst the other two algorithms.

22.5 Conclusion

As per our study, there is vast potential for application of data science in the agricultural field to further improve prediction results. The current work can be extended to

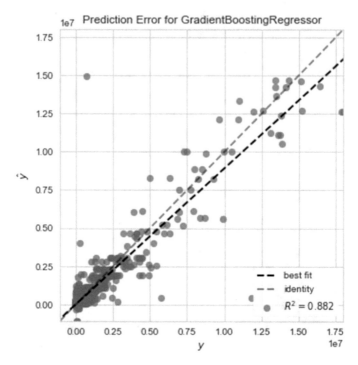

Fig. 22.5 Performance of gradient boosting algorithm

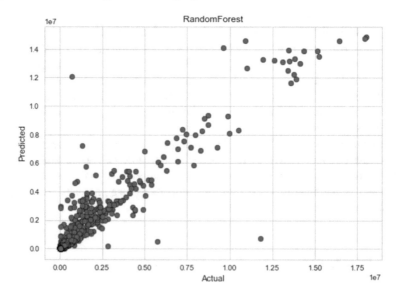

Fig. 22.6 Performance of random forest

include more features influencing crop yield. Hence, there is scope for better accuracy. The use of bulkier datasets can be done to better evaluate the performance of each technique. Also, the results of various algorithms obtained clearly show random forest-based crop yield prediction produces at par results than the other algorithms. Therefore by considering the above additional features in the future, there is an change of getting more error free results.

References

1. Ruß, G.: Data mining of agricultural yield data: a comparison of regression models. **5633**, 24–37 (2009). https://doi.org/10.1007/978-3-642-03067-3_3
2. Suraparaju, V., Mishra, B., Singh, C.D.: Soybean productivity modelling using decision tree algorithms. Int. J. Comput. Appl. **27**, 975–8887 (2011). https://doi.org/10.5120/3314-4549
3. Vagh, Y. Xiao, J.: Mining temperature profile data for shire-level crop yield prediction. In: 2012 International Conference on Machine Learning and Cybernetics, Xian, pp. 77–83 (2012). https://doi.org/10.1109/ICMLC.2012.6358890
4. Ananthara, M.G., Arunkumar, T., Hemavathy, R.: CRY—an improved crop yield prediction model using bee hive clustering approach for agricultural data sets. In: 2013 International Conference on Pattern Recognition, Informatics and Mobile Engineering, Salem, pp. 473–478 (2013). https://doi.org/10.1109/ICPRIME.2013.6496717
5. Ramesh, D., Vishnu Vardhan, B.: Region specific crop yield analysis: a data mining approach. UACEE Int. J. Adv. Comput. Sci. Appl. IJCSIA 3(2) (2013)
6. Veenadhari, S., Misra, B, Singh, C.D.: Machine learning approach for forecasting crop yield based on climatic parameters. 1–5. https://doi.org/10.1109/ICCCI.2014.6921718 (2014)
7. Ahamed, A.T.M.S., Mahmood, N., Hossain, N., Kabir, M. Das, K., Rahman, F., Rahman, M.: Applying data mining techniques to predict annual yield of major crops and recommend planting different crops in different districts in Bangladesh, pp. 1–6 (2015). https://doi.org/10.1109/SNPD.2015.7176185
8. Paul, M., Vishwakarma, S., Verma, A.: Analysis of Soil Behaviour and Prediction of Crop Yield Using Data Mining Approach, pp. 766–771 (2015). https://doi.org/10.1109/CICN.2015.156
9. Gandhi, N., Armstrong, L.J., Nandawadekar, M.: Application of data mining techniques for predicting rice crop yield in semi-arid climatic zone of India. In: 2017 IEEE Technological Innovations in ICT for Agriculture and Rural Development (TIAR), Chennai, pp. 116–120 (2017), https://doi.org/10.1109/TIAR.2017.8273697
10. Priya, P., Muthaiah, U., Balamurugan, M.: Predicting yield of the crop using machine learning algorithm. Int. J. Eng. Sci. Res. Technol. 7(1), 1–7 (2018)
11. Kumar, A., Kumar, N., Vats, V.: Efficient crop yield prediction using machine learning algorithms. Int. Res. J. Eng. Technol. **05**(06), 3151–3159 (2018)
12. Medar, R., Rajpurohit, V.S., Shweta, S.: Crop yield prediction using machine learning techniques. In: 2019 IEEE 5th International Conference for Convergence in Technology (I2CT), pp. 1–5 (2019). https://doi.org/10.1109/I2CT45611.2019.9033611
13. Sangeeta, S.G.: Design and implementation of crop yield prediction model in agriculture. Int. J. Sci. Technol. Res. **8**(01) (2020)
14. Josephine, M.B., Ramya, R.K., Rao, R., Kuchibhotla, S., Kishore, V.B., Rahamathulla.: Crop yield prediction using machine learning. Int. J. Sci. Technol. Res. **09**(02), 2102–2106 (2020)
15. Vogiety, A.: Smart agricultural techniques using machine learning. Int. J. Innov. Res. Sci. Eng. Technol. **9**, 8061–8064 (2020)
16. Narasimhamurthy, V.: Rice crop yield forecasting using random forest algorithm SML. Int. J. Res. Appl. Sci. Eng. Technol. V. 1220–1225 (2017). https://doi.org/10.22214/ijraset.2017.10176

Chapter 23
Music Genre Classification Using Federated Learning

Lakshya Gupta, Gowri Namratha Meedinti, Anannya Popat, and Boominathan Perumal

Abstract Federated learning (FL) is a decentralized privacy-preserving machine learning technique that allows models to be trained using input from multiple clients without requiring each client to send all of their data to a central server. In audio, FL and other privacy-preserving approaches have received comparatively little attention. A federated approach is implemented to preserve the copyright claims in the music industry and for music corporations to ensure discretion while using their sensitive data for training purposes in large-scale collaborative machine learning projects. We use audio from the GTZAN dataset to study the use of FL for the music genre classification task in this paper using convolutional neural networks.

23.1 Introduction

Consumers and legislators are putting more effort and focus on improving privacy in relation to data collection and use. Europe witnessed the adoption of the general data protection regular (GDPR) in 2018, affecting businesses operating in the European Union. GDPR requires businesses to pay closer attention to how they gather, store, use, and transfer user data. Similarly, the California Consumer Privacy Act (CCPA) went into force in the United States. With the CCPA, citizens in the United States now have the right to ask businesses to reveal what kind of data they hold about them and to request that it be removed. A large amount of data is required to train AI models (machine learning and deep learning), which frequently involves sharing

L. Gupta (✉) · G. N. Meedinti · A. Popat · B. Perumal
School of Computer Science and Engineering, Vellore Institute Of Technology, Vellore, India
e-mail: lakshya.gupta2019@vitstudent.ac.in

G. N. Meedinti
e-mail: gowri.namratha2019@vitstudent.ac.in

A. Popat
e-mail: anannyarajesh.popat2019@vitstudent.ac.in

B. Perumal
e-mail: boominathan.p@vit.ac.in

personal data. In the survey carried out in [1], an overview of launched attacks using various computer participants in order to create a unique threat classification, highlighting the most important attacks, such as poisoning, inference, and generative adversarial networks (GAN) attacks was analyzed. They demonstrate in their research that existing protocols do not always provide adequate security when it comes to controlling various assaults from both clients and servers, but there still exists a gap between today's status of federated AI and a future when mainstream adoption is possible [2].

In their survey [3], Ho Bae et al. stressed that AI-based applications have grown common in a variety of industries because of advances in deep learning technologies, but existing deep learning models are subject to a variety of security and privacy risks.

Consumer data is crucial to the music industry. For years, music festivals have used fan data to offer sponsorships to companies looking to get their trademarks on the right stages in front of the right audiences.

Many artists use "geo-fencing" location data to determine which songs entice audiences to the stage and which push them away in different parts of the world, countries, or even regions. Music publishers and song pluggers have utilized social media figures to offer songs for automobile advertisements, while record labels have always used data to decide which performers to pursue and which to dismiss. New regulations may put a stop to it. With the fast-rising pace of the need for a more secure methodology, mentioned new domains of legislation and policies, the machine learning methodology federated learning (FL) can be used to address such challenges.

FL has received a lot of attention owing to the way it maintains users' privacy by segregating the local data present at each end-user device and carrying out the aggregation of machine learning model parameters, such as the weights and biases of neural networks. FL's sole purpose is to train a global model collaboratively without losing data privacy. When compared to data center training on a dataset, FL offers significant privacy benefits. Client privacy can be jeopardized even when "anonymized" data is stored on a server since it is linked to other datasets. The information sent for FL, on the other hand, comprises little adjustments (weights) to improve the correctness of a machine learning model.

Federated learning has found applications in various domains, from fields such as IoT [4], healthcare industries [5, 6], and to autonomous driving [7]. There have been a lot of advancements in the security of the FL framework as well over the past few years [8, 9] In addition, much has been accomplished in achieving a fairer contribution from the training of clients in the architecture [10, 11].

Johnson et al. in [12] presented a study employing synthesized soundscape data to investigate FL for sound event detection. The amount of training data was evenly shared across customers (300 ten-second clips per client) in their study, which used a total of 25 event classes to model both IID and non-IID distributions. FL was able to build models that performed almost as well as centralized training when utilizing IID data, but when using non-IID data, performance was drastically lowered. Their findings show that FL is a viable solution for sound event detection (SED), but it runs into issues with the divergent data distributions that come with scattered client edge devices. In [13], Tao Sun Dongsheng Li and Bao Wang carried out extensive numerical experiments on training deep neural networks (DNNs) on numerous datasets in both IID and non-IID contexts. The results suggest that (quantized) DFedAvgM is effective for training ML models, reducing communication costs, and ensuring the privacy of training data participants.

Companies may be unable to merge datasets from diverse sources due to challenges other than data security, such as network unavailability in edge devices, another aspect where federated learning is superior. Federated learning makes it easier to access diverse data, even when data sources can only interact at particular periods. In FL, there is no need to aggregate data for continuous learning because models are constantly upgraded using client data. Because federated learning models do not require a single complex central server to evaluate data, the federated learning technique requires less complex hardware.

The prime contributions of this study are summarized as follows.

- Classifying the music genre with a decentralized approach, i.e., training the model without having the data at a single location
- Providing clients autonomy over their audio data, allowing for music labels to participate in more machine learning collaborations where their data is requested for training purpose.

23.2 Background

23.2.1 Federated Learning

Federated learning (also known as decentralized learning) is a relatively new technique in the field of machine learning which was introduced by Google in 2016. It allows end-users to exercise discretion over the data they have collected by not requiring them to upload it on a central location like the Google Cloud. Instead, only the global model is stored on the central server, which is distributed to multiple client devices in every communication round. In contrast, traditional machine learning approaches (also known as centralized learning) can only begin the training process after all the data is collected in a single location, which not only poses threat to users' privacy, but also increases the expense of storing this data. Clients and servers communicate on a regular basis, with clients sending the locally derived model parameters back to the server. For our model, we primarily based the federated learning on the federated average learning algorithm.

The incorporation of these parameters from the client models into the global model by the server is a critical component of FL. Several methods have been developed, including federated stochastic variance reduced gradient (FSRVG) [14] and CO-OP. The FedAvg algorithm, the most popular algorithm, has been proved to perform better than the others [15].

The federated averaging (FedAvg) algorithm is a relatively simple aggregation algorithm, which maintains the shared global model by computing the weighted sum of all the weight updates from the training clients in each communication round. A central server manages the training and hosts the shared global model weights. On the other hand, real optimization is carried out on the client's end locally using a plethora of optimization algorithms like SGD, Adam, or Adagrad.

The proportion of training clients (C), number of epochs E, the batch size for client data (B), and learning rate α are all hyperparameters in the FedAvg algorithm. When training with SGD, B and α are extensively utilized. E is the number of iterations that are carried out in total over the same data while training is carried out on the client's federated data before the global model is modified, which is also a widely used parameter in most optimization algorithms like SGD, Adagrad, and Adam [15].

The FedAvg algorithm initially begins with a random assignment of weights wt to the global model. Every communication round between the clients and server comprises the same process. Firstly, a random subset of training clients St, $|St| = C \cdot K \geq 1$, is selected by the server for global model update purposes. These clients are provided with the weights w_t of the global model, who thereafter update their local weights w^k to the global weights, $w^k \leftarrow w_t$. Thereafter, each of these clients splits their federated data into batches of size B and trains their local models E a number of times on the total number of batches. Lastly, the clients communicate the updated weights, w^k, from the training process back to the server, where aggregation of the weight updates is carried out by calculating their weighted sum, which subsequently leads to an updated global model, w_{t+1}.

23.2.2 Dataset

The GTZAN dataset was employed in this study. The dataset consists of ten unique genres, where each genre has a similar distribution. These ten genres include rock, hip-hop, blues, country, classical, jazz, disco, reggae, pop, and metal. The original dataset consists of 1000 music files for each of these genres, each lasting a total of 30 seconds. The monophonic 16-bit audio files have a sampling rate of 22050 Hz in .wav format. We opted for this dataset since it is the most widely used publicly available dataset that has been particularly used extensively for music genre recognition (MGR). The dataset incorporates diverse recording settings for the audio files by collecting the audio from numerous sources like radios, personal CDs, and microphone recordings.

23.3   Methodology

23.3.1   Federated Learning Architecture

We used the centralized architecture which was the best forming architecture between hierarchical, regional, decentralized, and centralized [16]. It maintained its accuracy and consistency with multiple datasets [16]. Figure 23.1 shows the centralized architecture adopted, wherein all clients communicate their updates to a client–server. The client–server carries out the task of model updates aggregation and maintains the global model by using federated averaging. A centralized architecture is most suitable for a small-scale FL system, the likes of which have been proposed in this research. When compared to other options, centralized architecture in FL allows for simple setups and node administration for the simple reason that it has a single central point that is responsible for handling all the participating edge devices as well as computing model aggregation [16].

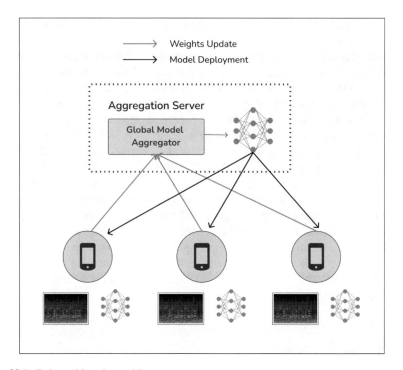

Fig. 23.1 Federated learning architecture

23.3.2 Data Pre-Processing and Feature Extraction

The audio from the GTZAN dataset is analyzed. The genre of a song or piece of music can be determined after only 4–5 s of listening. We split a single audio file into ten 3-s audio pieces because 30 s are a lot of time and information for the model to process at once. The number of training examples rose tenfold after pre-processing, with 1000 training instances in each genre and a total of 10,000 training examples employed.

The audio was converted into a spectrogram with 128 bands. The audio files are separated, and empty directories are created for each genre to contain the audio files (after they have been split) and their spectrograms. The photographs were reduced to 72 × 108 pixels and scaled down to 25% of their original size. Spectrograms were converted into grayscale images since a grayscale spectrogram has all the information that an RGB spectrogram has and it saves memory space to work with grayscale images. Using Keras, the individual instances were shuffled, sorted into batches, and the features were converted from NumPy arrays to tensors. We also use "repeat" to run numerous epochs across the data collection. After that, all of the audio files were label encoded into one of ten genres: rock, hip-hop, blues, country, classical, jazz, disco, reggae, pop, and metal.

The data for federated learning, in particular, has been altered to make it more suitable for FL. A federated dataset, or a collection of data from several users, is required for federated learning. Non-IID data is common with federated data, which presents a distinct set of issues [17]. We provided each audio file with a client ID to simulate a realistic federated environment. The federated data was then divided into two categories: training and testing. The first four clients were utilized for training, while the remaining 5,6 client IDs were used for testing. For our research and modeling with federated learning, we employed TensorFlow federated (TFF), an open-source framework for decentralized data machine learning and other computations. The training and testing datasets were created using *ClientData.from_clients and tf.fn.* Shuffling the data and employing repeats to boost data were the final stage in our pre-processing. Much like real-world working, a random subset of training clients is taken in every round for model weight aggregation and making a global model. In specific, we pick 4 random clients from the total of 6 training clients that we are working with. This ensures that the global model does not overfit to data of a particular set of clients.

23.3.3 Proposed Model

Figure 23.2 shows the architecture of the neural network model being implemented in our approach. The model has 4 convolutional layers, each of kernel size 3 × 3, and filters 8,16,32,64, respectively. For each convolutional layer, a MaxPool layer of size 2 × 2 is used with a stride of 2. At the end of the 4 convolutional layers, we flatten the array and pass it to a dense layer, with softmax activation, corresponding to the 9 different genres.

The following algorithm is used to extract the most significant features from the spectrograms pertaining to about each genre:

$$y^{j(r)} = \max\left(0, b^{j(r)} \sum_i k^{i,j(r)} * x^{i(r)}\right) \qquad (23.1)$$

The MaxPool layer incorporated in the proposed model uses the following algorithm to take the maximum value in the convolutional layer's kernel:

$$y_{i,j(k)} = \max\left(x_{i,j*s+m,k*s+n}\right) \qquad (23.2)$$

Lastly, the densely connected layer at the end uses the following mathematical formula:

$$y_j = \max\left(0, b_j + \sum_i x_i * w_{i,j}\right) \qquad (23.3)$$

The units in the last dense layer are all put through a "softmax" activation function, to indicate the probability of each of the ten genres. The formula for the same is

$$\sigma(\vec{z})_i = \frac{e^{z_i}}{\sum_{j=1}^{k} e^{z_j}}. \qquad (23.4)$$

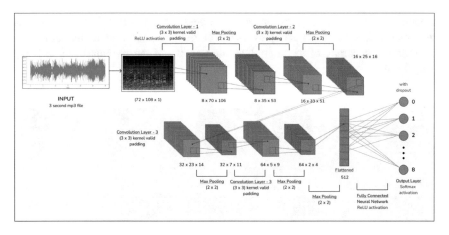

Fig. 23.2 Proposed CNN model

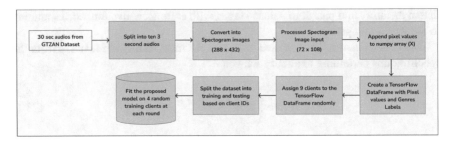

Fig. 23.3 Task flow

23.3.4 Task Flow

Figure 23.3 depicts the task flow diagram of the proposed federated framework for music genre classification.

23.4 Results

The graphs in Fig. 23.4 depict the average spectrogram image for each genre. As can be seen from the sixth graph in the sequence, it is observable that it is much brighter than the other ones. The reason being that the 6th graph corresponds to the metal genre which is the most powerful form of music among other genres and thus has the brightest plot.

Figure 23.5 gives a graphical representation of the genres' distribution for all training clients. Every client has a different distribution which is the characteristic property of federated learning wherein we deal with non-IID data. The centralized approach has the best accuracy, however, there is only a diminutive difference between the accuracy of the centralized approach and the decentralized approach (Fig. 23.6).

Fig. 23.4 Average spectrogram images

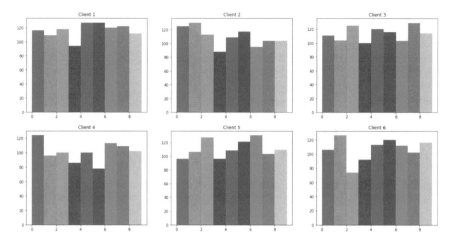

Fig. 23.5 Graphical representation of the genres' distribution for all training clients

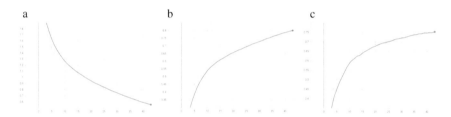

Fig. 23.6 FL on unscaled data **a** loss, **b** training accuracy, **c** testing accuracy

The loss, training accuracy, and testing accuracy numbers for training in the federated average algorithm using unscaled data are shown in Fig. 23.8. Note that when we trained the model, the accuracy values continued to rise, but the loss measure rapidly decreased. The loss, training accuracy, and testing accuracy numbers for training in the federated average algorithm using 25% scaled down data are shown in Fig. 23.7. As expected, the model performed better with unscaled images.

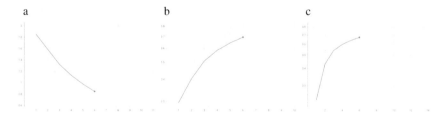

Fig. 23.7 FL on 25% scaled down data **a** loss, **b** training accuracy, **c** testing accuracy

a   b

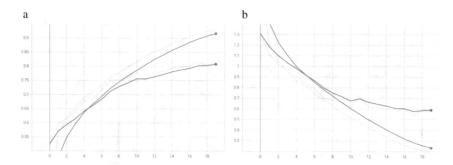

Fig. 23.8 Centralized **a** training versus testing accuracy, **b** training versus testing loss

Table 23.1 Model comparison

S. No.	Model comparison		
	Training approach	Training accuracy (%)	Testing accuracy
1	Proposed federated learning w/o scaling	80.2	76.02
2	Proposed federated learning w. 25% scaling	76.46	71.78
3	Centralized (neural network)	92.84	81.65

The proposed federated learning model outperformed the traditional centralized model which achieved an accuracy of 70% [18] (Table 23.1).

Discussions

The Hann window was the digital signal processing (DSP) window's default, because it has good all-around frequency resolution and dynamic-range features. The window function is used to convert the mp3 file to a spectrogram. The Bartlett, Hamming, Rectangular, and Raiser are other options for windows. Of these, other windows can be explored to test various accuracies.

Our experiment carried out this simulation on a single device set up. Because it was carried out on a single system, it took a heavy toll on the RAM. However, in actual production, FL is supposed to be integrated with separate devices. Carrying out the training on different systems could have yielded better and more realistic results for the federated learning. Since we are working with non-IID data, perhaps employing a fair contribution optimizer for each client could be investigated.

23.5   Conclusion

In this paper, we have presented a study in federated learning with audio data derived from the GTZAN dataset. The observation is to make a decentralized framework that can classify the music genre of a music file that a client has without the need for training the model with data in a centralized location. A federated learning framework is proposed wherein a server-client architecture (also known as centralized architecture) is used to train clients' data without the need for clients to upload the data to a single location.

To minimize overfitting and to imitate an actual federated environment, we used a variety of clients. Our findings show that when all client devices have significant amounts of data, federated learning may produce global models that perform similarly to centrally-trained models, without requiring access to any local data. We further compared the results with scaled down data, and rightly, the 25% scaled down data gave us less accuracy than the original data, albeit still performing better than traditional approaches to this problem statement.

The proposed framework is marginally less accurate than the centralized model. However, it provides the obvious benefit of user privacy and storage cost reduction which makes it more usable from an industrial feasibility aspect. With the added advantage of privacy protection, this technique for extracting new possibilities from a user's data without disclosing it has a lot of potential in future.

References

1. Zhang, J., Li, M., Zeng, S., Xie, B., Zhao, D.: A survey on security and privacy threats to federated learning. In: 2021 International Conference on Networking and Network Applications (NaNA), pp. 319–326. IEEE Access, Urumchi City, China (2021)
2. Mothukuri, V., Parizi, R.M., Pouriyeh, S., Huang, Y., et al.: A survey on security and privacy of federated learning. Future Gener. Comput. Syst. **115**, 619–640 (2021)
3. Bae, H., Jung, J., Jang, D., Ha, H., et al.: Security and Privacy Issues in Deep Learning. arXiv: 1807.11655 (2018)
4. Savazzi, S., Nicoli, M., Rampa, V.: Federated learning with cooperating devices: a consensus approach for massive IoT networks. IEEE Internet Things J. **7**(5), 4641–4654 (2020)
5. Sheller, M.J., Edwards, B., Reina, G.A., Martin, J., et al.: Federated learning in medicine: facilitating multi-institutional collaborations without sharing patient data. Sci. Rep. **10**, 12598 (2020)
6. Xu, J., Glicksberg, B.S., Su, C., et al.: Federated learning for healthcare informatics. J. Healthc. Inform. Res. **5**, 1–19 (2021)
7. Nguyen, A., Do, T., Tran, M., Nguyen, B.X., et al.: Deep Federated Learning for Autonomous Driving. arXiv: 2110.05754 (2021)
8. Xu, R., Baracaldo, N., Zhou, Y., et al.: HybridAlpha: an efficient approach for privacy-preserving federated learning. In: Proceedings of the 12th ACM Workshop on Artificial Intelligence and Security (AISec'19), pp. 13–23. Association for Computing Machinery, London, United Kingdom (2019)
9. Wei, K., Li, J., Ding, M., Ma, C., et al.: Federated learning with differential privacy: algorithms and performance analysis. IEEE Trans. Inf. Forensics Secur. **15**, 3454–3469 (2020)
10. Qi, T., Wu, F., Wu, C., Lyu, L., et al.: FairVFL: A Fair Vertical Federated Learning Framework with Contrastive Adversarial Learning. arXiv: 2206.03200 (2022)
11. Wei, S., Tong, Y., Zhou, Z., Song, T.: Efficient and fair data valuation for horizontal federated learning. In: Yang, Q., Fan, L., Yu, H. (eds.) Federated Learning. Lecture Notes in Computer Science, vol. 12500, pp. 139–152. Springer, Cham. (2020)
12. Johnson, D.S., Lorenz, W., Taenzer, M., Mimilakis, S., et al.: DESED-FL and URBAN-FL: Federated Learning Datasets for Sound Event Detection. arXiv: 2102.08833v3 (2021)

13. Sun, T., Li, D., Wang, B.: Decentralized Federated Averaging. arXiv: 2104.11375 (2021)
14. Konecný, J., McMahan, H,B., Yu, F,K., Richtárik, P., et al.: Federated learning: strategies for improving communication efficiency. In: NIPS Workshop on Private Multi-Party Machine Learning. arXiv: 1610.05492 (2016)
15. Nilsson, A., Smith, S., Ulm, G., et al.: A performance evaluation of federated learning algorithms. In: Proceedings of the Second Workshop on Distributed Infrastructures for Deep Learning, DIDL '18, pp. 1–8. Association for Computing Machinery, Rennes, France (2018)
16. Zhang, H., Bosch, J., Olsson, H.: Federated learning systems: architecture alternatives. In: 27th Asia-Pacific Software Engineering Conference (APSEC), pp. 385–394. IEEE, Singapore (2020)
17. Zhu, H., Xu, J., Liu, S., Jin, Y.: Federated Learning on Non-IID Data: A Survey. arXiv: 2106.06843 (2021)
18. Dong, M.: Convolutional Neural Network Achieves Human-level Accuracy in Music Genre Classification. arXiv: 1802.09697 (2018)

Chapter 24
Perception Mapping of Internet of Things in Augmenting Agroforestry: A Preliminary Exploration

David Lalrochunga⊙, **Adikanda Parida**⊙, **and Shibabrata Choudhury**⊙

Abstract Agroforestry practices have helped to balance the environment while also increasing agricultural productivity. Much research and development have improved the capabilities of computational technologies in realizing Internet of things (IoT)-based intelligent systems. The Internet of things has changed the way devices, sensors, and networks communicate, making them more versatile than the previous technologies. Devices have become communicable, and the connection made by IoT through use of stringent computational paradigm has found to significantly agricultural farmlands. With the introduction of IoT, information on the type of crops grown, the soil moisture content, and the amount of irrigation required for a specific type of crop has become smart. Even farmland policy has become feasible with all of the sustainable goals of the governing body that mandates agricultural farmland operations. The stakeholders' preconceptions have been mapped, with contributing factors such as technology, utility, and adaptability prospects of IoT paradigms in relation to agroforestry. As a result, this paper discloses the concerned respondents' perceptions on the promising beneficial intervention of IoT in the operation of agroforestry. The use of Likert scale for descriptive statistics and multiple regression analysis for predictive statistics also has been adopted for the perception mapping.

24.1 Introduction

Agroforestry, represented as the unification of trees and woody bushes with crop and livestock production systems, is an advanced and viable means of resolving conservation and development goals and objectives around the world, supporting in the attainment of the 2030 United Nations Sustainable Development Goals [1]. Research has indicated that cultivating trees on farms can preclude ecological degradation, promote sustainable efficiency, boost carbon sequestration, produce cleaner water, and encourage healthy soil and biological systems while providing stable

D. Lalrochunga (✉) · A. Parida · S. Choudhury
North Eastern Regional Institute of Science and Technology, Nirjuli, Itanagar,
Arunachal Pradesh 791109, India
e-mail: lalrochunga@gmail.com

earnings and other economic advantages to human government assistance [2]. The potential of trees on farms for livelihoods and environmental agroforestry in rural economy has shown promising results [3], laying the foundation for achievable effective interventions in agroforestry practices leading to ecological balance.

24.2   Challenges and Scopes

The challenge is comparatively high in adopting IoT technology in low- and middle-income nations [1] as well as the simplicity with which ecosystem services and human well-being may be implemented in high-income countries [2]. For IoT to be implemented, low- and middle-income countries will need more developed infrastructure, but high-income countries will find it easier to design prototypes and operationalize new relevant treatments. Traditional farming methods combined with trees and bushes have shown to be quite productive, but any IoT involvement would have to be tailored to the type of agroforestry practiced in a certain area or region.

The rigorous evidence of interventions designed to support and promote agroforestry through evidence gap map (EGM) identifying what evidence exists, rather than summarizing effect size estimates [1] has yielded another perspective where the Agriculture 4.0 supported by Internet of things (IoT) [4] adding further elements to efficient and productive agroforestry practices. Smart and precision farming are the future of farming rising at the horizon. Agroforestry has the scope of alleviating poverty in forests and tree-based systems [5], and incorporating IoT will enhance the spatial, geographical, and demographic information of the particular area, the stakeholders, surrounding biodiversity, and ecology as a whole.

24.3   IoT Aiding Agroforestry

Manuscripts discussing the agricultural Internet of things (IoT) have provided mass sensor data for agricultural production and scientific research [6], services of Internet of things (IoT), and smart service has helped the Eriocheir sinensis crab seedling breeding for successful quality dynamic traceability system [7] which was developed and tested.

The cooling effect in low carbon IoT greenhouse in summer [8], research and development on automated based on Internet of things for grading potted flowers [9] contributed much to the recent works that are available today.

Telegram-based IoT control system for nutrients in floating hydroponic system for water spinach (Ipomea reptans) [10], IoT for designing an automatic water quality monitoring system for aquaculture ponds [11], IoT-based decision support systems for agroforestry [12], agroforestry IoT system supporting sustainable agro-food management [13], low-cost sensing systems for woody plants supporting tree management [14], automatic smart irrigation based on the moistness of soil [15, 16], key structure of agricultural Internet of things [17] has set marks for the upcoming prospects on potential research and development which will soon play a key role in sustaining smart farming.

24.4 IoT in Agroforestry Perception Mapping

The perception mapping of IoT in agroforestry has been done in a variety of ways. The current study, on the other hand, is concerned with technology, utility, and adaptability. The IoT infrastructure in agricultural is structured on three layers: perceive, transportation, and application [18]. Other supporting technology, such as uninterrupted quality power and internet supply, real-time data capture, and so on, is required for this infrastructure to function. Similarly, IoT technology necessitates high-end tools for storing and retrieving a variety of data files utilizing computational tools such as cloud computing [4]. The issues, which range from day-to-day operations to disasters, entail IoT intervention for convenient utility services [15]. People in the other region, unlike those in developed economies, face difficulties in adopting to sophisticated IoT technologies [1, 2]. Horizontal federated learning model based on FedProLs for IoT perception data prediction [19] yielded promising outcomes and multidimensional graphic design-based voice perception model utilizing IoT [20] resulted in an esthetic evaluation system. Thus far, perception mapping on agroforestry with respect to IoT has not been observed. Hence, the current work would like to explore on the perception of the stakeholders regarding the Internet of things in augmenting agroforestry.

24.4.1 Methodology

To present the examined data in this study, a Google form questionnaire was disseminated to about 500 stakeholders from academics to industry professionals, out of which 234 responded. Basic information was integrated in the formation of questionnaire. Statement-based questions relating to the technology, utility, and adaptability were formed for perception mapping using the Likert scale for IoT. The Likert scale ranged from 1 to 5, with scale 1 indicating strongly disagree, scale 2 indicating disagree, scale 3 indicating neutral, scale 4 indicating agree, and scale 5 indicating strongly agree. Gender distribution Fig. 24.1a is being considered for demographic samples across the student and professional community, where the technical awareness of intermediate, undergraduate, postgraduate, and doctoral researchers Fig. 24.1b was collected.

The summarized data has been analyzed by descriptive statistics and predictive statistics.

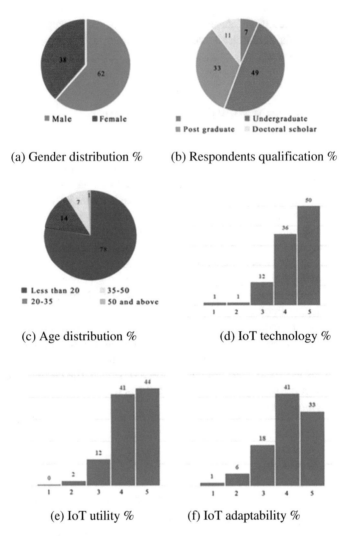

Fig. 24.1 **a** Gender distribution %, **b** respondents qualification %, **c** age distribution %, **d** IoT technology %, **e** IoT utility %, **f** IoT adaptability %

Descriptive Statistics

The respondents were asked questions regarding IoT infrastructure that would pilot the technology intervention and its utility in the large-scale promising adaptability of IoT in agroforestry. As indicated in Fig. 24.1c, 78% respondents were between the age group of 20–35 years followed by less than 20 years with 14 percentage respondents. Figure 24.1d 86% of the respondents 'agree' and 'strongly agree' see a promising scope of the IoT technology for precision and smart farming. 85% 'agree' and 'strongly agree' that IoT has the ability to empower intelligence in the methods employed in farming techniques as shown in Fig. 24.1e. While 74% 'agree' and 'strongly agree' and think that IoT paradigms will be well adopted in the farming sector for the generation of the full information of farmlands and the crops being grown enhancing knowledge on soil health and mitigating possible disasters as well as efficient management as shown in Fig. 24.1f.

According to the mean generated (Table 24.1), most respondents appear to be more aware of the technology (4.32), utility (4.26), and adaptability (4.00) of IoT applications. The Internet of things infrastructure has yet to be fully developed in high-income countries, and much more in low- and middle-income countries [1, 2]. The measurement of the data set between the data set does not vary that much but more inclination toward the technology, and utility can be seen from the variance generated from the mean. While the infrastructure and adaptability showed a varied behavior, indicating more practical implementation needs to be executed.

Predictive Analysis

The data in Table 24.2 shows that there is a strong positive correlation existing among all the variables. Utility is closely associated with 0.655 and 0.598 as correlation coefficient with respect to adaptability and utility, respectively. Technology and adaptability also have near to 0.5 correlation coefficient. Based upon the correlation, the predication has been calculated based on the regression analysis as shown in Tables 24.3 and 24.4.

The number of observations is 234. Multiple R is 0.67 which is considered to be strong linear regression. The adjusted R square is 0.44, and standard error of regression is 0.69. The two independent variable, namely 'technology' and 'utility', has p-value less than 0.05; hence, the result can be considered to be significant. Based on the coefficient obtained the regression line can be prepared as follows:

$$\text{Adaptability} = 0.42 + (0.18 * \text{Technology}) + (0.66 * \text{Utility}) + \text{error} \quad (24.1)$$

Table 24.1 Descriptive statistics of IoT in agroforestry perception mapping survey

Groups	Count	Sum	Mean	Variance
Technology	234	1013	4.32	0.69
Utility	234	998	4.26	0.62
Adaptability	234	936	4.00	0.86

Table 24.2 Correlation coefficient matrix of perception mapping survey

	Technology	Utility	Adaptability
Technology	1		
Utility	0.59	1	
Adaptability	0.49	0.66	1

Table 24.3 Regression statistics of IoT in agroforestry perception mapping survey

Regression statistics					
Multiple R	0.67				
R square	0.44				
Adjusted R square	0.44				
Standard error	0.69				
Observations	234				
ANOVA					
	df	SS	MS	F	Significance F
Regression	2	89	45	92.53	3.0E-30
Residual	231	111	0.5		
Total	233	200			

Table 24.4 Significance of technology and utility for application of IoT in agroforestry perception mapping survey

	Coefficients	Std. error	t Stat	*P*-value	Lower 95%	Upper 95%
Intercept	0.42	0.27	1.51	0.13	− 0.13	0.96
Technology	0.18	0.07	2.56	0.01	0.04	0.31
Utility	0.66	0.07	9.17	2.61E-17	0.52	0.8

24.5 Conclusion

This study was exploratory in nature, so it makes no grand claims. It has, however, provided some important insights into the perceptions of stakeholders/experts in the field of IoT applications for agroforestry.

In respect of future research agenda, it would be an upgrade for:

- IoT systems that would rely on renewable energy to power the entire network of devices related to IoT systems for smart agroforestry
- Drones that would use IoT paradigms for smart and precision farming.

This study contributes modestly to the development of theoretical models, empiricism, and policymaking in the area of technological interventions in agroforestry. However, the current analysis focused on stakeholders' preconceptions, with contributing factors such as infrastructure, technology, utility, and adaptability prospects of IoT paradigms in relation to agroforestry. Subsequent research will focus on adoption issues and barriers to adoption, as well as an assessment of the efficiency and productivity benefits of IoT paradigms in agroforestry practices.

Acknowledgements Authors would like to acknowledge funding support by the National Fellowship for Higher Education for Scheduled Tribe Students, Ministry of Tribal Affairs, Government of India, Award [202021-NFST-MIZ-00489].

References

1. Castle, S.E., Miller, D.C., Ordonez, P.J., Baylis, K., Hughes, K.: The impacts of agro-forestry interventions on agricultural productivity, ecosystem services, and human well-being in low- and middle-income countries: a systematic review. Campbell Syst. Rev. **17**(2), e1167 (2021)
2. Brown, S.E., Miller, D.C., Ordonez, P.J., Baylis, K.: Evidence for the impacts of agro-forestry on agricultural productivity, ecosystem services, and human well-being in high-income countries: a systematic map protocol. Environ. Evid. **7**(1), 1–16 (2018)
3. Nöldeke, B., Winter, E., Laumonier, Y., Simamora, T.: Simulating agroforestry adoption in rural Indonesia: the potential of trees on farms for livelihoods and environment. Land 2021, 10, 385. Ecosystem Services, Sustainable Rural Development and Protected Areas, 21 (2021)
4. Symeonaki, E., Arvanitis, K., Piromalis, D.: A context-aware middleware cloud approach for integrating precision farming facilities into the IoT toward agriculture 4.0. Appl. Sci. **10**(3), 813 (2020)
5. Hajjar, R., Newton, P., Ihalainen, M., Agrawal, A., Gabay, M., Alix-Garcia, J., Timko, J., et al.: Levers for alleviating poverty in forests and tree-based systems. forests, trees and the eradication of poverty: potential and limitations, 125 (2020)
6. Xiao, B., Guo, X., Wang, C., Wu, S., Lu, S., Wen, W.: Discussion on application of context aware computing technology in agricultural internet of things. J. Agri. Sci. Technol. (Beijing) **16**(5), 21–31 (2014)
7. Yu, L., Yang, J., Ling, P., Cao, S., Cheng, Y., Wang, C., Xia, J.: Research on dynamic quality traceability system of Eriocheir sinensis seedling based on IOT smart service. J. Fish. China **37**(8), 1262–1269 (2013)
8. Dong, W., Zhou, Z., Bu, Y., Lan, L., Chen, Y., Li, D.: On cooling effect in low carbon IOT greenhouse in summer. J. Shenyang Agri. Univ. **44**(5), 565–569 (2013)
9. Sun, J., Zhou, Z., Bu, Y., Zhuo, J., Chen, Y., Li, D.: Research and development for potted flowers automated grading system based on internet of things. J. Shenyang Agri. Univ. **44**(5), 687–691 (2013)
10. Anri, K., Hanis, A.L.: Control system of nutrient in floating hydroponic system for water spinach (Ipomea reptans) using telegram-based IoT. Jurnal Teknik Pertanian Lampung **9**(4), 326–335 (2020)
11. Nguyen, Q.H., Vu, T.T.G., Le Vu, Q.: Application of the Internet of Things technology (IoT) in designing an automatic water quality monitoring system for aquaculture ponds. Tap Chi Khoa Hoc Nong Nghiep Viet Nam/Vietnam J. Agri. Sci. **3**(2), 624–635 (2020)
12. Cavaliere, D., Senatore, S.: A multi-agent knowledge-enhanced model for decision-sup-porting agroforestry systems. In: 2021 IEEE Symposium Series on Computational Intelligence (SSCI), pp. 01–08. IEEE (2021)
13. Laksono, G.P.B.: Sustainable agrifood management with agroforestry system. SAMI2020 33 (2020)
14. Putra, B.T.W.: A new low-cost sensing system for rapid ring estimation of woody plants to support tree management. Inf. Process. Agri. **7**(3), 369–374 (2020)
15. Choudhari, N.K., Mayuri, H.: Automated plant irrigation system based on soil moisture and monitoring over IoT. Int. J. Res. Appl. Sci. Eng. Technol. **5**(6), 2551–2555 (2017)
16. Kulkarni, S.S., Shweta, A., Sukanya, W., Umavati, B., Satyam, P.: IOT based smart agro system. Int. J. Res. Appl. Sci. Eng. Technol. **6**(4), 323–326 (2018)
17. Yue, Y., Yue, X., Zhong, Y.: Research progress on system structure and key technology of agricultural Internet of things. J. Agri. Sci. Technol. (Beijing) **21**(4), 79–87 (2019)
18. He, Y., Nie, P., Liu, F.: Advancement and trend of internet of things in agriculture and sensing instrument. Nongye Jixie Xuebao Trans. Chinese Soc. Agri. Mach. **44**(10), 216–226 (2013)
19. Zeng, Q., Lv, Z., Li, C., Shi, Y., Lin, Z., Liu, C., Song, G.: FedProLs: federated learning for IoT perception data prediction. Appl. Intell. 1–13 (2022)
20. Wang, Z.: Aesthetic evaluation of multidimensional graphic design based on voice perception model and internet of things. Int. J. Syst. Assur. Eng. Manage. **13**(3), 1485–1496 (2022)

Chapter 25
Student Placement Analyser and Predictor Using Machine Learning and Data Analytics

Rushabh Jangada, Anushka Kulkarni, Shweta Barge, and Vrushali Lad

Abstract Improving the placement success of the students is one of the main difficulties that higher education institutions confront nowadays. As educational entities become more complicated, so does the placement analyser and prediction. It is crucial that educational institutions strive for more effective technology that may support improved administration, decision-making processes and help them create new strategies. Introducing fresh information about procedures and entities that contribute to management education is one of the most efficient ways to handle these issues and improve quality. Information regarding historical student data is contained in the dataset for system implementation who may have graduated from that educational institute. The knowledge can be drawn out of the operational and archival data stored in the databases of the educational organisation using machine learning techniques. These data are utilised to train the rule identification model and test the classification model. This essay offers a categorization and prediction system that assigns students to one of the six placement classes—Classes A, B, C, D, E, and F—based on their academic performance. With the use of this model, an organisation's placement cell may recognise potential students and focus on and enhance both their necessary technical and interpersonal skills. This will allow the students to work on developing these talents. The institutions will be informed of the criteria used by the businesses when choosing their employees. Additionally, students can use this system to determine their particular placement status, where they stand, and what they are most likely to accomplish in the both their B. Tech. course's pre-final and final years. With this, people can work harder and focus on their weak areas in order to be hired by better firms.

R. Jangada (✉)
Vishwakarma Institute of Technology, Pune, India
e-mail: rushabh.jangada16@vit.edu

A. Kulkarni · S. Barge · V. Lad
Cummins College of Engineering For Women, Pune, India

25.1   Introduction

The main reason students enrol in professional programmes at universities or other schools of higher education is to get a lucrative position with an established business. Specialised training could be entirely managerial or entirely technical. A Bachelor of Technology (B. Tech) is a technical education degree that can be obtained in a number of disciplines, including Computer Science and Engineering, Electronics and Communication Engineering, Mechanical Engineering, and more. With this degree, students will become specialists in both theoretical and practical knowledge in a variety of engineering areas. The main reason students enrol in professional programmes at universities or other higher education institutions is to get a well-paying job with a reputable company. Professional education could be entirely managerial or entirely technical. A Bachelor of Technology (B. Tech) is a technical education degree that can be obtained in a number of disciplines, including Computer Science and Engineering, Electronics and Communication Engineering, Mechanical Engineering, and more. With this degree, students will become specialists in both theoretical and practical knowledge in a variety of engineering areas. For the system's implementation, we used the Scikit-learn machine learning module, which has simple and effective data mining and data analytics capabilities, along with the random forest classifier, decision tree classifier, support vector machine, kernel support vector machine, K-nearest neighbours, and Naive Bayes algorithms.

25.2   Materials and Method

25.2.1   Machine Learning

Machine learning is concerned with the creation, examination, and study of algorithms that can automatically identify patterns in previous data, utilise those patterns to forecast future data, and make decisions in the light of those predictions [1]. In order to make this prediction, machine learning builds models by changing a few parameters for a certain method based on accuracy [2]. The fields of bioinformatics, computer vision, robot locomotion, computational finance, search engines, and many others are all impacted by the booming and emerging subject of machine learning.

25.2.2   Decision Tree

Observations are made on in real-world issues the problem's parameters in order to draw conclusions or forecast the parameter values that should be achieved. With the use of decision trees, this mapping is included and used in a prediction model. Decision tree learning is the name given to this learning strategy. This is just one of the several predictive modelling techniques used in data mining [3], machine learning, and statistics. We employed classification trees in this model; typically, a decision tree only allows a finite set of categorical values for the predictor variable (target variable) (like here, we have categorical values like classes A, B, C, D, E, F). The leaves of these trees stand in for labels for the classes, and branches reflect the paths that decisions take as they move from the tree's root to its leaves.

25.2.3   Naïve Bayes

The Bayes theorem serves as the foundation for this categorization strategy. It is based on the premise that the predictors are self-sufficient. Simply defined a Naive Bayes classifier thinks that the existence of one feature or set of parameters in a class has no bearing on the existence of any other feature. Even though they are dependent on one another or the presence of other variables, each of these characteristics increases the chance on its own. The Naive Bayes model is simple to construct, extremely helpful, and appropriate for very big data sets. In addition to being straightforward, Naive Bayes is known to outperform even extremely complex classification techniques [4].

25.2.4   Random Forests

Regression, categorization, and other issues, random forests, usually ensemble learning is referred to as random decision forests approaches. These techniques are based on building a large number of decision trees during the training phase and then producing the class that represents the mean of the predictions (in the case of regression) or the mode of the output classes (classification) of each tree. The problem of the decision tree algorithm overfitting to the training dataset is overcome by the random decision forest technique [4].

25.2.5   K-Nearest Neighbour

A classification algorithm is one that uses the k-nearest neighbours method. It fits into the category of supervised learning, i.e., it uses a large number of labelled points as a resource to teach itself how to label new ones. To label a new point (the value that we need to predict), it looks at the labelled points closest to it (those are its nearest neighbours) and asks those neighbours to cast their votes. As a result, the label for the new point will be determined by the mode of the value of the neighbours (the "k" is the number of neighbours it checks) [4].

25.2.6 Support Vector Machine

The training data for support vector machines is represented as points in a space that is as clearly divided into multiple categories as possible. Then, new examples or the values we need to forecast are mapped into that same space, and as a consequence, depending on whose side of the divide they are on, they are anticipated to fall into one of several categories. This method works well in high-dimensional spaces and only employs a small portion of the decision function's training points, making it memory-efficient. Probability estimates are derived by means of an expensive five-fold cross-validation, rather than directly or potentially being provided by the algorithm.

25.2.7 Scikit-Learn

Scikit-learn is a free and open-source machine learning package for Python [4] that includes a variety of regression, grouping, and classification techniques. This module includes a number of important algorithms, including Naive Bayes, decision tree, random forest, support vector machines, logistic regression, gradient boosting, K-means, and DBSCAN. The primary goal of this module is to resolve issues related to supervised and unsupervised learning. By offering an abstraction utilising a multi-purpose high-level language, it aims to make machine learning approachable for beginners. This module's main attributes include simplicity of use, documentation, performance, and API consistency [5].

25.2.8 Background and Related Work

In order to generate new, useful knowledge in the sphere of education and help students do better in their placements, machine learning techniques play a very important role. To establish the methodology for performance analysis and placement, numerous data scientists from across the world have invested a lot of time in research and development. To offer an indication of what has previously been done and what future growth is anticipated in this field of work, some of the pertinent work in this area is set forth. Hijazi and Naqvi [5] performed a survey to identify the variables influencing students' academic success. They employed surveys to extract information from the children, highlighting elements including the family's size, the frequency of teachers, the subjects the teachers were interested in teaching, and the students' interest in extracurricular activities. They used the Pearson correlation coefficient to identify and emphasise the key elements that contributed to this, and they discovered that the educational level of the mother and the family's income significantly influenced students' academic achievement. A classification model was suggested in a study by Pal and Pal [6] using student data that included details about their academic achievements in order to identify a reliable and effective way to forecast student placements. They came to the conclusion that, when compared to J48 and multilayer perceptron algorithms, the Naive Bayes classifier was the most effective classification technique that could be applied in placements. In a study they did, Ramanathan et al. [7] employed the sum of difference approach to forecast where students will be placed. They utilised a variety of criteria, including age, academic performance, achievements, and others, to make their prediction. They came to the conclusion that higher education institutions may provide their students with a better education based on their findings. Arora and Badal [8] undertook a study to use data mining to forecast where students will be placed. Based on data from MCA students in Ghaziabad, Uttar Pradesh, they developed forecasts by taking into account variables like MCA result, communication skills, programming abilities, co-curricular activity involvement, gender, 12th result, and graduation result. They concluded that their decision tree algorithm-based model can help the placement cell, and faculties identify the group of students who are likely to experience difficulties during final placements. Elayidom et al. [9] construction of a generalised data mining framework for placement probability prediction questions made a contribution to this topic. They employed decision trees and neural networks to determine for each student the branch of study that will be excellent, good, average, or poor based on their reservation category, sector, gender, and entry rank. Before allowing students to enrol in an MCA course, Naik and Purohit [10] developed a study that used data mining and prediction techniques to gather information about potential pupils.

25.2.9 Data Preparation

By generating a Google form and asking people to fill it out over the phone, we were able to collect the dataset that was utilised for both training and testing. 400 + records of students who graduated from the university during the educational year 2016–2017 make up the data sample.

25.2.9.1 Procedure

Students from the institution's many departments received a google form. Following responses, the information was divided into a number of columns, some of which included CPI, project domain, and internship domain. Some columns were re-spit into other domains, such as the project domain column. Machine learning, deep learning, and artificial intelligence all have their own columns, as do embedded systems and image processing, VLSI and signal processing, the Internet of things and Android, mechanical, and research. We would assign a "1" to the columns for machine learning/deep learning/data analytics, "1" to the columns for signal

Table 25.1 Parameters

Description	Possible values
GR number	Integer
Department	Comp, E&TC, electronics, mechanical, IT, instrumentation
Domain of project	ML, DL, AI, signal processing, image processing, research, IoT, Android, mechanical, data mining, data analytics, communication, theory of machines, gears
Domain of internship	ML, DL, AI, signal processing, image processing, research, IoT, Android, mechanical, data mining, data analytics, communication
CPI	Integer

processing and VLSI, and "−1" to the rest of the columns if a student had signal processing for his third-year project and machine learning for his final year project.

When all columns have received this treatment, we proceed to applying various classifiers and determining which classifier will work best with our data. *P* values were utilised to identify which column was the most significant factor in establishing the company's class and how it related to other factors.

To achieve greater precision, data manipulation was carried out in a certain way. Each internship type was given a unique code word since Python and Spyder consider machine learning and ML to be two distinct types of internships. 2 or 3 projects with similar topics were combined.

For projects, domains clubbed together were as follows:

- Machine learning, deep learning, data analytics.
- Artificial intelligence, blockchain, data mining
- VLSI, signal processing
- Embedded, image processing
- IoT, Android, communication
- Mechanical, theory of machines
- Gears
- Research (Table 25.1).

25.2.9.2   Machine Learning Model Implementation

Amongst the optimum data analytics languages, Python is widely employed in the sector. It is the greatest practical language for creating a variety of goods and has extensive and advanced machine learning and data mining abilities. Python is a favourite and a good choice for data processing because of this. Python was created as a compromise because in data processing, scalability and complexity are frequently trade-offs. For simpler tasks, we can utilise NumPy as a scratchpad and Anaconda Navigator (Spyder) notebook, whilst Python is an effective tool for processing medium-sized amounts of data. Additionally, Python has several benefits like the large data community, which provides a huge selection of toolkits and

Table 25.2 CS + IT + E&TC results

Algorithm	Accuracy (%)
Random forest	81.03
Decision tree	67.24
KNN	56.8
SVM	55
Kernel SVM	67.3
Naïve Bayes	46.5

Table 25.3 CS + IT + E&TC + Elex + Instru results

Algorithm	Accuracy (%)
Random forest	75.3
Decision tree	70
KNN	60
SVM	53.42
Kernel SVM	56.1
Naïve Bayes	19.1

functionalities. Scikit-learn is a complex Python module that includes practically all of the most popular machine learning techniques. The 2014 pass out batch's placement data is included in the training dataset, which is loaded into the Python code, followed by the addition of macros to the variables for quick processing and the fitting of using a decision tree classifier Scikit libraries. The test data is transferred to Python when the model creation is finished, where the predict function is then used to read variables using macros provided by Scikit-learn. In relation to the training data, this generates the macro output that corresponds to the placement status class. Eventually, the macro output is transferred back to the model's recruitment status variable.

25.3   Experiments and Result

See the Tables 25.2, 25.3, and 25.4.

25.4   Conclusion

As a result, we were successful in achieving our goal of predicting the placement status that Btech students will probably have at the conclusion of their final year

Table 25.4 CS + IT + E&TC + Elex + Instru + Mech results

Algorithm	Accuracy (%)
Random forest	79
Decision tree	72
KNN	53
SVM	6
Kernel SVM	61
Naïve Bayes	27

placements. Using the p value, we were able to determine the parameters that affect placements the most. Any parameter over the significance level of 0.05 was eliminated. In the end, the most significant factors were CPI, department, whether or not the applicant has worked on any ML/DL/AI projects, VLSI/Signal processing, embedded computing, or image processing, and internship. The accuracy of 79% when using tested data from actual life shows the system's dependability for attaining its primary goal, function is to support educators at an institution's placement cell in finding potential students and give them proper coaching so they can succeed in recruitment processes held by various organisations. The approach aids in increasing an institution's placement rate and can therefore play a significant role in enhancing the institution's reputation. It is evident from the analysis that the system's implementation methodology is effective enough to significantly advance the sophisticated classification techniques now used in the placement sector.

References

1. Kohavi, R., Provost, F.: Glossary of term. Mach. Learn. **30**, 271–274 (1998)
2. Bishop, C.M.: Pattern Recognition and Machine Learning. Springer, ISBN0: 387-31073-8 (2006)
3. Rokach, L., Maimon, O.: Data mining with decision trees: theory and applications. World Scientific Pub Co Inc. ISBN 978-98127717711 (2008)
4. Pedregosa, F., Varoquax, G., Gramfort, A., Michel, V., Thrion, B., Grisel, O., Blondel, M., Prettenhofer, P., Weiss, R., Dubourg, V., Vanderplas, J., Passos, A., Cournaeau, D.: Scikit-learn: machine learning in Python. J. Mach. Learn. Res. **12**, 2825–2830 (2011)
5. Hijazi, S.T., Naqvi, R.S.M.M.: Factors affecting student's performance: a case of private colleges. Bangladesh e-J. Sociol. **3**(1) (2006)
6. Pal, A.K., Pal, S.: Analysis and mining of educational data for predicting the performance of students. (IJECCE) Int. J. Electron. Commun. Comput. Eng. **4**(5), 1560–1565, ISSN: 2278-4209 (2013)
7. Ramanathan, L., Swarnalathat, P., Gopal, G.D.: Mining educational data for students' placement prediction using sum of difference method. Int. J. Comput. Appl. **99**(18), 36–39 (2014)
8. Arora, R.K., Badal, D.: Placement prediction through data mining. Int. J. Adv. Res. Comput. Sci. Softw. Eng. **4**(7) (2014)

9. Elayidom, S., Idikkulaand, S.M., Alexander, J.: A generalized data mining framework for placement chance prediction problems. Int. J. Comput. Appl. **31**(3), 0975–8887 (2011)
10. Naik, N., Purohit, S.: Prediction of final result and placement of students using classification algorithm. Int. J. Comput. Appl. **56**(12), 0975–8887 (2012)

Chapter 26
Difference Equations, Stationary and Non-stationary Discrete Systems in Block Ciphers

Ruslan Skuratovskii and Aled Williams

Abstract In this article, for Markov ciphers, we prove that they are resistant to differential cryptanalysis and some statements made for MS are obtained. The upper estimates of the probabilities of integer differentials are significantly improved when compared to previously known results. Our differential cryptanalytic algorithm finds weak subkeys that have more than 80 bits and 128 bits for 128-bit keys.

26.1 Introduction

We obtain an analytical estimation of the upper boundary of the Feistel-like block ciphers differential probabilities, resistance characteristics of unbalanced Feistel circuits to differential and linear cryptanalysis. Further, a formalized description and method of study of non-Markov symmetric block ciphers resistance to differential cryptanalysis are developed. New schemes of cascade block encryption are investigated and, in this case, we develop a method which is used for evaluating the stability non-Markov ciphers. The estimates of R-block encryption schemes resistance to differential cryptanalysis are obtained. In addition, the crypto stability of the national standard of symmetric encryption GOST 28147: 2009 to fault attacks is considered. We both consider and compare different cryptanalysis methods [10] and discover that our algorithm can find weak subkeys that have more than 80 bits and 128 bits for 128-bit keys.

R. Skuratovskii (✉)
National Aviation University, Kiev, Ukraine
e-mail: ruslan.skuratovskii@nau.edu.ua; ruslan@imath.kiev.ua

Interregional Academy of Personnel Management, Kiev, Ukraine

V. I. Vernadsky National Taurida University, Kiev, Ukraine

A. Williams
Department of Mathematics, London School of Economics and Political Science, London, UK
e-mail: a.e.williams1@lse.ac.uk

The focus of the research is block ciphers [9, 13, 15] with a round function of the form

$$G_k(x) = L_m(S(x \oplus k)).$$

These ciphers are considered from the view of their belonging to the class of Markov or the generalized Markov: all mathematical calculations and proofs belong to Skuratovskii R., examples and programs necessary for calculations were developed by the authors jointly.

The subject of the research is the study of the above ciphers by the method of difference analysis, finding their properties, constructing estimates of the probabilities of integer differentials for round functions of the form that was mentioned earlier, processing and systematizing results.

The research methods are the construction of a model which is used to describe the concepts or statements that are being analyzed. In discrete systems, both input and output signals are discrete signals. The variables in the discrete systems vary with time. In this type of system, the changes are predominantly discontinuous. The state of variables in discrete system changes only at a discrete set of points in time. Note that by a discrete system, we mean a technical device or program that transforms a discrete sequence $x(n)$ into a discrete sequence $y(n)$ according to the determined algorithm. The algorithm for transforming the input sequence $x(n)$ to the output sequence $y(n)$ is described by the relation

$$R_y[y(n)] = R_x[x(n)],$$

where R_x and R_y are operators. Considering the type of operator, discrete systems can be divided into

- linear or nonlinear,
- stationary or non-stationary,
- physically realizable (causal) or unrealizable (non-causal).

Definition 1 A discrete system is called linear if and only if its operator R satisfies additivity and homogeneity properties, namely if:

1. $R[x_1(n) + x_2(n)] = R[x_1(n)] + R[x_2(n)]$ for any $x_1(n)$ and $x_2(n)$, and
2. $R[\alpha \cdot x(n)] = \alpha \cdot R[x(n)]$ for any α and $x(n)$.

These properties may be expressed as the single condition

$$R[\alpha \cdot x_1(n) + b \times x_2(n)] = \alpha \cdot R[x_1(n)] + \beta \cdot R[x_2(n)].$$

Note that the last condition implies the reaction of a linear system to a complex action is equal to the sum of reactions to individual actions taken with the same coefficients α and β.

Definition 2 A discrete system is called stationary (invariant in time) if its parameters do not change in time. In this case, the action applied to the input of the system will always lead to the same reaction, regardless of when the action is applied.

We introduce the notation $M = (\mu_0, \ldots, \mu_r)$, where $\mu_i : G \times G \to G$ with commutative group operations on the group G, $\mu_i(a, b) = a \circ_i b$ where $a, b \in G$ and $i = \overline{0, r}$.

The magnitude of the input differences ω_0 and ω_1 is the differences which appear in the first and second rounds, respectively.

Definition 3 The generalized differential characteristic (GDC) of cipher (26.1) is the sequence

$$(\Omega, M) = ((\omega_0, \mu_0), (\omega_1, \mu_1), \ldots, (\omega_{r+1}, \mu_{r+1})),$$

where $\omega_i \in G \backslash \{0_i\}$ and $i = \overline{1, r[2]}$.

26.2 Main Result

26.2.1 Introduction and Definitions

For the Data Encryption Standard (DES) algorithm, it is known [3, 19] that after finding 48 bits of the key of the last round, the remaining 8 bits are found via a complete search. The following condition is necessary for a successful application of an attack by the RK method:

$$\exists \Delta x, \Delta y \in V_m \forall K \in (V_n)^2 \quad \forall x \in V_m :$$
$$P\left(E_K^{(r-1)}(x \oplus \Delta x) \oplus E_K^{(r-1)}(x) = \Delta y\right) = p,$$

where $p \gg 2^{-m}$ and the probability is taken for $x \in V_m$.

To describe the essence of the RK method [7], we make use of the following notation. Let the encrypting key $E_k(x)$ be a function determined by the equality

$$E_k(x) = E(K, x), \tag{26.1}$$

where $x \in V_m$, $K \in (V_n)^r$, $K = (k_1, \ldots, k_r)$, $k_i \in V_n$, $E_K(x)$ is the r-th block ciper and for any $K \in (V_n)^r$ display E_K, where $V_m \to V_n$ is a bijection.

Denote by

$$f_k(x) = f(k, x), x \in V_m, k \in V_n \tag{26.2}$$

the round function of the cipher E. Then, using our notation, we have that

$$E_K(x) = f_{k_r} \circ f_{k_{r-1}} \circ \ldots \circ f_{k_1}(x) \tag{26.3}$$

holds, where $k_1, \ldots, k_r$ is a sequence of round keys. In addition, for $l = \overline{1, n}$, we let

$$E_K^{(l)}(x) = f_{k_l} \circ f_{k_{l-1}} \circ \ldots \circ f_{k_1}(x). \tag{26.4}$$

The definitions of the Markov cipher (MC) were given for the first time in [11].

Definition 4 *(from [11]).* The cipher is a Markov cipher if

$$P(\Delta Y = \beta \mid \Delta X = \alpha, X = Z) \tag{26.5}$$

does not depend on Z, provided that the subkeys are randomly distributed.

If the cipher is Markov, then ΔY then almost all of them are the same, they do not change with a change in the subkeys Z so the dependence cannot be established. This makes them resistant to Differential Cryptanalysis.

Using our notation, where for convenience, we set γ for Z and then the definition has the form

$$P(f_k(\gamma \oplus \alpha) \oplus f_k(\gamma) = \beta) = 2^{-n} \sum_{k \in V_m} \delta(f_k(\gamma \oplus \alpha) \oplus f_k(\gamma), \beta). \tag{26.6}$$

It is worth emphasizing that this probability does not depend on γ. If instead the probability did depend on γ, then the cipher is a non-Markov cipher. The γ in (26.6) can consequently be treated as an arbitrary element with V_m and, if $\gamma = 0$, then we obtain the expression

$$2^{-n} \sum_{k \in V_m} \delta(f_k(\alpha) \oplus f_k(0), \beta).$$

The definition of MC can in such case be rewritten as follows.

Definition 5 A block G cipher [9, 13] with round function $f_k : V_n \rightarrow V_m$ and $k \in V_n$ is MC if

$$\forall \alpha, \beta \in V_m : 2^{-n} \sum_{k \in V_m} \delta(f_k(\gamma \oplus \alpha) \oplus f_k(\gamma), \beta) = 2^{-n} \sum_{k \in V_m} \delta(f_k(\alpha) \oplus f_k(0), \beta).$$

Corollary 1 *If the cipher is Markovian, then ΔY then almost all of them are the same, they do not change with a change in the subkeys so the dependence cannot be established. This makes them resistant to Differential Cryptanalysis.*

Further, the possible differential attack on the 16-round DES requires 2^{58} chosen plaintexts [14, 16].

Table 26.1 Dependence of the complexity of hacking DES by the method of RK on the number of rounds

Number of Rounds	Complexity
1	2
4	2^4
6	2^8
8	2^{16}
9	2^{26}
10	2^{35}
11	2^{36}
12	2^{43}
13	2^{44}
14	2^{51}
15	2^{52}
16	2^{58}

26.2.2 Generalized Markov Ciphers and Their Properties

Suppose that some mapping $f : V_n \times G \to G$ is given such that for each $k \in V_n$, the mapping $f(k, x) := f_k(x)$ is a bijection on G. We associate the set M_x of matrices of dimension $|G| \times |G|$, where $x \in G$ with this mapping. The elements of the matrix M_x are $a^x_{\alpha,\beta} \in [0, 1]$, $\alpha, \beta \in G$, where $a^x_{\alpha,\beta} \in [0, 1] = d^f_{\mu_1,\mu_2}(x; \alpha, \beta)$. It is assumed that some linear order is fixed on the group G [16, 17]. If $G = V_m$, then the bit-vectors naturally correspond to the integers from 0 to $2^m - 1$. We denote by P the set of substitution matrices of dimension $|G| \times |G|$ (Table 26.1).

Definition 6 The mapping $f : V_n \times G \to G$ will be called a generalized Markov mapping, relative to operations μ_1, μ_2, if $\forall x, x' \in G$ and $\exists \pi, \pi' \in P$ such that

$$\pi_x \cdot M_x = \pi'_{x'} \cdot M_{x'}, \tag{26.7}$$

where multiplication is standard usual matrix multiplication and, in this case, is reduced to simply permutation of rows of the matrices M_x and M_y.

Definition 7 A block cipher E will be called a generalized Markovian cipher (GMC) in the restricted sense if their round functions $f_k(x) = f_k(k, x)$, with $x \in V_m$ and $k \in V_n$ are generalized Markov mappings (GMM) f_1 and f_2 with corresponding operations μ_0 and μ_1, respectively.

Lemma 1 (Property of GMC). *For a GMC f, the equation*

$$\forall \beta \in G \quad \max_{\substack{x,\alpha \in V_n \\ \alpha \neq 0}} d^f_{\mu_1,\mu_2}(x; \alpha, \beta) = \max_{\substack{\alpha \in V_n \\ \alpha \neq 0}} d^f_{\mu_1,\mu_2}(0; \alpha, \beta).$$

holds.

The proof of this Lemma follows directly from the definition of GMC. In particular, since the columns of M_x and M_0 and the number β differ only by some permutation of their elements, where the maximum element in the columns of M_x and M_0 is the same, as stated in the Lemma.

Remarks

1. If $G = V_m$, $\mu_1 = \mu_2 = XOR$, $\pi = \pi' = \mathrm{Id}$ and $\forall x, x' \in G$, then the definition coincides with the classical definition of Markov BC (see e.g., [3]).
2. Definition 5 is equivalent to stating $\forall i = \overline{1, r}$ and $\forall x \in G$ $\exists \sigma_{x, \mu_{i-1}}$, i.e., a permutation on G such that $\forall \alpha, \beta \in G$ we have

$$d^f_{\mu_{i-1}, \mu_i}(x; \alpha, \beta) = d^f_{\mu_{i-1}, \mu_i}\left(0_{i-1}; \sigma_{x, \mu_i}(\alpha), \beta\right). \tag{26.8}$$

In particular, if $\mu_{i-1} = \mu_i = \mu$, then

$$d^f(x; \alpha, \beta) = d^f(0; \sigma_x(\alpha), \beta). \tag{26.9}$$

The following Theorem demonstrates the performance of the GMC for some estimates like those previously obtained for the MC.

Theorem 1 *For any GMC (with respect to operations M), the following statements hold:*

1. $\forall x, \omega' \in G$,

$$\max_{\omega \in G} d^f_{\mu_{i-1}, \mu_i}(x; \omega, \omega') = \max_{\omega \in G} d^f_{\mu_{i-1}, \mu_i}(0; \omega', \omega'), \quad \forall i = \overline{1, r}. \tag{26.10}$$

2. $\forall x, \omega' \in G$,

$$
\begin{aligned}
\max_{\omega \in G} d^f_{\mu_{i-1}, \mu_i}(\omega, \omega') &\leq \max_{\omega \in G} d^f_{\mu_{i-1}, \mu_i}(x; \omega, \omega') \\
&= \max_{\omega \in G} d^f_{\mu_{i-1}, \mu_i}(0; \omega, \omega'), \forall i = \overline{1, r}.
\end{aligned}
\tag{26.11}
$$

3. $\forall x, \omega' \in G$,

$$EDP(\Omega, M) \leq \prod_{i=1}^{r} \max_{\omega \in G} d^f_{\mu_{i-1}, \mu_i}(0; \omega, \omega_i). \tag{26.12}$$

4. $\forall x, \omega' \in G$,

$$\max_{\Omega} EDP(\Omega, M) \leq \prod_{\substack{i=1}}^{} \max_{\substack{\omega, \omega' \in G \\ \omega' \neq 0_i}} d^f_{\mu_{i-1}, \mu_i}(0; \omega, \omega'). \tag{26.13}$$

Proof Firstly, (26.12) follows directly from the definition of the GMC and by applying Lemma 1 since

$$\max_{\omega \in G} d^f_{\mu_{i-1}, \mu_i} \left(x; \omega, \omega' \right) = \max_{\omega_0} d^f_{\mu_{i-1}, \mu_i} \left(0; \omega_0, \omega' \right),$$

where $\omega_0 = \sigma_{x, \mu_{i-1}}$ for $x \in G$ and $i = \overline{1, r}$.

Next, note that (26.13) follows from (26.12) since

$$\max_{\omega \in G} d^f_{\mu_{i-1}, \mu_i} \left(\omega, \omega' \right) = \max_{\omega \in G} \frac{1}{|G|} \sum_{x \in G} d^f_{\mu_{i-1}, \mu_i} \left(x; \omega, \omega' \right)$$

$$\leq \frac{1}{|G|} \sum_{x \in G} \max_{\omega \in G} d^f_{\mu_{i-1}, \mu_i} \left(x; \omega, \omega' \right) = \max_{\omega \in G} d^f_{\mu_{i-1}, \mu_i} \left(0; \omega, \omega' \right)$$

holds. Finally, (26.12) follows from Lemma 1 and then (26.13) is an immediate consequence of (26.12), which concludes the proof.

26.2.3   Non-Markov Ciphers and Examples

Two examples of non-Markov ciphers are the old GOST 211428 and the new Belarusian BelT 34.101.31-2007. It is interesting to evaluate their resistance to Differential Cryptanalysis.

At the current time, the general theory of evaluating the practical stability of Markov ciphers with respect to difference (or linear) cryptanalysis has been developed. It is worth noting that some of the fundamental works in this direction are [3, 6, 11, 13, 19]. As a rule, when constructing estimates, several consequences of formula (26.17) are used, namely

$$\max_{\Omega} \text{EDP}(\Omega) \leq \max_{\Omega_1} \text{EDP}\,(\Omega_1) \max_{\Omega_2} \text{EDP}\,(\Omega_2), \tag{26.14}$$

where $\Omega = (\Omega_1, \Omega_2)$ and

$$\max_{\Omega} \text{EDP}(\Omega) \leq \left(\max_{\omega_1, \omega_2 \neq 0} d^f\,(\omega_1, \omega_2) \right)^r. \tag{26.15}$$

Similarly, we have

$$\max_{\Omega} \text{EDP}(\Omega) \leq \max_{\Omega} p_s^{\#\Omega}, \tag{26.16}$$

where $\#\Omega$ is the minimum number of possible active S-boxes in Ω, $p_s = \max_{s \in S} \max_{\omega_1, \omega_2} d^s\,(\omega_1, \omega_2)$ and where S is the set of S-blocks of the cipher [8, 14, 16] (if its round function is a composition of linear transformations and a block of substitutions). As for non-Markov BC [1, 8, 12], the property of Theorem 1 for them does not hold, which makes it difficult to obtain estimates of the form (26.14)–(26.16) by analogous methods. Instead, when constructing analogs of these estimates, it is necessary to consider the dependence in (26.8) on x.

Theorem 2 *(about the estimate for non-Markov block ciphers). For the value* $EDP(\Omega, M)$ *the following inequalities hold:*

$$EDP(\Omega, M) \leq \prod_{i=1}^{r} \max_{x \in G} d^{f}_{\mu_{i-1}, \mu_i} (x; \omega_{i-1}, \omega_i), \qquad (26.17)$$

$$\max_{\Omega, M} EDP(\Omega, M) \leq \prod_{i=1}^{r} \max_{x \in G} \max_{\substack{\omega_{i-1}, \omega_i \\ \omega_{i-1} \neq 0_{i-1} \\ \omega_i \neq 0_i}} d^{f}_{\mu_{i-1}, \mu_i} (x; \omega_{i-1}, \omega_i). \qquad (26.18)$$

Proof For simplicity, we instead prove (26.17) for the two round characteristic (Ω, M)
$= ((\omega_0, \mu_0), (\omega_1, \mu_1), (\omega_2, \mu_2))$ and then deduce (26.18) as a direct consequence to (26.17). Note that

$$\text{EDP}(\Omega, M) = \frac{1}{|G|} \sum_{x_0 \in G} d^{f}_{\mu_1, \mu_2} (x_0; \omega_1, \omega_2) d^{f}_{\mu_2, \mu_3} (x_1; \omega_2, \omega_3),$$

where $x_1 = f_{k_1}(x_0)$, $k_1 \in V_n$ is the key of the first round, then

$$\text{EDP}(\Omega, M) \leq \frac{1}{|G|} \sum_{x_0 \in G} \max_{x \in G} d^{f}_{\mu_1, \mu_2} (x; \omega_1, \omega_2) \cdot \max_{x \in G} d^{f}_{\mu_2, \mu_3} (x; \omega_2, \omega_3)$$

$$= \max_{x \in G} d^{f}_{\mu_1, \mu_2} (x; \omega_1, \omega_2) \cdot \max_{x \in G} d^{f}_{\mu_2, \mu_3} (x; \omega_2, \omega_3)$$

holds, which concludes the proof.

It is worth noting that the presence of an additional parameter $x \in G$ from (26.17) and (26.18) significantly complicates the construction of numerous estimates and, at the same time, makes the estimates obtained rougher, which in some cases may become trivial. Because of this one cannot generally use this approach in practice.

26.3 Construction of Upper Estimates for the Probabilities of the Integer Differential of the Round Functions Module 2

26.3.1 Conventions and Approval

Firstly, let us introduce some notation. For any $n \in N$, let $V_n = \{0, 1\}^n$ an array of n-dimensional vectors. If $n = pu$ with $p \geq 2$, then $\forall x \in V_n$ we can represent such an x as

$$x = \left(x^{(p)}, \ldots, x^{(1)} \right),$$

where $x^{(i)} \in V_n$ and $i = \overline{1, p}$.

Denote by $L_m : V_n \to V_n$ the mapping which produces a left shift by m-bits of the vector V_n. On the set V_n, we define the subsets

$$\Gamma_m(\gamma) = \{\beta \in V_n : \exists k \in V_n : L_m(k \oplus \gamma) - L_m(k) = \beta\} \text{ and}$$

$$\Gamma_m^{-1}(\beta) = \{\gamma \in V_n \mid \exists k \in V_n : L_m(k \oplus \gamma) - L_m(k) = \beta\},$$

then a bijective mapping $S : V_n \to V_n$ is defined

$$\forall x \in V_n : S(x) = \left(S^{(p)}\left(x^{(p)} \right), \ldots, S^{(1)}\left(x^{(1)} \right) \right), x^{(i)} \in V_u, i = \overline{1, p}.$$

We also denote

$$\tilde{x} = \left(x^{(p)}, \ldots, x^{(2)} \right) \in V_{n-u}; \tilde{S} : V_{n-u} \to V_{n-u},$$

where $S(x) = \left(S^{(p)}(x^{(p)}), \ldots, S^{(2)}(x^{(2)}) \right)$. Further, we introduce

$$\tau(k, \alpha) = \begin{cases} 0, & \text{if } S^{(1)}(k \oplus \alpha) \geq S^{(1)}(k) \\ 1, & \text{else} \end{cases}$$

Let

$$\Delta_{\oplus+}^{(1)} = \max_{\alpha, \gamma \in V_n \setminus \{0\}} \max \{I_1, I_2\} =$$

$$\max_{\alpha, \gamma \in V_n \setminus \{0\}} \max \left\{ 2^{-u} \sum_{k^{(1)} \in V_u : \tau\left(k^{(1)}, \alpha^{(1)}\right) = 0} \delta\left(S^{(1)}(k^{(1)} \oplus \alpha^{(1)})\right) - S^{(1)}(k^{(1)}), \right.$$

$$\left. 2^{-u} \sum_{k^{(1)} \in V_u : \tau\left(k^{(1)}, \alpha^{(1)}\right) = 1} \delta\left(S^{(1)}(k^{(1)} \oplus \alpha^{(1)})\right) - S^{(1)}(k^{(1)}, \gamma_j^{(1)}) \right\}$$

and, in addition, for any $\beta \in V_n$ with $\beta = q \cdot 2^m + r, 0 \leq q < 2^t - 1$ and $0 \leq r < 2^m - 1$ we introduce the following notation to work with elements of the set $\Gamma_m^{-1}(\beta)$:

$$\gamma_1 = \gamma_1(\beta) = \beta \cdot 2^t + q, \gamma_2 = \gamma_2(\beta) = \gamma_1 + 1 \text{ and}$$
$$\gamma_3 = \gamma_3(\beta) = \gamma_1 - 2^t, \gamma_4 = \gamma_4(\beta) = \gamma_1 - 2^t + 1$$

$\forall j = \overline{1, p}$ and assuming that

$$d_{\oplus+}^{S^{(j)}} = \max_{\alpha, \beta \in V_n \setminus \{0\}} 2^{-u} \sum \delta\left(S^{(j)}(k \oplus \alpha) - S^{(j)}(k), \beta\right)$$

then $\Delta_{\oplus+} = \max_{i=1,p} d_{\oplus+}^{S^{(j)}}$ holds.

Finally, we will use round functions, which are the composition of a key adder, a substitution block and a shift operator with form

$$G_k(x) = L_m(S(x \oplus k)). \tag{26.19}$$

26.3.2 Berson's Result

When obtaining further results, we will use the main result from [4], which we reformulate here using our notation. Hence, upon making use of our notation, the following Theorem holds.

Theorem 3 *For any* $m \in N$, $\gamma \in V_n$, $\gamma = q \cdot 2^t + r$, *with* $0 \le r < 2^t - 1$, *we have*

$$\Gamma_m(\gamma) \subset \{\beta, \beta + 1, \beta - 2^m, \beta - 2^m + 1\},$$

with $\beta = q + r \cdot 2^m$, *where all operations are performed* mod 2^n.

26.3.3 Construction of Upper Bounds for the Probabilities of Integer Differentials of Round Functions Containing an Adder Module 2

Theorem 4 *Let* $t \ge u$ *and* $p \ge 2$*If the round function has the form* (26.19) *[5], then the inequality*

$$\forall \alpha, \beta \in V_n \setminus \{0\} : d_{\oplus+}^G(\alpha, \beta) \le \max \left\{ 2\Delta_{\oplus+}, 4\Delta_{\oplus+}^{(1)} \right\}$$

holds.

Proof Average probabilities of integer round differentials for functions of the form (26.19) have the form

$$d_+^G(x; \alpha; \beta) = 2^{-n} \sum_{k \in V_n} \delta \Big(L_m(S((x + \alpha) \oplus k)) - L_m(S(x \oplus k)), \beta \Big). \tag{26.20}$$

It is the mean (behind the keys) probability of the differential of the mapping at the point x

$$d_+^G(\alpha; \beta) = 2^{-2n} \sum_{x,k \in V_n} \delta \Big(L_m(S((x + \alpha) \oplus k)) - L_m(S(x \oplus k)), \beta \Big). \tag{26.21}$$

Let $\mu(x; \alpha) = (x + \alpha) \oplus x \oplus \alpha$, then

$$d_+^G(x; \alpha; \beta) = 2^{-n} \sum_{k \in V_n} \delta\Big(L_m(S((x + \alpha) \oplus k)) - L_m(S(x \oplus k)), \beta\Big)$$

$$= 2^{-n} \sum_{k \in V_n} \delta\Big(L_m(S(x \oplus \alpha \oplus k \oplus \mu(x; \alpha)))\Big).$$

Let us introduce further notation to simplify, namely $x \oplus k = k'$, $\alpha \oplus \mu(x; \alpha) = \alpha' = \alpha'(x; \alpha)$ and $k = k'$. Then our expression becomes

$$d_+^G(x; \alpha; \beta) = 2^{-n} \sum_{k \in V_n} \delta\Big(L_m\big(S\left(\alpha' \oplus k\right)\big) - L_m(S(k), \beta)\Big) = d_{\oplus+}^G(0; \alpha; \beta).$$

We deduce that

$$\max_{\alpha \in V_n \setminus \{0\}} d_+^G(x; \alpha; \beta) = \max_{\alpha' \in V_n \setminus \{0\}} d_{\oplus+}^G(0; \alpha; \beta),$$

which can be simplified to

$$d_{\oplus+}^G(0; \alpha; \beta)$$

$$= 2^{-n} \sum_{k \in V_n} \sum_{\gamma \in V_n} \delta\Big(L_m(S(k) + \gamma) - L_m(S(k), \beta)\Big) \times \delta\Big(S(k \oplus \beta) - S(k), \gamma\Big)$$

$$\leq 2^{-n} \sum_{k \in V_n} \sum_{\gamma \in \Gamma^{-1}(\beta)} \delta\Big(S(k \oplus \alpha') - S(k), \gamma\Big)$$

$$= 2^{-n} \sum_{k \in V_n} \sum_{j=1}^{4} \delta\Big(S(k \oplus \alpha) - S(k), \gamma_i\Big) = \sum_{\gamma \in \Gamma_\beta^{-1}} d_{\oplus+}^G.$$

There are now two cases, namely $\alpha^{(1)} \neq 0$ and $\alpha^{(1)} = 0$.

If $\alpha^{(1)} \neq 0$ and because $p \geq 2$, then for any $j = \overline{1, 4}$ the differential $d_{\oplus+}^S$ $(0; \alpha; \gamma_j(\beta))$ can be represented as

$$d_{\oplus+}^S\left(0; \alpha; \gamma_j(\beta)\right) = 2^{-n} \sum_{k^{(1)} \in V_u} \delta\Big(S^{(1)}(k^{(1)} \oplus \alpha^{(1)}) - S^{(1)}(k^{(1)}), \gamma_j^{(1)}\Big)$$

$$\times \sum_{\tilde{k} \in V_{n-u}} \delta\Big(\tilde{S}(\tilde{k} \oplus \tilde{\alpha}) - \tilde{S}(\tilde{k}) - \tau(k^{(1)}, \alpha^{(1)}), \tilde{\gamma}_j\Big) \qquad (26.22)$$

We now represent (26.22) in a more convenient form, namely

$$2^{-(n-u)} \cdot 2^{-u} \sum_{k^{(1)} \in V_u} \delta\left(S^{(1)}(k^{(1)} \oplus \alpha^{(1)}) - S^{(1)}(k^{(1)}), \gamma_j^{(1)}\right)$$

$$\times \sum_{\tilde{k} \in V_{n-u}} \delta\left(\tilde{S}(\tilde{k} \oplus \tilde{\alpha}) - \tilde{S}(\tilde{k}) - \tau(k^{(1)}, \alpha^{(1)}), \tilde{\gamma}_j^{(1)}\right)$$

$$= 2^{-(n-u)} \sum_{\tilde{k} \in V_{n-u}} 2^{-u} \sum_{k^{(1)} \in V_u : \tau(k^{(1)}, \alpha^{(1)}) = 0} \delta\left(S^{(1)}(k^{(1)} \oplus \alpha^{(1)})\right.$$

$$\left. - S^{(1)}(k^{(1)}), \gamma_j^{(1)}\right) \times \delta\left(\tilde{S}(\tilde{k} \oplus \tilde{\alpha}) - \tilde{S}(\tilde{k}), \gamma_j^{(1)}\right).$$

The following transformation yields

$$2^{-u} \sum_{k^{(1)} \in V_u : \tau(k^{(1)}, \alpha^{(1)}) = 0} \delta\left(S^{(1)}(k^{(1)} \oplus \alpha^{(1)}) - S^{(1)}(k^{(1)}), \gamma_i^{(1)}\right)$$

$$\times 2^{-(n-u)} \sum_{\tilde{k} \in V_{n-u}} \delta\left(\tilde{S}(\tilde{k} \oplus \tilde{\alpha}) - \tilde{S}(\tilde{k}), \tilde{\gamma}_i\right)$$

$$+ 2^{-u} \sum_{k^{(1)} \in V_u : \tau(k^{(1)}, \alpha^{(1)}) = 1} \delta\left(S^{(1)}(k^{(1)} \oplus \alpha^{(1)}) - S^{(1)}(k^{(1)}), \gamma_i^{(1)}\right)$$

$$\times 2^{-(n-u)} \sum_{\tilde{k} \in V_{n-u}} \delta\left(\tilde{S}(\tilde{k} \oplus \tilde{\alpha}) - \tilde{S}(\tilde{k}) - 1, \tilde{\gamma}_i\right)$$

then by transforming (26.22) we obtain

$$d_{\oplus+}^S\left(0; \alpha; \gamma_j\right) \le \max\left(I_1, I_2\right)$$

$$\times 2^{-(n-u)} \sum_{\tilde{k} \in V_{n-u}} \left\{\delta\left(\tilde{S}(\tilde{k} \oplus \tilde{\alpha}) - \tilde{S}(\tilde{k}), \tilde{\gamma}_i\right)\right.$$

$$\left. + \delta\left(\tilde{S}(\tilde{k} \oplus \tilde{\alpha}) - \tilde{S}(\tilde{k}) - 1, \tilde{\gamma}_i\right)\right\}$$

We emphasize that $\forall \tilde{k} \in V_{n-u}$, if $\delta(S(k \oplus \alpha) - S(k), \gamma_i) = 1$, then $\delta(S(k \oplus \alpha) - S(k) - 1, \gamma_i) = 0$ and vice versa. In consequence, we deduce that

$$\delta\left(S(k \oplus \alpha) - S(k), \gamma_i\right) + \delta\left(S(k \oplus \alpha) - S(k) - 1, \gamma_i\right) \le 1,$$

so further

$$d_{\oplus+}^S\left(0; \alpha; \gamma_j\right) \le \max\left(I_1, I_2\right) \cdot 2^{-(n-u)} \sum_{k \in V_{n=\mu}} 1 = \max\left(I_1, I_2\right) \le \Delta^{(1)}.$$

In other words, for $\alpha^{(1)} \neq 0$, we have $d_{\oplus+}^{(S)}\left(0; \alpha; \gamma_j\right) \le \Delta^{(1)}$ and therefore $d_{\oplus+}^G(\alpha, \beta) \le 4\Delta_{\oplus+}^{(1)}$ holds.

Consider now when $\alpha^{(1)} = 0$. In such case, it is clear that

$$d_{\oplus+}^{S}(\alpha, \gamma) \leq d_{\oplus+}^{S^{(1)}}(\alpha^{(1)}, \gamma^{(1)})$$

holds and, if the condition $\gamma^{(1)} \neq 0$ is also met, the condition $d_{\oplus+}^{S^{(1)}}(\alpha^{(1)}, \gamma^{(1)}) = 0$ follows as required. In this case, the condition $d_{\oplus+}^{S}(\alpha, \gamma) \neq 0$ is met only if $\gamma^{(1)} = 0$ and therefore, in our case, we have $d_{\oplus+}^{S}(\alpha, \gamma) \neq 0$ which yields $\gamma^{(1)} = 0$ and so

$$d_{\oplus+}^{G}(\alpha, \beta) = \sum_{\substack{\gamma \in \Gamma_m^{-1}(\beta): \\ \gamma^{(1)}=0}} d_{\oplus+}^{S}(0; \alpha, \gamma)$$

and hence $\Gamma_m^{-1}(\beta) = \{\gamma, \gamma + 1, \gamma - 2^t, \gamma - 2^t + 1\} = \{\gamma_1, \gamma_2, \gamma_3, \gamma_4\}$ with $\gamma = q + \beta \cdot 2^t$. Note that the set $\{\gamma \in \Gamma_m^{-1}(\beta) : \gamma^{(1)} = 0\}$ contains no greater than two elements, namely either $\gamma^{(1)}$ and $\gamma^{(3)}$ or $\gamma^{(2)}$ and $\gamma^{(4)}$. Therefore,

$$d_{\oplus+}^{G}(\alpha, \beta) \leq 2 \max_{\alpha, \gamma \in V_m \{\{0\}}} d_{\oplus+}^{S}(0; \alpha, \gamma) \leq 2 \max_{i=2, p} d_{\oplus+}^{S^{(i)}}(0; \alpha, \gamma)$$

$$\leq 2 \max_{i=2, p} \left\{ \max_{\alpha, \gamma \in V_m \setminus \{0\}} d_{\oplus+}^{S^{(i)}}(0; \alpha, \gamma) \right\} = 2 \max_{i=2, p} d_{\oplus+}^{S^{(i)}} = 2\Delta_{\oplus+}$$

holds. Since for arbitrary values of the index $i = \overline{2, p}$ and $\alpha, \gamma \in V_n$ we have that $d_{\oplus+}^{S}(0; \alpha, \gamma) \leq d_{\oplus+}^{S^{(i)}}(0; \alpha^{(i)}, \gamma^{(i)})$ holds and hence $d_{\oplus+}^{S} \leq d_{\oplus+}^{S^{(i)}}$, which concludes the proof.

Theorem 5 *For any UMC, each of the following statements are true:*

1. $\forall \alpha, \beta \in G$,

$$\max_{x \in G} d^f(x; \alpha, \beta) \leq \max_{x' \in G'} d^\phi\left(0; \alpha'(\alpha, \beta), \beta'(\alpha, \beta)\right). \tag{26.23}$$

2. $\forall \alpha, \beta \in G$,

$$\max_{x \in G} d^f(x; \alpha, \beta) \leq \max_{\alpha' \in G'} d^\phi\left(0; \alpha', \beta'(\alpha, \beta)\right) = \max_{\alpha' \in G'} d^\phi\left(\alpha', \beta'(\alpha, \beta)\right). \tag{26.24}$$

3.

$$EDP(\Omega) \leq \prod_{i=1}^{r} \psi\left(\omega_{i-1}, \omega_i\right) \max_{\alpha' \in G'} d^\phi\left(0; \alpha', \beta'\left(\omega_{i-1}, \omega_i\right)\right)$$

$$= \prod_{i=1}^{r} \psi\left(\omega_{i-1}, \omega_i\right) \max_{\alpha' \in G'} d^\phi\left(\alpha', \beta'\left(\omega_{i-1}, \omega_i\right)\right) \tag{26.25}$$

$$\leq \prod_{i=1}^{r} \max_{\alpha' \in G'} d^\phi\left(\alpha', \beta'\left(\omega_{i-1}, \omega_i\right)\right).$$

4. *If* $EDP(\Omega) \neq 0$, $\alpha'(\omega_{i-1}, \omega_i) \in U$, $\beta'(\omega_{i-1}, \omega_i) \in V$, $i \in I$ *and some* $I \subset \{1, \ldots, r\}$ *with* $U, V \subset G'$, *then*

$$\max_{\Omega} EDP(\Omega) \leq \prod_{i \in I} \max_{\alpha' \in U, \beta' \in V} d^{\phi}\left(0; \alpha', \beta'\right) = \prod_{i \notin I} \max_{\alpha', \beta' \in G'} d^{\phi}\left(0; \alpha', \beta'\right).$$

(26.26)

The proof of the Theorem is carried out similarly to Theorem 1 by simply using the definition of UMC.

Remark the last statement of the theorem can be generalized to the case of several subsets $\{1, \ldots, k\}$ and several G'.

26.3.4 Examples of GMC in a Broad Sense

The national standard for block ciphering GOST 28147-89 is UMC in a broad sense is related to the bitwise addition operation. Indeed, in this case with $G = V_{64}$ and $G' = V_{32}$,

$$d^f(x; \alpha, \beta) = \psi(\alpha, \beta) d^{\phi}\left(x'; \alpha', \beta'\right)$$

where $\psi(\alpha, \beta) = \delta(\alpha_2, \beta_1)$, $x'(x) = x_2$, $\alpha'(\alpha, \beta) = \alpha_2$, $\beta'(\alpha, \beta) = \alpha_1 \oplus \beta_2$, $x = (x_1, x_2)$, $\alpha = (\alpha_1, \alpha_2)$, $\beta = (\beta_1, \beta_2)$ and ϕ_k is a round transformation, which is a generalized Markov mapping (see e.g., [2, 4, 16, 18]). We check the possibility of a differential attack on AES and show the dependence of number of differetials in Table 26.2.

This means we get better method even in [10–13].

Table 26.2 Dependence of the number of zero differentials depending on the round number

Number of Rounds	Number of Differentials
2	56180
3	12780
4	880
5	0
6	0
7	0
8	0
9	0

26.3.5   Conclusion to Sect. 26.3

In this section, an upper estimate of the probability of integer differential of round functions has been found. The proof is based on Berson's result. An estimation is found which is important because it improves an existing result and describes all foreseeable cases. This result can be implemented for analysis of crypto stability of block cipher in relation to round crypto analysis.

References

1. Avanzi, R.: The qarma block cipher family. Almost mds matrices over rings with zero divisors, nearly symmetric even-mansour constructions with non-involutory central rounds, and search heuristics for low-latency s-boxes. In: IACR Transactions on Symmetric Cryptology, pp. 4–44 (2017)
2. Berson, T.A.: Differential cryptanalysis mod $2^3 2$ with applications to md5. In: Workshop on the Theory and Application of of Cryptographic Techniques, pp. 71–80. Springer, Heidelberg (1992)
3. Biham, E., Shamir, A.: Differential fault analysis of secret key cryptosystems. In: Annual International Cryptology Conference, pp. 513–525. Springer, Heidelberg (1997)
4. Bogdanov, A., Boura, C., Rijmen, V., Wang, M., Wen, L., Zhao, J.: Key difference invariant bias in block ciphers. In: International Conference on the Theory and Application of Cryptology and Information Security, pp. 357–376. Springer, Heidelberg (2013)
5. Daemen, J., Rijmen, V.: Statistics of correlation and differentials in block ciphers. IACR ePrint archive **212**, 2005 (2005)
6. Giraud, C.: Dfa on aes. In: International Conference on Advanced Encryption Standard, pp. 27–41. Springer, Heidelberg (2004)
7. Gnatyuk, V.A.: Mechanism of laser damage of transparent semiconductors. Physica B: Condensed Matter **308–310**, 935–938 (2001)
8. Iatsyshyn, A.V., Kovach, V.O., Romanenko, Y.O., Iatsyshyn, A.V.: Cloud services application ways for preparation of future phd. In: Proceedings of the 6th Workshop on Cloud Technologies in Education, vol. 2433. Arnold E. Kiv, Vladimir N. Soloviev (2018)
9. Lai, X., Massey, J.L., Murphy, S.: Markov ciphers and differential cryptanalysis. In: Workshop on the Theory and Application of of Cryptographic Techniques, pp. 17–38. Springer, Heidelberg (1991)
10. Langford, S.K., Hellman, M.E.: Differential-linear cryptanalysis. In: Annual International Cryptology Conference, pp. 17–25. Springer, Heidelberg (1994)
11. NIST Fips Pub: 197: Advanced Encryption Standard (AES). Federal information processing standards publication **197**(441), 0311 (2001)
12. Romanenko, Y.O.: Place and role of communication in public policy. Actual Probl. Econ. **176**(2), 25–31 (2016)
13. Skuratovskii, R.: An application of metacyclic and miller-moreno p-groups to generalization of diffie-hellman protocol. In: Proceedings of the Future Technologies Conference, pp. 869–876. Springer, Heidelberg (2020)
14. Skuratovskii, R., Osadchyy, V., Osadchyy, Y.: The timer inremental compression of data and information. WSEAS Trans. Math. **19**, 398–406 (2020)
15. Skuratovskii, R., Osadchyy, Y., Osadchyy, V.: The timer compression of data and information. In: 2020 IEEE Third International Conference on Data Stream Mining & Processing (DSMP), pp. 455–459. IEEE (2020)
16. Skuratovskii, R.V.: A method for fast timer coding of texts. Cybern. Syst. Analy. **49**(1), 133–138 (2013)

17. Skuratovskii, R.V., Williams, A.: Irreducible bases and subgroups of a wreath product in applying to diffeomorphism groups acting on the möbius band. Rendiconti del Circolo Matematico di Palermo Series 2, **70**(2), 721–739 (2021)
18. Skuratovskii, R.V.: Employment of minimal generating sets and structure of sylow 2-subgroups alternating groups in block ciphers. In: Advances in Computer Communication and Computational Sciences, pp. 351–364. Springer, Singapore (2019)
19. Tunstall, M., Mukhopadhyay, D., Ali, S.: Differential fault analysis of the advanced encryption standard using a single fault. In: IFIP International Workshop on Information Security Theory and Practices, pp. 224–233. Springer, Heidelberg (2011)

Chapter 27
Promotion of Sustainable Entrepreneurship Through Executive MBA Education

Pradnya Vishwas Chitrao, Pravin Kumar Bhoyar, and Rajiv Divekar

Abstract An MBA degree in India equips one with good management skills. Today, preparing executives to be business leaders is important, and so, MBA programs offer this leadership training. Executive MBA education helps working professionals progress in the organizational hierarchy of their company (Taj, Firms introduce unique executive education programmes. Economic Times (online) (2011) 14). Today, Executive MBA programs help working professionals start their own enterprises based on the knowledge gained during the course. The Symbiosis Institute of Management Studies' (SIMS) Executive MBA has inspired professionals to start their own ventures and successfully sustain them during the COVID-19 lockdowns. The research studies how Executive MBA can be used for entrepreneurial inputs.

Usually, a Master's course is expected to advance specific knowledge and skill sets [9]. The MBA program should ideally develop skills and knowledge that create leadership potential in postgraduates [13]. Management education needs to be revamped in order to improve socioeconomic condition of any given society and to ensure sustainability especially post COVID-19 business losses. It needs to impart knowledge and build competencies for creating and running sustainable businesses. It must inculcate responsibility in the executives who are prepared for the corporate world. Management education must now adopt a paradigm shift in terms of its research system, its knowledge dissemination system, as also its knowledge utilization system in terms of its learning and consultancy and industry projects. Only then it can result in instilling responsibility in the executives who can think of ways of establishing and running sustainable businesses.

P. V. Chitrao (✉) · P. K. Bhoyar · R. Divekar
Symbiosis Institute of Management Studies (SIMS), Khadki, Pune, India
e-mail: pradnyac@sims.edu

P. V. Chitrao
A Constituent of Symbiosis International Deemed University, Pune, India

The United Nations came up with the initiative of Principles for Responsible Management Education (PRME) for preparing professionals for conducting responsible and sustainable businesses besides running profit making enterprises. The initiative is a global call to incorporate in business education the international values that are embedded in the Global Compact framework on human rights, labor, anti-corruption, and the environment. It seeks to upgrade business education on account of changing ideas about corporate citizenship, corporate social responsibility, and sustainability. Its purpose is to provide the framework required to mold management education to the new after crisis realities—in terms of curriculum, research, and learning methodologies [12].

The educational system in India has rigid boundaries that compel students to take up traditional employment rather than dare to do what they like. A Gallup study by two of its senior executives, Daniela Yu and Yamii Arora [3], discovered that only about 22% of people aspiring for entrepreneurship have access to proper training. Another study found that more than 80% of students in developed countries learn a skill before the age of 14, whereas in India, only 4% learn a skill. This makes us realize the need for education that will lay the foundation for entrepreneurial activity. Entrepreneurship does play a critical role in the development of any given society. Education is one of the most important tools for developing entrepreneurship and an entrepreneurial culture.

Management education influences greatly the way in which future generations learn business practices [2]. The Master of Business Administration (MBA) program is probably the world's best known and most well recognized post graduate program. The MBA is a professional program, people working in all aspects of business. Now offered globally, the MBA program usually offers general and specialist curriculum in core areas such as accounting and finance, economics, organizational behavior, marketing, general and strategic management, and human resource management. The training provided in the MBA program is based on a strong theoretical basis of what is effective management along with an understanding of the internal operations of a firm and its interface with society and environment.

MBA courses have been found to develop the candidates' creativity and innovativeness through courses like design thinking [10].

Entrepreneurs, and the new businesses they set up, are critical for the development and well-being of societies. Today, increasing acknowledgment is being made of the role played by new and small businesses in an economy (Global Entrepreneurship Monitor (GEM) 2017).

Research Methodology

The paper will seek to establish through two main case studies and a few caselets of alumni of Symbiosis Institute of Management Studies' (SIMS) Executive MBA Program of how management education especially for working professionals can prove useful for setting up a sustainable enterprise that survived the hardships of COVID-19's lockdown, and its losses especially in the manufacturing sector. It will also try to find out whether it develops the innate intrapreneurial skillsets of the candidates. The first case is that of AFY Technologies set up by SIMS' two alumni of the executive MBA program. The second case is that of Sarvadnya Electro Tech founded by Mr. Jayesh Patil. In both cases, the entrepreneurs give credit for their success to the management lessons that they learned during their Executive MBA course that they did at Symbiosis Institute of Management Studies (SIMS), a Constituent of Symbiosis International Deemed University. So, the primary research is primarily qualitative in nature through a structured interview of the entrepreneurs.

The paper will be backed by secondary research.

Relevance of the Study

The paper will help entrepreneurs understand the importance of MBA education for setting up and growing a business even during difficult times like COVID-19. It will motivate working professionals wishing to start their own business to enroll for Executive MBA programs and thereby simultaneously learn management and entrepreneurial tenets from the experts. It will also encourage business professionals to develop their intrapreneurial skillsets so that they will contribute meaningfully to the organization for which they are working.

27.1   Introduction

Sustainability is becoming very important in the business sector all over the world. In India, there are very few opportunities for building the capacity for sustainability in the business sector. Also, there is a need to grow businesses in the face of a growing population and increasing unemployment. In fact, entrepreneurship is responsible for the growth and vitality of any given economy. At the same time, most of the new ventures do not last especially during tough times like the COVID-19 pandemic. This is because new ventures are vulnerable and require care and nurture in the first few years of their existence. In normal circumstances also, starting a new business requires working for long hours, coping with a lot of stress, and being prepared for a low chance of success. Funding is another major issue though today there are many funding options available that can partly or greatly finance your enterprise. Having one's own business, however, helps one to be in control of one's own life on both professional and personal levels.

Start-ups and SMEs require vital business skills and expertise in order to execute and implement the vision of the business in a logistic and economical manner. To attain these skillsets, the Executive MBA is a safe platform for bringing in new professional perspectives and putting theoretical concepts into practice.

27.2   Literature Review

A Master's program tries basically to impart specialized knowledge and skillsets [9]. MBA courses were first introduced in the US in the early 20th C to meet industry demand [7].

The Harvard MBA became a success because of the active participation of many industries, and it was later offered by other American universities. Most of the participants were mature professionals who were successful at their work. So, their commitment was the key to the MBA's success. US universities wanted to ensure the receipt of 'quality intakes' for MBA, and so introduced GMAT, namely the Graduate Management Admission Test [1]. Today's organizations are constantly in a state of flux. Business leaders today need to reflect on these changes and handle them appropriately. It is especially important because of the influence of organizations in society today [6]. Also, when businesses become complex and global in size, managers need to hone their leadership and managerial skills as also develop their decision-making ability and strategic vision [16]. This is the reason why students in good MBA programs are given the resources to support their big dreams and are exhorted to do big things and are at the same time permitted to fail [8]. Such Executive MBA programs focus generally on building leadership, team work, a global outlook, stress management, social strategy, and networking.

However, many universities are not equipped to impart these transformation skills. In fact, most MBA programs emphasize traditional, disciplinary knowledge like marketing, accounting/finance, human resources, operations management, and information systems, even when complex transformations call for cross-disciplinary skills. It is difficult to find such integrated, cross-disciplinary courses equipping the students to handle competently change even at many of the renowned business [15].

Entrepreneurship today is essential for the construction and growth of any economy. Entrepreneurship is regarded as one of the main reasons for the fourth Industrial Revolution and the basis for the partnership between the World Economic Forum and the Global Entrepreneurship Monitor [17]. In fact, the relationship between entrepreneurial activities and economic growth has been established long ago. An entrepreneur finds opportunities and develops resources to convert them into a sustainable business [4]. Since entrepreneurship has a lot of economic, social, and educational benefits, it has resulted in the growth of business education programs at colleges and universities across the globe [5].

The dynamic global business environment calls upon employees to be innovative and entrepreneurial. Consequently, there is a strong interest in the emerging entrepreneurial mindset [11]. Government in its search for alternative and innovative approaches is motivating the youth to become entrepreneurs in order to create job opportunities for themselves and their peers. Many B-Schools as a result are offering separate entrepreneurship courses. These institutes are an appropriate vehicle for developing management graduates into entrepreneurs and intrapreneurs who have a lot of integrity, ethical standards, a strong sense of social responsibility, who are committed to the progress of their organizations, and who are very clear about their role as drivers of positive change in the way business is done.

27.3 Research Objective

The paper will try to find out whether MBA education and in particular Executive MBA encourage and guide people in starting and running their own enterprises. It will also find out whether the education also helps develop the candidates' intrapreneurial capabilities so that they are in a position to shoulder important responsibilities and contribute constructively to their company and to society.

27.4 Research Methodology

The researchers will interview two Symbiosis Institute of Management Studies' (SIMS) Executive MBA alumni. They will also examine the opinions of other alumni of SIMS Executive MBA to find out whether the course helped them start and run successfully their enterprises, and whether it developed the intrapreneurial capabilities of the candidates, thereby enabling them to rise to responsible positions. The paper will also be based on secondary sources.

27.5 Relevance of the Study

The study is important as working professionals will be motivated to take up Executive MBA courses. Also, universities and management institutes can mold their Executive MBA courses in a way that will encourage working professionals to start their own enterprises. The more the number of successful and sustainable enterprises, the better it is for a nation's economy and employment scenario. Entrepreneurship being the foundation for the robustness of any economy, it is essential to see whether the Executive MBA courses being offered are contributing toward the encouragement of the entrepreneurial and intrapreneurial spirit or are simply obsessed with the imparting of traditional business management concepts.

27.6 Findings

The Executive MBA at Symbiosis Institute of Management Studies (SIMS) is a 20 weeks program. The program covers both learning periods at the institute as well as project work. The students take an actual project usually in the company where they are employed and submit a report on the same. The program is either in the evenings from Monday to Friday or full day on Saturdays and Sundays. Faculty with shared vision and common background work with the students and guide them in their projects. Experienced practitioners are part of the faculty that take sessions with them.

Interviews with some of the alumni of the Executive MBA program were conducted.

27.7 AFY Technologies

Pravin Oswal and Sudhir Kalkar both from pharma background started this company with two more partners, namely Ganesh Jamdar and Ravi Parab in 2019. The company manufactures pharmaceutical equipment, accessories, and attachments and also provides support for equipment upgradation and automation systems. The Symbiosis Institute of Management Studies (SIMS) education was the starting point of this journey. Both Mr. Oswal and Mr. Kalkar knew each other for fifteen years. But the MBA course sets them thinking about starting their own venture. Earlier, they were good engineers who had an overview of the pharma industry. Knowledge of marketing, legal aspects, operations and especially finance, and discussion of practical issues with the respective faculty and inviting them to visit their factory and gives inputs helped them a great deal.

Initially, the company started with technical consultancy to pharmaceutical companies and allied industries like equipment manufacturers. Slowly, it moved to automation and upgradation of process equipment as per regulatory requirement, safety upgrades, and efficacy enhancements. Now, Afy is a pioneer in these fields.

This start-up has a unique combination of experience for pharmaceutical equipment manufacturing as well as pharmaceutical process know how. It has a team of more than 100 years of combined experience of directors. It helped to understand customer requirements and support their needs through knowledge and good engineering solutions. Each partner contributed Rs. 2 lacs as the initial investment to start this start-up. Company has a registered office at Narayan Peth and the manufacturing unit in Khed, Shivapur. This young start-up began its operations with 6 employees including partners in 2019 and currently employing more than 22 people and that too engineers.

Initially for the first one year, all four partners did the marketing work. But after COVID and lockdown of 2020, execution became a problem. They could not visit customers onsite. Again, the cost of materials, especially new materials, increased. Cost of steel increased by 40%. They started using all indigenous materials. They bore all the losses. During the peak of the COVID period, the manufacturing unit was closed for only two months. They faced difficulties in the travel arrangements of employees after opening. They then employed mostly people from the Khed, Shivapur areas. Again, they asked the employees to come on rotation basis. They installed hand-free sanitizing machines, went in for regular fumigation, and such other measures to ensure the safety of the employees. The owners motivate the employees by offering them a lot of learning opportunities, a good work environment, a good salary, and good training as also Corona Kavacch Insurance policy for them and their family members (which gives them a sense of security). Their trips to the clients are sponsored by the company, and they are usually sent by air. They are treated to monthly dinners in restaurants and annual picnics. They are given a dress code and a uniform as also petrol allowances. The employees are mostly young engineers and are given good accommodation near the factory. Around four to five engineers are from Pune and are given company bikes for commuting purposes. Technical training is mandatorily given. A consultant has been hired for corporate governance and ethics compliances purposes.

The company for the last two and a half years has been exporting products to America, Malaysia, Bangladesh, South Africa, Vietnam, and others. Initially, Mr. Oswal and Mr. Kalkar used to travel. Later, employees also were given the opportunity to travel. Now, meetings with customers are also conducted online especially for foreign clients. The owners admitted that the Executive MBA taught them the importance of video presentations as also updated Websites. They also said that the marketing knowledge shared by the concerned faculty helped them a great deal in growing the business. They agreed that the MBA program gave them insights about how to coordinate with customers and their banks and one's own banks. They admitted that the HR courses taught them how to handle their employees. With knowledge from the MBA course, Mr. Oswal and Mr. Kalkar at present prepare the business proposals with inputs of course from the other two partners. They also claimed that the theoretical knowledge imparted to them along with examples about vision and mission helped them a great deal as also guided them regarding communicating with customers.

Afy is currently working with many large pharmaceutical companies like Cipla, Sun, Zydus, Abbott, Unichem, Indoco, and Markson. They have clients not only in India but in the USA, South Africa, Vietnam, Malaysia, Bangladesh, etc. With more than 25 Crore orders in hand now, the company's revenue is more than 15 Crore for the financial year 2021–22 though it is yet to get over. Professionally managed company's ambition is to achieve a 50 Crore milestone in next 2 years, and its manufacturing unit is moving to new premises due to expansion. The new premise is much more spacious with the state-of-the-art amenities. This technology start-up is happy to contribute to nation building through charity.

These alumni entrepreneurs believe that the Executive MBA program not only imparted advanced knowledge for the business start-up but actively supported, appreciated, and motivated them to do their best.

27.8 Sarvadnya Electrotech Private Limited

The company was started in 2010 by Mr. Jayesh Patil, an engineer who worked earlier for ten years in various companies like Cipla, Tatas, and John Deere India

Pvt. Ltd. He started with an initial capital of only Rs. 50,000/-. His aim was to provide employment for people. He started in a 100 sq ft room under the name of Sanika

System Services. After one year, he got a major breakthrough Sahyadri Enterprises in Gujarat. Sahyadri was into manufacturing roof sheets. It had two plants in Pune that were facing a lot of issues. Their workers were not using additives, and so their asbestos sheets had problems. So Mr. Patil got the first big order of Rs. 90 lakhs from Sahyadri Enterprises and with the thirty percent advance, completed the order successfully. He got a repeat order in 2018 from the same company. Mr. Patil now moved the unit to a 3000 sq ft area with the advance that he got from this order. This was necessary as electrical panels require a lot of space. Mr. Patil provided Sahyadri Enterprises electrical panels and electrical solutions. The company is now getting orders from Cipla, Thermax, and other big companies. In 2017, name was changed to Sarvadnya Electrotech Private Limited. Purchased in 2017 land from farmer and converted it to industrial status. Mr. Patil got big orders from HPCL and Honeywell and completed orders in twelve locations. Now got thirteenth order last year. The company now does complete electrical work right form designing to commissioning. Doing turnkey solutions for Sudarshan Chemicals and a 7.4-MW project for Thermax. Mr. Patil now started realizing the need for acquiring managerial skills. So, he joined the weekend Executive MBA course of SIMS. He told the researchers that he imbibed a lot of knowledge from the course like the review system of five years for employing people, information about the share market (which he plans to use for diversifying into the share market), creating a sales strategy for the next five years, importance of talking to gatekeepers, importance of talking to the decision makers, etc.

Mr. Patil has realized that management education is as important as technical knowledge. He plans to cross the hundred crores mark in the coming five years. He, now, is planning the second phase of expansion of his business. He has three persons in sales department. For pre-engineering, i.e., estimations, there are four persons. Design department, accounts department, and production department have again three persons each. He has around eighteen to twenty executives plus workers, and the total number of employees working for him is seventy.

Mr. Patil had to shut down in March 2020 due to COVID. But the business started working again from May. Cost of materials like steel and copper increased which resulted in increased cost of production. This created a challenge of selling at the previously agreed price to Government and companies like Thermax. Consequently, he had to bear a loss of around Rs. 50–60 lakhs. Again, there was the challenge of getting materials as main supply used to come from China which stopped which resulted in increase in prices. But six months later with the relaxation of the lockdown norms, he could secure new orders. Today, Honeywell Automation, ABB India Ltd., Yokogawa India Ltd., Serum Institute of India, Cipla Company, Thermax, and Government agencies like

Rashtriya Fertilizers, Bhabha Automic Research Center are some of Mr. Patil's main customers. He is employing some rural people in his company. He has hired some consultants especially for RCM activities on the basis of the inputs of his SIMS' MBA course inputs. He hires workers trained from ITI colleges as apprentices and trains them. In January 2022, the Business Development department will be started with recruits from MBA colleges. He did not lay off any one during lockdown. He only gave people working from home eighty percent of their salary. His office is in Keshav Nagar, Pune. Seventy percent of the office staff was working from home, while everyone form the manufacturing unit had to come to factory.

27.9 Other Alumni of SIMS' Executive MBA

Mr. Abhay Ghosalkar is presently working in one of the Fortune 500 companies as a Senior Vice President. He joined the SIMS Executive MA 2021–23 batch. He claims that the knowledge he got from SIMS' faculty has greatly helped him in his professional growth. Ms Priya Jacob, an Executive MBA (2013–15 batch) alumnus, is now COO of Nanded City Development & Construction Company Ltd., Pune. She started her entry level career with Magarpatta Township. She has exhibited an acumen in negotiation skills, business planning, and customer relationship management which has made her a top notch in real estate industry. She has given credit for her growth to the Executive MBA Program. Four employees of Kirloskar Oil Engines Ltd. (KOEL) who completed the SIMS Executive MBA program in 2020 have been promoted in 2021. One became GM Marketing, two became AGM-SCM, and one became Senior Manager, Finance after doing the Executive MBA Course.

Mr. Ahishek Jonnalagadda, of SIMS Executive MBA 2021–23, became a CEO of Saiprasad Group and Jyotiee Engineering Works in 2020. He claimed that enrolling for this course was his life shaping decision as it has helped him shape his career better. According to him, the course caters to professional objectives and has helped him build up different concepts that were highly productive. Ms Sakshi Mahale, General Manager, Townships, Paranjape Schemes Construction Ltd., Pune said in her interview that enrolling in the Executive MBA was a perfect decision on her part as it helped her flourish in her current role. She came to understand the financial aspects of her company. All the subjects according to her took her deep into the different concepts of business environment, culture, strategies, branding, sales, learning, motivation, training, communication, and helped her understand the importance of CEM-measuring tools, and how customer complaint iceberg is important.

Mr. Saravanan Gandhi of the SIMS' Executive MBA batch of 2014–16 emphatically stated in his interview that the program is designed in such a way that there is no need for professionals to get additional certifications like PMP as they get all the necessary knowledge from this education itself. Dr. Abhigyan Upadhyay of SIMS'.

Executive MBA 2020–22 was honored with the Pillars of India 2021 Award instituted by Tamil Nadu Government in memory of Dr. APJ Abdul Kalam. Mr. Omprakash Maurya, alumnus of Executive MBA 2011–13, started Diacto Technologies Pvt. Ltd. in 2018 with the vision pf making it the world's most trusted BI brand for speed, quality, and customer success by 2030. The company provides business leaders with 360 degrees view of their organization and truly actionable insights to drive better business decisions and stay ahead of the competition. Its team has experience of deploying across industries like manufacturing, logistics, construction, marketing, and retail, to name a few. Before starting the company, he headed Emerson Global operations for two and a half years. He started the initiative with an initial capital of Rs. 25 lakhs. Now, the company's turnover is Rs. 15 crores, and it employs sixty people. He is of the opinion that an entrepreneur must identify the market value which can be done after acquiring some relevant work experience. All entrepreneurs need bootstrapping and capital light, and the same money can be used to hire the most potential team. The focus according to him has to be on not earning money but on quality, delivery, and not on volume. In Maurya's opinion, the Executive MBA course taught him how a business is run, and how advanced technology leads to paradigm shift in business.

Now, Mr. Arun Bharat, Principal Commissioner Income Tax, Pune, has joined the SIMS' Executive 2022–24 batch. He may join the private sector based on this education three years from now after retiring from service.

27.10   Conclusions and Recommendation

We realize that while Executive MBA education may not make entrepreneurs out of everyone, it certainly develops the entrepreneurial abilities of working professionals and helps them contribute constructively to their work and their organization.

The Executive MBA audience requires a more target-oriented entrepreneurship education approach. This is because participants seeking professional education are older and far more experienced. It is therefore essential to introduce team-based content that utilizes experiential learning methods. B schools should also develop culturally-based teaching materials as also a flexible curriculum. Institutes should aim for the imparting of life-long learning while teaching business skills. One way this can be done is by linking incubators to B-schools. Practical inputs on financing and marshaling resources, marketing, idea and opportunity identification, business planning, managing growth, organization building, new venture creation, and SME management should be given to the candidates. The design and delivery of the EMBA course should be based on questioning assumptions and promoting critical thinking. The program should aim to foster the spirit of enterprise and should give strategic guidelines and models for practicing corporate entrepreneurship either in an existing private business, governmental, or non-governmental setting. Ultimately, it should see that candidates become responsible entrepreneurs and intrapreneurs.

References

1. Alam, G.M., Parvin, M., Roslan, S.: Growth of private university business following "oligopoly" and "SME" approaches: an impact on the concept of university and on society.

Soc. Bus. Rev. **16**(2), 306–327 (2020)
2. Bryant, A.N., Gayles, J.G., Davis, H.A.: The relationship between civic behavior and civic values: a conceptual model. Res. High. Educ. **53**(1), 76–93 (2012)
3. Business Today-In Nov 24 2021: Indians have the Talent and Attitude to become Successful Entrepreneurs
4. Cantino, D., Cortese, R., Longo: Place-based "EMBA as an Entrepreneurship inductor: the ISCTE Executive Education case" 76 network organizations and embedded entrepreneurial learning: emerging paths to sustainability. Int. J. Entrepreneurial Behav. Res. **23**(3), 504–523 (2017). https://doi.org/10.1108/IJEBR-12-2015-0303
5. Canziani, B., Welsh, D.H.B., Hsieh, Y.J., Tullar, W.: What pedagogical methods impact students' entrepreneurial propensity? J. Small Bus. Strateg. **25**(2), 97–113 (2015)
6. Drucker, P.F.: A functioning society: selections from sixtyfive years of writing on community, society, and polity. Transaction Publishers, New Brunswick, NJ (2003)
7. Kaplan, A.: European management and European business schools: insights from the history of business schools. Eur. Manag. J. **32**(4), 529–534 (2014)
8. Kuratko, D.F., Morris: Examining the future trajectory of entrepreneurship. J. Small Bus. Manage. **56**(1), 11–23 (2018). https://doi.org/10.1111/jsbm.12364
9. Krishnamurthy, S.: The future of business education: a commentary in the shadow of the Covid-19 pandemic. J. Bus. Res. **111**(10), 1–15 (2020)
10. Larso, D., Saphiranti, D.: The role of creative courses in entrepreneurship education: a case study in Indonesia. Int. J. Bus. **21**(3), 216–225 (2016)
11. Obschonka, M., Hakkarainen, K., Lonka, K., Salmela-Aro, K.: Entrepreneurship as a twenty-first century skill: entrepreneurial alertness and intention in the transition to adulthood. Small Bus. Econ. **48**(3), 487–501 (2017). https://doi.org/10.1007/s11187-016-9798-6
12. PRME.: (Principles for Responsible Management Education). Outcome Statement of the 1st Global Forum for Responsible Management Education. New York City (2008)
13. Saleh, A., Drennan, J.: An empirical investigation on the motivational factors for pursuing an MBA. Int. J.of Bus. Res. **13**(4), 133–140 (2013)
14. Taj Y.: Firms introduce unique executive education programmes. Economic Times (online) (2011)
15. Team Denzler.: Executive MBA: Zukunft des NDU St.Gallen. Internal Report, University of St. Gallen (2002)
16. Vaudrev, S.: The Importance of Teamwork in an Executive MBA Program (2015)
17. World Economic Forum.: Europe's Hidden Entrepreneurs—Entrepreneurial Employee Activity and Competitiveness in Europe. World Economic Forum (2016)

Chapter 28
An Overview of Self-Organizing Network (SON) as Network Management System in Mobile Telecommunication System

Kennedy Okokpujie, Grace Chinyere Kennedy, Sunkanmi Oluwaleye, Samuel N. John, and Imhade P. Okokpujie

Abstract The rapid advancement in technologies employed in mobile telecommunication industries has improved the sector over the years. However, it has also introduced another problem of ensuring backwards compatibility between newer and older generations. Furthermore, as the technology evolved from the older generation to newer ones, configurable parameters increased, making it more complex to manage manually during installation. This situation worsens when the mobile network operator integrates network elements from different Original Equipment Manufacturers (OEMs). As a result, the Self-Organizing Network (SON) management system was developed. However, with increasing data traffic supplemented by new and developing technologies and correspondingly bigger networks, it is clear that network operations must be redefined in order to ensure optimal performance. For device installations, configurations, resetting network settings, and general network administration, a manual configuration technique necessitates specialized skills. This is a time-consuming and expensive operation. In today's wireless technology, using this strategy results in poor network quality. As a result, the emergence of enhanced

K. Okokpujie (✉) · S. Oluwaleye
Department of Electrical and Information Engineering, Covenant University, Ota, Ogun State, Nigeria
e-mail: kennedy.okokpujie@covenantuniversity.edu.ng

K. Okokpujie
Africa Centre of Excellence for Innovative & Transformative STEM Education, Lagos State University, Ojo, Lagos State, Nigeria

G. C. Kennedy
Department of Computer Science and Engineering, Kyungdong University, Gangwon-do, Korea

S. N. John
Department of Electrical and Electronic Engineering, Nigerian Defence Academy, Kaduna, Nigeria

I. P. Okokpujie
Department of Mechanical and Mechatronics Engineering, Afe Babalola University, Ado Ekiti, Ekiti State, Nigeria

Department of Mechanical and Industrial Engineering Technology, University of Johannesburg, Johannesburg 2028, South Africa

309
C. So-In et al. (eds.), *Information Systems for Intelligent Systems*, Smart Innovation, Systems and Technologies 324, https://doi.org/10.1007/978-981-19-7447-2_28

mobile networks has brought attention to the need of automation. SON enables operating effectiveness and next-generation simplified network monitoring for a mobile wireless network by automating the process. As a result of the introduction of SON in LTE, network performance is improved, end-user Quality of Experience (QoE) is improved, and operational and capital expenses are reduced (OPEX). This paper highlights the SON techniques in the mobile wireless network and briefly describes SON architecture.

28.1 Introduction

Over the years, a dramatic development in mobile telecommunication technologies has critically demanded the need for a robust network management system such as the Self-Organizing Network (SON), a set of functional algorithms for configuring, optimizing, and repairing mobile communications networks automatically [1–3]. The breakthrough started with Global System for mobile communication, the Second Generation (2G) technologies. 2G uses a digital telecommunication approach based on Time Division Multiple Access (TDMA), which reduces data transmission errors and better optimizes the bandwidth to serve more users. With 2G, voice calls, data transfer up to 9.6kbps, short message service (SMS), and SIM card were part of the services. In addition, both circuit switching and packet switching data transport were supported in 2G. The need for a higher data rate increased as Internet technology rapidly improved, leading to the development of Universal Mobile Telecommunication System (UMTS), and approved as third-generation mobile technology (3G). 3G used UMTS Terrestrial Radio Access (UTRA) that supported frequency and time division duplex. The Wideband Code Division Multiple Access (WCDMA), High-Speed Downlink Packet Access (HSDPA), and Evolved High-Speed Packet Access (HSPA+) deliver better data rates than 2G mobile technology. To further meet the need of ever-growing demand for the higher data rate required in Internet technology, Long Term Evolution Advanced (LTE-A) was developed and approved as the fourth-generation (4G) mobile technology [4, 5]. 4G used the Orthogonal Frequency Division Multiple Access (OFDMA) and smart antenna techniques to deliver a much higher data rate. To meet the need for required Internet services for several industries, fifth-generation (5G) mobile technology was developed. 5G provides enhanced Mobile Broadband (eMBB), ultra-Reliable and Low Latency Communication (uRLLC), and massive Machine Type Communication (mMTC) as user scenarios [6, 7]. This continuous development in mobile telecommunication has brought more complexity to the deployment and maintenance of telecommunication networks, and this has led to the development of Self-Organizing Networks (SON) by 3rd Generation Partnership Project (3GPP), a standard organization that develops protocols for mobile telecommunication technologies [8–10].

28.2   Benefits of SON

The number of configurable parameters increases as the technology evolves from one generation to another, making it more difficult to manage manually. Furthermore, because the evolution from one technology to the other is so fast, there is a need for backwards compatibility, which requires that the newer mobile technology should be able to integrate well with the older ones. Finally, different OEMs manufacture mobile technology equipment, and this equipment is met to work together seamlessly in a telecommunication network. All these three essential factors motivated the development of SON. The significant benefits of SON can be broadly categorized as a reduction in Capital Expenditure (CAPEX) and Operation Expenditure.

(OPEX) Figure 28.1 shows a typical network deployment and maintenance process. The CAPEX cost includes planning and deployment of the network equipment, while the OPEX cost focuses on the operation and maintenance of the networks [11, 12]. SON aims to reduce these costs and improve quality of service (QoS) by using its functional features such as Self-Configuration, Self-Optimization, and Self-Healing [13–15].

When SON is enabled in a mobile network, apart from the cost reduction in both CAPEX and OPEX, user experience and overall network performance are improved. Figure 28.2 shows the need for human intervention in a mobile network which is significantly reduced in a SON-enabled network.

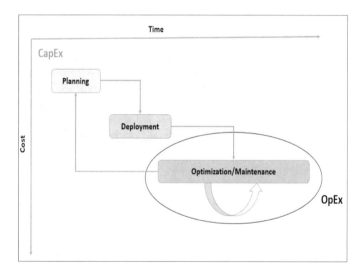

Fig. 28.1 Mobile telecommunication deployment process

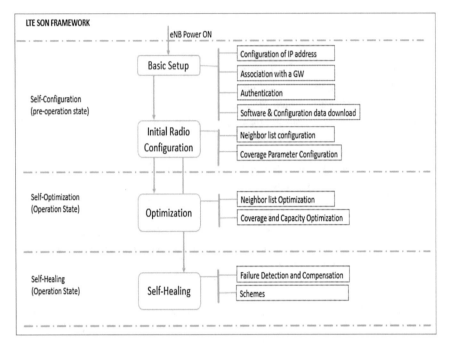

Fig. 28.2 Features of SON

28.3 Features of SON

SON was first released with LTE mobile technologies but has been extended to older technologies such as 2G and 3G in its later release. The various functionalities of SON can be grouped under these three categorized features: Self-Configuration, Self-Optimization, and Self-Healing. Each of these features has a set of functionalities seamless work together to achieve the overall aim of the SON's feature. Figure 28.2 shows the interoperate relationship of these features.

28.3.1 Self-Configuration

These functional algorithms set up newly installed eNB for ordinary network operation. The aim is to achieve a "plug and play" system that reduces human errors during deployment [16–18]. These functions were among the first standardized SON features by 3GPP (Release 8), and the stepwise algorithm implemented during the Self-Configurations of eNB as shown in Fig. 28.3.

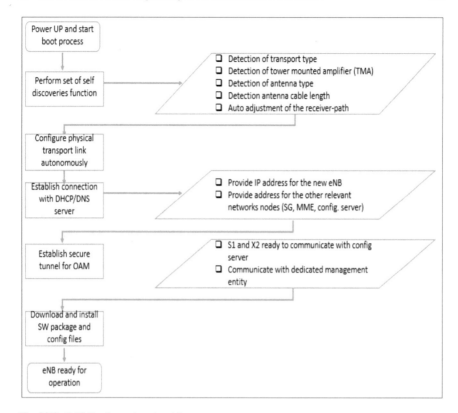

Fig. 28.3 Self-Configuration algorithm

28.3.2   Self-Optimisation

Several-Configuration settings in the network ground station regulate various characteristics of the cell site. Due to observations at both the ground station (BS) and observations at the mobile terminal or smartphone (ME), this may be changed to affect the network's behaviour [19–21]. Others enhance random access parameters or mobility robustness for handover oscillations, whereas SON automatically creates neighbour relations (ANR). The automated switch off of specific base stations during the night hours is a particularly illustrative use case. The settings of the neighbouring base stations would then be re-configured to ensure that the signal reached the entire region. The "sleeping" communication systems "wake up" fairly instantly in the event of a sudden increase in connection demand for any reason [22]. For the network provider, this process results in huge energy savings. Self-Optimisation functions include the following:

(a) Mobility load balancing (MLB)
(b) Mobility robustness optimization (MRO)
(c) Minimization of drive testing (MDT)
(d) Fast and proactive parameters optimization
(e) Increased network performance.

28.3.3 Self-Healing

Self-Healing is a set of SON processes that identifies faults and fixes or mitigates them in order to minimize user impact and maintenance expenses [23–25].

The SON standards were taken into releases 9 and 10 by 3GPP TS 32.541 [26].

The Self-Healing procedure in SON is usually applied in the following areas:

(a) Self-Diagnosis: develops a diagnostic model based on previous experiences.
(b) Self-Healing: initiates remedial activities to address the problem automatically.

The performance of Self-Healing procedures heavily depends on some outputs obtained from the Self-Optimization procedure that involves analysis of the network performance.

Some of the Self-healing functions available are

(i)   Cell outage: The procedure detects an outage in the network, as shown in Fig. 28.4, using various analysis techniques available in the SON protocol.
(ii) Self-recovery of network element (NE) software: After a fault is detected, the SON procedure maximizes the capacities of the neighboring cells to compensate for the outage, as shown in Fig. 28.5, in other to reduce the impact of the affected network elements on the user experience and overall network performance.

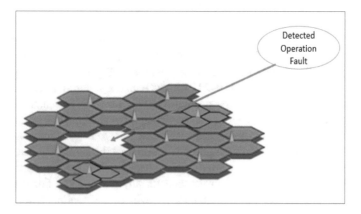

Fig. 28.4 Fault detection

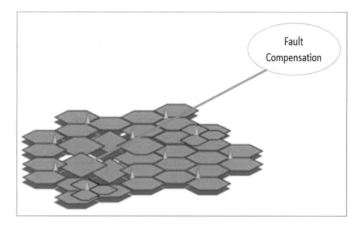

Fig. 28.5 Fault compensation

   Self-Healing of board faults. While the failure is temporarily compensated for, the SON procedure attempts to repair the network element using the available self-recovery software functionality or alerts the Operation Administration and Management (OAM) team if this process fails. Afterwards, the decision is taken to repair and replace the network element to restore the network to optimal operation.

28.4   SON Architecture in Mobile Telecommunication Network

As shown in Fig. 28.6, SON uses three main design choices in cellular networks, with network management system (NMS), element management system (EMS), and operational support system (OSS) as the essential components of the network architecture. Centralized, distributed, and mixed structures are the three alternatives [27, 28]. Distinct structural layouts in a comparable network can achieve different SON functionalities.

28.5   Conclusion

In conclusion, because of the increased need for data traffic compensated by new and developing technologies with matching bigger networks, the necessity for SON develops in Network Maintenance. It has been understood that SON comes with a lot of benefits that improve the network services in the telecommunication industries. This include: improvement in end-user Quality of Experience, reduction in capital expenditure and operational expenditure, efficient resource utilization, increased network performance, and Faster Network Maintenance.

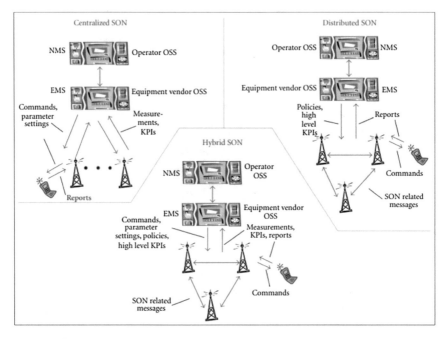

Fig. 28.6 Three approved standard options of SON architecture

Acknowledgements This paper is partly sponsored by Covenant University Center of Research, Innovation, and Discovery (CUCRID) Covenant University, Ota, Ogun State, Nigeria.

References

1. Moysen, J., Giupponi, L.: From 4G to 5G: self-organised network management meets machine learning. Comput. Commun. **129**, 248–268 (2018). https://doi.org/10.1016/j.comcom.2018.07.015. (In: Conference 2016, LNCS, vol. 9999, pp. 1–13. Springer, Heidelberg (2016))
2. du Jardin, P.: Forecasting corporate failure using ensemble of self-organising neural networks. Eur. J. Oper. Res. **288**(3), 869–885 (2021), ISSN 0377-2217, https://doi.org/10.1016/j.ejor.2020.06.020
3. Li, S., Gao, T., Ye, Z., Wang, Y.: Comparative research on the formation of backbone media of wireless self-organising network by DLA-GF algorithm and ant colony algorithm. Alexandria Eng. J. (2021), ISSN 1110-0168, https://doi.org/10.1016/j.aej.2021.06.003
4. Agboje, O., Nkordeh, N,. Idiake, S, Oladoyin, O., Okokpujie, K., Bob-Manuel, I.: MIMO channels: optimising throughput and reducing outage by increasing multiplexing gain. Int. J. Appl. Eng. Res. (2020), ISSN 0973-4562
5. Olabode. I., Okokpujie, K., Husbands, R., Adedokun, M.: 5G wireless communication network architecture and its key enabling technologies. Int. Rev. Aerosp. Eng. (I. RE. AS. E) **12**(2), 70–82 (2019)
6. Oshin, O., Luka, M., Atayero, A.: From 3GPP LTE to 5G: an evolution. Trans. Eng. Technol. 485–502 (2016). https://doi.org/10.1007/978-981-10-1088-0_36

7. Wiwatcharakoses, C., Berrar, D.: SOINN+, a self-organising incremental neural network for unsupervised learning from noisy data streams. Expert Syst. Appl. **143**, 113069 (2020), ISSN 0957-4174, https://doi.org/10.1016/j.eswa.2019.113069

8. Osemwegie, O., John, S., Adeyinka, A., Noma-Osaghae, E., Okokpujie, K.: Comparative analysis of routing techniques in chord overlay network. Int. J. Electr. Comput. Eng. **11**(5), 4361–4372 (2021)

9. Bayazeed, A., Khorzom, K., Aljnidi, M.: A survey of self-coordination in self-organising network. Comput. Netw. **196**, 08222 (2021), ISSN 1389-1286, https://doi.org/10.1016/j.com net.2021.108222

10. Bhattacharyya, S., Pal, P., Bhowmick, S.: Binary image denoising using a quantum multilayer self-organising neural network. Appl. Soft Comput. **24**, 717–729 (2014), ISSN 1568-4946, https://doi.org/10.1016/j.asoc.2014.08.027

11. Osterbo, O., Grondalen, O.: Benefits of self-organising networks (SON) for mobile operators. J. Comput. Networks Commun. **2012** (2012), https://doi.org/10.1155/2012/862527

12. Atayero, A.A., Adu, O.I., Alatishe, A.A.: Self organising networks for 3GPP LTE. Lect. Notes Comput. Sci. (including Subser. Lect. Notes Artif. Intell. Lect. Notes Bioinformatics), **8583**(5), 242–254 (2014), https://doi.org/10.1007/978-3-319-09156-3_18

13. Belisle, J., Clayton, M.: Coherence and the merging of relational classes in self-organising networks: extending relational density theory. J. Contextual Behav Sci **20**, 118–128 (2021), ISSN 2212-1447, https://doi.org/10.1016/j.jcbs.2021.03.008

14. Huang, K., Ma, X., Song, R., Rong, X., Li, Y.: Autonomous cognition development with lifelong learning: a self-organising and reflecting cognitive network. Neurocomputing **421**, 66–83 (2021), ISSN 0925-2312, https://doi.org/10.1016/j.neucom.2020.09.027

15. Pan, W., Sun, Y., Turrin, M., Louter, C., Sariyildiz, S.: Design exploration of quantitative performance and geometry typology for indoor arena based on self-organising map and multi-layered perceptron neural network. Autom. Constr. **114**, 103163 (2020), ISSN 0926-5805, https://doi.org/10.1016/j.autcon.2020.103163

16. Li, W., Li, M., Zhang, J., Qiao, J: Design of a self-organising reciprocal modular neural network for nonlinear system modelling. Neurocomputing **411**, 327–339 (2020), ISSN 0925-2312, https://doi.org/10.1016/j.neucom.2020.06.056

17. Wiwatcharakoses, C., Berrar, D.: A self-organising incremental neural network for continual supervised learning. Expert Syst. Appl. **185**, 115662 (2021), ISSN 0957-4174, https://doi.org/10.1016/j.eswa.2021.115662

18. Qin, Z., Lu, Y.: Self-organising manufacturing network: a paradigm towards smart manufacturing in mass personalisation. J. Manuf. Syst. **60**, 35–47 (2021), ISSN 0278–6125, https://doi.org/10.1016/j.jmsy.2021.04.016

19. Kamboh, U.R., Yang, Q., Qin, M.: Impact of self-organizing networks deployment on wireless service provider businesses in China. Int. J. Commun. Netw. Syst. Sci. **10**(05), 78–89 (2017). https://doi.org/10.4236/ijcns.2017.105b008

20. Okokpujie, K., Chukwu, E., Olamilekan, S., Noma-Osaghae, E., Okokpujie, I.P.: Comparative analysis of the performance of various active queue management techniques to varying wireless network conditions. Int. J. Elec. Comp. Eng. **9**(1), 359–68 (2019)

21. Balaji, K., Lavanya, K., Geetha Mary, A.: Clustering algorithm for mixed datasets using density peaks and self-organising generative adversarial networks. Chemometr. Intell Lab. Syst. **203**, 104070 (2020), ISSN 0169-7439, https://doi.org/10.1016/j.chemolab.2020.104070

22. 3GPP TR 36. 902.: Self-configuring and self-optimising network (SON) use cases and solutions (Release 9). v.9.3.1 (2011)

23. Ng, R.W., Begam, K.M., Rajkumar, R.K., Wong, Y.W., Chong, L.W.: An improved self-organising incremental neural network model for short-term time-series load prediction. Appl. Energy **292**, 116912 (2021), ISSN 0306-2619, https://doi.org/10.1016/j.apenergy.2021.116912

24. Qiao, X., Guo, W., Li.: An online self-organising modular neural network for nonlinear system modelling. Appl. Soft Comput. **97**, Part A, 106777 (2020), ISSN 1568-4946, https://doi.org/10.1016/j.asoc.2020.106777

25. Kebonye, N.M., Eze, P.N., John, K, Gholizadeh, A, Dajčl, J., Drábek, O., Němeček, K., Borůvka, L.: Self-organising map artificial neural networks and sequential Gaussian simulation technique for mapping potentially toxic element hotspots in polluted mining soils. J. Geochem. Explor. **222**, 106680 (2021), ISSN 0375-6742, https://doi.org/10.1016/j.gexplo.2020.106680
26. 3GPP TS 32. 541.: Telecommunication management, Self-Organising Networks (SON), Self-healing concepts and requirements. v.10.0.0 (2011)
27. Østerbø, O., Grøndalen, O.: Benefits of self-organising networks (SON) for mobile operators. J. Comput. Netw. Commun. **2012**(862527), 16. https://doi.org/10.1155/2012/86252
28. 3GPP TS 32. 501.: Telecommunication Management, Self-Organizing Networks (SON), Concepts and requirements (Release 11). v.11.1.0 (2011)

Chapter 29
Deep and Transfer Learning in Malignant Cell Classification for Colorectal Cancer

Gauraw F. Jumnake, Parikshit N. Mahalle, Gitanjali R. Shinde, and Pravin A. Thakre

Abstract Colorectal cancer has shown wide spread over a decade, projected number of cancer cases in 2022 will be almost 71% as per ICMR and NCBI data (Kather et al., 100,000 histological images of human colorectal cancer and healthy tissue (Version v0.1) (2018)) due to lifestyle and changing dietary habits. If diagnosis in its early stages, then will significantly boost survival rate of patient. Computer integrated system had positive influence on smoothing out the process of detection or classification. Furthermore, learning methods added more accuracy and details in this process. In this paper, deep learning and transfer learning methods were experimented and analyzed to know the impact of various parameters and model-related factors in identification and classification of malignant cells for colorectal cancer on whole slide stained tissue image samples.

29.1 Introduction

India is seeing notable changes in cancer incidences from last few years; ICMR and NCBI report showed projected number of cancer cases in 2023 will be almost 79%. In male tobacco and alcohol-related cancer incidence are more while in case of female breast and cervical cancer incidences are reported. Beside this due to vastly changing lifestyle and eating habits, colorectal cancer is steadily spreading its legs in India, resulting in a substantial rise in CRC patients over the past few decades [1]. The detection of cancer can be avoided or extended in its early stages and with proper diagnosis. Efficient care can be taken with the use of suitable screening procedures. In early screening for CRC using histopathological images obtained through the

G. F. Jumnake (✉)
SKNCOE Savitribai Phule Pune University, Pune, Maharashtra, India
e-mail: g.f.jumnake83@gmail.com

P. N. Mahalle · G. R. Shinde
VIIT Savitribai Phule Pune University, Pune, Maharashtra, India

P. A. Thakre
ZCOER Savitribai Phule Pune University, Pune, Maharashtra, India

C. So-In et al. (eds.), *Information Systems for Intelligent Systems*, Smart Innovation, Systems and Technologies 324, https://doi.org/10.1007/978-981-19-7447-2_29

relevant/appropriate screening methods will allow the learning algorithm to be used for classification and segregation of malignant and non-malignant CRC cells [2].

Use of computer aided methods makes it possible to analyze histopathological image samples in details for malignant cells. A learning model should be capable of classification, prediction and estimation or similar tasks. A CNN model with simple vector machine works with hyper tuning works fine. Pre-trained networks like VGG16 and VGG19 with tuning meet up to expectations and predicted all samples right other model like InceptionV3, ResNet50, AlexNet on fine tuning gave promising results. A fine classification model should be capable of fitting given training set well and should be able to classify all the instances. One should also consider model over-fitting of test error rate is more than training error rate which can be avoided by making ML model more intricate. Once the classification model is prepared, performance of model is measured in terms of precision, recall, and accuracy [3]. In this research, these learning models were evaluated and tested for different hyperparameters and depth of convolutional layers to improve accuracy of classification.

29.2 Literature Review

Wan et al. [4] discussed about cfDNA which has some characteristics like cancer associated mutation, translocation, large number of chromosomal copy variants, and DNA fragments are comparatively shorter. Mostly, there are less no of cfDNA in early stage of cancer these factors put question mark on technique used. These limitations are managed by studying cfDNA from different perspective like other changes which occurs at early stage which include blood analysis there enough evidences showing relation between cancerous cell and non-cancerous cells, i.e., fibroblast, platelets, and immune cells. ML model includes cross-validation process to get accuracy. Different CV models are tested based on definition with k-fold. Input features are obtained from whole genome data with the help of given classification model.

Lee et al. [5] studied somatic alterations in cancerous cells. As it is known that cancer is critical phenomenon due to irregular cellular proliferation in cells and blood. To get accuracy of ML classifiers, various mutation features selection is applied. In reviewed method cancer predictor using ensemble model classifier is applied which was based on ensemble of deep neural network and random forest classifier. Three hidden layer DNN is applied for predictions and Adam optimizer to train network. To maximize accuracy, tenfold-cross-validation is used to get features from input data. Most optimal working features are collected using different algorithms like extra tree-based, LASSO, and LSVM.

Blanes-Vida et al. [6] proposed method based on results obtained through colonoscopy. Research focuses on polyps where researcher tried to identify accurate size of polyps, which is an influential factor for better CRC predicament. Factor which characterize polyps are size, location, and morphology. In given study,

ML method is used to test samples obtained by fecal immunochemical test by getting similarity index between polyps using Gower's coefficient.

Luyon Ai et al. [7] researched on gut microbial as sometimes blood-based test may lead to limited sensitivity and specificity and also stool may not contain considerable amount of blood or it may not be detectable in a single stool sample. Gut microbiome in a healthy person is different from gut microbiome of CRC patients; it is reported from number of bacterial species which are involved in CRC carcinogenesis. Applied Naive Bayes/BayesNet, Random Forest, Logistic Regression and LMT classification models out of which Naive Bayes and random forest gave most promising results still random forest had more negative false rate than Naive Bayes. Another algorithm J48 is also applied to show most discriminative bacterial species. J48 only analyzes several known attribute of bacteria to learn decision tree giving good interpretations of classification model. Samples are obtained from two different populations and tested for CRC microbial activity. Fecal microbiome with FOBT gave good results with AUC 0.93 and 0.94, respectively, over both models. WEKA packages deployed for classification and analysis as it contains vast collection of different machine learning models.

Hornbrook et al. [8] studied CRC patient's blood count and evaluated for presence of cancer. These samples are tested on decision tree and cross-validation-based tool ColonFlag. Tool was developed to test Maccabi Healthcare Services (MHS) and Israel National Cancer Register (INCR) data which further used to test samples from UK-based Healthcare Information Network. FOBT can be applied when there is sufficient presence of blood samples in fecal material which is not possible in every case. So, by counting patient's blood count for factors like haemoglobin, RBC, etc., will improvise result up to very large extent and will make early detection possible.

In a research conducted by Bibault et al. [9], machine learning model is constructed based on demographic data which includes features like TNM staging, patient's medical history, work habits, and cancer-related treatments taken if any. Model is designed based on gradient boosting with XGBoost hyperparameters were selected by nested cross validator by applying Bayesian optimizer. To access the performance of model, nonparametric bootstrap procedure is applied. Model was trained on data obtain from NCBI repository, around 2359 samples used to training and testing. Accuracy of 84% and precision was 65% while recall rate of 82% was obtained. Result can be enhancing by balancing data and putting more complex and accurate system.

29.3 Background Work

29.3.1 Deep Neural Network

It is a kind if neural network, where a network composed of minimum two layers. Deep neural network (DNL) is consisting of input, output, and multiple hidden layers.

Fig. 29.1 Generalized deep learning model (DL-CNN)

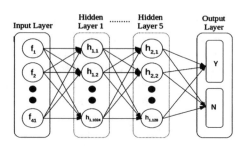

Here, layers in DNL takes output of previous layer as an input to the current layer and trains distinct features. During training, every hidden layer learns features, correlates, executes own logic (ordering, sorting, etc.) and rebuilds input from the sample feature dataset. Figure 29.1 shows a generalized DLN model that has five hidden layers. An input layer takes 41 features as an input to the model and has two features in the output dataset [10].

Convolution Neural Network (CNN): It is a deep learning algorithm which takes image as an input to the network, assigns weight, bias to the objects in the input image, and has ability to differentiate objects in input image. The architecture of the CNN is similar to the human brains neuron patterns. The generalized CNN model has input layer, convolutions layer, pooling layer, and fully connected layer.

Convolution Layer: This layer uses filters which executes convolution operations on input with respect to the input dimension vector. It uses hyperparameters filter size and stride S. The output of this layer is termed as activation map.

Pooling Layer: This layer performs down sampling operations on the output of convolution layer. This layer applies spatial invariance on the convolution layer output. It has maximum, minimum, or average pooling operation during down sampling process.

Fully Connected Layer: It operates on the flattens input, where each neuron is connected with input. If it is present in the CNN, it is available at the end of CNN, where it performs optimize operation such as class score. Figure 29.2 shows the generalized view of the CNN model [10].

Evaluation Metrics: The evaluation of the proposed convolution neural network model is done by finding accuracy, F1 score, and precision value of the proposed model. Equation (29.1) is used to find accuracy of the proposed model. Precision value of the model is computed using Eq. (29.2), recall value of the model is computed

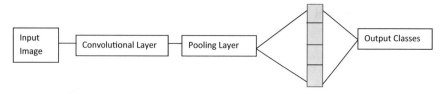

Fig. 29.2 Layers of CNN model

using Eq. (29.3), and the Eq. (29.4) is used to compute the F1 score of the proposed model. Achieving high accuracy is always not useful; many times model shows 98 or 99% accuracy but when it comes to correct class prediction results are quire discouraging mostly when data is skewed or unbalanced class distribution. So, one more parameter interest metric (29.5) is included to monitor class prediction which is ratio of correctly predicted samples and smoothing factor to total no of samples and classes.

$$\text{Accuracy} = \frac{\text{TP} + \text{TN}}{\text{TP} + \text{TN} + \text{FN} + \text{FP}} \tag{29.1}$$

$$\text{Precision} = \frac{\text{TP}}{\text{TP} + \text{FP}} \tag{29.2}$$

$$\text{Recall} = \frac{\text{TP}}{\text{TP} + \text{FN}} \tag{29.3}$$

$$\text{F-1 Score} = 2 * \frac{\text{Precision} * \text{Recall}}{\text{Precision} + \text{Recall}} \tag{29.4}$$

$$\text{Interrest Metric} = \frac{N_{tc}}{N_c + n} \tag{29.5}$$

N_{tc}—Number of correctly predicted samples t from class c
N_c—No of samples with class c
n—Total no of classes

The 80% of images are used for training model and remaining 20% for testing model. Upon model creation, the proposed model has been tested with image datasets [11].

AlexNet was first CNN model with gpu boosted performance metrics. It is consisting of five convolutional layers with three maxpooling layers, and two fully connected layers with one dense layer for image classification. Model was executed on ImageNet dataset of 1.4 million images with 1000 different classes [12].

VGG16 and VGG19 are CNN models which are also trained on ImageNet dataset. These deep convolutional neural networks are used for image recognition and clas-sification. Here, 16 and 19 represent number of weighted convolutional layers with varying filter size of 3×3, 5×5, and 1X1 which are then incorporated with maxpooling layers. Three fully connected layers, two dense, and one layer for output with softmax activation are applied for final image classification [12].

In this research for experimentation 42 layer deep sparsely connected Inception V3 model is used. These layers are merged together with variable filter of sizes 5×5, 3×3, and 1×1 which are concatenated with output filters together before feeding it to next layer. Additional 1×1 layer with maxpooling layer is parallely incorporated for dimensionality reduction before connecting to previous layer. This makes it possible for internal layers to select specific filter size and in this way will help to learn essential information [13].

ResNet is residual, highly deep neural network which may incorporate layers up to 152. It manages issue of vanishing gradient which is big concern while working with deep neural networks. In general, gradient is back propagated to adjacent layer and after multiplication it becomes very small, as the network goes deeper it starts saturating or sometimes starts vanishing. Convolutional layers are stacked together with original input and output of other convolutional block which makes it possible to handle said issue and also ensure higher layer will perform as good as previous one atleast not worst [14].

29.4 Proposed Methodology

Dataset used for proposed work consists of more than 90,000 images collected from mix sources divided into nine classes including lymph cells, normal cells, and tumor cells. Images were extracted from stained tissue samples of patients containing primary tumor cells and normal cells. Images are color normalized and resized into 224×224 non-overlapping samples. Results on this dataset are proven by several researchers [10, 15]. Identifying cancerous cells is not the only motivation of this experiment. It should be able to generate parameters which will be helpful for clinical analysis and will lead to better diagnosis. Also to evaluate system which is capable of differentiating between malignant and non-malignant cells and able to classify them [16].

Purpose of this experimentation is to examine importance of network depth and other hyperparameters in medical image classifications. An investigation is conducted on deep learning networks with CNN, AlexNet, InceptionV3, VGG16, VGG19, and ResNet50. These CNN are of different abilities which are tested to extract useful information from images modalities. A simple deep learning network is trained on mixed image dataset which is fine tuned for parameters like number of epochs, batches per epochs, RMSprop, and Adam optimizer with learning rate [17, 18]. Mean square error between the image pixels is analyzed higher the value of MSE less similar images are

$$\text{MSE} = \frac{1}{mn} \sum_{i=0}^{m} \sum_{j=0}^{n} [A(i, j) - P(i, j)]^2 \qquad (29.6)$$

i—values in rows
j—values in columns
$A(I, j)$—target image pixel values
$P(I, j)$—Generated image pixel values

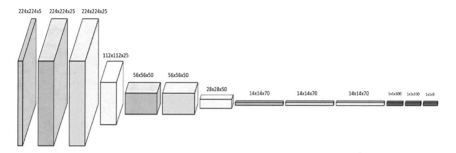

Fig. 29.3 Schematic representation of DL-CNN

A deep learning model with 11 layers is incorporated which has two convolutional layers of 5×5 and 2×2 filter size. These filters are applied to reduce the size of input image. Each convolutional layer in the model follows batch of ReLU activation function, which is used to degrade vanishing issue. Here, maxpool has been deployed, which is used to extract maximum information layer of size 2×2. Three fully connected layers with dropout layer in between added to avoid the over fitting in the network. Schematic representation of proposed model is shown in Fig. 29.3.

Proposed network had good accuracy over train and test dataset but was lagging in class prediction. So, transfer learning frameworks are tested VGG16, VGG19, InceptionV3, and Resnet50 which are pre-trained on ImageNet dataset, last layer was removed and replaced from these networks keeping same weights, Fig. 29.4 represents schematic of transfer learning model layers. While working with VGG16 and VGG19, fully connected layer was added with 9 neurons and softmax classifier as dataset contains images belonging to nine different classes but in case of inception and ResNet, an another dense layer is added before final classification layer to control feature [19]. Table 29.1 gives details about selected pretrained model with their layer details and parameters. Researchers have proven their work on used dataset in several ways but to make model more concrete and flexible, images are collected from different repositories, analyzed, and mixed together which are categorized in nine different classes based on tissue samples collected and cellular structures; those are tumor cells, normal cells, muscles, mucus, adipose, stroma, lymph nodes, and other mix structures.

29.5   Result

In this part, we have evaluated investigative results on six CNN models which are hyper-tuned for various parameters. Whole image dataset is converted into data frames where column holds pixel information about images and their label. These data frames are stored in csv format of future use. Kernel initializers are kept smaller and constant trained and tested on both Glorot normal and XGBoost initializes. Furthermore, categorical cross entropy was applied as loss function. To get prediction

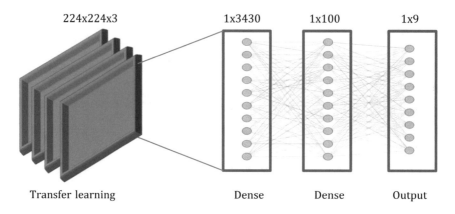

Fig. 29.4 Schematic representation of transfer learning models

Table 29.1 Learning models with details

Model name	Layers details	Actual parameters	Trainable parameters
DL-CNN	4 CNN, 3 maxpooling, 3 dense	419,359	419,199
AlexNet	5 CNN, 3 maxpooling, 2 dense	503,220	402,362
VGG16	16 layers, 1 dense (output)	14,940,489	225,801
VGG19	19 layers, 1 dense (output)	20,590,560	305,999
InceptionV3	22 layers, 2 dense, 1 output	74,241,833	2,439,049
ResNet50	50 layers, 2 dense, 1 output	75,275,273	5,710,473

accuracy, set of 68 differently processed images samples were taken. Experimental results obtained are show in table Comparative analysis of results obtained by testing these pretrained model are shown in chart (Fig. 29.5).

As we can see, DL-CNN model produces good result identified 54 images belongs to six classes very well, AlexNet also manages to do the same but when VGG16 and VGG19 was evaluated predicted all nine classes of 66 and 67 images, respectively. In case of InceptionV3 and ResNet50, extra dense layer was added to reduce loss due to very small gradient values and improved accuracy is obtained able to identify 55 and 61 images from nine different classes, respectively.

Interest metrics which focuses on class prediction how many samples were correctly classified to respective class is significant for VGG16 with 91% and least significant for custom network DL-CNN. Dataset used for above experimentation is obtained from cited repository created using National Center for Tumor diseases records [20].

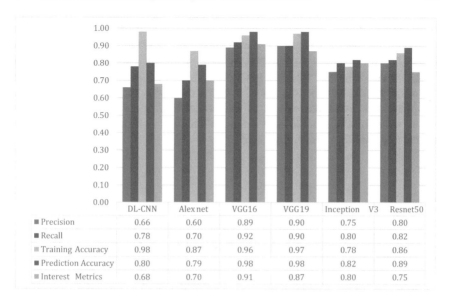

	DL-CNN	Alex net	VGG16	VGG19	Inception V3	Resnet50
■ Precision	0.66	0.60	0.89	0.90	0.75	0.80
■ Recall	0.78	0.70	0.92	0.90	0.80	0.82
■ Training Accuracy	0.98	0.87	0.96	0.97	0.78	0.86
■ Prediction Accuracy	0.80	0.79	0.98	0.98	0.82	0.89
■ Interest Metrics	0.68	0.70	0.91	0.87	0.80	0.75

Fig. 29.5 Comparative analysis of different learning models

29.6   Conclusion

In this research, deep learning and other transfer learning models are applied on mix dataset collected from different sources to evaluate depth of model and parameter tuning. We have examined how model performs if numbers of layers are increased and if parameters are adjusted. DL-CNN network worked fine identified features from six classes very well but inadequate to classify image belonging to adipose and mix cells. While VGG networks shown great improvement with tuned parameters, classified almost all the image and also able to identify features of all nine classes accurately. Inception and ResNet manage to extract features of seven classes accurately and also classified most of the images. In VGG model, layers are appended directly with output layer before training and gave accuracy of 98%, but in case Inception and ResNet two more dense layers were appended to boost feature extraction and reduce loss due very small values of gradient which allows us to get accuracy of 89%.

Adding more layers and making it denser helped model to learn well but looking at the result obtained in this experimentation, it is difficult to obtain high classification accuracy. It is due to high similarity index between the few classes. Images belonging to some classes may be visually different or in some classes may look same. So, applying same technique on individual class may not generate same result. In the future, combination of output from different layers can be explored to get better accuracy. Other deeper and wider networks like LSTM recurrent network, densely connected RNN, or inception residual network might be tested for better classification.

In regard to boost interest metrics which will insure correct prediction of class, more robust feature extraction and selection strategy shall be applied. Self-supervised interest point detection convolutional model which extracts features from full sized image by collectively computing pixel level interest points with multiscale multi-homography [21]. Another approach local features network (LF-Net) uses two stage neural network, first stage is dense network which returns keypoint locations, orientation, scales and in second stage it returns local descriptors for samples cropped around keypoint obtained in first stage [11].

References

1. Jumnake, G.F., Mahalle, P.N., Shinde, G.R.: Learning techniques for pre-malignancy detection in human cells a review. Int. J. Eng. Adv. Technol. **9**(6), 555–561 (2020). https://doi.org/10.35940/ijeat.f1622.089620
2. Ladabaum, U., Dominitz, J.A., Kahi, C., Schoen, R.E.: Strategies for colorectal cancer screening. Gastroenterology **158**(2), 418–432 (2020). https://doi.org/10.1053/j.gastro.2019.06.043
3. Bhatt, D.B., Emuakhagbon, V.S.: Current trends in colorectal cancer screening. Curr. Colorectal Cancer Rep. **15**(2), 45–52 (2019). https://doi.org/10.1007/s11888-019-00432-4
4. Wan, N., et al.: Machine learning enables detection of early-stage colorectal cancer by whole-genome sequencing of plasma cell-free DNA. BMC Cancer **19**(1), 1–10 (2019). https://doi.org/10.1186/s12885-019-6003-8
5. Lee, K., oh Jeong, H., Lee, S., Jeong, W.K.: CPEM: accurate cancer type classification based on somatic alterations using an ensemble of a random forest and a deep neural network. Sci. Rep. **9**(1), 1–9 (2019). https://doi.org/10.1038/s41598-019-53034-3
6. Blanes-Vidal, V., Baatrup, G., Nadimi, E.S.: Machine learning-based colorectal cancer detection. In: Proceedings of 2018 Research Adaptation Converging System RACS 2018, pp. 43–46 (2018). https://doi.org/10.1145/3264746.3264785
7. Ai, L., Tian, H., Chen, Z., Chen, H., Xu, J., Fang, J.Y.: Systematic evaluation of supervised classifiers for fecal microbiota-based prediction of colorectal cancer. Oncotarget **8**(6), 9546–9556 (2017). https://doi.org/10.18632/oncotarget.14488
8. Hornbrook, M.C., et al.: Early colorectal cancer detected by machine learning model using gender, age, and complete blood count data. Dig. Dis. Sci. **62**(10), 2719–2727 (2017). https://doi.org/10.1007/s10620-017-4722-8
9. Bibault, J.E., Chang, D.T., Xing, L.: Development and validation of a model to predict survival in colorectal cancer using a gradient-boosted machine. Gut 1–6 (2020). https://doi.org/10.1136/gutjnl-2020-321799
10. Hu, Z., Tang, J., Wang, Z., Zhang, K., Zhang, L., Sun, Q.: Deep learning for image-based cancer detection and diagnosis—a survey. Pattern Recognit. **83**, 134–149 (2018). https://doi.org/10.1016/j.patcog.2018.05.014
11. Ono, Y., Fua, P., Trulls, E., Yi, K.M.: LF-Net: learning local features from images. Adv. Neural Inf. Process. Syst. **2018**(NeurIPS), 6234–6244 (2018)
12. Razzak, M.I., Naz, S., Zaib, A.: Deep learning for medical image processing: overview, challenges and the future. Lect. Notes Comput. Vis. Biomech. **26**, 323–350 (2018). https://doi.org/10.1007/978-3-319-65981-7_12
13. Szegedy, C., Vanhoucke, V., Ioffe, S., Shlens, J., Wojna, Z.: Rethinking the inception architecture for computer vision. In: Proceedings of IEEE Computer Society Conference Computing Vision Pattern Recognition, vol. 2016, pp. 2818–2826 (2016). https://doi.org/10.1109/CVPR.2016.308

14. He, K., Zhang, X., Ren, S., Sun, J.: Deep residual learning for image recognition. In: Proceedings IEEE Computer Society Conference Computer Vision Pattern Recognition, vol. 2016, pp. 770–778 (2016), https://doi.org/10.1109/CVPR.2016.90
15. Jiang, D., et al.: A machine learning-based prognostic predictor for stage III colon cancer. Sci. Rep. **10**(1), 1–9 (2020). https://doi.org/10.1038/s41598-020-67178-0
16. Misawa, D., Fukuyoshi, J., Sengoku, S.: Cancer prevention using machine learning, nudge theory and social impact bond. Int. J. Environ. Res. Public Health **17**(3), 1–11 (2020). https://doi.org/10.3390/ijerph17030790
17. Puttagunta, M., Ravi, S.: Medical image analysis based on deep learning approach. Multimed. Tools Appl. (2021). https://doi.org/10.1007/s11042-021-10707-4
18. Kainz, P., Pfeiffer, M., Urschler, M.: Segmentation and classification of colon glands with deep convolutional neural networks and total variation regularization. PeerJ **2017**(10), 1–28 (2017). https://doi.org/10.7717/peerj.3874
19. Simonyan, K., Zisserman, A.: Very deep convolutional networks for large-scale image recognition. In: 3rd International Conference on Learning Representation ICLR 2015—Conference Track Proceedings, pp. 1–14 (2015)
20. Kather, J.N., Halama, N., Marx, A.: 100,000 histological images of human colorectal cancer and healthy tissue (Version v0.1). Zenodo (2018). https://doi.org/10.5281/zenodo.1214456
21. Detone, D., Malisiewicz, T., Rabinovich, A.: SuperPoint: self-supervised interest point detection and description. IEEE Comput. Soc. Conf. Comput. Vis. Pattern Recognit. Work. **2018**, 337–349 (2018), https://doi.org/10.1109/CVPRW.2018.00060

Chapter 30
The Knowledge Management Model for Spa Business Entrepreneurship in the Upper Northern Thailand

Ploykwan Jedeejit⊙, Yingsak Witchkamonset⊙, and Pratya Nuankaew⊙

Abstract This study intends to raise the standard of spa services in the upper northern Thailand in the terms of research and development (R&D) based on the model of knowledge management in health spa operators. The research process focuses on managing knowledge, building new skills, and enhancing current capabilities to provide standard services for certain expectation of marketing and future customers. It found that the spa businesses in the upper northern Thailand are characteristic according to the traditional cultures, identities, and the style of "Lanna." It reflects the five senses of human beings: sight, taste, smell, sound, and touch which can satisfy spa customers. Furthermore, six spa business professionals reviewed and evaluated the proposed model under the concept of project assessment. It indicated that the recommendations for the spa operators must include the ability to build skills and knowledge in six categories that consist of effective knowledge management through adult learning and entrepreneurial readiness. Therefore, training, seminars, and workshops were conducted as tools for managing the knowledge to achieve the goal.

30.1 Background

According to the spa business, Thailand is ranked in the top five of the worlds. It is an aspect to promote the country. Besides the natural fascination, the characteristic of traditional massage attracts customers globally [1–4]. The service models of spas in Thailand vary depending on locations [5–7]. Hence, spa operators in Thailand can adopt their characteristic local identities such as local traditions, herbs, and knowledge to their spa businesses. Therefore, further innovations with appropriate support could be added to leverage the existing knowledge and understanding of spa operators

P. Jedeejit · Y. Witchkamonset
College of Art Media and Technology, Chiang Mai University, 50200, Chiang Mai, Thailand

P. Nuankaew (✉)
School of Information and Communication Technology, University of Phayao, Phayao 56000, Thailand
e-mail: pratya.nu@up.ac.th

in each local area of the country, especially applications of local resources. However, among countless wellness spas in Thailand, it found that there are only 521 spas meet international standards verified by the Ministry of Health. This is a relatively small number of spas in Thailand [8–11]. Nevertheless, engaging staffs with proper qualifications and professional skills, such as foreign language communication skills are the main areas of improvement.

Therefore, this research presented the results of creating the model of knowledge management in wellness spa operators in the upper northern Thailand. It showed how it performs in real situations through the activities of the selected participants. In addition, this research intends to support the health spa business operators in the upper northern region by creating a model of knowledge management along with new methods to raise the standard of spa services.

Moreover, the knowledge management and license renewal enhance the knowledge and skills of the selected operators in assessing the suitability and quality of a model of knowledge management [12, 13]. In this research, spa businesses recruited experts to discover new options and approaches for spa operators to provide quality and standard service to meet certain expectations of marketing and future customers. The model of knowledge management in health spa operators in the upper northern Thailand adopted the data from several studies on the differences between spa businesses in the upper northern region with the expectations of spa customers to synthesize and develop a model of knowledge management [14]. Spa businesses in the upper northern Thailand are distinctive in the traditional cultures, identities, and the style of "Lanna" [11]. It reflects the five senses of human beings: sight, taste, smell, sound, and touch, which can satisfy spa customers. In facility management, the operators were likely to have outstanding knowledge of spa business management in all six categories of knowledge specified in this research. Regarding the spa customer expectations, it also found that selected customers emphasized staff management the most regarding providing quality spa services [15, 16]. Therefore, the information was applied to support the spa business operators in the upper northern Thailand throughout the model of knowledge management.

Research Objective

This research intends to improve the health spa operators in the upper northern Thailand throughout a model of knowledge management.

30.2 Research Methodology

Several methods were used in this research to learn more about spa business situations in the upper northern Thailand. It was adopted to interview ten spa operators in the upper northern Thailand as given in Table 30.1. Questionnaires were applied to collect information from 373 spa operators and 400 spa customers regarding the expectations of spa businesses. It was applied to develop the model of knowledge

Table 30.1 Demographic characteristics of key informants ($n = 10$)

Demographic characteristics of key informants	N	Percent
Gender		
1. Male	3	30.00
2. Female	7	70.00
Age		
1. 30–40 years old	4	40.00
2. 41–50 years old	4	40.00
3. 50 years and older	2	20.00
Work experiences		
1. 1–2 year(s)	5	50.00
2. 3–4 years	2	20.00
3. 5 years or more	3	30.00
Business scales		
Small	7	70.00
Middle	3	30.00
Provinces		
1. Chiang Mai	4	40.00
2. Nan	3	30.00
3. Lampang	3	30.00

management in wellness spa operators in the upper northern Thailand. The created model was evaluated later by six spa business experts.

30.3 Research Results

According to the objective and the model of knowledge management in wellness spa businesses in the upper northern Thailand implemented in this study, the results can be summarized implementing the steps of the CIPP model [17, 18] as follows.

30.3.1 Context Evaluation: C

This research has two important factors, which are the main contexts. The details can be summarized as follows.

Upper northern Thailand is a region of relaxation

In this research, the upper northern Thailand includes Chiang Mai, Lampang, and Nan. The spa businesses in the upper northern Thailand are characteristic by their identities and cultures that reflect facilities, uniforms, dialects, and local knowledge, such as applications of local herbs, local massage styles, and other "Lanna" identities. They are measured through the five basic senses of human beings: sight, taste, smell, sound, and touch, including natural environments that promote delights and relaxation.

Sense of Sight

Sense of sight includes visual perceptions of landscapes, nature, forests, and relaxation. It also consists of the Lanna architecture styles of upper northern decorations, such as the upper northern style woven fabrics, craved woods, local-lifestyle paintings, local-musical instruments, local flowers, uniforms, and services of local staff.

Sense of Smell

Sense of smell includes applications of natural fragrant plants used to decorate facilities. Some are made into essential oils, incents, candles, etc. Moreover, it can be used in upper northern spas to relax and serve as a natural way of treatment.

Sense of Taste

Sense of taste includes local herbs and vegetables adopted as refreshing herbal drinks and seasonings for customers.

Sense of Sound

Sense of sound includes Lanna's music played with local instruments. It makes customers feel relaxing while they are receiving treatments.

Sense of Touch

Sense of touch includes characteristic massage skills applying the local upper northern herbs along with physical touches that can relax muscles and make customers feel more comfortable.

Factors in the Tourism Businesses

Tourism businesses are factors determining the success of tourism businesses. It consists of the following factors.

Attractions

The main attraction of the spa business in Thailand, especially the upper north, is the advantageous climate and natural environment that attracts tourists from all around the world. Besides, implementing local resources in the health spa business along with other hospitality factors makes visitors feel relaxed and impressed.

Accessibility

Spa businesses in the upper northern Thailand are easily accessible by transportation or places that can be reached by personal car since the transportation routes in Chiang Mai have been developed for several years.

Amenity

Spa businesses in the upper northern Thailand provide various amenities to facilitate their customers whether parking lots, shuttle buses, basic quality infrastructures, and decorations on facilities. It can support customers feeling about the characteristic of Lanna's cultures, identities, and nature. Other amenities include clean spa rooms, Wi-Fi access, etc.

Accommodation

Some spa businesses in the upper northern Thailand are close to hotels and shuttle bus services. Most spas also have lobbies where customers can enjoy and relax the environment with books, magazines, drinks, and refreshments while they are waiting for treatments.

Activities

Spa businesses in the upper northern Thailand provide customers with additional enjoyable activities. During the waiting period, they can enjoy several books, magazines, computers with Internet access, spacious lobbies for families, etc.

30.3.2 Input Evaluation: I

There are six inputs collaborated with the seven standards of ASEAN Spa Service Standards. It comprises place, service, people, products, equipment, management, and environmental practices. They were chosen to be input factors in creating the model of knowledge management in spa operators in the upper northern Thailand as the following declaration:

Human Resources Activities

The spa business in the upper northern Thailand adequately understands human resources. The reason is that operators can evaluate and hire skilled and experienced staff in the spa business. It provides an idea of how many teams are needed to join the company. Also, it covers the suitability of each job position, provision of reasonable wages and welfare management, and inspiration of teamwork. Moreover, they can track the satisfaction of employees in the workplace to maintain service standards.

Management

The spa business in the upper northern Thailand understands the management appropriately. They can plan strategies and keep the company up to date. Its main goal is to provide quality and commercial services to effectively discover how to develop marketing strategies, information, and design attractive campaigns, and financial and accounting situations based on economic conditions at the present.

Products and Services

Spa businesses in the upper northern Thailand can offer good products and services by applying local materials and ingredients to create characteristic effects. Important examples are the application of rare folk herbs, local wisdom, and the design of promotional materials that impress customers.

Physical Environments

Spa businesses in the upper northern Thailand can design excellent facilities and accommodation where operators can consider the characteristic positioning and design of the areas to reflect the Lanna culture. In addition, it also meets the spa standards set by the Ministry of Public Health and provides customers with the highest quality services.

Service Standards

Spa businesses in the upper northern Thailand can design effective services implementing their existing knowledge and control the quality of their services to satisfy customers. They also have plans to support their staff efficiencies in learning better skills, knowledge, and performance to benefit and gain customer trust.

Foreign Language Skills

Spa businesses in the upper northern Thailand mostly use Thai as a primary communication language. Nonetheless, the operators support their staff to learn new alternative significant languages, such as English, Chinese, Korean, and Japanese. This implementation relies on the ASEAN businesses to facilitate visitors and support tourism in the country.

30.3.3 Process Evaluation: P

Development of knowledge management processes to drive knowledge management for spa operators in the upper northern Thailand is the Adult Learning Andragogy (ALA). ALA is a learning process for adults and a lifelong learning style with many different learning management processes in childhood. Generally, learning management for children consists of primary and secondary education. Teachers are responsible to set the goals and objectives and organize learning activities structured according to a predetermined lesson plan. On the other hand, education for adults focuses on promoting the skills and knowledge needed to be applied in daily life.

Spa businesses manage internal knowledge within the organization by learning from real-life situations. Exchanging experiences and knowledge from the previous generation of employees with more knowledge to another generation of employees is necessary. Furthermore, it includes learning from the example of the spa business and inviting experts to improve their knowledge, skills, and expertise.

Nevertheless, there are programs for other staff members in other parts, such as front desk staff and information staff. These programs support them to learn new skills and modern business approaches in the current situation. It also establishes learning environments for the staff to learn and exchange experiences with each other to meet the expectations of customers.

Driving Mechanisms

The key concept used in the CIPP model is the stakeholder assessment. 400 stakeholders may be affected by the evaluation and are expected to participate in the appraisal. The revenue comprises spa operators and clients who are visited or are interested in spa operators in the upper northern Thailand.

The researchers applied the CIPP model (Context, Input, Process, and Product) as a primary concept in developing this research framework. The CIPP model needs the specific components that related a functional framework in an appropriate context of knowledge management and spa business. Input, process, and product are included to extend the desire outputs.

Several contexts refer to the academic, wellness, and spa as the central problems of the process in the knowledge management. Input comprises of five components. The spa business operators and staffs as the beneficial knowledge. In contrast, the providers of information and knowledge likely become information knowledge recipients.

The beneficial elements are presented in a form of learning contents in various questionable issues that the reliable information from learners can be obtaining by the suitable information data sources.

However, other elements with the "Input" aspects were the physical environment and language necessary for the spa business.

First of all, this research focused on finding the knowledge contents under several issues of the spa business entrepreneur as the vital input element. It was determined to formulate the appropriate knowledge management process within the CIPP model. This points out the relevance of research objectives.

Then, a model of knowledge management in spa business operators was developed with its procedural techniques of information acquisition and interpretation to conduct their business expected. This focus would be relevant to the research objective by applying the results of their existing knowledge perceived by themselves.

The model's product was a set of expected competencies necessary for the spa business as the result of the model of knowledge management application. It was judged by spa business experts against the knowledge process and some related ASEAN spa standard criteria. Afterward, the outcome should be a knowledge management process for the spa business in the upper northern Thailand.

The driving mechanisms of the CIPP model used in the spa business in the upper northern Thailand are presented as follows:

Adult Learner Experience

Staff in spa businesses are primarily adults with some experience in their field of work and are interested in improving their current skills and solving problems. Therefore, adult learning requires considering specific factors like physical conditions, needs, desires to know, past experiences, and desires to improve. It also includes the learning readiness and the learning abilities of each learner. In this research, the selected staffs were actively ready to achieve better skills, knowledge, and experiences in specific areas to encourage them in providing better services to customers.

Readiness to Learn

The selected participants in this research are accounted to acquire knowledge, develop skills, and gain experiences by learning from others. They are likely to be sharing their experiences with others for improving the performance of their work and teams.

30.3.4 Product Evaluation: P

Knowledge management manuals evaluated by experts are highly effective. This model contains knowledge management methods with six categories of knowledge for any spas in the upper northern Thailand. Another category of knowledge recommended by experts is northern Thailand's cultural adaptation and awareness.

The experts found that the proposed knowledge management manual created matches with learners and instructors in the spa business in the upper north of Thailand. It can improve the cognition of operators and individuals in a long-term operation if properly adapted with Andragogy.

It also found that the six categories of knowledge presented were required for several spa businesses, such as Human Resources. The experts suggested that physical circumstances and environments can play a significant role for spa businesses since spa businesses require workforces to be effective. Moreover, the main factors attracting the customers are treatments and relaxation.

Moreover, experts also recommend the seventh input for the creation of model since all experts have agreed that cultural adaptation and awareness of Thailand can encourage spas in upper northern Thailand to find the characteristic to compete internationally. An overview summary of the spa business model analysis of the northern region is shown in Fig. 30.1.

A comparison of the existing knowledge of spa operators in the upper north of Thailand, spa operators are intended to learn across the six categories, is shown in Fig. 30.2.

According to Figs. 30.1 and 30.2, it shows that the existing knowledge of spa operators in upper northern Thailand in six categories including Human Resources, Management, Products and Services, Physical Environments, Service Standards, and Foreign Language Skills was lower than the additional knowledge in the same categories that the spa operators desired to learn. This study applied a congruence

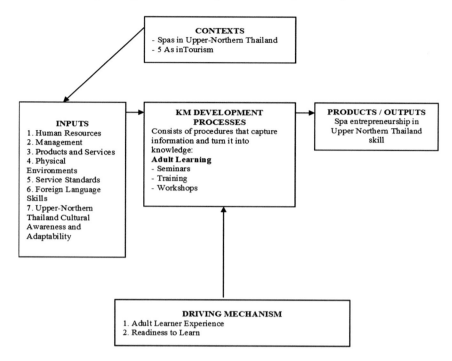

Fig. 30.1 Knowledge management model for spa businesses in the upper northern Thailand with the 7th input suggested by experts

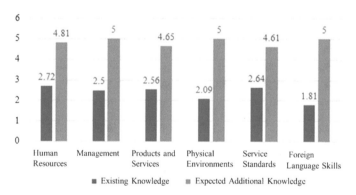

Fig. 30.2 Comparisons between the existing knowledge of spa entrepreneurs in upper northern Thailand in the six knowledge categories and the additional knowledge that the spa entrepreneurs expected to learn

test of the existing knowledge of the spa operators over the empirical data to test the reliability through accuracy of the data.

30.4 Discussion and Conclusion

In this research, the model of knowledge management in spa operators in the upper northern Thailand was adopted through the implementation of Stufflebeam's CIPP model [17, 18]. It was applied to define the contexts of spas in the upper northern Thailand, which are traditionally distinctive in the local wisdom, identities, and Thai "Lanna" cultures. These contexts have been inherited from previous generations; therefore, they cannot be discovered elsewhere. The spas in the upper northern Thailand indicated that the consumption of products as a service respond to the five senses of human beings: sight, taste, smell, sound, touch, and positive environments adjusted in the "Lanna" style blending with nature. It also includes satisfied service charges at a proper price compared with the different domestic regions and other countries.

The proposed main factors were obtained from Stufflebeam's CIPP model. It consisted of six categories of necessary knowledge including management, foreign language skills, physical environments, products and services standards, and human resources. It was used to create a suitable model of knowledge management regarding knowledge integrations, experiences, and skills in activities based on the learning readiness of adults in spa businesses. Moreover, there were training, seminars, and workshops improving the outcomes and products. It can enhance the system operation of knowledge management for any spa operators in the upper northern Thailand.

Besides the proposed six inputs, the experts suggested that the current model with six categories of knowledge for spa operators was an essential category for spa operators in the upper northern Thailand according to cultural awareness and adaptability.

Acknowledgements This research project was supported by the Thailand Science Research and Innovation Fund and the University of Phayao (Grant No. FF65-UoE006). In addition, this research was supported by many advisors, academics, researchers, students, and academic staff from two organizations including the College of Art Media and Technology, Chiang Mai University, Chiang Mai, Thailand, and the School of Information and Communication Technology, University of Phayao, Phayao, Thailand. The authors would like to thank all of them for their support and collaboration.

Conflict of Interest The authors declare no conflict of interest.

References

1. Hirankitti, P., Mechinda, P., Manjing, S.: Marketing strategies of Thai spa operators in bangkok metropolitan (2009)
2. Noviello, M., Smętkiewicz, K.: The revitalisation of thermal areas in the bagnoli district (Naples) as a chance for tourism development in the campania region in the context of selected European experiences. Quaestiones Geographicae **38**, 119–131 (2019)
3. Peric, L.G., Stojiljković, M., Gašić, M., Ivanović, P.V.: Perspectives of development of spa tourism in Serbia. J. Awareness **2**, 597–614 (2017)
4. Szromek, A.R.: An analytical model of tourist destination development and characteristics of the development stages: example of the Island of Bornholm. Sustainability **11**, 6989 (2019)
5. Smith, M.K., Ferrari, S., Puczkó, L.: Service innovations and experience creation in spas, wellness and medical tourism. In: The Handbook of Managing and Marketing Tourism Experiences. Emerald Group Publishing Limited (2016)
6. Lalitnuntikul, P.: Integrated management model of day spas in Thailand (2014)
7. Boleloucka, E., Wright, A.: Spa destinations in the Czech Republic: an empirical evaluation. Int. J. Spa Wellness **3**, 117–144 (2020)
8. Chundasutathanakul, S., Chirapanda, S.: Thailand 4.0: a new value-based economy and its implication on wellness business. Content Page. **113** (2021)
9. Chantaburee, S.: Opportunity and competitiveness in the tourism business in Thailand. Univ. Thai Chamber Commerce J. Humanit. Soc. Sci. **36**, 129–148 (2016)
10. Chanthanawan, S., Fongthanakit, R.: Factors affecting decision making on implementing spa services towards consumers in Chatuchak district, Bangkok. J. Assoc. Res. **24**(3), 190–204 (2019)
11. Chaoprayoon, P., Madhyamapurush, W., Panyadee, C., Awirothananon, T.: The application of Lanna wisdom for spa business in the upper northern Thailand. Mekong-Salween Civil. Stud. J. **7**, 113–124 (2016)
12. Downe-Wamboldt, B.: Content analysis: method, applications, and issues. Health Care Women Int. **13**, 313–321 (1992). https://doi.org/10.1080/07399339209516006
13. Ashforth, B.E., Harrison, S.H., Corley, K.G.: Identification in organizations: an examination of four fundamental questions. J. Manag. **34**, 325–374 (2008)
14. Sama, R.: Impact of media advertisements on consumer behaviour. J. Creat. Commun. **14**, 54–68 (2019)
15. Holloway, I., Galvin, K.: Qualitative Research in Nursing and Healthcare. John Wiley & Sons (2016)
16. Bessant, J., Tidd, J.: Innovation and Operatorship. John Wiley & Sons (2007)
17. Stufflebeam, D.L.: The CIPP model for evaluation. In: Evaluation Models. pp. 279–317. Springer (2000)
18. Stufflebeam, D.L.: The 21st Century CIPP Model. Evaluat. Roots. 245–266 (2004)

Chapter 31
Traffic Analysis Using Deep Learning and DeepSORT Algorithm

Aizat Zafri Zainodin, Alice Lee, Sharifah Saon⊙, Abd Kadir Mahamad⊙, Mohd Anuaruddin Ahmadon, and Shingo Yamaguchi⊙

Abstract With the help of advanced technology available, intelligent traffic has been implemented on roads to improve traffic congestion. Traffic analysis is essential to improve traffic flow in traffic light junctions. Therefore, this project analyzes vehicle speed, type, and count. The Jupyter notebook, Google Colaboratory, and Visual Studio Code are utilized to create live code. For multiple object tracking, Simple Online Real-time Tracking with Deep Association Metric (DeepSORT) algorithm is used with the help of the state-of-the-art object detection model You Only Look Once version 4 (YOLOv4). YOLOv4 is chosen as it is a simpler approach for object detection as compared to the regional proposal method, and it often takes far less processing time. Other than that, TensorFlow acts as an open-source platform for machine learning. It is a useful tool and library to provide workflows with high-level APIs. The average accuracy for each parameter is taken from the tabulated accuracy in different conditions: daytime, nighttime, and rainy weather. As a result, the system is able to detect vehicles with 83.65% average accuracy. There is a slight error in bad weather, nighttime, and traffic congestion for the vehicle count. Therefore, the average accuracy for vehicle count is 65.98%.

31.1 Introduction

People living in a city have to face their daily life with traffic congestion. According to the statistic provided by the Ministry of Transport Malaysia in 2015, Kuala Lumpur had the highest number of vehicles on the road, which is 4,805,029 [1]. Moreover, the density of traffic is increasing from year to year. Supposedly, the increasing number

---

A. Z. Zainodin · A. Lee · S. Saon (✉)
Faculty of Electrical and Electronic Engineering, Universiti Tun Hussein Onn Malaysia, 86400 Batu Pahat, Malaysia
e-mail: sharifa@uthm.edu.my

A. K. Mahamad · M. A. Ahmadon · S. Yamaguchi
Graduate School of Sciences and Technology for Innovation, Yamaguchi University, Yamaguchi, Japan

of vehicles on the road should be collateral with smart traffic. Smart traffic can be categorized as a system that effectively manages traffic to avoid or reduce congestion.

In this era, deep learning techniques are becoming more popular. It is widely used to identify moving objects in videos. Recently, with the great success of deep convolutional neural networks (CNN) in object detection/recognition [2], the CNN-based approaches have the advantage of deep architectures to create non-handcrafted features accurately to represent objects. Deep learning is also commonly used for traffic analysis [3]. However, the existing system for analysis is developed with high costs.

The main target of this project is to analyze vehicle speed, vehicle type, and vehicle count. These traffic data are essential in transportation systems for more effective utilization of resources, improving environmental sustainability, and estimating future transportation needs.

31.2 Materials and Methods

This section explains the methodology used to design this system. It focuses on the architecture design of vehicle detection, vehicle count, and vehicle speed estimation in traffic light junctions.

31.2.1 Evaluation Model for Object Detection

Average Precision (AP) is the most used metric to measure object detection accuracy [4]. Other terms that need to be known are true positive (TP), false positive (FP), and false negative (FN). The definition of TP is a correct detection of the ground-truth bounding box, and FP means an incorrect decision of a non-existent object. Lastly, FN defines the undetected ground-truth bounding box [5]. To decide, the correct and incorrect detection is based on IoU calculation as shown in (31.1).

$$\text{IoU} = \frac{\text{area of overlap}}{\text{area of union}} \tag{31.1}$$

IoU is compared to a threshold value to determine the correctness of vehicle detection. If IoU > threshold, the detection is correct, but if IoU < threshold, then it is incorrect. IoU is used to evaluate the object detection model and applied in the system.

Equation (31.2) shows the calculation of AP with precision and recall value. Precision is the percentage of correct positive predictions, while recall is the percentage of correct positive predictions among all given ground truths.

$$AP = \sum_n (R_{n+1} - R_n) P_{\text{interp}}(R_{n+1}) \tag{31.2}$$

The mean Average Precision (mAP) measures the accuracy of object detectors over all classes. The formula to calculate mAP is shown in (31.3).

$$\text{mAP} = \frac{1}{N} \sum_{i=1}^{N} AP_i \tag{31.3}$$

where N = total number of classes evaluated.

Confusion matrix represents a table layout of the different outcome of prediction and result of the classification problem and help visualize its outcomes. It is a tool for predictive analysis in machine learning. Four elements inside a confusion matrix are TP, FP, FN, and TN.

TP is true positive; its actual value was positive and the model predicted a positive value. While FP is false positive it is our prediction is positive, and it is false. It also calls as the type 1 error. FN is false negative which is it our prediction is negative, and result it is also false. It also calls as type 2 error. Lastly, TN is true negative, which is the actual value was negative and the model predicted a negative value. Four main matrices are:

Accuracy. It uses to find portion of the correctly classified values. It shows how often our system is right. It is the sum of all true values divided by the total values as shown at (31.4).

$$\text{Accuracy} = \frac{\text{TP} + \text{TN}}{\text{TP} + \text{TN} + \text{FP} + \text{FN}} \tag{31.4}$$

Precision. Precision is used to calculate the model ability to classify positive values correctly. When the model predicts a positive value, how often is it right. It is true positive divided by the total number of predicted positive values as shown at (31.5).

$$\text{Precision} = \frac{\text{TP}}{\text{TP} + \text{FP}} \tag{31.5}$$

Recall. Recall is used to calculate the model ability to predict positive values. How often the model predicts the correct positive values. It true positive divided by total number of actual positive values as shown at (31.6).

$$\text{Recall} = \frac{\text{TP}}{\text{TP} + \text{FN}} \tag{31.6}$$

F1-Score. It is the harmonics mean of recall and precision. It is useful when we need to take both precision and recall into account. Equation (31.7) shows F1-Score.

$$F1\text{-Score} = \frac{2(\text{Pecision})(\text{Recall})}{(\text{Pecision} + \text{Recall})} \tag{31.7}$$

31.2.2 System Design

YOLOv4 is used to make this parameter because, based on previous research, YOLOv4 has accuracy as the region-based algorithms at a higher speed. For the detection and counting, we will use YOLOv4 technique.

The detection will have classification such as cars, buses, vans, and trucks in this parameter. In order to detect all the vehicle, a bounding box in pixels need to be created to diagnose the passing vehicle belong to which of the vehicle classes. Each bounding area will show which class should be allocated for the vehicle. Each vehicle can be divided by different colors of rectangle box. Vehicle counting is referred to as the number of passing vehicles at the cross-section zone detection zone. The vehicle is classified by the number. A log text giving is created by the following details: (1) number of counted cars, (2) number of counted buses, (3) number of counted vans, (4) number of counted trucks, (5) number of counted vehicles, and (6) time and date of recording. Figure 31.1a shows the flowchart of vehicle detection and counting.

OpenCV/image processing is used to build this parameter for speed detection, in which the video acquisition is part of defining the region of the interest. This is a small portion of the road to measure movement and the vehicles.

In image recognition and region of interest monitored, the first step involved image subtraction, which helps find where the vehicle is. Masking is used to make the image binary that will make it black and white or one or zeros which is the vehicle would be white, and the background will be black, and that will be able to pinpoint where the vehicle is. Erosion is basically used to distinguish the foreground from the background because sometimes, what happens is when two vehicles are going close, they will have detected together. Thus, the erosion will be able to find the difference between the object. Contour detection is the method that finds the shape of the vehicle that is coming, and it saves the detail of where it is and draws the bounding box accordingly. Object tracking is very important because it is different from object recognition, where object recognition is just finding where the object but object tracking is specifying an 'id' to a particular vehicle, and it follows that vehicle around.

For speed calculation, we start the timer and end the timer when a car crosses a particular segment of road, and based on that we can calculate the speed. Lastly, we save the image into a file. Figure 31.1b shows block diagram of the vehicle speed estimation.

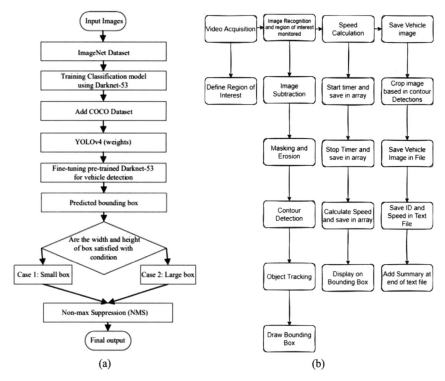

Fig. 31.1 a Flowchart of vehicle detection and counting system using YOLO and **b** block diagram of the speed detection

31.3 Results and Discussion

31.3.1 Vehicle Detection Database

For image classification and object detection, a dataset needs to be used. This project uses Common Object in Context (COCO) dataset format as it is much simpler and widespread usage. Another option of the dataset would be Pascal VOC, but since COCO has larger instances than Pascal, thus COCO dataset format is chosen. COCO dataset contains 91 common object categories, with 82 of them having more than 5000 labelled instances [6].

For object detection, the weight file is also an important parameter. The needed COCO dataset is then imported to the weight file. Then, with the dataset, YOLOv4 algorithm is trained to recognize vehicles in this project. YOLOv4 has been use because it having improvement in the mean Average Precision and at the number of frame per second. However, the performance YOLOv4 model depends on the GPU running on different architecture.

Research in [7] shows how different GPU affects the performance of YOLOv4 model. This research applied common GPU such as Maxwell GPU, GTX Titan, Tesla M40 GPU, Pascal GPU, and GTX 1080 Ti.

31.3.2 Analysis and Result of Vehicle Detection

Google Colaboratory is used to run the Python code. The system is implemented with Python on Google Colaboratory inside a Laptop with 2.5 GHz speed, under 12 GB, and Intel Core i5-7200U processor. To measure the system's effectiveness, we report the number of vehicle detection and the actual number of vehicles available in the 20-s recorded video. Some part of the video is shown in Fig. 31.2. The average frame per second run by Google Colaboratory is 24 frames per second. Other than that, the total frame for this recorded video is 621 frames. Table 31.1 shows the number of vehicles detection by the algorithm with the actual number of the available vehicle with the average accuracy for vehicle detection. On rainy weather, the system detects a truck as bus. This is due to the heavy rain thus the video appears unclear. The average is calculated based on the detection accuracy from daytime, nighttime, and rainy weather results.

Fig. 31.2 Vehicle detection during daytime

Table 31.1 Average accuracy for vehicle detection in different condition

Condition	% error				Average accuracy (%)
	Car	Motorcycles	Truck	Bus	
Daytime	0	0	0	0	100
Nighttime	0	0	0	0	100
Rainy	0	0	25	100	50.95
Total average					83.65

31.3.3 Analysis and Result of Vehicle Count

To measure the systems effectiveness, we report the number of vehicles count and the actual number of vehicles available in the video. Some part of the video is shown in Fig. 31.3. The average frame per second run by Google Colaboratory is 24 frames per second. Other than that, the total frame for this recorded video is 621 frames. Table 31.2 shows the number of vehicles counted by the algorithm with the actual number of the available vehicle, with the average accuracy for vehicle counting. The algorithm detected a several errors in all condition. This is because it detected a van and a small lorry as a car. Meanwhile, for buses, in the video, there are no buses, but the system still detects the number of the bus as 1. This is due to it detecting a long lorry as a bus. In rainy condition, the percentage error is 15.55% it is because this is due to the heavy rain, thus the video appears unclear. The average is calculated based on the counting accuracy from daytime, nighttime, and rainy weather results.

Fig. 31.3 Vehicle counting system during daytime

Table 31.2 Average accuracy for vehicle count in different condition

Condition	% error				Average accuracy (%)
	Car	Motorcycles	Truck	Bus	
Daytime	12.5	0	100	100	52.60
Nighttime	8.33	12.5	0	100	49.75
Rainy	15.55	0	25	100	95.60
Total average					65.98

Fig. 31.4 Vehicle speed estimation system during daytime

31.3.4   Analysis and Result of Vehicle Speed Estimation

As can be shown in Fig. 31.4, the speed is ranging from 7 to 29 km/h. However, since this is a highway road, the speed shown is irrelevant. This happened because a recorded video is used for speed detection. Therefore, the calculation is based on pixel per meter and frame rate. The limitation to get accurate result is the usage of graphics processing unit (GPU).

31.4   Conclusion and Recommendation

To conclude, traffic light system has improved greatly, starting from conventional system, up to sensor-based traffic light. Now, the use of deep learning is implemented into the system. The main purpose of this project is to analyze parameters to measure traffic at road. With the help of various deep learning techniques such as RCNN and YOLO, traffic analysis is performed on vehicle speed, vehicle type, and vehicle count. In conclusion, this vehicle detection method utilized YOLOv4 algorithm with the help of Google Colaboratory to write and execute Python code. Designing the system requires DeepSORT technique because it is more organized than using bounding box technique for object detection. From the result, the accuracy of vehicle detection that obtain is 100% for the daytime and nighttime. Meanwhile, at rainy weather is almost 100%. This is because the system got affected by the droplets of rain. Therefore, it is hard for the system to recognize the object. Overall, for a smaller sample size, the detection system almost obtained high accuracy and precision result. However, due to the limitation on PC processor, it affects the accuracy of speed detection.

This system can be improved by using a good GPU to do the video processing smoothly. Other than that, video enhancement for this system can be improved in the future. For instance, during rainy weather, the video will become blurry, resulting in the system having a slight error in detecting the object. With future research on video enhancement, this issue can be mitigated.

References

1. Bilangan Kenderaan di atas Jalan Raya Mengikut Negeri (Number of Vehicles On The Road by State)—Set Data—MAMPU. https://www.data.gov.my/data/ms_MY/dataset/bilangan-ken deraan-di-atas-jalan-raya-mengikut-negeri. Accessed 28 Dec 2021
2. Zhang, H., Liptrott, M., Bessis, N., Cheng, J.: Real-time traffic analysis using deep learning techniques and UAV based video. In: 16th IEEE International Conference on Advance Video Signal Based Surveillance, AVSS 2019, pp. 1–5 (2019). https://doi.org/10.1109/AVSS.2019.890 9879.
3. Won, M., Sahu, S., Park, K.J.: DeepWiTraffic: low cost WiFi-based traffic monitoring system using deep learning. In: Proceedings of IEEE 16th International Conference Mobile Ad Hoc Smart Syst. MASS 2019, pp. 476–484 (2019). https://doi.org/10.1109/MASS.2019.00062
4. Padilla, R., Netto, S.L., Da Silva, E.A.B.: A survey on performance metrics for object-detection algorithms. In: International Conference on System Signals, Image Processing, vol. 2020-July, pp. 237–242 (2020). https://doi.org/10.1109/IWSSIP48289.2020.9145130
5. Kim, C.E., Dar Oghaz, M.M., Fajtl, J., Argyriou, V., Remagnino, P.: A comparison of embedded deep learning methods for person detection. In: VISIGRAPP 2019—Proceedings of 14th International Joint Conference on Computing Vision, Imaging Computer Graphics Theory Application, vol. 5, pp. 459–465 (2019). https://doi.org/10.5220/0007386304590465.
6. Colleges, G.T.U.A., et al.: Microsoft COCO. Eccv, no. June, pp. 740–755 (2014)
7. Bochkovskiy, A., Wang, C.-Y., Liao, H.-Y.M.: YOLOv4: Optimal Speed and Accuracy of Object Detection. April, 2020, [Online]. Available: http://arxiv.org/abs/2004.10934

Chapter 32
Blockchain-Based Certificate Verification

Alice Lee, Aizat Zafri Zainodin, Abd Kadir Mahamad◉, Sharifah Saon◉, Mohd Anuaruddin Ahmadon, and Shingo Yamaguchi◉

Abstract Academic certificates are essential to acknowledge one's academic achievement. With the credentials, it can be used to develop a career in a desired industry based on the qualifications. However, nowadays, fake academic certificates are used by irresponsible individuals to apply for a job. Therefore, it is crucial for the employer to validate one's academic certificate before offering them the job position. To validate, a certificate may be too troublesome and require a long time for the verification process. In addition, the employer must check with the certificate's issuer, which can involve a lot of processes. Thus, to solve this problem, blockchain-based certificate verification is introduced where the issuer can store the certificates details inside a blockchain network with the help of a smart contract feature that is immutable and anti-counterfeit. Smart contract is written in Solidity programming language and executed in Remix IDE. In terms of security for this smart contract, only the certificate's issuer is able to add, edit, modify, and delete the details stored. Not only that, the contract's owner also unable to add data into the blockchain due to the decentralized feature of a blockchain network. The average execution time for the smart contract is 8.3 s. All changes made to the certificate's details can be tracked; thus, changes to the previous recording are impossible. This system will benefit the university because the students' data can be stored in a system and can be retrieved anytime. For a third-party user such as the employer, it can save time to check the validity of the academic certificates they received because the system generates the confirmation of transaction in less than five seconds.

A. Lee · A. Z. Zainodin · S. Saon
Faculty of Electrical and Electronic Engineering, Universiti Tun Hussein Onn Malaysia, 86400 Batu Pahat, Malaysia

A. K. Mahamad (✉) · M. A. Ahmadon · S. Yamaguchi
Graduate School of Sciences and Technology for Innovation, Yamaguchi University, Yamaguchi, Japan
e-mail: kadir@uthm.edu.my

32.1   Introduction

Blockchain technology started to get popular when Satoshi Nakamoto used it in 2009 for the digital cryptocurrency Bitcoin. Now, blockchain can be used for broad purposes, such as education, agriculture, and supply chain. Blockchain is a distributed ledger that is open to anyone. As the name implies, the data is stored in a block. Each block has an identity which is called a hash value. Hash is completely unique or can be described as a "fingerprint." Another unique feature of this hash will be that if the data inside the block changes, it will also affect the hash. In other words, the block is no longer the same.

The key advantage of blockchain is widely considered to be decentralization. It can help establish disintermediate peer-to-peer (P2P) transactions [1, 2], coordination, and cooperation in distributed systems without mutual trust and centralized control among individual nodes, based on data encryption, time-stamping, and distributed consensus algorithms, and economic incentive mechanism [3].

Nowadays, there are seen increasing cases of fraud in academic certificates. Some called "experts" are using fake certificates to be recognized by the community. These fake certificates can be easily obtained illegally. According to New Straits Times, it reported that each certificate could be obtained ranging from RM 6500 up to RM 10,000 [4].

Verifying the validation of a certificate usually takes some time and needs to be done manually by looking at the university database according to the serial number provided in the certificates. With blockchain technology, certificates are stored in digital form and can be verified by anybody, anywhere, and anytime.

32.2   Methods

32.2.1   Smart Contract

The development of blockchain is based on smart contract. The smart contract architecture requires the parties to the smart contract to agree to interact with each other on certain conditions. The general idea of the working mechanism of smart contract deployment in Remix IDE can be seen in Fig. 32.1.

The smart contract is written on Remix IDE. Remix IDE is an open-source software that is used to construct smart contract in Solidity language. Another common software used by blockchain developer is Visual Studio Code. The methods of using these two softwares are almost the same. However, Remix IDE did not require any installation because it is based on the browser platform. With Remix, testing, debugging, and deploying of the smart contract are performed.

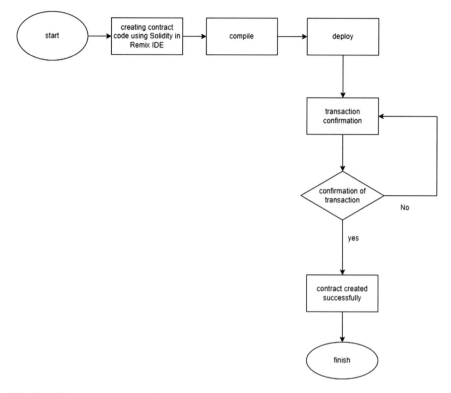

Fig. 32.1 Smart contract execution process

32.2.2 System Design

An overview of the system design needs to be developed to provide a better understanding of the project.

The certificate contains all the details of its holder. For instance, an academic certificate contains the students' details such as the student's name, student's identification number, and other important details. These details are key in before deploying the contract. Next, hash values are generated and stored in blocks. The blocks are linked together to form a blockchain. Any change in the actual information will change the corresponding hash also, thus making it hard for somebody to manipulate the document.

The hash values generated when deploying the contract using third-party address should be verified and compared to the hash values stored in the blockchain to validate the certificate [5–10]. If both are the same, it means the certificate provided by the student is valid. The general framework is illustrated in Fig. 32.2.

From Fig. 32.2, the flowchart of the coding can be detailed as in Fig. 32.3. Registrar, shown in Fig. 32.2, refers to the certificate's issuer. In this case, it is referring to the university. The trusted third party is the individual that wants to use the system, which is a hiring manager.

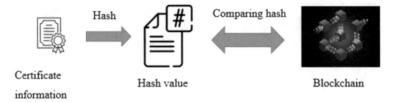

Certificate
information   Hash value   Blockchain

Fig. 32.2 General illustration of system design

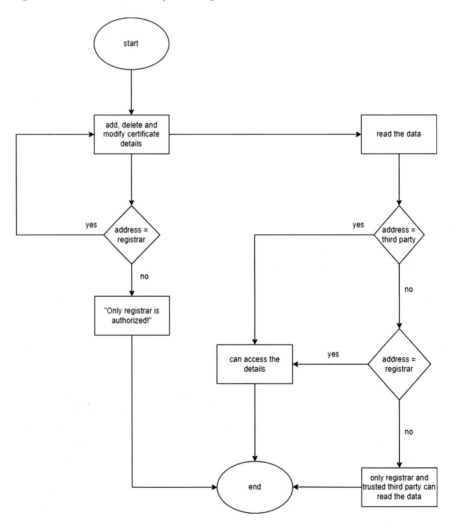

Fig. 32.3 Flowchart of smart contract coding

When the user wanted to perform transaction, the account address needs to be set in Remix IDE. If the address matches registrar, then the user is able to add, delete, and modify the data. However, if the address does not match, then it will show "only registrar is authorized."

32.2.3   Ethereum Wallet

Remix IDE is one of the most popular Ethereum platform used by developers to write, compile and debug smart contract [11]. To perform the operations, a cryptocurrency token known as Ether is used. Therefore, a browser-based plugin known as MetaMask is utilized as a cryptocurrency wallet that enables users to store Ether.

For this project, Ropsten Test Network is used. Ropsten is known as a public blockchain network in which users can deploy transactions without any real cost associated for testing purposes. MetaMask plugin serves as a connection of identifying user's Ethereum accounts inside various configured blockchain network (Ropsten network), through the identification of their account's private key or either their validated account hash and the password. After the correct association of the user account, the smart contract is now ready to be deployed onto the blockchain network. To get Ether for Ropsten, developers could request testnet ETH on the Ropsten testnet using a Ropsten faucet.

32.2.4   Ganache

Other than using test network inside MetaMask, Ganache account can be imported to MetaMask account. Ganache is a personal Ethereum blockchain that is used to run tests, execute commands, and inspect states while controlling how the chain operates.

Ganache gives the user ten different account addresses. Each account has 100 ETH. To import Ganache account to MetaMask wallet, first need to copy private key of one of the account in Ganache. To add the local blockchain network in MetaMask, entered the RPC URL and Chain ID (Default Value: http://127.0.0.1:8545 and 1337, respectively). Then, a new account is created by entering the private key. Take note that the Ganache account given is for testing purposes only.

32.3   Results and Discussion

Three parties are involved in this smart contract. The first one is owner address to represent the smart contract's owner. Owner only can deploy and destroy the contract. Secondly, the registrar. Registrar is the person that are able to add, delete, and modify the students' details. Lastly, the third-party user. It is the employer that wishes to

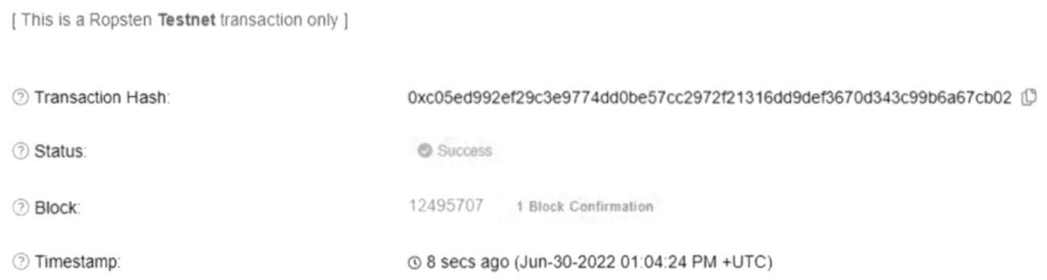

[This is a Ropsten **Testnet** transaction only]

⑦ Transaction Hash:	0xc05ed992ef29c3e9774dd0be57cc2972f21316dd9def3670d343c99b6a67cb02 🗐
⑦ Status:	⊘ Success
⑦ Block:	12495707 1 Block Confirmation
⑦ Timestamp:	⊙ 8 secs ago (Jun-30-2022 01:04:24 PM +UTC)

Fig. 32.4 Etherscan shows the time stamp of each transaction in smart contract

Table 32.1 Average time taken and gas consumption for smart contract execution to add certificate details

Certificate number	Time taken for the execution of smart contract (s)	Gas consumption (Gwei)
1	8	2.500000007
2	5	2.500000007
3	17	2.500000007
4	3	2.500000007
5	10	2.500000009
6	6	2.500000009
7	7	2.500000009
8	5	2.500000008
9	9	2.500000007
10	13	2.500000007
Average	8.3	2.5000000077

read certificate's details and check its validity therefore the third-party user are only able to read the details.

The time taken for MetaMask transaction to add certificate's details into the network is approximately five seconds. The transaction details are viewed on Etherscan as shown in Fig. 32.4. The average time taken is calculated in seconds for the execution of smart contract is tabulated in Table 32.1.

The time taken for the smart contract to be executed depends on the gas limit that indicates the maximum amount of gas to be spend for the transaction.

32.4   Conclusion and Recommendations

To conclude, this system is focused on smart contract development that acts as back end application. For testing purposes, the smart contract only displayed how the students' details are stored into the blockchain network. For future works, a front end application written in HTML can be built to create user interface of the application. To do this, a certificate needs to be in PDF file and uploaded into the application. Since storing data in the blockchain network is expensive, another alternative that can be taken into consideration is to integrate InterPlanetary File System (IPFS) into the network.

References

1. Crosby, M., Pattanayak, P., Nachiappan, Verma, S., Kalyanaraman, V.: BlockChain Technology: Beyond Bitcoin. Controlling, **27**(4–5), 222–228 (2015). doi: https://doi.org/10.15358/0935-0381-2015-4-5-222
2. Croman, K., et al.: On scaling decentralized blockchains (A position paper). Lecture Notes Computer Science (including Subseries Lecture Notes Artificial Intelligence Lecture Notes Bioinformatics) 9604. pp. 106–125 (2016). https://doi.org/10.1007/978-3-662-53357-4_8
3. Bokariya, P.P.: Motwani, D: Decentralization of credential verification system using blockchain. Int. J. Innov. Technol. Explor. Eng. **10**(11), 113–117 (2021). https://doi.org/10.35940/ijitee.k9514.09101121
4. Imam, I.T., Arafat, Y., Alam, K.S., Aki, S.: DOC-BLOCK: A blockchain based authentication system for digital documents. Proc. 3rd Int. Conf. Intell. Commun. Technol. Virtual Mob. Networks, ICICV 2021, pp. 1262–1267 (2021). doi: https://doi.org/10.1109/ICICV50876.2021.9388428
5. Ghani, R.F., Salman, A.A., Khudhair, A.B., Aljobouri, L.: Blockchain-based student certificate management and system sharing using hyperledger fabric platform. Periodicals of Engineering and Natural Sciences **10**(2), 207–218 (2022). https://doi.org/10.21533/pen.v10i2.2839
6. Shawon, M.S.K., Ahammad, H., Shetu, S.Z., Rahman, M. M., Hossain, S.A.: Diucerts DAPP: A blockchain-based solution for verification of educational certificates. Paper presented at the 2021 12th international conference on computing communication and networking technologies, ICCCNT 2021, (2021). doi:https://doi.org/10.1109/ICCCNT51525.2021.9579533
7. Malsa, N., Vyas, V., Gautam, J., Ghosh, A., Shaw, R.N.: CERTbchain: A step by step approach towards building A blockchain based distributed appliaction for certificate verification system. Paper presented at the 2021 IEEE 6th international conference on computing, communication and automation, ICCCA 2021, pp. 800–806 (2021). doi:https://doi.org/10.1109/ICCCA52192.2021.9666311
8. Kumutha, K., Jayalakshmi, S.: The impact of the blockchain on academic certificate verification system-review. EAI Endorsed Transactions on Energy Web **8**(36), 1–8 (2021). https://doi.org/10.4108/eai.29-4-2021.169426
9. Taha, A., Zakaria, A.: Truver: A blockchain for verifying credentials: Poster. Paper presented at the proceedings of the ACM symposium on applied computing, pp. 346–348 (2020). doi:https://doi.org/10.1145/3341105.3374067
10. Nguyen, B.M., Dao, T., Do, B.: Towards a blockchain-based certificate authentication system in Vietnam. Peer J Computer Science, **2020**(3), (2020). doi:https://doi.org/10.7717/peerj-cs.266
11. Liu, L., Han, M., Zhou, Y., Parizi, R. M., Korayem, M.: Blockchain-based certification for education, employment, and skill with incentive mechanism, (2020). doi:https://doi.org/10.1007/978-3-030-38181-3_14

Chapter 33
Optical Wireless Communication for 6G Networks

Rudresh Deepak Shirwaikar, H. M. Shama, and Kruthika Ramesh

Abstract New high data rate multimedia services and applications are evolving continuously and exponentially increasing the demand for wireless capacity of fifth generation (5G) and beyond. Radio Frequency (RF), which was the most used form of communication, cannot satisfy the need for 6G. Optical wireless communication (OWC), which uses an ultra-wide range of unregulated spectrum, has emerged as a promising solution to overcome the RF spectrum crisis. The few important and common issues related to the service quality of 5G and 6G communication systems come with great capacity, massive connectivity, low reliable latency, high security, low-energy consumption, high quality of experience, and reliable connectivity. Of course, 6G communication will provide several-fold improved performances compared to the 5G communication regarding these issues. This chapter will represent how OWC technologies, such as visible light communication, light fidelity, optical camera communication, and free space optics communication will be an effective solution for successful deployment of 6G.

33.1 Introduction

With the brief knowledge on Optical Wireless Communication (OWC), lets further understand the role of OWC in the deployment of 6G. The most used wireless communication till date was the Radio Frequency also known as RF. It provided a range of 3 kHz to 10 GHz which proved to be exhausting and insufficient to the given advanced version of 6G. Only a small fraction of optical communication is in usage,

R. D. Shirwaikar (✉)
Department of Computer Engineering, Agnel Institute of Technology and Design, Goa University, Assagao, Goa, India
e-mail: rudreshshirwaikar@gmail.com

H. M. Shama
Innovation Centre, BMS Institute of Technology, VTU, Bangalore, Karnataka, India

K. Ramesh
Deparment of ISE, BMS Institute of Technology, VTU, Bangalore, Karnataka, India

i.e. visible, infrared, and ultraviolet. Future research and development will lead to furthermore exploitation of the spectrum. Terahertz communication falling in the range of 100 GHz to 10 THz, provides the main requirement for the 6G network which is enabling high data rates and frequency connectivity. Its frequency can be enhanced by exploring frequency bands that do not fall in the category of molecular absorption. The main setback for this communication system is the propagation loss. After much awaited research on Visible Light Communication (VLC), it proved to be more mature than terahertz and it can be conventionally used indoors because of the implementation of cellular coverage over short distances. Outdoor and space communication can be implemented by Free Space Optical communication (FSO). As the name suggests, it is used as a free space medium. For short distances FSO can be implemented by LEDs following VLC. FSO can also be implemented point to point using infrared which is its most simple form. However, FSO has failed to be reliable over short ranges because of its dropping packets. Further the ultraviolet-B (UV-B) communications have proven to provide extremely high data rates as compared to the visible light. Similarly, UVC band communications are also used wildly, these being solar blind can provide accurate ground-based photon detectors. Their ideal transmission is provided by NLOS along with a large Field of View (FOV). The major drawback of this band is its requirement for dense network configuration and having a high channel attenuation leads to reduction in data rate as well as transmission range [1, 2].

Given some of the technologies used by the OWC for the implementation of the 6G, the fruit of further development can be added to this basket of this knowledge. Further development of Terahertz is to make it available in terms of an inexpensive small laser diode, using detectors which can absorb well are also used to enhance its transmission. The overall data rate expansion for the VLC is done by techniques such as Orthogonal Frequency Division Multiplexing (OFDM), Discrete Multitone Modulation (DMM), and Multiple Input Multiple-Output (MIMO), and Wavelength Division Multiplexing (WDM), and so on. The performance of FSO can be improved by implementation of SAC OCDMA and data rate is increased further by Low Pass Filter (LPF). The UVC can be further developed to implement UV-B-LED which will provide extraordinary data rate using QAM. These enhancements are likely to be done this decade owing to improvisation and successful launch of 6G for the coming decade [3, 4].

33.2 Background

The only constant known to mankind is evolution, we have constantly evolved in terms of appearance, culture, food habits, technology along with everything that can be further developed. In the twenty-first century, the most important aspect to a modern human being is communication technology. Communication has come a long way with its rich history whilst taking several turns culturally and politically. Passing along Telegraph, to Landline Telephones, to Phonograph and to reaching the medieval age of radio and television. Picking from here it expanded to mobile technology and to the most known technology worldwide "Internet". The use of the Internet as a global mass communication is vastly practised with wireless communication. The wireless communication started with 1G FM Cellular systems in 1981 developing further to 5G in 2020 [2].

The services offered by 5G are an extravagant amount of high multi Gbps data speeds, ultra-reliable and low latency communications (uRLLC), additional reliability, massive network capacity, improvised and increased availability, massive Machine type communications (mMTC) and a more uniform user experience. With this massive improvement in 5G globally it has provided goods and services worth $13 trillion. With the industry growing upfront and increased exploitation of its services has raised concern due to incompetency of the old devices, development of the infrastructure not being economical. With the blooming development and increased commercialization of 5G, it's the right time to expand further to the 6G. A tentative timeline has been drafted for the implementation of 5G, B5G, and 6G, respectively, by the international standardization bodies. They reveal that the 6G will take the world by storm by 2030 [5]. The development regarding the same has begun in various countries. 6G being developed will send a rippling effect of high data speeds provided by the technologies that will be playing a keen role in the development of this wireless network. The operating power of 6G is expected to be 100 GHz to 10THz and the data speed will boost up to 1000 gigabits/seconds. 6G is expected to be 1000 times faster as compared to other networks. Having an advanced network makes everyday problems like connectivity, infrastructure, and limitation of mobile phones to be banished away forever. The technologies which will be the main act for 6G are AI, 3D networking, Optical wireless communication, THz band, unmanned aerial vehicles, etc. [6].

Optical wireless communication is deployed for the 6G network. Optical wireless communication (OWC) as the name suggests is a form of communication based on optical nature like ultraviolet, infrared, and visible light. These optical mediums carry the signal. When the signal is sent in the visible optical band, it is known as Visible Light Communication (VLC). VLC are used to pulsate signals at extreme speeds which makes it impossible to notice by a naked eye, this is mainly implemented by LEDs. These are used in wide range applications such as WLAN, vehicular networks along with the rest. Ultraviolet communication which uses UV spectrum is rather successful because the majority of the UV radiation in the sunlight is absorbed by the ozone layer, giving almost zero background noise to use this form of communication.

The massive scattering effect of the photon tends to be feasible for non-line of sight (NLOS) communication. The high absorption rate of the UV band ensures the communication is highly secured and protected making it useful for communication in the field of environment sensing, unattended ground sensor network amongst many others. Infrared unlike UVC requires signal transferring between the range of 430THz to 300 GHz and a wavelength of 980 nm falls under this radar [7]. There are mainly two types of infrared communication point to point and diffuse point. The transmitter and the receiver pointing at each other marks point to point and reflecting or bouncing of transmitted signal between the transmitter and the receiver, respectively, marks diffused point. It is proven to be highly secure and maintained to have high-speed data communication [8].

33.3 Radio Frequency (RF)

Wi-Fi is the most used form of wireless connection across the world. This network allows the user to travel to remote places without affecting the connection. This form of wireless connection was once satisfied by radio frequency. This Radio frequency provides a range of 3 kHz–10 GHz. This can only support some small amount for optical wireless communication. In this section we are going to learn about how RF is used for wireless communication and its limitations.

33.3.1 RF Wireless Communication

This form of communication has a rich history of over 100 years. In 1901 Marconi et al. [9] conducted experiments where alerts through radio were sent to the surrounding area. From this it further developed to being the most used wireless communication for over a century. It was the main broadcast nature, which was created by AM and FM modulation. After the discovery of the mobile phones in 1979, RF was used as the main communication network for verbal exchange.

33.3.2 The Structure of RF

RF systems are classified for both terrestrial and space-based systems. Terrestrial system is composed of WLAN which is point to point. One of them being mobile Phones. This system also constituted Line of sight (LOS) communication which is the reason for its limiting factor. Space-based systems, the communication between the satellites is done by uplink and downlink for up and down information transfer [9]. The system mainly constitutes of transmitter, channel and receiver as indicated in Fig. 33.1. The Transmitter transfers the signal which can be audio, video and data or sometimes it can contain the combination of all the three. The signal being transferred expires by time, it is a victim of atmospheric loss and the weather plays a huge role in the attenuation. Sometimes human inference also can damage the information being passed. The channel which is considered as the transmission medium plays a huge role. The type of channel which is mainly air or the atmosphere is used. Once the transmitter sends the data signals, it is transferred by the channel to the receiving end. The receiver which receives the signal converts it back to its original form and the information is further used. The main drawback of the RF is its security issue.

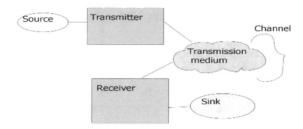

Fig. 33.1 A simplified version of RF system

The signal can be easily inferred, and the hacker can then gain entry to the imitation network and consume the data.

Propagation of the signal or wave can be done in various methods. Free Space Propagation, this form of propagation takes place when the transmitter and receiver are separated by a distance, the waves covering those distances have a transmitting power. The second form of propagation is through reflection and refraction. This happens when the waves are reflected by a vacant spot which has a greater wavelength, this mainly takes place because of sudden interruption. The following Fig. 33.2 can give an idea of reflection and refraction. The third form of propagation is through diffraction. This happens when the waves are interrupted by a partial object which changes its course of path. The fourth form of propagation is through inference, this form generally slows down the transmission over long distances, as it happens when an obstacle comes across the path. It causes alteration to the waves and can lead to time-varying results [10].

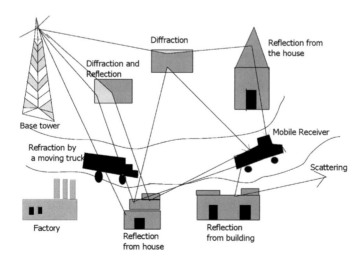

Fig. 33.2 Propagation through reflection and refraction

33.3.3 Security Issues and Challenges of RF

According to the National Institute standards or Technology. The following have been considered as a threat to wireless communication by RF networks.

(a) Theft of information leading to forgery.
(b) Hacking and gaining access to the data being transferred.
(c) The code is easily hackable.
(d) Not considered ideal for transferring personal and professional details.
(e) The omission of mistakes leads to attenuation and low efficiency.
(f) Poor infrastructural support.
(g) Active attacks that lead to loss of privacy.
(h) Damage of network service.

The reasons for RF not used as a technology for 6G.

(a) It breaches security and privacy shown in Fig. 33.3.
(b) It has a short range of frequency.
(c) It does not support the complexity of the newly advanced electronic gadgets.
(d) It cannot support transferring high data per sec.

33.4 Terahertz Communication

After the radio frequency failed to supply the necessary frequency range. The closest required frequency range is provided by the Terahertz. Terahertz is the substitute name for the spectrum range that falls into the category of 0.1–10 Thz. This has proven to have the required technology to support the 6G networks. Terahertz has provided the base for nano communication as well. To match up to the level of the nano devices, a special antenna is installed to achieve its high range frequency. Similarly, the frequency used by the terahertz can penetrate through a lot of materials with a minor attenuation making it very feasible to be used around opaque objects. Terahertz frequency is very similar to radio frequency, they are quite different when it comes to diffraction, antenna properties, transmission, and absorption by materials.

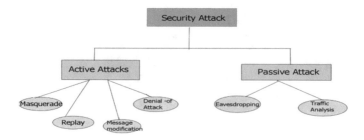

Fig. 33.3 Types of security attacks

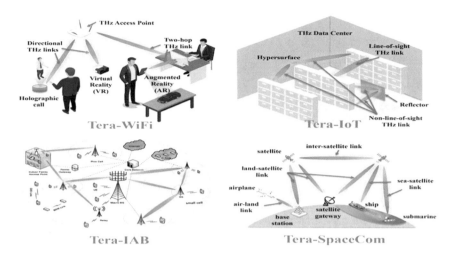

Fig. 33.4 Various application of terahertz communication

Some of the applications of the Terahertz WLAN system include Tera-IOT, Tera-space Com, Tera-Wi-Fi as depicted in Fig. 33.4. Despite having all these features terahertz has its own flaws which include, low antenna gains at the receiver end, given its high frequencies and attenuation it may lead to gain reduction. Power consumption of the devices using the terahertz has proven to be challenging. A few more challenges include link stability, beam pointing control, and so on. After this brief introduction, the topics discussed further will contain the Propagation through terahertz channel, Application of terahertz for 6G communication and challenges of terahertz communication.

33.4.1 Terahertz Channel

The Terahertz system consists of a transmitter, channel and receiver. Given these aspects, the propagation characteristics are modelled between the transmitter and the receiver. The atmosphere plays an important role in the absorption of the frequency and it cannot be neglected like that of a RF. There are various methods to calculate the absorption which is provided by the International Telecommunication Union (ITU-R). This loss is referred to as atmospheric loss [11]. Figure 33.5 shows the plots of atmospheric attenuation for the range 0–1 THZ. Terahertz supports non-line of sight communication (NLOS) which leads to multipath diversion. The scattering effect of the terahertz is a major concern. Despite this disadvantage, research once done in 2007 showed that a surprisingly good performance by the multipath diversion given the transmitter and the receiver are optimally aligned. The loss occurred in the above passage is called as loss due to scattering and diffusion. There is another loss which has arisen worries that being the spreading loss. This occurs due to antenna aperture

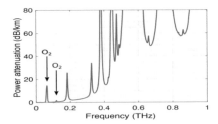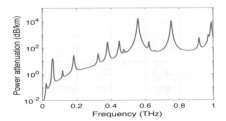

Fig. 33.5 Terahertz wave atmospheric power attenuation

which varies the frequency at the end links. This can be reduced if the antenna manages to have a constant physical area which will keep the frequency gains equal at both the end links. Apart from this weather also influences a lot to attenuation, if the particles present in the air is lesser than the wavelength of the frequency then it leads to strong attenuation in non-clear skies. Similarly, there are many more factors that cause these attenuations. Terahertz communication is used in system-wide channel sounding. This is implemented using Channel impulse response (CIR) [12].

33.4.2 Applications and Challenges of Terahertz Communication in 6G Network

UM-MIMO—To reduce the spreading loss, the antenna performance details are carefully fed to the UN-MIMO. This can also increase the spatial multiplexing, make it more energy efficient and distance can be enhanced. Three unique architectures of terahertz are taken and combined to form a hybrid which provides good spectrum efficiency, low hardware complexity, and energy efficiency. Fully connected (FC), Array of subarray (AoSA), and dynamic array of subarray (DAoSA) are the three distinct architectures used in UM-MIMO [11].

Medium Access Control (MAC)—There are various challenges faced by the MaC of Thz some of them being deafness problem, transmission problem due to avoidance of the deafness problem, the received signal strength (RSS). These challenges are overcome by MAC protocols in the THz wireless networks which opens a path to the distinctive spectrum features [11].

Modulation and coding—Modulation is designed on the distance between the transmitter and the receiver, for short range pluses can be modulated using pulse amplitude modulation (PAM), OOK. The details of this are discussed in the next section. The coding schemes developed are with low complexity and to support THz communication. This must be further implemented.

The challenges include the following—(a) The attenuation is greater than expected, the main being is already discussed, future methods to avoid the same need to be discussed. (b) Due to frequency inference and the total range of the frequency has proven to not add up to the requirement of the 6G communication. (c)

The lack of coding application which still is a tedious task must be taken care of. (d) The shift of appliances from 5 to 6G is not economical.

33.5   Visible Light Communication (VLC)

The range and data transmission offered by the Radio frequency do not meet the bar required for the new advancement in 6G networks. It also fails to reach the potential transmission speed and the transmission medium. With the improvement in technology and advancement in gadgets, an average person uses at least 2 or more electronic gadgets which requires this network making it impossible for the previously discussed technologies to catch up. Penetration through walls and inside a room is the main reason for the security breach in RF technology. The inference from the electronic devices disrupts the frequency. These given limitations are overcome by VLC. It offers unlimited bandwidth, it acts as a shield for any external inference, it has a better privacy as light cannot penetrate through opaque objects and is more economically feasible. Solid state lighting (SSL) has been the main reason for the development and deployment of the VLC communication. As the name suggests, this communication is mainly based on the visible light region. There are two broad approaches to generate VLC, one of them includes generating white light through LEDs and the other one includes RGB LEDs. The second approach can also create white light with the right mixture of the three basic colours (red, green, blue). The first approach is more economically feasible and user friendly making it the most used method in VLC [13]. The Fig. 33.6 depicts the two approaches.

33.5.1   Indoor Visible Light Communication System

With the increase in electronic devices usage per specific area, it is more convenient to use indoor VLC as the devices used per person on an average has increased to 2 or more. The use of VLC has been more accurate because wired communication

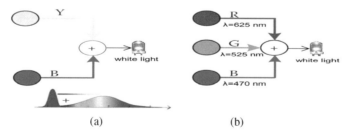

(a)   (b)

Fig. 33.6 Two approaches to produce white light from LEDs. **a** Phosphor LED method, **b** RGB method

makes it impossible to use given its high feasibility and complexity and use of Radio frequency has its own limitation in terms of bandwidth. Similarly using Infrared Technology has an adverse health effect. The VLC system consists of a transmitter, a channel and receiver. As shown in Fig. 33.7, the transmitter and the receiver use Digital and analogue components, digital to analogue converter (DAC) and analogue to digital converter (ADC), respectively. The transmitter, which is the light source, is mainly the LEDs. The LEDs are installed in the places of need such as workspace, home, and so on. The most common transmission medium for VLC is the air. The light particles penetrate through the air medium to reach the receiver. The receiver of VLC is the photodiode, it first has a filtering process to remove all the additional noise and converted to DC power after amplification to the required need [14]. Figure 33.7 shows a basic block diagram of Indoor VLC.

The transmitters are mainly LEDs, which convert DC to AC and emit light into free space. The LEDs are considered DAC because they are beneficial towards health, they yield high efficiency from low power consumption, and so on. The while light emitted from the LEDs is fed into free space. From the free space, the signal is then received by the receiver. The connection between the transmitter and the receiver in this system is called the VLC link. The link is mainly of two types, first one is based on the degree of directionality between the transmitter and the reviewer also called non-line of sight communication and the second one is the presence of a direct path between the two also called line of sight communication. The VLC link is chosen based on the nature of the transmitter and the receiver. The LOS is more efficient but setting up the required units is a tedious task. On the contrary the NLOS depends on reflection by an opaque object and uses multipath diversion to satisfy its need [13]. Figure 33.8 illustrates the types of VLC Link.

The receiver, in this case is the photodiode along with the pre-amplifier circuit. The sole purpose of the photodiode is to convert the analogue input to a digital medium which is the electric current.

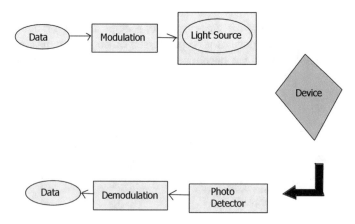

Fig. 33.7 Indoor visible light communication system structure

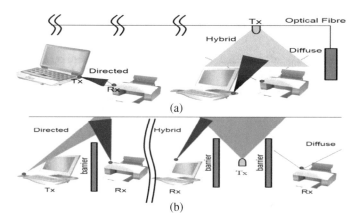

Fig. 33.8 **a** LOS communication, **b** NLOS communication

33.5.2 Challenges and Applications

Indoor VLC was a substitute to solve all the problems of the radio frequency, which has proven its worth making it more efficient than the latter. However, it does generate its own flaws and a few of those have grown to be a little challenging. The main challenges include: (a) Low bandwidth Modulation of LEDs—The modulation of the bandwidth of the LEDs are shown to be comparatively lower than its channel which has resulted in the low data rates. Various methods such as usage of blue optical fibre must be deployed as a filter for the LED to increase its data rates. Thus, making it more non economical. (b) Space for an uplink for the VLC system -Due to low compatibility of the mobile phone and the harmful emission from the light to the eye, fitting an uplink for the VLC system is a very challenging task. To solve this issue various methods have been induced, some of them being RF-VLC combination, bidirectional indoor communication, and so on. (c) Multipath diversion—The main delay in the transmission to the receiver is the multipath diversion which causes the signal to reflect in its own pace and reach the receiver. (d) Signal Modulation techniques—They are mainly four different techniques namely, On–Off Keying (OOK), Pulse position modulation (PPM) Orthogonal Frequency Division Multiplex, and Colour Shift keying (CSK).

Some of the challenges of VLC include: (a) LEDs can be shifted from one to another, it leads to loss of data and connection. (b) LEDs have very low Bandwidth modulation which is a major challenge to be solved further without compromising on the feasibility. (c) Noise that is added and inferred from other light sources causes decrease in the efficiency of the VLC system. (d) The multipath diversion or the scattering effect of the light causes delay in transferring the signal and setting up light units for a direct path leads to decrease in the feasibility. (e) Signal installation of the light units at the required places causes loss in transmission. (f) The increased exposure to LED light has also had a drastic impact on the human eye.

33.6   Free Space Optical Communication System (FSO)

Free space optical communication as the name suggests uses an open space such as vacuum and air. They are mainly indoor called indoor optical wireless communication and outdoor also called outdoor optical wireless communication. The signal in the form of light is transferred through the air medium. FSO has vast applications in the different fields such as LAN, underwater communication, satellite communication. History on FSO dates to the World War 1906–1060, where transmitters and receivers were developed for OWC systems. Since then, a vast amount of improvement has been made with regards to LEDs, it expanded to several thousand kilometres and by using optical telescopes as beam expanders. This led to expansion of the FSO range. Majority of the current companies work on the design and manufacturing of FSO systems for outdoor communication. Some of the companies include Cassidian, Novasol, etc. The structure mainly contains transmitter, channel and receiver. Figure 33.9 contains the basic structure of the FSO System. The information originates in the source which is then given to the transmitter, this transmitter which contains a modulator that produces the data input towards the transmitting medium, from then it is directed towards the receiver which is then received by the receiver. The receiver then contains a demodulator which is then amplified and given to the required destination. The modulation techniques used here are OOK, AM, FM etc. The transmitter used here is also LEDs. The receiver is mainly of two types, Coherent and Non-coherent receivers. The receivers used here are photo detectors [15]. During the propagation through the channel, a lot of attenuation takes place. Some of these losses are caused due to the following. (a) Geometric and Misalignment Losses—This loss is mainly due to the off-course path of the beam, it is calculated based on the angle of change, the properties of the receiver. This can be rectified by increasing the beam divergence at the transmitter and to improve the link capacity. (b) Atmospheric loss -The scattering effects of the signal due to the particles and absorption of the same can lead to loss of signal and cause attenuation. Since it is being transferred through air, it acts similar to visible light. The particles that cause smog, pollution and so on cause reflection and absorption of the light source leading to loss of signal. (c) Turbulence of Atmosphere- Even under clear atmospheric conditions where the atmospheric loss associated can be neglected but it causes another major problem called fading. This happens because of all the properties such as temperature, pressure, humidity, and so on. d) Radiation- The radiation caused due to background noise can degrade the performance of the FSO links. This noise can be inevitable like the sunlight [16].

Issues caused by FSO (a) The range of signal still stays insufficient and must be expanded. (b) The loss of signal during propagation in the channel due to the reasons mentioned above (c) security issues are also caused due to easy inference by another device or person. (d) The use of atmosphere as the channel leads to various physical challenges is the major setback. (e) Single beam causes a lot of attenuation loss which leads to usage of multiple beams making it not feasible.

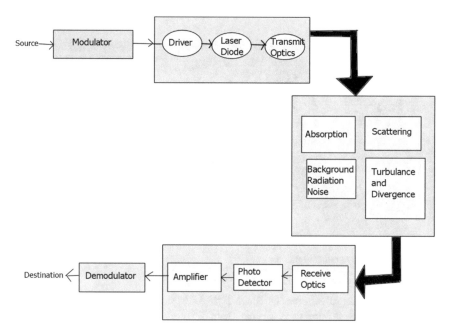

Fig. 33.9 Basic structure of FSO

33.7   Visible Free Space Optical Communication System (FSO)

UV communication system is used for short range which achieve high data rate. RF, Terahertz, VLC has limitations due to medium of transmission such as scattering for NLOS and obstacle for LOS. UV waves have high scattering effects which makes it useful for LOS when no other alternative is available. To use it for NLOS Ultraviolet C band is used, C band is used because it is solar blind hence not inferred by the sunlight. The research in UVC dates to 1960s Similarly Ultraviolet a Band has comparatively low propagation losses as compared to UVC despite it being solar blind. The main challenge of the UVC is the data loss due to the environment.

33.7.1   UVC Structure

The main source for UVC is the UV LEDS, which produces UV radiation through vaporized mercury. Sometimes even Laser diodes are used. UV lamps are not used because they release very high heat as compared to the earlier option. There are a lot of commercially available LEDs with varying energy sources. The UV led application area is shown in Fig. 33.10.

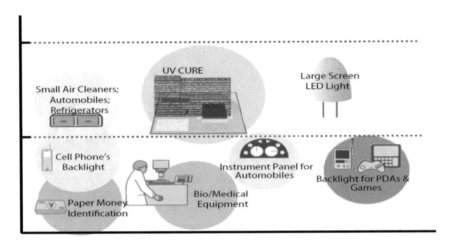

Fig. 33.10 UV LED application areas

To utilize economical sources Black light UV lamps can be used. The range of length of these back lamps are varying from 320 to 400 nm which is UV-A region. They are mainly used in networks since they are low cost and produce good efficiency. They have one major drawback, i.e. Filters of its inside layers, which does not allow the output with original energy. The main usage of these is for dermatology, oil lamps, and so on. UV LEDs are better because of them being highly efficient and low heat generation. Apart from these UV Lasers can also be used UV light Detectors and channels. Detectors are also called receivers, the main of them being photodetectors. These photodetectors collect all the UV light and convert them to electrical voltage and current signal. There are various types of photodetectors, some of them are PIN photodetectors, APD photodetectors, PMT photodetectors, and so on. The channel is the medium of communication, here the mainly used medium is the atmosphere. The atmosphere creates a link between the transmitter and the receiver, the connection between these two is called the link. The link can be direct as in case of LOS. As in NLOS an overlapping area is mentioned which is the region for multipath diversion [17]. Figure 33.11 shows the different types of links.

33.7.2 UVC Applications and Challenges

UV communication is highly secured and so can be used for both civil and defence purposes. The following shows an image on how UV can be used as Defence communication. UV communication is used for civil purposes, which include day to day activity. Some of them being our cell coverage, data transmission from one device to the other. Encrypted message transversal between devices. UVC is also used in Civil purposes [17]. Another major application is the Lidar, this is the combination

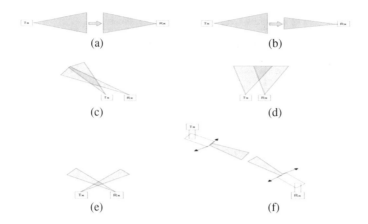

Fig. 33.11 **a** Direct LOS, **b** Diffused LOS, **c** NLOS, **d** NLOS, **e** NLOS, **f** Tracked system

of laser beam and telescopes field of view to create a new form of communication over large surface areas. UVC is also used for Aircraft Landing under low visibility conditions.

The challenges include: (a) The NLOs communication has its own limitation of power and bandwidth. (b) Turbulence of the atmosphere requires a good channel coding, if not it leads to disintegration of the system. (c) A subject of quantum mechanics must be applied to understand the scattering effect of the light. (d) There is a high limitation to the signal detection methods and are the source for further development.

33.8   Summary and Conclusion

Given some of the technologies used by the OWC for the implementation of the 6G, the fruit of further development can be added to this basket of this knowledge. Further development of Terahertz is to make it available in terms of an inexpensive small laser diode, using detectors which can absorb well are also used to enhance its transmission. The overall data rate expansion for the VLC is done by techniques such as orthogonal frequency division multiplexing (OFDM), discrete multitone modulation (DMT), and multiple-input multiple-output (MIMO), and wavelength division multiplexing (WDM), and so on. The performance of FSO can be improved by implementation of SAC OCDMA and data rate is increased further by low pass filter (LPF). The UVC can be further developed to implement UV-B-LED which will provide extraordinary data rate using QAM. These enhancements are likely to be done this decade owing to improvisation and successful launch of 6G for the coming decade, Table 33.1 describes the summary of networks with their range, applications and limitations.

Table 33.1 Summary of networks, range, application and limitation

No.	Name of the network	Range	Application	Limitation
1	Radio frequency [5, 6]	3 kHz–10 GHz	• Used in broadcast • Used in one-to-one communication in 1900's • Main source of early mobile network	• Leads to theft • Poor infrastructure • Damage of network service • Hacking and accessing data during transmission
2	Terahertz communication [15]	0.1–10 THz	• Tera IOT • Tera-Space • Tera-Wi-Fi	• Low antenna gains at the receiver • Power consumption is challenging • Link instability and low beam pointing control
3	Visible light communication [15, 18]	400–800 THz	• Health care institution • Defence communication • Air travel service • Underground communication	• Loss of data and communication • Noise inference • Multi scattering causes decrease in efficiency
4	Free Space optical Communication [17, 19]	Above 300 GHz	• Used in WLAN • Used for LAN building to building connectivity • Used in LAW firm	• Range is inefficient • Loss of signal during propagation • Multiple beam cause attenuation
5	Ultraviolet communication [17, 20]	UV Spectrum (200–280 nm)	• Defence communication • Civil communication • Aircraft landing monitoring	• Limitation of power and bandwidth • Turbulence leads to disintegration of the network • Limitation in signal detection

References

1. Huang, T., Yang K, Wu J, Jin ma and Zhang D.: A survey on green 6G Network: architecture and technologies. IEEE Special Section on Green Internet of Things **7**, 175758–175768 (2019)
2. Alsharif, M.H., Kelechi, A.H., Albreem, M.A., Chaudhry, S.A., Zia, M.S., Kim, S.: Sixth generation (6G) wireless networks: vision, research activities, challenges and potential solutions. Symmetry **12**(4), 676 (2020)
3. Uysal, M., Hatef N.: Optical wireless communications—an emerging technology. In: 2014 16th International Conference on Transparent Optical Networks (ICTON), IEEE, pp. 1–7 (2014)
4. Giordani, M., Polese, M., Mezzavilla, M., Rangan, S., Zorzi, M.: Towards 6G networks: use cases and technologies. IEEE Commun. Mag. **58**(3), 55–61 (2020)
5. Chowdhury, M.Z., Shahjalal, A.S., Jang, Y.M.: A comparative survey of optical wireless technologies: architectures and applications. IEEE (2018)

6. Chowdhury, M.Z., Shahjalal, A.S., Jang, Y.M.: 6G wireless communication systems: applications, requirements, technologies, challenges, and research directions. IEEE Open Journal of the Communications Society 99, 1–1 (2020)
7. Mohsan, H.A.S, Mazinani, A., Malik, W., Younas, I.: 6G: envisioning the key technologies, applications and challenges. International Journal of Advanced Computer Science and Applications 11(9) (2020)
8. Akhtar, M.W., Hassan, S.A., Ghafar, S., Jung, H., Garg, S., Hossain, M.S.: The shift to 6G communications: vision and requirements. Human-Centric Computing and Information Sciences 53 (2020)
9. Abood, B., Fahad, J., Rida, A.: Survey of improved performance radio frequency channels in wireless communication systems. International Journal of Civil Engineering and Technology (IJCIET) 10(2), 70–83 (2019)
10. Aastha, Gulia, P.: Review on security of radio frequency identification technology. Research India Publications 10(8), 2427–2433 (2017)
11. Han, C., Yongzhi, W., Chen, Z., Wang, X.: Terahertz communications (TeraCom): challenges and impact on 6G wireless systems. IEEE 1(1) (2019)
12. John, F., O'Hara, Ekin, S., Choi, W., Son, I.: A perspective on terahertz next-generation wireless communications. MDPI 7(2) (2019)
13. Hussein, A.T.: Visible light communication system. Doctoral dissertation, University of Leeds (2016)
14. Yadav, N., Kundu, P.: Literature study of visible light communication techniques. IJIET 7(1), 275–280 (2016)
15. Lemic, F., Abadal, S., Tavernier, W., Stroobant, P., Colle, D., Alarcon, E., Barja, J.M., Famaey, J.: Survey on terahertz nanocommunication and networking: a top-down perspective. Internet Technology and Data Science Lab 1(1) (2021)
16. Khalighi, M.A., Uysal, M.: Survey on free space optical communication: a communication theory perspective. IEEE Communications Surveys Tutorials 16, 2231–2258 (2014)
17. Patmal, M.H.: A study on ultraviolet optical wireless communication employing WDM. Doctoral dissertation, Waseda University (2019)
18. Saif, N.I., Muataz, S.: A review of visible light communication (VLC) technology. In: 2nd International Conference on Materials Engineering and Science. AIP Conference Proceedings, vol. 2213(1) (2020)
19. Sharma, D., Khan, S.A., Singh, S.: Literature survey and issue on free space optical communication system. IJERT 4(2), 561–567 (2015)
20. Vavoulas, A., Sandalidis, H.G., Chatzidiamantis, N.D., Xu, Z., Karagiannidis, G.K.: A survey on ultraviolet C-band (UV-C) communications. IEEE Communications Surveys Tutorials 21(3), 2111–2133 (2019)

Chapter 34
Workpiece Recognition Technology Based on Improved SIFT Algorithm

Meng Wang, Jasmin Niguidula, and Ronaldo Juanatas

Abstract To solve the problem of workpiece image matching and recognition in complex environments such as rotation, scaling, and partial occlusion, and to solve the problems of slow running speed and low matching accuracy of SIFT algorithm, a scheme of reducing the dimension of feature point descriptor and selecting the best feature point matching is proposed. A concentric ring with a radius of 2 pixels, 4 pixels, and 6 pixels is constructed with the feature point as the center. According to the rule, the pixels close to the feature points significantly impact the features. The pixels far from the feature points have a minor impact on the features. They are divided into 4, 8, and 16 partitions, respectively, generating 56-dimensional descriptors. The feature points are divided into two sets in image matching: maximum and minimum. The distance between the feature points of the same typeset between the two images is calculated to select the best matching points, which reduces the amount of image matching calculation and saves the algorithm time consuming. The experimental results of workpiece image recognition show that the improved algorithm of feature point descriptor dimensionality reduction and image matching can effectively improve the accuracy and reduce the image matching time.

34.1 Introduction

Workpiece recognition technology is essentially image processing and recognition. Image recognition refers to the technology of processing images to identify different targets and objects. Generally, industrial cameras are used to take pictures, and then processing and image matching is carried out according to the pictures. Therefore, the

M. Wang (✉) · J. Niguidula · R. Juanatas
Technological University of the Philippines, Manila, Philippines
e-mail: ahjdwm@ahcme.edu.cn

M. Wang
Anhui Technical College of Mechanical and Electrical Engineering, Wuhu, China

critical step of image recognition is image matching. There are many image matching methods. The feature-based matching method is a hot research topic because of its good noise resistance, high matching accuracy, and better robustness to object rotation, occlusion, and other problems.

At present, feature-based matching methods include the SIFT algorithm, surf algorithm, etc. LOWE DG proposed the SIFT algorithm in 2004, can adapt to different scales, illumination, and rotation angles, and has affine invariance and high matching accuracy. However, the scale-invariant feature transform (SIFT) algorithm also has apparent shortcomings: the descriptor dimension of feature points is too large, which leads to an increase in data volume and time consuming, which seriously affects the algorithm's efficiency.

The literature proposed the PCA–SIFT algorithm for principal component analysis (PCA) of SIFT algorithm. Compared with SIFT, the performance of this algorithm in running time and affine change is improved. However, it is more sensitive to scale transformation and smooth blur, and the amount of calculation is more significant [1]. Using the idea of integral graph and Harr wavelet features, the scale space and feature description are simplified, and the SURF algorithm is proposed. The real-time performance is greatly improved, but its rotation invariance is worse than SIFT algorithm [2]. Aiming at the poor real-time performance of the scale-invariant feature transformation (SIFT) algorithm and the loss of information when generating gray images, a SIFT feature extraction algorithm with contrast enhancement and Daisy descriptor is proposed. The algorithm can obtain more feature points and reduce the time of descriptor generation and feature matching while ensuring accuracy [3]. Aiming at the problems of a large amount of calculation, poor real-time performance, and high error matching rate in scale-invariant feature transformation (SIFT) feature matching algorithm, a method based on the distance ratio criterion is proposed to remove the error matching in SIFT feature matching. Using the distance ratio criterion, the method has high matching accuracy, reduces the matching time, and improves the real-time performance [4]. The image matching process is divided into coarse matching and good matching to reduce the time consuming of the algorithm. In addition, according to the needs of the algorithm, the error removal algorithm used after coarse and fine matching is improved. The matching speed is improved and guaranteed matching accuracy [5].

In industry, robots are required to recognize workpieces in real time, efficiently, and accurately. This paper optimizes and improves the SIFT algorithm based on analyzing the basic principle of SIFT algorithm. Firstly, a new feature descriptor is designed to reduce the dimension of the high-dimensional descriptor in the original algorithm; then, the feature points are divided into maximum and minimum. The distance between the feature points of the same typeset between the two images is calculated to reduce the amount of calculation of image matching.

34.2 Methodology

34.2.1 SIFT Algorithm

SIFT algorithm is mainly used to describe and detect the local features in the image. It looks for the extreme points in the spatial scale and extracts their position, scale, and rotation invariants [6]. The basic steps of SIFT image recognition include feature extraction, feature matching; determination of registration model; image resampling. Feature extraction and feature matching mainly include key point detection, local feature descriptor generation, and feature point matching [7].

Key point detection. SIFT algorithm is a point-based matching algorithm, so we must first find the critical points with representativeness and stability. The purpose of SIFT algorithm key point detection is to find out the points' exact position that keeps the image scaling, rotation, and a certain degree of noise unchanged in the image scale space and screen these points to collect enough information for subsequent processing [8].

Local feature descriptor generation. After obtaining the key points that keep the image scaling and rotation unchanged in the scale space, sift generates a descriptor for the main direction and each auxiliary direction of each key point and makes these descriptors as invariant as possible [9].

Feature point matching. SIFT algorithm judges the similarity according to the Euclidean distance between descriptors and matches the feature points according to the similarity. The ratio purification method is usually used to purify the matching results to reduce false matching. In addition to querying the descriptor of the nearest neighbor, we should also consider the next nearest neighbor; When the ratio of the following nearest neighbor distance to the nearest neighbor distance is less than the threshold, it is considered that the basic process of SIFT image registration is not significant enough. The matching pair is eliminated [10].

34.2.2 Algorithm Improvement

Dimensionality reduction of feature descriptor. The original SIFT algorithm uses 256 pixels in the rectangular area 16*16 around the feature points to determine the descriptor of the feature points. Finally, each feature point generates a $4*4*8 = 128$-dimensional descriptor. The amount of calculation is relatively large, and the image matching is time consuming [11]. In this paper, the algorithm simplifies the feature descriptor, and the specific steps are as follows:

(1) Taking the feature points as the circle and taking 2 pixels, 4 pixels, and 6 pixels as the radius, the field is divided into three rings. According to the fact that the pixels closer to the feature points have a greater impact on the features, and the pixels farther from the feature points have a smaller impact on the features,

this algorithm divides the rings with different distances into regions of different sizes to reduce the impact of peripheral pixels on the features. The ring with a radius of 6 pixels is equally divided into four regions. The ring with a radius of 4 pixels is equally divided into eight regions. The ring with a radius of 2 pixels is equally divided into 16 regions.

(2) The gray level of pixels in each region is normalized, and the average gray level is calculated to keep the illumination unchanged. The calculation formula is as follows:

$$\overline{g}_i = \frac{g_i}{\sqrt{\sum_{j=1}^{8} g_i}} \quad i = 1, 2, \ldots, 8 \tag{34.1}$$

(3) Calculate the gray difference value for each subregion, and the calculation formula is as follows:

$$d\overline{g}_i = \begin{cases} \overline{g}_i - \overline{g}_{i+1} & i = 1 \\ 2g_i - g_{i-1} - g_{i+1} & 1 < i < 8 \\ \overline{g}_i - \overline{g}_{i-1} & i = 8 \end{cases} \tag{34.2}$$

(4) In this algorithm, each feature point descriptor comprises a gray mean value and gray difference value of all partitions. The format is as follows:

$$g = (g_1 + g_2 + \cdots + g_{28} + \overline{g}_1 + \overline{g}_2 + \cdots + \overline{g}_{28}) \tag{34.3}$$

Each feature descriptor has $(4 + 8 + 16) * 2 = 56$ dimensions. Compared with SIFT algorithm, each feature descriptor reduces data by 56%.

The feature descriptor established by this algorithm makes full use of the gray information and difference information in the field of feature points, reduces the dimension of the feature descriptor, and retains the essential characteristics of images such as translation, scale, and rotation.

Feature point matching and selection. For matching two image feature points in the SIFT algorithm, the distance between the feature points of the two images needs to be calculated. Due to the large number of feature points in the image, the calculation amount of the distance between the feature points is relatively large. If the number of feature points of two images is m and n, respectively, the number of feature point distances is calculated as *mn* times in theory.

In SIFT algorithm, the feature points are the extreme points in the corresponding DoG. When the image changes in illumination, scale, and rotation, the feature points are generally still extreme and will not change. Aiming at the property of feature points, this paper optimizes the distance calculation of two images as follows:

The maximum and minimum values are stored in two data sets when the algorithm judges the extreme points according to DoG. When calculating the distance between two image feature points, it only needs to calculate the distance between two maximum sets and two minimum sets. In the best case, the maximum and minimum sets account for 50%, respectively, and the distance times of feature points are 1/2 mn times.

34.3   Results and Discussion

This algorithm is based on Intel corei7-4510, 2G 4-core CPU and 8 GB memory windows10 operating system, OpenCV 4.5 platform, and Python 3.06 language development and implementation. The experimental verification is carried out by testing the data set of the algorithm for identifying specific objects.

Figure 1a shows the target workpiece graph, and Fig. 1b shows the workpiece heap graph. The experiment uses SIFT algorithm and improved algorithm to find the target workpiece from the workpiece heap graph.

Figures 34.2, 34.3, 34.4, 34.5, 34.6 and 34.7 shows the detection results of scaling, rotating, and occlusion of the target workpiece image using the SIFT algorithm and the improved SIFT method.

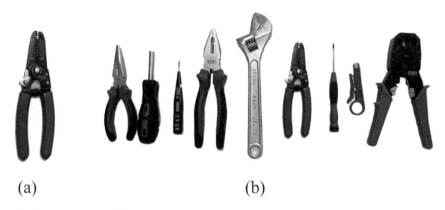

(a)   (b)

Fig. 34.1   a Target workpiece graph and **b** workpiece heap graph

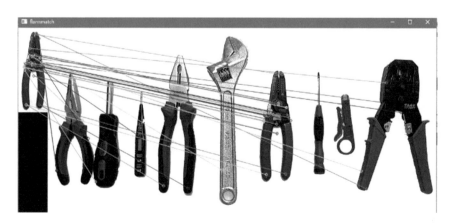

Fig. 34.2 Recognition of scaling workpiece graph by SIFT algorithm

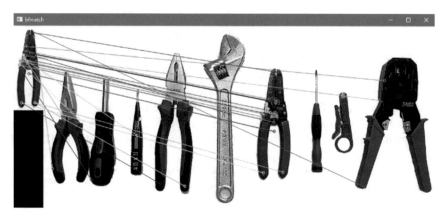

Fig. 34.3 Recognition of scaling workpiece graph by improved algorithm

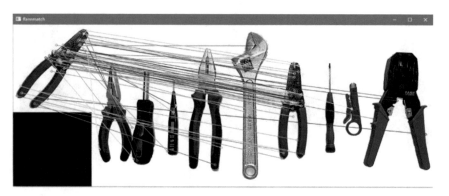

Fig. 34.4 Recognition of rotating workpiece graph by SIFT algorithm

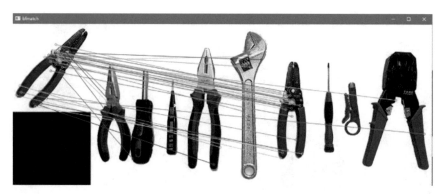

Fig. 34.5 Recognition of rotating workpiece graph by improved algorithm

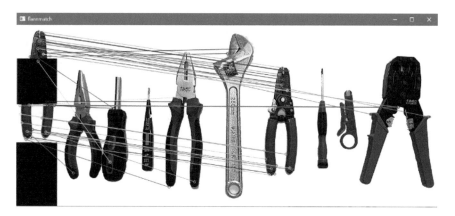

Fig. 34.6   Recognition of the occluded workpiece graph by SIFT algorithm

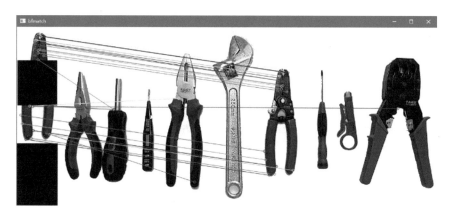

Fig. 34.7   Recognition of occluded workpiece graph by improved algorithm

The original SIFT algorithm and the improved algorithm are compared and analyzed in terms of the number of feature points, the number of correct matching points, the correct matching rate, and the time consuming of the algorithm. The specific data are listed in Tables 34.1 and 34.2.

Because the algorithm in this paper adopts the ring grouping domain dimension reduction method to generate descriptors, it retains the gray information of the domain and the difference information of the domain. All better describe the local features of feature points. The correct matching rate of the image is improved compared with the original SIFT algorithm. From the data in Table 34.1, we can see that the correct rate is improved by about 7%.

This algorithm effectively reduces the dimension of descriptors. Each feature point is reduced from 128-dimensional description data of the original SIFT algorithm to 56-dimensional description data. The matching time can also be reduced through the calculation method of feature point grouping distance. It can be seen from the data

Table 34.1 Comparison of feature points between improved algorithm and SIFT algorithm

Image change	Query image resolution	Number of feature points		Correctly match the number of points		Correct matching rate	
		SIFT algorithm	Improved algorithm	SIFT algorithm	Improved algorithm	SIFT algorithm (%)	Improved algorithm (%)
30% reduction	75*210	37	30	24	22	64.86	73.33
Rotate 45 degrees	230*250	79	51	61	42	77.22	82.35
50% occlusion	110*300	41	34	23	19	67.65	82.61

Table 34.2 Table captions should be placed above the tables

Image change	Time-consuming (MS)		Lift time ratio (%)
	SIFT algorithm	Improved algorithm	
30% reduction	190	176	7.37
Rotate 45°	210	184	12.38
50% occlusion	232	208	10.34

in Table 34.2 that the running time of this algorithm is reduced by 10% on average compared with the original SIFT algorithm.

34.4 Conclusions

The SIFT algorithm is improved based on the study of the SIFT algorithm to improve the accuracy of workpiece recognition and reduce the recognition time. Because of the large dimension of SIFT algorithm feature descriptor and a large amount of calculation of image matching feature point distance, a scheme to reduce the dimension of the feature point descriptor is proposed. The algorithm takes the feature point as the center to construct concentric rings with a radius of 2 pixels, 4 pixels, and 6 pixels. Different rings are divided into different zones to reflect the weight of different fields. The ring with a radius of 6 pixels, 4 pixels, and 2 pixels is divided into 4, 8, and 16 partitions, respectively, and finally generates a 56-dimensional descriptor.

Because of the large amount of distance calculation between feature points in image matching, dividing the feature points into two sets of maximum and minimum is proposed. The calculation distance of feature points between the same type sets between two images can reduce the amount of calculation by 50% in the best case.

The experimental results of workpiece image recognition show that the improved method of feature point dimensionality reduction and image matching can effectively improve the accuracy and reduce the image matching time. According to the experimental results, the correct matching rate is improved by 7%, and the time consumption of the algorithm is reduced by 10%. Later, we can consider parallelizing the implementation and operation of the algorithm to improve the real-time performance of the algorithm further.

Acknowledgements Key project of natural science research in colleges and universities of Anhui Province (KJ2020A1102).

Anhui teachers' teaching innovation team (2021jxtd065).

References

1. Ke, N.Y., Sukthankar R.: PCA-SIFT: a more distinctive representation for local image descriptors. In: IEEE Computer Society Conference on Computer Vision & Pattern Recognition (2004)
2. Bay H., Ess, A.: Speeded-up robust features (SURF). In: European Conference on Computer Vision (2008)
3. Hang, X.I.: A method for improving matching accuracy of SIFT features. Optical Instruments **38**(6), 497–500 (2016)
4. Li, H.Y.: A real-time SIFT feature extraction algorithm. Journal of Astronautics **38**(8), 865–871 (2017)
5. Jintao, D.: Improvement of real-time performance of image matching based on SIFT. Electron. Opt. Control. **27**(3), 80–88 (2020)
6. Yufeng, L.: Image mosaic algorithm based on area blocking and SIFT. Opt. Precis. Eng. **24**(5), 1197–1205 (2016)
7. Hai-Bo, M.A.: Fast SIFT image matching method based on two-dimensional entropy. Journal of North University of China (Natural Science Edition) **40**(1), 63–69 (2019)
8. Sunan, D.: Image registration method based on improved SIFT algorithm. Transducer and Microsystem Technologies **39**(10), 45–50 (2020)
9. Fuyu, L.I.: Summarization of SIFT-based remote sensing image registration techniques. Remote Sensing for Land and Resources **28**(2), 14–20 (2016)
10. Huang, H.B.: A survey of image registration Based on SIFT. Software Guide **18**(1), 1–4 (2019)
11. Zhao, S.S.: Railway fastener state detection algorithm based on SIFT feature. Transducer and Microsystem Technologies **37**(11), 148–154 (2018)

Chapter 35
An Analysis of COVID Effected Patients

**R. Sivarama Prasad, D. Bujji Babu, G. Srilatha, Y. N. V. Lakshmi,
Y. Lakshmi Prasanna, V. Vani, and Sk. Anjaneyulu Babu**

Abstract Due to COVID-19 pandemic, public health emergency was created throughout the world. So, we took the base data and perform analysis on how the effect of vaccination on the human lives in terms of recovery, severity, side effects, and deaths on the globe. We also analyzed the country wise vaccination to understand the scenarios in the world, because the COVID virus is transforming in different countries in different ways, therefore the understanding the mutations of the virus and the use of the drug analysis also very much important for the future generations and also useful to face the future COVID virus mutations.

35.1 Introduction

In china, a new virus called "corona" (COVID-19) was identified in Wuhan city, and it was spreader all over the world. The COVID-19 vaccination analysis of research serious of the critical cases and visualizing active cases, cured and deaths cases. Due to COVID-19 pandemic, public health emergency was created throughout the world.

The COVID-19 pandemic impacts heavily on every citizen of the world till now, even though the effort of intervention and vaccination. COVID-19 is still there

R. Sivarama Prasad (✉) · V. Vani
Acharya Nagarjuna University, Nagarjuna Nagar, Guntur, India

D. Bujji Babu · G. Srilatha · Y. N. V. Lakshmi · Y. Lakshmi Prasanna · Sk. Anjaneyulu Babu
Department of MCA, QIS College of Engineering & Technology, Ongole, Andhra Pradesh, India
e-mail: mcahod@qiscet.edu.in

G. Srilatha
e-mail: 19491F0002@qiscet.edu.in

Y. N. V. Lakshmi
e-mail: 19491F0008@qiscet.edu.in

Y. Lakshmi Prasanna
e-mail: 19491F0009@qiscet.edu.in

Sk. Anjaneyulu Babu
e-mail: anjaneyulu.sk@qiscet.edu.in

resulting more number of cases and deaths. More number of articles reviewed and recommended for publication throughout the world.

This resource is to provide guidance for countries which are receives, stores, distribute, and manage COVID-19 vaccines. Distribute COVID-19 vaccines to remote vaccination sites to ensure efficacy, quality, tracking, and reporting of vaccine utilization. To protect the environment and populations, we need to assess, develop, and implement appropriate safety and waste management mechanisms.

35.2   Literature Survey

COVID-19 is a novel disease caused by respiratory syndrome [1]. In two manners, the citizens can be protected by increasing antibodies. The relationship between population immunity and transmission of highly infected diseases like COVID-19 [2]. This study will help to protect the people by implementing best practices.

Understand the efficacy and potency of different vaccines available in the market to challenge new viruses are explained [3]. To improve herd immunity, vaccines are very important [4]. In this paper, author stated that vaccines are not suitable for elder people sometimes.

35.2.1   Security and Privacy Analysis Methodology

Any study should focus on sequence of steps that are performed on any data. It requires hypothetical analysis to understand the outcome of the pattern. This leads to extract useful information for good decision making.

Dataset was taken from Kaggle.com [5] and WorldinData.com [6]. The changes in COVID-19 were identified by drawing statistical graphs in Python.

35.2.2   Sentiment Analysis and Stance Detection

Opinion mining plays important role to determine view points toward goal of Internet, which can be done using computational methods [7]. To find whether a text has positive, negative, or neutral, we need to apply different sentiment analysis [8]. Uncover emotions can be detecting by using emotion identification [9]. Determine the text is subject can objective can be done by using subjective detection [10].

Stance detection is different from polarity detection, which determines the agreement or disagreement in relation to a specific target [11]. State of art results are obtained from recurrent neural networks (RNN) and convolutional neural network (CNN) for machine translation [12], document generation [13], and syntactic parsing

[14]. Open-AI GPT and bidirectional encoder representations from transformers (BERT) are easily fine-tuned for natural language processing (NLP) tasks [15, 16].

Polarity detection from football-specific tweets using several machine learning algorithms [17]. Polarity analysis and stance detection various approaches explained in lexical-based methods [18] and machine learning methods [19].

35.2.3   Twitter Sentiment Analysis on COVID-19 Data

Social media plays vital role to maintain social distance during lockdown period [20].

35.3   Methodology

Algorithms: Genetic algorithms show great performance in many domains. There are three main operations, which are selection, crossover, and mutation. Genetic algorithms are repetitive algorithms which repeated until condition met.

Based on population size, GA generates random set of solutions. Fitness function used to evaluate each solution. Then, crossover function performed after selection function. Based on fitness value, the current population is updated.

Input:

-nP: Size of base population
-nI: Total number of iterations
-rC: Crossover rate
-rM: Mutation rate
-cI: Current iteration

Method:

Generate initial population of size nP
According to fitness function, evaluate initial population

While ($cI \leq nI$)

// Breed $rC \times nP$ new solutions.

From current population select two parent solution

From offspring's solutions via crossover.

IF (rand (0.0, 1.0) < rM) Mutate the offspring's solutions.

end IF

According to the fitness function, evaluate each child solution.

Add offspring to population.

Remove the rC × nP least—fit solutions from population.

end While

output:
  Best solution.

E: Overall error rate obtained from ANN,

β: Predetermined fixed value (β=5),

|R|: Number of the selected features

|N|: Total.

Fitness

It is a measure of the body's ability.

Fitness is more specific for sport. Fitness has two main components, health—related fitness.

Formula:

$$\text{Fitness} = E * \left(1 + \beta * \frac{|R|}{|N|}\right) \qquad (35.1)$$

E: Over all error rate.

R: The number of selected features count.

N: Original features count.

Accuracy

Accuracy is ratio of correct predicted observation to total number of observations. If accuracy is more than model is best.

Formula:

$$\text{Accuracy} = \frac{TP + TN}{TP + FP + FN + TN} \qquad (35.2)$$

TP: Number of real positive.

TN: Number of negative.

FP: Number of real negatives.

FN: Number of real positive.

Precision

Precision is rate of correctly predicted positive observation to total predictive positive operations.

Formula

$$\text{Precision} = \frac{TP}{TP + FP} \tag{35.3}$$

TP: Number of real positive.

FP: Number of real negatives.

Recall

Recall is the ratio of correctly predicted positive observations to all observations.

Formula

$$\text{Recall} = \frac{TP}{TP + FN} \tag{35.4}$$

TP: Number of real positive.

FN: Number of real positive.

F-measure

It is weighted average of precision and recall.

Formula

$$\text{F-measure} = \frac{2 * (Recall * Precision)}{Recall + Precision} \tag{35.5}$$

I. Computation of base prevalence and herd immunity

Base prevalence is the ratio of sum of number of people vaccinated and people who are recovered to total population size.

$$BP = \frac{VC + VE + PR}{P} * 100\% \tag{35.6}$$

where

BP: Base prevalence metric value

VC: Number of people vaccinated

PR: Number of people who recovered

P is the total size of the population

$$\text{Herd immunity} = \left(1 - \frac{1}{R_0}\right) \times 100 \qquad (35.7)$$

where

R_0: Reproduction number of the virus COVID-19

Herd immunity calculation for the countries UAE and Bahrain is given below.

UAE:

$$
\begin{aligned}
\text{Herd immunity} &= \left(1 - \frac{1}{R_0}\right) \times 100 \\
&= \left(1 - \frac{1}{2 \cdot 75}\right) \times 100 \\
&= 63.63\%
\end{aligned}
$$

Bahrain:

$$
\begin{aligned}
\text{Herd immunity} &= \left(1 - \frac{1}{R_0}\right) \times 100 \\
&= \left(1 - \frac{1}{3 \cdot 39}\right) \times 100 \\
&= 70.5\%
\end{aligned}
$$

Herd immunity of above countries is minimum so that the disease is an epidemic. The formulae from (35.1) to (35.7) are used in various computations for analytics.

35.4 Implementation and Result

For the implementation, we have COVID-19 vaccination analysis data set features used the CSV file means comma separated values, each line of the file is a data record. Each row and column in a csv file represent a set of values delimited with a particular delimiter. Each row has the same number of values from all rows with the column. Each column in a csv file by print the contents vertically in the output file. Given column is written the subsequent column and NumPy in the Jupiter note book in windows operating and the system configuration of RAM is 4.00 GB local disk(c) 3.70.

Figure 35.1, In chart, X-axis represented by millions and Y-axis represented by total countries, specifies that is where data for full vaccinations is available, it shows how many people have been fully vaccinated (which many require more than 1 dose).

People Vaccinated

Individuals who received the first dose of the vaccine

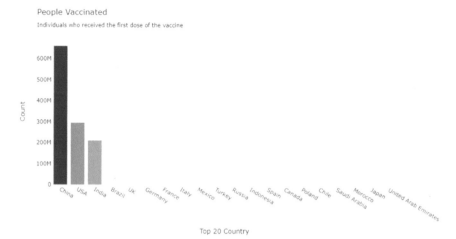

Fig. 35.1 People vaccinated

Percentage Vaccinated

Percentage of the total population that have received the first dose

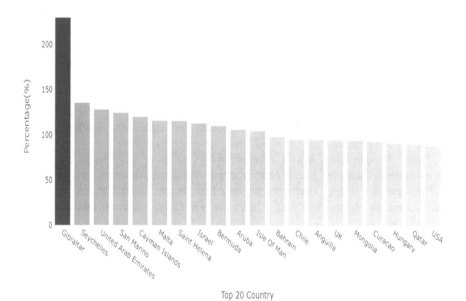

Fig. 35.2 Percentage vaccinated

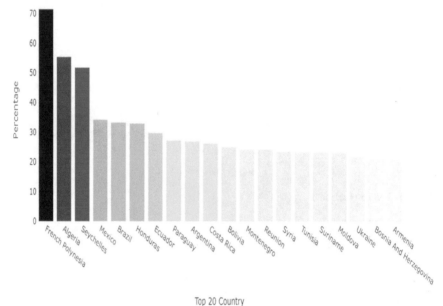

Fig. 35.3 Tested positive

Figure 35.2, In chart, X-axis represented by percentage of the total population and Y-axis represented by countries. Full vaccination means infected with virus and 1 dose of the protocol.

Figure 35.3, In chart, X-axis represented by percentage and Y-axis represented by countries. Fraction of people that tested positive among those that were tested.

The analysis is very useful to the common person to understand the use of COVID vaccine, at the same time, it helps the public to create awareness on COVID precautions.

35.5   Conclusion

Due to COVID-19 pandemic, public health emergency was created throughout the world. The COVID-19 vaccination analysis of research serious of the critical cases and visualizing active cases, cured, and deaths cases.

The proposed block chain management system deal with vaccine supply, vaccine expiry data, and fraud recording using smart contracts. IOT and block chain techniques provide smart system for different domains in health care.

To improve vaccine coverage, the COVID vaccine is to teck patient location, humidity, and temperature. The Gcoin block chain is proposed for drug data flow. The COVID-19 vaccine distribution chain helps in track the vaccine which are not wasted.

Acknowledgements We acknowledge all people who are directly and indirectly motivate us to complete this task. We sincerely thank all the authors and co-authors of the papers which are mentioned in references section.

References

1. Wikipedia: Severe acute respiratory syndrome coronavirus 2
2. Anderson, R.M., May, R.M.: Vaccination and heard immunity to infectious diseases. Nature **318**(6044), 323–329 (1985)
3. Rashid, H., Khandaker, G., Booy, R.: Vaccination and herd immunity: what more do we know? Curr. Opin. Infect. Dis. **25**(3), 243–249 (2012)
4. Lipsitch, M., Dean, N.E.: Understanding COIVD-19 vaccine efficacy. Science **370**(6518), 763–765 (2020)
5. Coronavirus (COVID-19) Vaccination—Statistics and Research—Our World in Data
6. COVID-19 World Vaccination Progress | Kaggle
7. Liu, B.: Sentiment Analysis: Mining Opinions, Sentiments, and Emotions, 1st edn. Cambridge University Press, Cambridge (2015)
8. Liu, B.: Sentiment Analysis and Opinion Mining, vol. 5. Morgan & Claypool, San Rafael (2012)
9. Cotfas, L.-A., Delcea, C., Segault, A., Roxin, I.: Semantic web-based social media analysis. In: Nguyen, N.T., Kowalczyk, R. (eds.) Transactions on Computational Collective Intelligent XXII, vol. 9655, pp. 147–166. Springer, Berlin (2016)
10. Chaturvedi, I., Ragusa, E., Gastaldo, P., Zunino, R., Cambria, E.: Bayesian network based extreme learning machine for subjectivity detection. J Franklin Inst. 355(4), 1780–1797 (2018). 10. 1016/j.franklin.2017.06.007.
11. D' Andrea, E., Ducange, P., Bechini, A., Renda, A., Marcelloni, F.: Monitoring the public opinion an out the vaccination topic from tweets analysis. Expert Syst. Appl. 116, 209–226 (2019).https://doi.org/10.1016/j.eswa.2018.09.009
12. Vaswani, A., Shazeer, N., Parmar, N., Uszkoreit, J., Jones, L., Gomez, A.N., Kaiser, L., Polosukhin, I.: Attention is all you need. In: Proceedings of 31st International Conference on Neural Information. Long Beach, CA, USA, pp. 6000–6010, (2017). Accessed 24 Jan 2021 [Online]
13. Liu, P.J. Saleh, M., Pot, E., Goodrich, B., Sepassi, R., Kaiser, L., Shazeer, N.: Generating Wikipedia by summarizing long sequences. Presented at the 6th international conference on learn (2018). Available: https://openreview.net/forum?id=Hyg0vbWC- [Online]
14. Kitaev, N., Klein, D.: Constituency parsing with a self-attentive encoder. In: Proceedings of 56th Annual Meeting Associate Computed Linguistics, Melbourne, VIC, Australia, pp. 2676–2686 (2018). https://doi.org/10.18653/v1/p18-1249.
15. Radford, A., Narasimhan, K., Salimans, T., Sutskever, I.: Improving language understanding by generative pre-traning (2018). Available: https://cdn.openai.com/research-covers/language unsupervised/language understanding paper.pdf [Online]
16. Devlin, J. Chang, M.-W., Lee, K., Toutanova, K.: BERT: pretraining of deep bidirectional transformers for language understanding. In: Proceedings of Conference on North American Chapter Associate Computed Linguistics, Human Language Technology, Minneapolis, MN, USA, pp. 4171–4186 (2019). https://doi.org/10.18653/v1/n19-1423

17. Aloufi, S., Saddik, A.E.: Sentiment identification in football specific tweets. IEEE Access **6**, 78609–78621 (2018). https://doi.org/10.1109/ACCESS.2018.2885117

18. Cotfas, L.-A., Delcea, C., Roxin, I.: Grey sentiment analysis using multiple lexicons. In: Proceedings of 15th International Conference on Informant Economy (IE), Cluj-Napoca, Romania, pp. 428–433 (2016)

19. Cotfas, L.-A., Delcea, C., Nica, I.: Analysing customer' opinions towards product characteristics using social media. In: Eurasian Business perspectives. Springer, Cham, pp. 129–138 (2020). https://doi.org/10.1007/978-3-030-48505-4_9

20. Merchant, R.M., Lurie, N.: Social media and emergency preparedness in response to novel coronavirus. J. Amer. Med. Assoc. **323**(20), 2011–2012 (2020). https://doi.org/10.1001/jama.2020.4469

Chapter 36
Keyword-Based Global Search to Understand the Impact of the Scenario of the Keyword

D. Bujji Babu, K. Guruprasanna, Y. Narasimha Rao, K. Jayakrishna, G. Dayanandam, P. Govinda Reddy, and T. Chandirika

Abstract Due to globalization, the world became a global village, the computer-based search playing a vital role in understanding the scenarios around the globe. Only one solution to explore search engine data is Google Trends, which is at free of cost and potential to predict the official data immediately. In this paper, we experimented with a keyword-based search to understand various scenarios in different countries. We, analyzed the Google Trends using various techniques like long short-term memory in feed forward networks in deep learning, multi-linear regression and elastic net, SVR and time series algorithms for analysis.

36.1 Introduction

Google search engines have major impact on people's daily life. Retrieval of data from search engine is very important. Search engines are act like filters for information

D. B. Babu (✉) · K. Guruprasanna · K. Jayakrishna · P. G. Reddy · T. Chandirika
Department of MCA, QIS College of Engineering and Technology Ongole, Ongole, Andhra Pradesh, India
e-mail: mcahod@qiscet.edu.in

K. Guruprasanna
e-mail: 19491F0004@qiscet.edu.in

K. Jayakrishna
e-mail: Jayakrishna.k@qiscet.edu.in

P. G. Reddy
e-mail: 19491F0006@qiscet.edu.in

T. Chandirika
e-mail: 19491F0007@qiscet.edu.in

Y. N. Rao
School of Computer Science and Engineering-SCOPE VIT, AP University, Amaravati, India

G. Dayanandam
Departement of Computer Science, Government College for Men(A), Kadapa, Andhra Pradesh, India

© The Author(s), under exclusive license to Springer Nature Singapore Pte Ltd. 2023
C. So-In et al. (eds.), *Information Systems for Intelligent Systems*, Smart Innovation,
Systems and Technologies 324, https://doi.org/10.1007/978-981-19-7447-2_36

available on the Internet. They are used to find information as genuine interest without need to walk through irrelevant web pages.

Nowadays, only one solution to explore search engine data is Google Trends, which is at free of cost and potential to predict the official data immediately. It has explained in more indexed papers database.

Search engines provide users search results with relevant information that are available on high quality websites. According to user searches, search engines deliver results to attain market share in online searches.

Google search is a famous tool for explore the data because it is at free of cost, easy to access and potential to predict the data immediately using Google Trends data.

The paper was organized with seven sections. Section 36.1 describes the introduction concepts which helps to understand the work. Section 36.2 focused on literature survey, we Sur versed around 26 papers clearly defined the problem statement; in Sect. 36.3, the section methodology web explained. In Sect. 36.4, the implementation environment and result are placed; in Sect. 36.5, we presented the conclusion of the work as well as the feature work also mentioned, and finally, expressed the acknowledgment and references.

36.2   Literature Survey

In big data research, the data analysis of search engines will be interesting research area.

Google search engine is a tool for searching information in various fields. It deals an immediate response for need of its users. It traffic is voluminous, and search engines requests can be tracked [1]. According to this theory [2], as compared to most qualified decision maker, better forecasts can be produced by diverse people will produce better and more solid forecasts. As practice shows, the group's guesses mean individual estimation will not precise than groups guess. The wisdom of the crowd gives us the right answer, it will show, it is good answer than other answers. [3] participant independence, and network devolution. Variety deals with each member's crowd has some personal information and a unique interpretation. Independence means individuals opinions are not decided those around them. Devolution means that crowd member can learn from their sources and their own specializations data as predictive tool for academics and practitioners in various fields. Ginsberg et al. [4] noticed these exact features in scientific research. They found that Google search engine tracking queries gives the possibility of populations' flu established Google Flu Trends, which is epidemic tracking tool.

36.2.1 Google Search Volume Research with Data

Up to 2006, finally, SVI is an indicator of the flu virus spread launched. An advanced technology, i.e., Google insights launched for providing data of search query is private until Google Trends latest after two years. GT data of previous [4, 5] has explained that data search in Google can help [6] flu spread and rate of unemployment there increase in sales if there is search property of real estate using Google Trends in a specific location.

Explored search engine data is Google Trends, which is at free of cost and potential to predict the official data immediately. It has explained in lot of indexed papers.

Google Trends is a significant tool in housing market, which predict real estate price changes. Due to differences in technology, the reliability cannot be widespread in web and social media.

36.2.2 Google Search Data

Google Trends filtered data will give users querying volume of aggregated search category, the data collected through Google search engine is anonymous and normal. The CSV file of search term index at national and local levels. GT gives 0 (zero) index of query if search volume is less for certain amount of time [7].

Previously, several studies explored on various online data streams They are articles related to news [8–10], websites and blogs of health [11], Wikipedia [12, 13] search engines [14] and Twitter data. Data on Twitter provides spatiotemporal observations for Twitter data. The sets related to infection reflects actual illness accurately [15], false positives and negatives pursuing while doing data collection, preprocessing and classification. Moreover, there is limited usage for Twitter [16].

Online tracking of websites, browser attacks and phishing attacks are classified based on substantial literature of the web browsers privacy and security.

Study of online tracking focused on cookies and related ecosystem, browser fingerprinting used to prevent tracking and interested for long time to preventing tracking using browser IP address. Tor is most prominent technology in this area related to attacks and defenses/mitigations.

Timing-based side channels are against browser which leak information such as user browsing history [17–19] and website. JavaScript is executed against memory-based attacks [20]. Reference [21] focused on user generated data for public health surveillance. Reference [22] suggests Twitter is used for public health. Mogo [23] due to changes in inclinations of technology cultural and economic, we cannot generalize developed countries [24] and [25].

36.3 Methodology

Algorithm:

1. Seasonal ARIMA with or without external regressions.
2. MLR means multiple linear regression.
3. EN stands elastic net.
4. Support vector machine regression.
5. Feedforward neural network (FNN).
6. Long shot-term memory (LST).

1. Seasonal ARIMA with or without external regressions.

2. Multiple linear regression.
   Response variable predictors by multiple regression.
   It is an extension of linear regression which used only one explanatory variable.
   In the area of econometrics and financial inference, we use MLR.
   MLR formula
   $$yi = \beta o + \beta_{1xi}1 + \beta_{2x}i2 + \beta_{3x}i3 + \ldots + \beta_{pxip}$$
   where n observation for i value
   Dependent variable is y_i
   Explanatory variables is x_i
   y intercept is βo
   Each slope coefficient variable is β_p
   Models error term is ϵ

3. Elastic net (EN)
   It is a regularized regression method. It combines L1 and L2 penalty functions. Based on the sum of squared coefficient values, one penalty is penalized. It is L2 penalty. L2 prevents the coefficient that is removed from the model. L2 minimizes the all-coefficients size.
   $l2 - penalty = sum\ j = 0\ to\ p\ beta - j^2$
   L1 penalty is sum of absolute coefficient values. L1 allows some coefficient to zero. L1 removes predictor from the model.

   • $L1 - penalty = sum\ j = 0\ top\ abs\ (beta - j)$

   It includes L1 and L2 penalties during training and it is penalized linear regression model.

4. Support vector regression
   It is used to predict discrete values. It works as similar to support vector machine principle. Finding the best-fit line is main objective of SVR. SVR is best-fit line that has the maximum number of points.
   $y = wx + b$ is hyperplane.

5. Feedforward neural networks
   It approximate function

y = f* (X) is a formula for classifier assigns X to Y.

Ө FNN will map y = f (x; Ө). Ө will memorize which is appropriate function.

6. Long short-term memory

In AI and deep learning areas, we use this artificial neural network method, i.e., LSTM. It is based short-term memory processes to create longer-term memory and LSTMs a complex area of deep learning.

36.4 Implementation and Result

For the implementation, we directly used the pandas, Pytrends in the Jupiter note book in windows 10 operating system and we executed on the machine is the configuration of RAM is 12.0 GB local disk (c) 145 GB, we executed on high end configuration system with 12 GB RAM and 145 GB hard disk.

Figure 36.1: Specifies that analytics has been the focus of so many companies and students for the last 3–4 years, so let's have a look at the trend of search queries based on "analytics" increased or decreased on Google.

Figure 36.2: In graph, X-axis represented by years and Y-axis represented by total count of Google searches for keyword analytics. So we can see that searches based

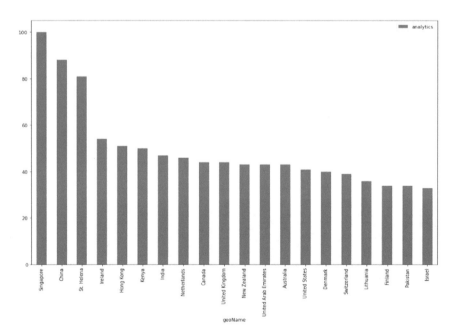

Fig. 36.1 Geo name analytics

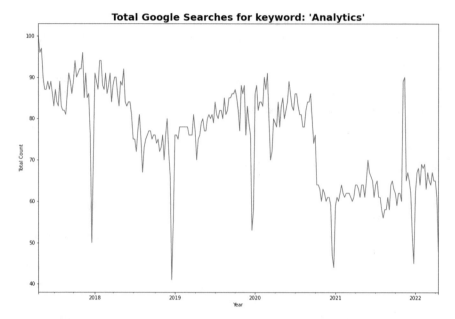

Fig. 36.2 Total Google searches for keyword: "analytics"

"analytics" Google started to decrease in 2018 and highest searches were done in 2022. This is how we can analyze Google searches based on any keyword.

Figure 36.3: This figure shows date by the date dimension and getting the sum of clicks for each of them, it is a type of summarization.

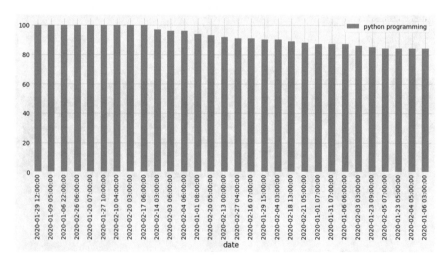

Fig. 36.3 Python programming

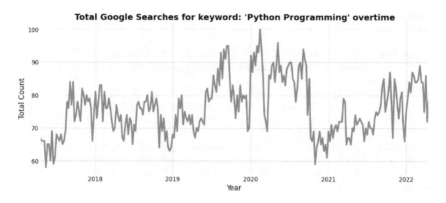

Fig. 36.4 Total Google searches for keyword: "Python programming" overtime

Figure 36.4: In this figure, X-axis shows over the period of five years, it has stayed pretty stable with the highest peak around 2020. Y-axis shows over the total count.

36.5   Conclusion

The methodology of this research data of was tested on data about user's activity in Twitter and Reddit and also data other than real estate. This will give next stage of development and improvement.

Google Trends plays vital role in understanding of housing market. Google Trends is a significant tool for predict real estate price changes.

Online users of the web predict price trends in housing market. Government reports are crucial for understanding of current housing market and trends of future market forecasting.

Google search data is more reliable for monitoring flu spread and produce comparison of forecasting accuracy to data of real-life ILI. The models forecasting capability enhanced the GT and ILI data is freely available without any delay can be used for addressing problems associated with traditional systems. To achieve sustainable surveillance, it will allow for preparation of better epidemic.

Acknowledgements We acknowledge all people who are directly and indirectly motivate us to complete this task. We sincerely thank all the authors and co-authors of the papers which are mentioned in references section.

References

1. Bordion, I., Battistion, S., Caldarelli, G., Cristelli, M., Ukkonen, A., Weber, I.: Web search queries can predict stock market volume. PLos ONE **7**(7), Art. No. e40014 (2012). 10.1371/journal.Pone.0040014
2. Surowiecki, J.: The Wisdom of Crowds. Anchor Book, New York (2005)
3. Hong, H., Ye, Q., Du, Q., Wang, G.A., Fan, W.: Crowd characteristics and crowd wisdom: evidence from an online investment community. J. Assoc. Inf. Sci. Technol. **71**(4), 423–435 (2020). https://doi.org/10.100/asi.24255
4. Ginsberg, J., Mohebbi, M.H., Patel, R.S., Brammer, L., Somlinski, M.S., Brilliant, L.: Detecting influenza epidemics using search engine query data. Nature **457**(7232), 1012–1014 (2009)
5. Choi, H., Varian, H.: Predicting initial claims for unemployment benefits. Menlo Park, CA, USA: Google (2012)
6. Choi, H., Varian, H.: Predicting the present with Google trends. Econ. Rec. **88**(1), 2–9 (2012)
7. Google Trends: Understanding the data. Accessed: 1 Aug 2019. Available http://storage.goo gleleapis.com/gwebnewsinitiativetraining.appspot.com/upload/GO802NewsInitiativeLesso nsFundamentals-L04-GoogleTrends1saYVCP.pdf; Liu, B.: Sentiment analysis and opinion mining, vol. 5. Morgan & Claypool, San Rafael, Ca, USA (2012)
8. Grishman, R., Huttunen, S., Yang Arber, R.: Information extraction for enhanced access to disease outbreak reports. J. Biomed. Information **35**(4), 236–246 (2002). https://doi.org/10.1016/S1532-0464(03)000013-3
9. Abla, M., Blench, M.: Global public health intelligence network (GPHIN). In: Proceedings of 7th conference on assoc. Mach. Transl. Amer. pp. 8–12 (2006). Accessed 9 March 2018. Available: http://pdfs.Semticscholar.Org/7d88/e623aa6ca78510e0093e17e2e00db39bd ad5.pdf
10. Reilly, A.R., Iarocci, E.A., Jung, C.M., Hartley, D.M., Nelson, N.P.: Indications and warning of pandemic influenza compare to seasonal influenza. Inf. Syst. **b9**(8), 2008 (2008)
11. Hulth, A., Rydevika, G., Linde, A.: Web queries as a source for syndromic surveillance. PLoS ONE **4**(2), e4378 (2009). https://doi.org/10.1371/journal.Pone.0004378
12. Bardak, B., Tan, M.: Prediction of influenza outbreaks by integrating Wikipedia article access logs and Google flu trend data. In: Proceedings of IEEE 15th BIBE, pp. 1–6 (2015). https://doi.org/10.1109/BIBE.2015.7367640
13. Hickmann, K.S., Fairchild, G., Priedhorsky, R., Generous, N., Hyman, J.M., Deshpande, A., Valle, S.Y.D.: Forecasting the 2013–2014 influenza season using Wikipedia. PLOS Comput. Biol. **11**(5), Art. No. e1004239 (2015). https://doi.org/10.1372/journal.pcbi.1004239
14. Eysenbach, G.: Infodemiology: tracking flu-related searches on the web for syndromic surveillance. In: Proceedings of AMIA annual symposium, pp. 244–248 (2006)
15. Polgreen, P.M., Chen, Y., Pennock, D.M., Nelson, F.D., Weinstein, R.A.: Using internet searches for influenza surveillance. Clin. Infectious Diseases **47**(11), 1443 (2008). https://doi.org/10.1086/5939098
16. Moss, R., Zarebski, A., Dawson, P., McCaw, J.M.: Forecasting influenza outbreak dynamics in Melbourne from internet search query surveillance data. Influenza Other Respiratory Viruses **10**(4), 314–323 (2016). https://doi.org/10.1111/irv.12376
17. Pelat, C., Turbelin, C., Bar-Hen, A., Flahault, A., Valleron, A.J.: More diseases tracked by using Google trends. Emerg Infectious **15**(8), 1327–1328 (2009). https://doi.org/10.320/eid 1508.090299
18. Schootman, M., Toor, A., Cavazos-Rehg, P., Jeffe, D.B., McQueen, A., Eberth, J., Davidson, N.O.: The utility of Google trends data to examine interest cancer screening. BMJ Open **5**(6), Art. No. e006678 (2015). https://doi.org/10.1136/bmjopen-2014-006678
19. Teng, Y., Bi, D., Xie, G., Jin, Y., Huang, Y., Lin, B., An, X., Feng, D., Tong, Y.: Dynamic forecasting of Zika epidemics using Google trends. PLos ONE **12**(1), Art. no. e0165085 (2017). https://doi.org/10.1371/journal.Pone.0165085

20. Li, C., Chen, L.J., Chen, X., Zhang, M., Pang, C.P., Chen, H.: Retrospective analysis of the possibility of predicting the COVID-19 outbreak from internet searches and social media data, China, 2020, Euro surveillance, vol. 25, no. 10, Art, no. 2000199 (2020). https://doi.org/10.2807/1560-7917.ES.2020.25.10.2000199

21. Shakeri Hossein Abab, Z., Kline, A., Sultana, M., Noaeen, M., Nurmambetiva, E., Lucini, F., Al-Jefri, M., Lee, J.: Digital public health surveillance: a systematic scoping review. NPJ Digital. Med. 4(1), 1–13 (2021). 10. 1038/s41746-021-00407-6

22. Edo-Osagie, O., de La Iglesia, B., Lake, I., Edeghere, O.: A scoping review of the use of Twitter for public health research. Comput. Biol. Med. 122, Arr. no. 103770 (2020). 10. 1016/j.compbiomed.2020.103770

23. Mogo, E.: Social media as a public health surveillance tool: evidence and prospects. Accessed: 2 Nov 2018. Available: http://www.enterprise.sickweather.com/downloads/SWSocialMedia_Whitepaper.pdf

24. Nsoesie, E.O., Olaseni, O., Abah, A.S.A., Ndeffo-Mbah, M.L.: Forecasting influenza-like illness trends in Cameroon using Google search data. Sci. Rep. 11(1), 6713 (2021). https://doi.org/10.1038/s41598-021-85987-9

25. Bragazzi, N.L., Mahroum, N.: Google trends predicts present and future plague cases during the plague outbreak in madagascar: infodemiological study. JMIIR Public Health Surveill. 5(1), Art. no. e13142 (2019). https://doi.org/10.2196/13142

Chapter 37
Internet of Things: Innovation in Evaluation Techniques

Devapriya Chatterjee and Devarshi Chatterjee

Abstract The purpose of the innovation in evaluation techniques in IoT is to reduce the risks within the ever-changing IoT landscape and to improve the security of the IoT applications and systems. An important aspect to IoT security is compliance. The absence of compliance in the governments, industries, and other authorities augments the risk of security, in terms of exposure to lawsuits, fines, and penalties. However, it is observed that IoT security comprises of only one element of risk. The innovative IoT evaluation techniques are designed to be compliant with mandated compliance IoT framework, and are expected to reduce the security-related risks, and improve the posture of IoT security. The paper discusses the building of an evaluation system for the deployment of IoT that ensures the improvement of the posture of IoT security. The paper further discusses the tools for maintaining and managing the IoT evaluation framework. The evaluation techniques are designed to set up an IoT compliance schedule with innovative methods for continuous monitoring, addressing the issues of impacts to IoT to frequently utilized evaluation techniques, and making an analysis of the challenges posed by IoT devices for the introduction of the IoT techniques. The paper encourages to build, adapt, and tailor the IoT evaluation techniques, in accordance with the demands of the ever-changing IoT landscape.

37.1 Introduction

A deployed and integrated IoT system that is made up of people, processes, and techniques, is compliant with some sets of best practices and regulations. The innovative evaluation techniques, related to the IoT systems, address the challenges of easy scanning for vulnerabilities in IoT systems, frequent usage of alternative networking in

D. Chatterjee (✉)
Ex Director (MBA), Shankara Group of Institutions Jaipur, Jaipur, India
e-mail: drdpchatterjee@gmail.com

Management Consultant and Chartered Engineer (India), Salt Lake City 700064, Kolkata, India

D. Chatterjee
Indian Institute of Management Nagpur India, Mihan, Nagpur 441108, India

IoT systems that are not found in the existing enterprises, availability of limited documentation for IoT system operations, implementation of a diverse array of hardware computing platforms by IoT systems, difficulty in installation of firmware/software updates to IoT components, and usage of functionally limited operating systems by the IoT systems.

37.2   Objectives

The innovation in IoT evaluation techniques is aimed to ensure that IoT systems are implemented in a compliant manner, with the following considerations:

(a) Setting up the IoT systems in a test environment before operational deployment for ensuring the run of rigorous IoT functional and security tests, against the systems for the identification of vulnerabilities and defects, and also examining the behavior of the devices on the network.
(b) Setting up regular IoT evaluation schedules for the review of operation procedures, configuration and documentation, as well as maintain scanned results for audit, to ensure continued compliance.
(c) Setting up proper management approaches for all IoT devices.
(d) Setting up procedures for IoT incident response for dictating responses to malicious events and natural failure.
(e) Appointing partners for sharing the IoT compliance and audit records.
(f) Setting up and documenting the roles and groups that are authorized to interact with the IoT system.
(g) Documenting records of additional IoT device characterization, like physical security, self-identification of device, configuration limitations, and upgradability of device, for configuring monitoring solutions.
(h) Setting up documentation procedure for integration of each IoT system into the network, and maintaining diagrams and change control procedures on record for regular IoT audits.
(i) Setting up documentation for identification of areas of functionality of the IoT devices and the type of requirement of portals and gateways.
(j) Setting up the authorities for approval, responsible for approving the operation of IoT systems in the organizational environment.
(k) Setting up documentation for storage of sensitive information, points of interconnection to other systems, all ports and protocol used.

37.3   Methodology and Data Collection

A video-conferencing was made with the managers of compliance evaluation of IBM Watson IoT platform, Microsoft Azure IoT suite, Cisco fog computing, and AWS IoT of Amazon.

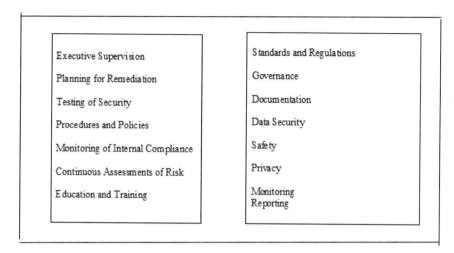

Fig. 37.1 IoT compliance program in an organization

The innovative evaluation initiative for the compliance of IoT necessitates a number of activities that should, at a minimum extent, be a sequence of the IoT compliance program. The activities are concurrent and ongoing functions that involve different stakeholders in the organization. In order to implement the innovative IoT evaluation systems, it is needed to ensure that each aspect of the IoT compliance program is in order. Figure 37.1 provides an idea about these activities.

37.4 Analyses and Results

As a critical business function, the innovative evaluation of IoT compliance, requires governance and executive supervision from multiple departments. Organizations that do not have supervision from executive level, and monitoring of policy mandates, put the stakeholders at a greater risk. The governance model for the innovative evaluation of IoT compliance needs to include the following organizational functions as well as departments:

(a) Operations
(b) Safety engineering
(c) Information security
(d) Privacy and legal representation

The governance needs to include some type of approval authority for operating the innovative evaluation system of IoT compliance. In absence of the same, there is the risk of potentially high-risk devices, being installed in the network. The governance team needs to be well versed in the security policies, and the technical standards, to which the IoT system needs to comply.

The policies and procedures for the sound operation of the innovative evaluation techniques for the IoT compliance systems need to inform how to safeguard the operation of the IoT systems and the data, and provide the details of actions for non-compliance.

The potential impact of misuse of an IoT compliance system could be avoided by organizing a comprehensive training program that would focus on the details of the skills needed to securely design, implement, and operate the IoT compliance systems. The knowledge and understanding that are to be imparted, need to include:

(a) Procedures of safety for IoT compliance systems
(b) Security tools that are IoT compliance specific
(c) Data security of IoT
(d) Privacy of IoT

The topics that are needed to be addressed in a broad way, in the training programs, need to include:

(a) Certification
(b) Tools for cybersecurity
(c) Privacy
(d) Network, cloud, and IoT
(e) Assessments of skill
(f) Security of data
(g) In-depth defense
(h) Threats

Certification would be necessary, focusing on privacy of data, and the understanding ability of the complex cloud environment, for powering the IoT implementation. Necessary training is needed to be imparted regarding the effective usage of the tools that provide regular inputs into the IoT compliance systems, and the tools that are used for the scanning of these systems. Adoption of measures of privacy pertaining to the underlying technology that drives the IoT evaluation systems in the organization, and the manner in which data is stored, transferred, and processed, need to be emphasized. The training programs need to consider the new architectures of network, that support the IoT deployment better, like the inclusion of software development networking (SDN) and network function virtualization (NFV) that are more adaptable, scalable, and dynamically responsible. The training program further needs to consider the secure configuration of the cloud-based, backend storage of data, and analytics systems for the prevention of malicious and non-malicious leakage of sensitive data. There could be unanticipated privacy and security risks due to the diversity in levels of sensitivity and types of data. The in-depth defense comprises of layering security mechanisms, such that the effect of failure of any of the layers could be minimized. Training for reinforcing this concept would enable to secure more robust IoT implementations and IoT security systems. The latest threats and cybersecurity alerts need to be addressed by adaptable and responsive in-depth defense approaches to the design of the IoT system that are conceptualized through training programs.

37.5 Discussion

The evaluation test beds are vital for IoT implementations, prior to the deployment in production environment. The functional testing in the test beds requires the ability to scale to the number of devices that would be deployed in the enterprise. During the initial test events, it may not be feasible to physically implement the exact numbers. As such, the requirement of a virtual test lab solution is made. These solutions upload and test the virtual machines in a simulated and realistic environment. The use of containers is leveraged to support the creation of the baselines of the IoT environment that could be tested with both security and functional tools. IoT deployments of higher assurance need to have rigorous security and safety regression tests, for the validation of proper system and device responses to error-state recoveries, security or safety related shut-offs, basic functional behavior, and sensor error conditions.

The organizations need to formulate a methodology for continuous testing of internal compliance, and evaluating the real-time security posture of IoT systems. The practical approach would be to adopt IoT integrated deployments. A six-step process has been formulated for continuous testing of the IoT systems in commercial organizations. These steps enable to identify continuously the new IoT security issues, even during the prioritization of resources, against pressing issues at any instant of time. The handling within an IoT system, warrants exploration for the purpose of adaptation (Fig. 37.2).

There is an additional step that focuses on the cause of the necessary IoT implementation and the necessary updating of the IoT system design. For an effective security management process, a feedback loop is designed between the necessary architectural update of the design of the IoT system, and the identification of the flaws.

The installation or updating of sensors that act as monitoring agents in IoT deployment, may not be feasible. However, the same could still be instrumented in the IoT systems, and a brief examination of the architectural fragment is discussed (Fig. 37.3).

The innovative evaluation is made of the data collected from the IoT architectural fragment of wireless sensor networks endpoints, while transmission of the data to a

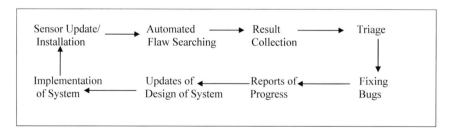

Fig. 37.2 Six-step IoT compliance testing process

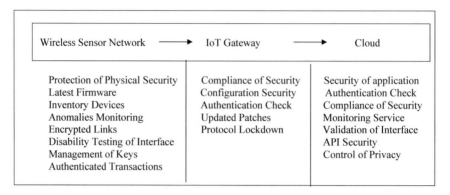

Fig. 37.3 Security-relevant data for architectural fragment of IoT deployment

protocol gateway. The data is then passed to the cloud by the gateway. The capabilities of the cloud service provider could then be leveraged for the capture of the data between the application endpoints that support the IoT sensors. The power of storage and processing is vested with the protocol gateway that is necessary for the installation of the IoT security tools. The components send back data on a scheduled or demand basis, for the support of the continuous IoT system monitoring from a cloud-based support structure.

The wireless sensor networks comprise of resource-limited IoT devices that lack memory, operating system, and processing support for instrumenting with audit and security agents. However, they play an important part in the security posture of the IoT systems.

The IoT devices, located in the diverse ecosystems, are needed to be subjected to automatic search for flaws, with the available protocols on the endpoints. These include gateways, interfaces, desktop applications, web servers, and mobile applications that are hosted in the cloud, for supporting the augmented analyses, collection, and reporting of data, that typically characterize IoT. The tools used in the automated search for flaws provide reports that allow for triage. The resources assigned to each flaw depend on its severity. A severity rating is assigned to each flaw, based on its impact on the IoT security, and a prioritization is made in accordance with the norms of fixing of high-severity findings on top priority. For fixing of bugs, inputs of disaster recovery need to be made in the IoT system, for clearing any backlog, and prioritizing them to the next sprint. For extreme severity, exceptions could be made by sole focusing on closing the security flaw, and stalling the development of the new feature in the IoT system. A regression testing is needed to be incorporated after every disaster recovery is completed, to ensure that unintentional flaws are not made in the IoT system, during the fixing by disaster recovery. The IoT compliance tools have their own reporting framework, and there are dashboards for reporting compliance to the evaluation team. As far as the updates of the design of system are concerned, it is needed to identify if the flaws necessitate configuration or design changes in the IoT systems and networks, or if the device is needed to be withdrawn. Period review meetings could identify the required changes in IoT architectures or baselines, and in the event of a severe vulnerability in any IoT device, a simple change of configuration could be made in the IoT network.

37.6 Findings

It is necessary for the organizations that deploy IoT solutions to have comprehensive penetration test programs. The innovative IoT evaluation technique would include a mix of white box and black box testing, along with fuzz testing against the IoT application protocols.

(a) White Box Technique of Evaluation

The white box or glass box technique of evaluation requires the evaluators to have full access to configuration information and design of the IoT system. A brief outline of the activities and descriptions of the white box technique of evaluation are discussed.

Activity	Description
Analysis of components hardware	Fingerprint devices may be used to ensure that components of hardware are not from unknown sources or not clones. The trustworthiness of the hardware components is determined from a supply chain perspective
Analyses of fault and attack tree	The fault trees comprise of a model framework, for analyzing how a system or device could fail from a set of unrelated leaf node events or conditions. With the updating of IoT system or product, fault tree models are updated to provide the available visibility in the safety risk management of the IoT system. The attack trees address system and device security. These need to be created as normal risk management activities for the understanding of the attacker's sequenced activities that would compromise the security of the IoT system or device. It is necessary to conduct a combined fault and failure tree modeling for a better understanding of the combined security and safety posture, for higher assurance IoT deployments, like life-critical medical systems. There is complex trade-off between safety and security as certain security controls could reduce safety of the IoT deployments

(continued)

(continued)

Activity	Description
Analysis of code	In order to identify vulnerabilities in any software of IoT system, it is needed to conduct static application security testing and dynamic application security testing
Interviews of staff	An evaluator needs to be fully aware of the deployment and integration points, the gamut of technologies used in the implementation, storage of critical data and processing of sensitive information, by way of conducting interviews of operational and developmental staff
Reviews of configuration documentation and design of system	It is necessary for the identification of the gaps in documentation and the areas of inconsistencies. It is necessary to create a security test plan by leveraging the documentation review, and also the review of designs of IoT systems and all documentation
Reverse engineering	In order to identify the possibility of any new development, based on the prevailing state of device firmware, reverse engineering of IoT device firmware is necessary to be performed whenever necessary

(b) Black Box Technique of Evaluation

The black box technique of evaluation requires the evaluators to break into the device without any prior knowledge of the technology of the implementation of the IoT device. This type of evaluation could be made at low costs, with third parties performing the tests, against the infrastructure as well as the devices. These evaluations are required annually for each IoT system, and more frequently, if there is change of systems. For IoT systems residing in the cloud, it is necessary to perform the application penetration testing against the representative virtual machines, deployed in the cloud containers. Valuable information could be derived from penetration testing against the deployed IoT system, if there is a test infrastructure mock-up of the system. This technique includes a characterization of the IoT system to enable the evaluator to understand the details that could be identified without authorization. A brief outline of the activities and descriptions of the black box technique of evaluation are discussed.

Activity	Description
Evaluation of mobile application	An evaluation is needed to be conducted of the mobile devices as communication is made by IoT devices with either gateways or mobile devices. The evaluation includes the characterization of the IoT technologies, capabilities, and features of the mobile application, along with the attempts of breaking the interfaces, connecting with the IoT devices through gateways or directly. Necessary investigations need to be made for alternative methods for replacing and overriding trust relationships between IoT devices and mobile applications
Evaluation of physical security	This technique characterizes the needs of physical security that is relative to the IoT deployment environment. This includes justification of tamper protections or embedded protections, and queries related to unprotected logical and physical interface
Analysis of cloud security	A detailed investigation is made into the protocols of communication used by the cloud hosted services, IoT devices, and mobile application. An analysis is also made regarding the deployment of secured communication, and the authentication of the IoT device and the mobile application to the cloud service. The infrastructure of the communication of the endpoint, need to be thoroughly tested, irrespective of the location, in cloud or on-premises. The evaluation technique needs to include that for management applications, with web servers having known vulnerabilities, and are public-facing
Analyses of software or firmware update process	The analyses include the queries related to performing manual updates or using an update server, the procedure of loading the software or firmware into the device at the initial stage, if the interface of loading software images are accessible, if there is any protection for the software and firmware during download, loading into memory or at rest, and if the integrity is protected at the file level. The analyses further include the authentication, the downloading of software patches in chunks, and the effect of the halting of the installation process midway, for some technical issue

(continued)

(continued)

Activity	Description
Evaluation of configuration security	In order to ensure that no unnecessary services are running in the IoT system, this process focuses on the optimal configuration of the IoT devices within a system. It ensures that only authorized protocols are enabled, and also evaluates the least privilege checking
Analyses of interface	The analyses map the IoT system services and all device applications, and then identify all the hidden and exposed interfaces. These determine the means of accessing each function or service. The analyses further include queries related to authentication, like the procedure of following a single authentication during accessing the device or initializing a session, or making the authentication on a per call basis. The function calls or services that are not authenticated, and require additional steps beyond authentication and prior to the initialization of services, are also analyzed. The circumstances in which sensitive operations are performed without authorization, are thoroughly investigated, and probed if the area of the operations is highly secured, with only authorized technicians
Evaluation of wireless security	The main purpose of this type of evaluation is to identify the known vulnerabilities with the protocols and figure what wireless protocols are used by the IoT device. An analysis is made regarding the usage of cryptography by the wireless protocol, and the nature of updating the default keys. It further evaluates the suitability of the default protocols that are used in conjunction with the wireless protocols. In these cases, the operating environments are analyzed, along with the sensitivity of the deployment environment

(c) Fuzz Testing Technique of Evaluation

The fuzz testing technique of evaluation is an advanced and specialized field that is exploited by the attackers of applications, through manipulation of states and abnormal usage of protocol. A brief outline of the activities and descriptions of the fuzz testing technique of evaluation are discussed.

Activity	Description
Processing of in header	This technique requires the implanting of unexpected fields the headers or header extensions of the protocols of IoT communication
Integration of analyzer	For the technique to be most effective, usage of various automated fuzzers is essential. These fuzzers need to have on the endpoint's behavior, an analysis engine, as it is being fuzzed. The fuzzed application's responses to various inputs are observed by a specially created feedback loop. It is also used for devising and altering valuable and new test cases that make the endpoint disabled and compromise it completely
Changing states of turning on and turning off power	This technique performs the detailed analyses for identifying the responses of IoT devices to unexpected and different inputs in various states. The technique also involves the sending of unexpected data, during changing states, to the IoT devices
Attacks on data validation	Improperly formatted data and random inputs are sent to the IoT gateways and endpoints. In other words, it is needed to send messages that do not conform to the application-acceptable message structures or the conforming message syntax
Value fields and length of protocol	This technique requires that unexpected values are implanted in the protocol fields of IoT communications. The possible examples include unexpected encodings and characters, as well as non-standard lengths of field inputs

37.7 Limitations and Further Scope of Research

The tracking of compliance with cybersecurity and regulations, as well as standards of data privacy, is the present practices of the industry. The embedded communications capabilities, introduced by IoT devices in the physical assets of the organizations, need to focus on compliance with safety regulations. The line between several regulatory IoT frameworks are blurred by the IoT devices. There are gaps due to the challenges posed by the fast changing pace of technology. In view of the same, the following IoT standard gaps need to be included in the guidance document:

(a) Big data and privacy
(b) Security and assurance of open source
(c) Security gateway

(d)   Measurement and management of IoT security
(e)   Risk assessment techniques of IoT devices
(f)   Incident response and guidance of IoT
(g)   Virtualization of network security
(h)   IoT application security guidance

Additional challenges are faced by health organizations, due to the IoT transition to connected healthcare equipment and other smart medical devices. There needs to be new IoT system deployments and device implementation in the organizations of the retail industry, and the industry of payment processors. These include

(a)   Ordering technologies that support automated delivery
(b)   Check-out in automatic mode
(c)   Smart rooms for fitting
(d)   Smart machines for vending
(e)   Radio frequency identification tags for inventory control
(f)   Proximity advertising

As some aspects of financial payment are involved, care needs to be taken to ensure that the prevalent requirements of payment processing are adhered to. The IoT risk management framework provisions for a set of continuous risk management activities that is needed to be followed by IoT system implementations. These comprise of:

(a)   System authorization for use
(b)   Implementation of only selected security controls
(c)   Continuous monitoring of system security framework
(d)   Assessment of security control implementation
(e)   Categorization on the basis of importance and sensitivity
(f)   Appropriate security control selection

It is a flexible process and could be adapted and applied to any IoT system implementation. There are still many gaps in the IoT frameworks and standards, but the developments are significant among the standard bodies, who are committed to closing the gaps.

Bibliography

1. Alansari, Z., Anuar, N.B., Kamsin, A., Belgaum, M.R., Alshaer, J., Soomro, S., Miraz. M.H.: Internet of things : infrastructure, architecture, security and privacy. In: 2018 international conference on computing, electronics communication engineering (ICCECOME), pp. 150–155 (2018). https://doi.org/10.1109/ICCECOME.2018.8658516
2. Ansari, D.B., Rehman, A.U., Ali, R.: Internet of things (IoT) protocols: a brief exploration of MQTT and CoAP. International Journal of Computer Applications **179**, 9–14 (2018)
3. Borgia, E., Gomes, D.G., Lagesse, B., Lea, R., Puccinelli, D.: Special issue on internet of things: research challenges and solutions. Comput. Commun. **89**(90), 1–4 (2016)
4. Cooper, J., James, A.: Challenges for database management in the internet of things. IETE Tech. Rev. **26**(5), 320–329 (2009)

5. Mahmud, S.H., Assan, L., Islam, R.: Potentials of internet of things (IoT) in Malaysian construction industry. Annals of Emerging Technologies in Computing (AETiC), vol. 2, no. 1, pp. 44–52. International Association of Educators and Researchers (IAER), Print ISSN: 256-0282, Online ISSN: 2516–029X. https://doi.org/10.33166/AETiC.2018.04.004

6. Mano, Y., Faical, B.S., Nakamura, L., Gomes, P.G., Libralon, R., Meneguete, G., Filho, G., Giancristofaro, G., Pessin, G., Krishnamachari, B., Ueyama, J.: Exploiting IoT technologies for enhancing healthsmart homes through patient identification and emotion recognition. Computer Communication 178–190. https://doi.org/10.1016/j.comcom.2016.03.010

7. Mazayev, A., Martins, J.A., Correia, N.: Interoperability in IoT through the semantic profiling of objects. IEEE Access **6**, 19379–19385 (2018)

8. Miraz, M., Ali, M., Excell, P., Picking, R.: Internet of nano-things, things and everything: future growth and trends. Future Internet **10**(8), 68 (2018). https://doi.org/10.3390/fi10080068

9. Miraz, M.H., Ali, M., Excell, P.S., Picking, R.: A review on internet of things (IoT), internet of everything (IoE) and internet of nano-things (IoNT), internet technologies and applications (ITA), pp. 219–224 (2015). https://doi.org/10.1109/ITechA.2015.7317398

10. Miraz, M.H., Ali, M.: Blockchain enabled enhanced IoT ecosystem security. In: Proceedings of the international conference on emerging technologies in computing 2018, London Metropolitan University, UK, Part of the Lecture Notes of the Institute for Computer Sciences, Social Informatics and Telecommunication Engineering (LNICST), vol 200, pp. 38–46, Online ISBN: 978-3-319-95450-9, Print ISBN: 978-3-319-95449-3, Series Print ISSN: 1867-8211, Series Online ISSN: 1867-822X. https://doi.org/10.1007/978-3-319-95450-9_3

11. Miraz, M.H.: Blockchain of things (BCoT): the fusion of blockchain and IoT technologies. Advanced applications of blockchain technology, studies in big data (2019). https://doi.org/10.1007/978-981-13-8775-3_7

12. Patel, K.K., Patel, S.M., et al.: Internet of things IoT: definition, characteristics, architecture, enabling technologies, and application future challenges. International Journal of Engineering Science and Computing **6**(5), 6122–6131 (2016).

13. Rajguru, S., Kinhekar, S., Pati, S.: Analysis of internet of things in a smart environment. International Journal of Enhanced Research in Management and Computer Applications **4**(4), 40–43 (2015)

14. Soomro, S., Miraz, M.H., Prasanth, A., Abdullah M.: Artificial intelligence enabled IoT: traffic congestion reduction in smart cities. In: IET 2018 smart cities symposium, pp. 81–86. https://doi.org/10.1049/cp.2018.1381

15. Tadejko, P.: Application of internet of things in logistics-current challenges. Ekonomia i Zarz a dzanie **7**(4), 54–64 (2015)

Chapter 38
Discovering the Performance of MANET with Malicious and Non-malicious Node Using Newton–Raphson Method

A. Ganesan and A. Kumar Kompaiya

Abstract The behavior of malicious and non-malicious nodes in mobile adhoc networks (MANET) is examined in this study using the Newton–Raphson method. This research uses the Newton–Raphson method to analyze the behavior of mobile adhoc networks (MANET) with malicious and non-malicious nodes. To find the infiltration, this MANET performance data is helpful. The network simulator is used to assess MANET's performance. Network measures including energy consumption, packet delivery ratio, and delay are used in the Newton–Raphson approach to determine the performance of the network. Three alternative configurations of the Medium Access Control (MAC) protocol—low; medium, and high—were used to carry out the experiment.

38.1 Introduction

A wireless network is a mobile adhoc network. This network is made up of a number of mobile nodes that can communicate with one another without the need of centralized management or specified infrastructure. This wireless open channel environment is used as transferring medium between originating node to destination node. The MANET has more attention from the network performance researchers. Aim of this work to discover the mobile adhoc network performance [8, 10] with malicious and non-malicious node using Newton–Raphson method [1]. This method is used to solve non-linear equation [7]. The three separate equations in this work's coefficients are a, b, and c. Three network metrics—packet delivery ratio, energy consumption, and delay—are assessed in this experiment utilizing NS2 in three different configurations with MAC protocol values of low, medium, and high. Finally, results are verified and

A. Ganesan (✉)
PG and Research Department of Computer Applications, Hindusthan Arts College, Coimbatore 641 028, India
e-mail: ganesan102.a@gmail.com

A. K. Kompaiya
Department of Computer Science, Chikkanna Goverment Arts College, Tirupur 641 602, India

compared with malicious and non-malicious node of MANET by Newton–Raphson method [1].

38.2   Experiment Value of MAC Protocol

The Medium Access protocol's (MAC) parameter [2] value such as contention window (CW), bit rate (BR), and transmission power are used in this experiment to discover MANET performance with malicious and non-malicious node. These MAC protocol parameters [13] effects on energy usage, packet delivery ratio, and transmission time are significant. The transmission power has an impact on the transmitter's energy consumption; if it is too high, it may shorten battery life and cause more interference with MANET nodes using the same frequency. The transmission power is the amount of electricity provided to the transmitter for data transmission (Tx). The maximum transmission power for a node between 930 and 1000 m away is 1584.89. Three transmission power values—low, medium, and high—are utilized in this work. That transmission powers are 500, 1000, and 1584.89 mW.

The second bit rate (BR) is a parameter of the MAC protocol that describes how much data (measured in bits) is sent in a given length of time. It is necessary to handle data transmission at 6, 12, and 24 Mbit/s. Since more data may be delivered at once at greater bit rates, network delays are reduced. Less time is required for the data transmission to be completed for improved performance.

The third MAC protocol's parameter is contention window (CW) [15, 16]. The duration of the network's operation in contention mode is defined as this parameter. It appears that the CW will have greater throughput under heavy load for messages of lower priority if its size is kept bigger. The collision rate will be lessened by the higher CW size [3]. In order to get lower delay for higher priority message in a heavy load, smaller CW size is better [3]. Thus, the smaller CW size, the less number of packet are waiting to be served. Then, it reduces delay. The minimum CW size is 31 and maximum CW size is 255 are used in this work (Table 38.1).

The selected three parameters of MAC protocol used for three different configuration for this experiment they are

Low: 500mW (Tx), 31(CW), and 6 Mbps (BR)
Medium: 1000mW (Tx), 128 (CW), and 12 Mbps (BR)

Table 38.1 Priority classes parameter source [3]

Parameter	Data	Audio	Video
CW min	31	7	15
CW max	1024	255	511
Packet size	1500 bytes	160 bytes	1280 bytes
Packet interval	12.5 ms	20 ms	10 ms
Flow rate	120 KBps	8 KBps	128 KBps

Table 38.2 Parameter used in simulator

Parameter	Value
Channel type	Wireless channel
Number of nodes	100, 200, 300, 400, and 500
Pulse time	20 s
Traffic type	CBR
MAC type	802_11
Node placement	Random
Mobility	Random way
Area simulation	1000 m*1000 m

High: 1584.89mW (Tx), 255 (CW), and 24Mbps (BR)

38.3   Methodology

In order to execute the simulation for this experiment, network simulator 2 (NS2) with a DSR routing protocol environment was employed. In order to obtain reliable results, each simulation was run for 0, 10, 20, 30, and 40 m simulator time five times. The average outcome for this simulator was then calculated. For this work, the empirical function (Sect. 39.5) was computed using this result.

To obtain location-based routing information in MANET for this research project, DSR routing protocol is used [12]. Additionally, the User Data gram protocol (UDP) has been utilized for the transport protocol along with the IEEE 802.11p MAC protocol [5]. This transport protocol is incredibly quick, has little latency, and is not impacted by traffic. The simulation environment and parameters used in this simulation are displayed in Table 38.2.

38.4   MANET Metrics Evaluation

The various simulation types used in this study work are used to assess and examine the network performance [4]. The average result value of these simulations is determined.

Packet delivery ratio, energy use, and transmission time delay are the chosen network metrics for analyzing network performance.

Ratio of Delivered Packets: By dividing the total number of data packets sent from sources by the total number of data packets received at destinations, several network metrics can be determined. The ratio of packets supplied from the source to those received at the destination is known as the packet delivery ratio. It can be computed as

Packet delivery ratio

$$= \sum \text{Number of packets received}/ \sum \text{Number of packets sent}$$

The proportion of packets sent by a sender that reach their destination successfully as compared to the total number of packets transmitted [11].

Energy Consumption: The energy consumed by each individual node when sending a packet, as well as by switching, transmitting, and accessing networks. Based on the many types of network performance metrics used in this research, mobility and retransmissions are two elements that might contribute to excessive energy consumption [9]. The main problem of host which is in wireless network is energy consumption node will take more energy while its data transmission. It wastes its resources due to delay of packet transmission and packet dropping. The encryption of packet will take more energy consumption. If the distance is quite far, the transmission in one hop uses more energy. Retransmissions, collisions, idle listening, control packets, and overhearing all affect how much energy is used. The energy consumption will be increased when the malicious node is exists.

Transmission Time Delay: The packet's transmission time delay includes the local transmission delay experienced from one node to the next. Each packet's delay is the total of the delays encountered at a series of intermediate nodes en route to the destination. The local delay often grows with the number of hops. As a result, information coming from the wireless network that has more hops before reaching the sink is probably going to encounter a lengthier delay. In order to transport less traffic that has been forwarded, a protocol or technology that is more hops away from the sink is required. The route discovery procedure and the number of nodes in the path from the source to the destination have an impact on the transmission delay.

38.5 Empirical Function

The goal of this research is to develop products with low energy consumption, which increases sensor longevity and improves network performance. It also aims to build products with high packet delivery ratios (PDR) and short transmission time delays. If the network is able to accomplish this performance, it is performing well. As a result, to accomplish the objectives of this work, an empirical function $f(E)$ equation, as shown in Eq. (38.1), is applied.

$$f(\text{PDR, Energy} - \text{Cons, delay})$$
$$= (\text{Energy consumption} * \text{Delay})/\text{PDR} = \text{EF} \qquad (38.1)$$

In order to obtain the necessary value for this experiment, as well as to obtain a high value of PDR and a low value of energy consumption and delay at the same time, the Ef in Eq. (38.1) must be low. PDR is the rate of successful packet transmission in the equation above. The MANET uses a certain amount of electricity to complete the data transfer, which is known as the energy consumption. The values of energy are directly proportional to **EF** (empirical function).

Last but not least, by reducing the value of EF, it can reduce the value of PDR as well as reduce the value of energy consumption and delay. Using the Newton–Raphson approach stated in Eq. (38.2) [1, 6], the computed value of EF is utilize to determine the values of coefficients a, b, and c.

$$a^2 x_{1+b}^2 x_{2+c}^2 x_3 = (\text{Energy consumption} * \text{Delay})/\text{PDR} = \text{EF} \qquad (38.2)$$

In this above mentioned Eq. (38.2), x_1 represents the contention window, x_2 represents the bit rate, and x_3 is the presentation range taken for this research.

38.6 Result Analysis

As demonstrated in the accompanying table, the trials were carried out both with and without a malicious node, using five different transmitted intervals.

38.6.1 Without Malicious Node Experiment

Tables 38.3, 38.4, 38.5, 38.6, 38.7 and 38.8 show the result of empirical function (EF) for the different configuration without malicious node.

It can be seen from Tables 38.4, 38.5, 38.6, 38.7 and 38.8 that as the parameter value increases, the EF value decreases. It demonstrates that as the parameter's value increases, the PDR will be high without malicious MANET while the latency and energy consumption will be minimal.

The coefficients a, b, and c for equation are determined using the Newton–Raphson method [14] (38.2). There are three distinct formulae used to calculate these coefficients. These equations represent the low, medium, and high values of the MAC protocol parameters. The Newton–Raphson technique is used in the computation of the coefficient values.

$$EF_1 = a^2(31) + b^2(6) + c^2(0.5) \qquad (38.3)$$

$$EF_2 = a^2(128) + b^2(12) + c^2(1) \qquad (38.4)$$

$$EF_3 = a^2(255) + b^2(24) + c^2(1.58489) \qquad (38.5)$$

In Eqs. (38.3), (38.4), and (38.5), the values 31, 128, and 255 are contention window (CW), the values 6, 12, and 24 are bit rate (BR), and the values 0.5, 1, and

Table 38.3 Number of node: 60 Number of packets: 50 Source node: 1 Destination node: 55

Simulator time/network-metric	Queue-length	Packet delivery	Throughput	Packet-drop	Delay	Energy	Life-time	Latency
0	9.25	9.25	9.25	10	10	10	9.25	10
10	9.43	9.44	9.39	9.87	9.89	9.89	9.45	9.97
20	9.51	9.51	9.44	9.76	9.87	9.84	9.51	9.92
30	9.62	9.54	9.51	9.64	9.8	9.78	9.58	9.84
40	9.69	9.59	9.56	9.53	9.73	9.7	9.65	9.78

Table 38.4 Low value of MAC protocol parameter result

Simulator time (m)	Delay	Energy	Packet delivery	EF_1
0	10	10	9.25	10.81081
10	9.89	9.89	9.44	10.36145
20	9.87	9.84	9.51	10.21249
30	9.8	9.78	9.54	10.04654
40	9.73	9.7	9.59	9.841606
				10.25458

1.58489 are transmission power (Tx). As shown in the Eq. (38.3) is the set of low value parameters, the Eq. (38.4) is the set of medium value parameter, and Eq. (38.5) is the set of high value parameter and also EF_1, EF_2, and EF_3 are the average empirical functions. The Newton–Raphson method used for solving nonlinear equation requires the evaluation matrix defined by Jacobian of the system.

$$J = \begin{pmatrix} \frac{\partial EF_1}{\partial a} & \frac{\partial EF_1}{\partial b} & \frac{\partial EF_1}{\partial c} \\ \frac{\partial EF_2}{\partial a} & \frac{\partial EF_2}{\partial b} & \frac{\partial EF_2}{\partial c} \\ \frac{\partial EF_3}{\partial a} & \frac{\partial EF_3}{\partial b} & \frac{\partial EF_3}{\partial c} \end{pmatrix} = \begin{pmatrix} 62a & 12b & c \\ 256a & 24b & 2c \\ 510a & 48b & 3.168c \end{pmatrix}$$

$$\begin{pmatrix} a_{k+1} \\ b_{k+1} \\ c_{k+1} \end{pmatrix} = \begin{pmatrix} a_k \\ b_k \\ c_k \end{pmatrix} - \begin{pmatrix} \frac{\partial EF_1}{\partial a} & \frac{\partial EF_1}{\partial b} & \frac{\partial EF_1}{\partial c} \\ \frac{\partial EF_2}{\partial a} & \frac{\partial OF_2}{\partial b} & \frac{\partial EF_2}{\partial c} \\ \frac{\partial EF_3}{\partial a} & \frac{\partial EF_3}{\partial b} & \frac{\partial EF_3}{\partial c} \end{pmatrix}^{-1} \times \begin{pmatrix} EF_1 \\ EF_2 \\ EF_3 \end{pmatrix}$$

(38.6)

The coefficients of variables a, b, and c are calculated using Eq. (38.6), and the predefined value of EF is the average value of EF from Tables 38.4, 38.5, 38.6, 38.7 and 38.8. Where k represents the current level's solution and $k + 1$ represents the level beyond that. The updated set of EF values is now calculated as follows

$$EF_1 = (31)(1)^2 + (6)(1)^2 + (0.5)(1)^2 - 10.25458 = 27.24542$$

$$EF_2 = (128)(1)^2 + (12)(1)^2 + (1)(1)^2 - 10.33328 = 130.66672$$

$$EF_3 = (255)(1)^2 + (24)(1)^2 + (1.58489)(1)^2 - 10.25051 = 280.58489$$

$$\begin{pmatrix} a_{k+1} \\ b_{k+1} \\ c_{k+1} \end{pmatrix} = \begin{pmatrix} 1 \\ 1 \\ 1 \end{pmatrix} - \begin{pmatrix} 62(1) & 12(1) & 1(1) \\ 256(1) & 24(1) & 2(1) \\ 510(1) & 48(1) & 3.168(1) \end{pmatrix}^{-1} \times \begin{pmatrix} 27.24542 \\ 130.66672 \\ 270.33438 \end{pmatrix}$$

This step will be continued until the desired solution is found. Using the following Eq. (38.7), we can find Jacobian transpose matrix (JT).

Table 38.5 Number of node: 65 Number of packets: 45 Source node: 2 Destination node: 60

Simulator time/network-metric	Queue-length	Packet delivery	Throughput	Packet-drop	Delay	Energy	Life-time	Latency
0	9.25	9.25	9.25	10	10	10	9.25	10
10	9.43	9.44	9.39	9.88	9.97	9.93	9.45	9.97
20	9.52	9.51	9.44	9.77	9.94	9.88	9.51	9.92
30	9.62	9.54	9.51	9.64	9.83	9.82	9.58	9.84
40	9.69	9.59	9.56	9.53	9.77	9.74	9.65	9.78

Table 38.6 Medium value of MAC protocol parameter result

Simulator time (m)	Delay	Energy	Packet delivery	EF$_2$
0	10	10	9.25	10.81081
10	9.97	9.93	9.44	10.48751
20	9.94	9.88	9.51	10.32673
30	9.83	9.82	9.54	10.11851
40	9.77	9.74	9.59	9.922815
				10.33328

$$J^- = \begin{pmatrix} \frac{\partial EF_1}{\partial a} & \frac{\partial EF_1}{\partial b} & \frac{\partial EF_1}{\partial c} \\ \frac{\partial EF_2}{\partial a} & \frac{\partial EF_2}{\partial b} & \frac{\partial EF_2}{\partial c} \\ \frac{\partial EF_3}{\partial a} & \frac{\partial EF_3}{\partial b} & \frac{\partial EF_3}{\partial c} \end{pmatrix} = 1/\det(J) \times \mathrm{Adj}\ (J) \tag{38.7}$$

Thus, the coefficient of a, b and c values calculated.

$$\begin{pmatrix} 0.42291 \\ -0.30756 \\ 25.22490 \end{pmatrix}$$

The value of $a = 0.42291$, $b = -0.30756$, and $c = 25.22490$.

38.6.2   With Malicious Node Experiment

Similarly, the Tables 38.9, 38.10, 38.11, 38.12, 38.13 and 38.14 show the result of empirical function (EF) for the different configuration with malicious node. It can be seen from Tables 38.10, 38.11, 38.12, 38.13 and 38.14 that as the parameter value increases, the EF value decreases. It demonstrates that as the parameter's value increases, the PDR with malicious MANET will be high and the delay and energy consumption will be low.

The coefficients of variables a, b, and c are calculated using Eq. (38.6), and the predefined value of EF is the average value of EF from Table 38.10, 38.11, 38.12, 38.13 and 38.14. Now, the malicious node is used to compute the new set of EF values as seen below.

$$EF_1 = (31)(1)^2 + (6)\,(1)^2 + (0.5)\,(1)^2 - 10.25458 = 27.24542$$

$$EF_2 = (128)\,(1)^2 + (12)\,(1)^2 + (1)\,(1)^2 - 10.33328 = 130.66672$$

$$EF_3 = (255)\,(1)^2 + (24)\,(1)^2 + (1.58489)\,(1)^2 - 10.36317 = 280.58489$$

Table 38.7 Number of node: 70 Number of packet: 60 Source node: 2 Destination node: 65

Simulator time/network-metric	Queue-length	Packet delivery	Throughput	Packet-drop	Delay	Energy	Life-time	Latency
0	9.25	9.25	9.25	10	10	10	9.25	10
10	9.43	9.43	9.38	9.87	9.89	9.87	9.44	9.97
20	9.51	9.51	9.43	9.76	9.87	9.83	9.52	9.92
30	9.61	9.54	9.51	9.64	9.8	9.78	9.58	9.84
40	9.69	9.59	9.56	9.54	9.73	9.7	9.65	9.78

Table 38.8 High value of MAC protocol parameter result

Simulator time (m)	Delay	Energy	Packet delivery	EF$_3$
0	10	10	9.25	10.81081
10	9.89	9.87	9.43	10.35146
20	9.87	9.83	9.51	10.20211
30	9.8	9.78	9.54	10.04654
40	9.73	9.7	9.59	9.841606
				10.25051

$$\begin{pmatrix} a_{k+1} \\ b_{k+1} \\ c_{k+1} \end{pmatrix} = \begin{pmatrix} 1 \\ 1 \\ 1 \end{pmatrix} - \begin{pmatrix} 62(1) & 12(1) & 1(1) \\ 256(1) & 24(1) & 2(1) \\ 510(1) & 48(1) & 3.168(1) \end{pmatrix}^{-1} \times \begin{pmatrix} 27.24542 \\ 130.66672 \\ 280.58489 \end{pmatrix}$$

Then, the coefficient of a, b, and c values calculated with malicious node MANET as follows.

$$\begin{pmatrix} 0.42291 \\ -0.33265 \\ 25.52599 \end{pmatrix}$$

Thus, the coefficient value of a is 0.42291, b value is -0.33265, and value of c is 25.52599.

38.7 Conclusion

The malicious node either will attack or will disrupt the performance of MANET. Therefore, detection of malicious node and solation of malicious node will improve the performance of MANET. The Newton–Raphson method is used to efficiently find the coefficient value of the a, b, and c values as a result of this research. According to the experiment, measurements for MANET with malicious nodes are more valuable than those for MANET without malicious nodes. As a result, this network uses a lot of power.

Table 38.9 Number of node:60 Number of packet:50 Source: node 1 Destination:58

Simulator time/network-metric	Queue-length	Packet delivery	Throughput	Packet-drop	Delay	Energy	Life-time	Latency
0	9.25	9.25	9.25	10	10	10	9.25	10
10	9.4	9.39	9.3	9.91	9.94	9.92	9.39	9.93
20	9.48	9.46	9.4	9.79	9.92	9.87	9.48	9.9
30	9.57	9.49	9.49	9.67	9.82	9.81	9.54	9.83
40	9.65	9.56	9.5	9.58	9.76	9.74	9.63	9.77

Table 38.10 Low value of MAC protocol parameter result

Simulator time (m)	Delay	Energy	Packet delivery	EF1
0	10	10	9.25	10.81081
10	9.89	9.89	9.44	10.36145
20	9.87	9.84	9.51	10.21249
30	9.8	9.78	9.54	10.04654
40	9.73	9.7	9.59	9.841606
				10.25458

Table 38.11 Number of node: 65 Number of packet:50 Source node: 2 Destination node:60

Simulator time/network-metric	Queue-length	Packet delivery	Throughput	Packet-drop	Delay	Energy	Life-time	Latency
0	9.25	9.25	9.25	10	10	10	9.25	10
10	9.43	9.42	9.48	9.94	9.97	9.95	9.42	9.97
20	9.5	9.48	9.49	9.81	9.94	9.87	9.5	9.92
30	9.58	9.51	9.5	9.68	9.83	9.82	9.55	9.84
40	9.66	9.57	9.51	9.59	9.77	9.75	9.64	9.78

Table 38.12 Medium value of MAC protocol parameter result

Simulator time (m)	Delay	Energy	Packet delivery	EF2
0	10	10	9.25	10.81081
10	9.97	9.93	9.44	10.48751
20	9.94	9.88	9.51	10.32673
30	9.83	9.82	9.54	10.11851
40	9.77	9.74	9.59	9.922815
				10.33328

Table 38.13 Number of node: 70 Number of packet: 60 Source node: 2 Destination node: 65

Simulator time/network-metric	Queue-length	Packet delivery	Throughput	Packet-drop	Delay	Energy	Life-time	Latency
0	9.25	9.25	9.25	10	10	10	9.25	10
10	9.43	9.42	9.48	9.94	9.97	9.95	9.42	9.97
20	9.5	9.48	9.49	9.81	9.94	9.87	9.5	9.92
30	9.58	9.51	9.5	9.68	9.83	9.82	9.55	9.84
40	9.66	9.57	9.51	9.59	9.77	9.75	9.64	9.78

Table 38.14 High value of MAC protocol parameter result

Simulator time (m)	Delay	Energy	Packet delivery	EF3(70-M)
0	10	10	9.25	10.81081
10	9.97	9.95	9.42	10.53094
20	9.94	9.89	9.48	10.36989
30	9.83	9.82	9.51	10.15043
40	9.77	9.75	9.57	9.953762
				10.36317

References

1. Skaflestad, B.: Newton's method for systems of non-linear equations. 3 Oct 2006. https://www. math.ntnu.no›notater›nr-systems-a4
2. Garg, V.K.: Wireless local area networks. Research Gate (2007)
3. Khalaj, A., Yazdani, N., Rahgozar, M.: Effect of the contention window size on performance and fairness of the IEEE 802.11 standard. Wirel. Pers. Communication. **43**(4), 1267–1278 (2007)
4. Bansal, R., Goyal, H., Singh, P.: Analytical study the performance evaluation of mobile Ad Hoc networks using AODV protocol. International Journal of Computer Applications **14**(4), 0975–8887 (2011)
5. Dalal, K., Chaudhary, P., Dahiya, D.P.: Performance evaluation of TCP and UDP protocol in VANET. Ad Hoc Netw. **10**(2), 253–269 (2012)
6. VijiPriya, J.: Application of Newton Raphson algorithm for optimizing TCP Performance. In: Fourth international conference on advances in recent technologies in communication and computing (ARTCom2012)
7. Remani, C.: Numerical methods for solving of nonlinear equations, vol. 4301 (2013)
8. Adimalla, R.R., Priyanka, M.L., Harika, K.: Performance evaluation of mobile Ad-Hoc network (MANET) routing protocols (reactive) by using network simulator-2. **IJERT 02**(03) (2013)
9. Awatef, B.G., Nejeh, N., Abdennaceur, K.: Impact of topology on energy consumption in wireless sensor networks. ENIS 4 Aug 2014
10. Sikarwar, N.: Performance comparison of Adhoc network with NS2 simulator. International Journal of Advanced Computational Engineering and Networking **2**(11) (2014). ISSN: 2320-2106
11. Tyagi, S., Gopal, G., Garg, V.: Detecting malicious node in network using packet delivery ratio. In: Published 2016 computer science 2016 3rd international conference on computing for sustainable global development (INDIACom)
12. Fahad, A.M., Alani, S., Mahmood, S.N., Fahad, N.M.: NS2 based performance comparison study between DSR and AODV protocols. International Journal of Advanced Trends in Computer Science and Engineering **8**(1.4) (2019)
13. Malik, M., Sharma, M.: Design and analysis of energy efficient MAC protocol for wireless sensor networks. International Journal of Engineering and Advanced Technology (IJEAT) **8**(3) (2019). ISSN: 2249-8958
14. Adesina, L.M., Abdulkareem, A., Katende, J., Fakolujo, O.: Newton-Raphson algorithm as a power utility tool for network stability. Advances in Science, Technology and Engineering Systems Journal **5**(5), 444–451 (2020)
15. Li, F., Huang, G., Yang, Q., Xie, M.: Adaptive contention window MAC protocol in a global view for emerging trends networks. IEEE Access (IF3.367), Pub Date: 25 Jan 2021. https://doi.org/10.1109/access.2021.3054015
16. Reinders, R., van Eenennaam, M., Karagiannis, G., Heijenk, G.: Contention window analysis for beaconing in VANETs

Chapter 39
GAN to Produce New Faces and Detection Expression

Sidhant Singh, Souvik Sarkar, Pomesh Kumar Deshmukh, Rohit Kumar, and Debraj Chatterjee

Abstract In a world where data is the most significant item. Image has now set off to be the most valuable data. Along with the evolving technologies, image is being bifurcated and used in the field of machine learning for various operations. From the recent past, image is bifurcated into segments and could be seen to be used everywhere, whether it is predicting market patterns or imitating the real world in a form of virtual grid. Segmentation of images is the most critical part of developing a machine learning model. Better is the training of the model, better will be the results, which may result in a successful machine learning model. A recent innovative proceeding has been introduced related to image segmentation, generative adversarial network (GAN), which will surely bring huge achievements in this field. In this study, the function of GAN in producing new faces and subsequently detecting facial expressions is elaborated. This study can be divided into two phases, the first is to generate faces using GAN, and the second phase is to detect expressions of the faces generated. For the first phase, the aim is to train the generator model with the existing face dataset and after the successful training, the generator will be able to produce new faces. In the second phase, a model is prepared which can extract the facial features and detect expressions based on these features.

S. Singh (✉) · S. Sarkar · R. Kumar · D. Chatterjee
Techno International New Town, Kolkata, India
e-mail: singhsidhant2000@gmail.com

S. Sarkar
e-mail: souviksarkar.ronnie@gmail.com

R. Kumar
e-mail: rohit.kumar.2018@cse.tict.edu.in

D. Chatterjee
e-mail: debraj.chatterjee@tict.edu.in

P. K. Deshmukh
Shri Shankaracharya Technical Campus, Bhilai, India
e-mail: pomesh.onlineclass@gmail.com

39.1 Introduction

Machine learning [1] is a modern technology used by computers to learn from data and information that are fed to them. These data are fed to them in the form of observations and real-world interactions. This process helps them to behave like artificial humans and also auto-improve themselves spontaneously. These are performed on the basis of algorithms and without any interference of humans. These algorithms are designed to learn data by themselves along with improvement from past experiences. Machine learning is being used widely in various fields such as speech recognition, image recognition, traffic prediction, and GAN. This study is mainly based on supervised learning that can be defined as a method in which labelled training data is used by the model to learn the mapping function from the input variables (X) to the output variable (Y).

$$Y = f(X) \tag{39.1}$$

Generative adversarial network (GAN) [2, 3] is an advanced study that extends the limits of modern technologies. Introduced in 2014, it is emerging as a broad area in research and has an active scholars working in this field. Learning from dataset that has been fed to the model, it can generate new results. Two models, namely as generator and discriminator, engage with each other to analyse, capture and replicate the variations in the dataset. Generator is a deep network that learns to create fake data by incorporating feedback from the discriminator. The generator is trained on random input, generator network, discriminator network, and discriminator output and generator loss. On the other hand, discriminator is simply a classifier that tries to contrast real data from the data created by the generator. The discriminator is trained on real data instances based on real images of people and fake data instances created by the generator. An overview of the GAN structure can be visualized as shown in Fig. 39.1.

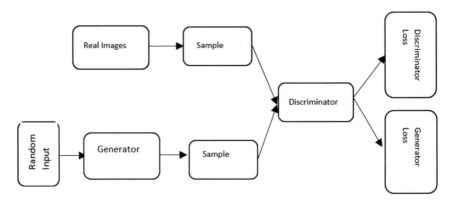

Fig. 39.1 Overview of GAN structure

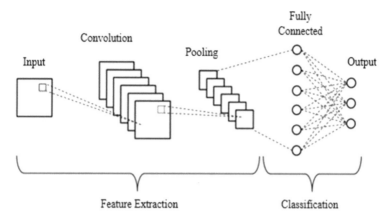

Fig. 39.2 Convolutional neural network

Convolutional neural network (CNN) [4] is an algorithm for classification of images in which objects/aspects are assigned with weights/biases from the images provided as input, and then, they are differentiated from other objects. This is majorly used for analysing visual images. In comparison with other classification algorithms, CNN requires much lower pre-processing of the images before training of the model. A simple CNN is shown in Fig. 39.2.

This network is made up of many neurons. A neuron has an input when after computing with weight and biases, then passed from an activation function [5] which will finally decide whether neuron will be fired or not. Machine learning models have various types of activation functions such as sigmoid, ReLU, softmax, and tanh. This study comprises of sigmoid, LeakyReLU, and ReLU as activation functions. ReLU activation function is the majorly used activation functions in the field of machine learning. It can be expressed mathematically as in Eq. (39.2).

$$f(x) = \max(0, x) \tag{39.2}$$

According to the above equation, any input less than 0 will simply be 0 and rest will return the number itself. It can be expressed easily as in Fig. 39.3.

Moreover, LeakyReLU function is another activation function that has been used in the model in the output layer. It is a variation of ReLU function which deals with the limitation of dying ReLU. LeakyReLU can be expressed mathematically as in Eq. (39.3).

$$f(x) = \max(0.2x, x) \tag{39.3}$$

The graphical representation of LeakyReLU function can is shown in Fig. 39.4.

OpenCV [6] is an open-source library that is implemented in computer vision, machine learning, and image-processing. Basically, it is a library for image and video

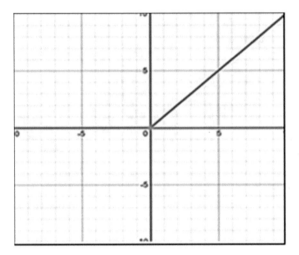

Fig. 39.3 Graphical representation of ReLU function

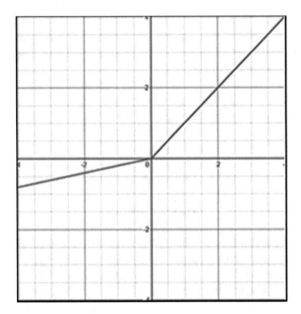

Fig. 39.4 Graphical representation of LeakyReLU function

analysis. Nowadays, it is playing a vital role in real-time operations. It can be used for processing images and videos to analyse faces, hand written scripts, etc.

Matplotlib [7] is a widely used library among all the libraries in python. It is an open-source plotting library used in machine learning to understand large data through proper visualizations. It is designed in such a way that most of the functions can be used in Python that are used for plotting in MATLAB.

39.2 Methodology

39.2.1 Collection and Pre-processing of Dataset

This study is segregated into two parts, generation of human faces and detection of expression. Till date, the progress of this study can be justified as successfully training a model that can generate human like faces based on the dataset provided. The generation step can be further divided into: collection of datasets, image pre-processing, preparation of required model, saving the state of model, generation of images, and conversion of images to .png format.

The dataset used in this project is taken from the "Yale Face Database" [8]. It comprises of 2378 images of male and female in .pgm format. To analyse these images, they are first converted to .png format and are stored in a directory. Then, the images are filtered according to the lighting conditions. This is done to ensure that only proper images could be feed into them. The images in the dataset are then scaled down to 64 × 64 pixel for faster computation. After that, it is converted to numpy array and is saved as a .npy file so that it does not require to compute the images again and again. On loading the numpy array for the training purpose, the whole array is clipped in the range [0,1] by dividing each value in the array by 255.0 as each number in the array represents the range of the pixel which will always in the range of [0, 255]. As the range of the input vectors reduces, it helps in better training and it subsequently results in better results. Image before pre-processing and after pre-processing had been shown in Fig. 39.5a and b, respectively.

39.2.2 Preparation of Face Generation Model

For the training of GAN, deep-convolutional GAN (DCGAN) is used in this project. According to the definition of GAN, a GAN is comprises of sequential layers of a discriminator and a generator. As stated, that DCGAN is used in this project, all the layers are majorly convolutional ones, i.e. Conv2D and Conv2D transpose. Starting with the discriminator, it comprises of repetition of convolutional, batch normalization, and LeakyReLU. The convolutional layers used are having filters 16, 32, 64, 128, 256, respectively, and with a kernel size of 3, in each interaction the activation function used in this, i.e. LeakyReLU is having a momentum of 0.2 in each iteration. The output generated from this is then passed to a series of dense layers with ReLU and sigmoid as activation functions. The discriminator is then compiled

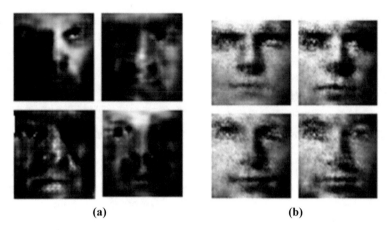

(a)   (b)

Fig. 39.5 **a** Face generation without filter. **b** Face generation with filter

in Adams optimizer as 0.0002 since learning rate and binary cross entropy as its loss function (Table 39.1).

So, in the discriminator, the size of each filter is getting decreased which is clear from the above table. But for the generator, a model is prepared that is complete opposite of the discriminator.

In the generator, the ultimate goal is to achieve a model that can generate the faces as in the training data. But what should be provided as an input to the generator? Usually in the generator, random noise is provided as an input. The noise which is used is an array of 100 random numbers. It is then passed to a dense layer of $(128 \times 4 \times 4)$ nodes and then reshaped to $(4, 4, 128)$. After that, it is passed to a repetition of sequence of conv2D transpose layers along with batch normalization with a momentum of 0.2 and LeakyReLU as activation function. The matrix so generated after passing from the repetition is in the shape of $(64, 64, 128)$. Since, the out image is in the shape of $(64, 64, 1)$, it is passed to a convolutional layer with 1 as a filter. Then, the generated image will have to pass the discriminator for the verification purpose. Table 39.2 can describe the above-mentioned process.

The last model that has been used in this project is GAN. This is a model with, generator as the first layer and the discriminator as the second layer. In the discriminator passed here, the property to train the model by a discriminator has been turned off. This means that while training the GAN, the vectors in the discriminator will not change and only the generator is able to train.

So, in the overall process, numpy image array will be imported and normalized. In this project, the batch size of 256 is decided and the training data having half of the array as the images from the numpy array will be assembled and labelled as 1 showing them as true and another half as generator generated images, which will be initially random noise and will be labelled as 0 showing them as false. After that, passing them to the discriminator so that the discriminator can learn the correct versions of both the given data present in the batch.

Table 39.1 Model description of the discriminator

Layers	Output shape
conv2d	(None, 32, 32, 16)
Batch_normalization	(None, 32, 32, 16)
Leaky_re_lu	(None, 32, 32, 16)
conv2d	(None, 16, 16, 32)
Batch_normalization	(None, 16, 16, 32)
Leaky_re_lu	(None, 16, 16, 32)
conv2d	(None, 8, 8, 64)
Batch_normalization	(None, 8, 8, 64)
Leaky_re_lu	(None, 8, 8, 64)
conv2d	(None, 4, 4, 128)
Batch_normalization	(None, 4, 4, 128)
Leaky_re_lu	(None, 4, 4, 128)
conv2d	(None, 2, 2, 256)
Batch_normalization	(None, 2, 2, 256)
Leaky_re_lu	(None, 2, 2, 256)
flatten	(None, 1024)
dense	(None, 128)
Re_lu	(None, 128)
dense	(None, 256)
Re_lu	(None, 256)
Dense	(None, 1)
Sigmoid	(None, 1)

Now, the random noise array will be passed in shape of (batch size, 100) and labelled as 1 which means true to the GAN so the input will first enter the generator and then to the discriminator. Here, it is being tried to trick the discriminator i.e., making the discriminator believe that the generated noise is the correct human face. The above steps are repeated for quite several iterations until suitable results are achieved.

39.2.3 Preparation of Expression Detection Model

For the next part, i.e. the emotion detection, "Emotion detection from Facial Expressions" [9] dataset from Kaggle is used to train the model. The dataset contains grayscale images of 48 × 48 pixel. Each pixel is of value 0–255, but to feed the value to the model every pixel needs to be in between 0 and 1. The model comprises of convolutional layers and dense layers followed by activation layers. The model

Table 39.2 Model description of generator

Layers	Output shape
Input	(None, 100)
Dense	(None, 2048)
Leaky_ReLU	(None, 2048)
Reshape	(None, 4, 4, 128)
Conv2D_transpose	(None, 8, 8, 16)
Batch_normalization	(None, 8, 8, 16)
Leaky_ReLU	(None, 8, 8, 16)
Conv2D_transpose	(None, 16, 16, 32)
Batch_normalization	(None, 16, 16, 32)
Leaky_ReLU	(None, 16, 16, 32)
Conv2D_transpose	(None, 32, 32, 64)
Batch_normalization	(None, 32, 32, 64)
Leaky_ReLU	(None, 32, 32, 64)
Conv2D_transpose	(None, 64, 64, 128)
Batch_normalization	(None, 64, 64, 128)
Leaky_ReLU	(None, 64, 64, 128)
Conv2D	(None, 64, 64, 1)

consists of following layers. The activation layers that has been used is mostly ReLU and softmax (Table 39.3).

After training the model for a batch size of 128 and for 40 epoch and with a loss function as categorical cross entropy, the accuracy achieved is 60%. After training the generator model, the output will be passed to the emotion model. The output of the generator model is of shape 60 × 60 pixel but the output have to be resized to 48 × 48 pixel to pass it as input of emotion detection model.

39.3   Result

39.3.1   Face Generation with GAN

The generator model generates new faces by learning from the dataset. Figure 39.6 shows few outcomes of the completely trained generator model after passing a random noise.

While the face generation by the generator, discriminator creates a loss. The discriminator loss against number of epochs and accuracy of the generator model against number of epochs are graphically presented in Fig. 39.7.

Table 39.3 Model description of emotion detection

Layers	Output size
Conv2D	(None, 48, 48, 64)
Batch Normalization	(None, 48, 48, 64)
Leaky Relu	(None, 48, 48, 64)
Max Pooling 2D	(None, 48, 48, 64)
Droupout	(None, 24, 24, 64)
Conv2D	(None, 24, 24, 128)
Batch Normalization	(None, 24, 24, 128)
Leaky Relu	(None, 24, 24, 128)
Max Pooling 2D	(None, 12, 12, 128)
Droupout	(None, 12, 12, 128)
Conv2D	(None, 12, 12, 256)
Batch Normalization	(None, 12, 12, 256)
Leaky Relu	(None, 12, 12, 256)
Max Pooling 2D	(None, 6, 6, 256)
Droupout	(None, 6, 6, 256)
Conv2D	(None, 6, 6, 512)
Batch Normalization	(None, 6, 6, 512)
Leaky Relu	(None, 6, 6, 512)
Max Pooling 2D	(None, 3, 3, 512)
Droupout	(None, 3, 3, 512)
Flatten	(None, 4608)
Dense	(None, 256)
Droupout	(None, 256)
Dense	(None, 512)
Relu	(None, 512)
Dense	(None, 5)
Softmax	(None, 5)

39.3.2 Expression Detection

The faces generated by the generator are fed into a model created for expression detection. Features are extracted from the image after which the model can differentiate between various expressions. Figure 39.8 shows the expression detected of some of the faces generated.

The model predicts the percentage of various expressions such as angry, fearful, happy, neutral, and sad. In Fig. 39.8, 'a', 'f', 'h', 'n', and 's' stands for angry, fearful, happy, neutral, and sad, respectively. Face having the highest percentage of an expressions is said to be having that particular expression among other expressions mentioned.

Fig. 39.6 Face generated by GAN

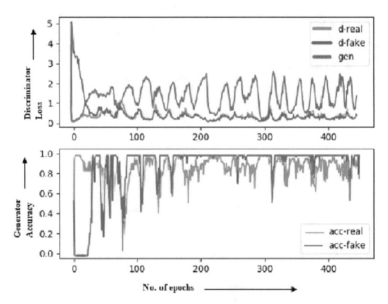

Fig. 39.7 Discriminator loss and generator accuracy graph

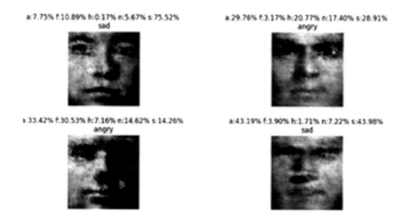

Fig. 39.8 Expression detection of the face generated

39.4   Conclusion

Technologies are brimming in the fields of computer vision and machine learning. This provided the source of motivation of this study. GAN is one of the achievement in the field of computer science which was introduced in 2014. It is a very vibrant field which has not been digged properly yet. New studies are being introduced in this field. Being early in this new tech, this study can take a strong stand on the novelty of its work. This study has successfully completed the generator model which is successfully trained. New images are being generated by the generator model which are satisfactory up to a good extent. The generated images are then fed into a model that has been prepared for expression detection. Now, the model can detect the expression of generated face with ease.

This study is favourable in way that a dataset itself could be generated using the generator of GAN to train and test the model prepared for expression detection. Though, it cannot detect a complex expression other than the mentioned expressions as it is totally dependent on the dataset being used for the model. The generator model could be implemented with a better dataset. Moreover, this study is generating faces in a single channel, i.e. grayscale, it can be further implemented to generate faces in RGB.

Acknowledgements We would like to express our heartfelt gratitude to Prof. Debraj Chatterjee of the Department of Computer Science Engineering, whose significant role as mentor was invaluable for this research work. We are extremely thankful for his encouraging and supportive efforts in advising and for the guidance and reference materials provided for moral support.

References

1. Margineantu, D., Wong, W.-K., Dash, D.: Machine learning algorithms for event detection. Machine Learning **79**(3) (2010)
2. Lata, K., Dave, M., KN, N.: Data augmentation using generative adversarial network. SSRN Electronic Journal (2019)
3. Wang, D., Dong, L., Wang, R., et al.: Targeted speech adversarial example generation with generative adversarial network. IEEE Access **8**, 124503–124513 (2020)
4. Samma, H., Lahasan, B.: Convolutional neural network for skull recognition. International Journal of Innovative Computing **12**, 55–58 (2021)
5. Yuen, B., Hoang, M., Dong, X., Lu, T.: Universal activation function for machine learning. Scientific Reports (2021)
6. SV, V., Katti, M., Khatawkar, A., Kulkarni, P.: Face detection and tracking using OpenCV. The SIJ Transactions on Computer Networks & Communication Engineering **04**(03), 01–06 (2016)
7. Nagesh, K., Nageswara, D., Choi, S.K.: Python and MatPlotLib based open source software system for simulating images with point light sources in attenuating and scattering media. International Journal of Computer Applications **131**(8), 15–21 (2015)
8. https://www.kaggle.com/datasets/olgabelitskaya/yale-face-database
9. https://www.kaggle.com/c/emotion-detection-from-facial-expressions

Chapter 40
Predicting Adverse Reaction of COVID-19 Vaccine with the Help of Machine Learning

Chintal Upendra Raval⬤, Ashwin Makwana⬤, Desai Vansh Brijesh⬤, and Aman H. Shah⬤

Abstract The COVID-19 epidemic demonstrated the importance of technology in the healthcare sector. A lack of ventilators and essential drugs results in a high mortality rate. The most important lesson from the pandemic is that we must use all available resources to alleviate the situation during the pandemic. In this paper, we combine pharmacovigilance and machine learning to predict the effect of an adverse reaction on a patient. We take VAERS data and preprocess it before feeding it to various machine learning algorithms. We assess our model using various parameters.

40.1 Introduction

The COVID-19 pandemic is a global pandemic that inspires us to think about high-quality medicines and better health care in the future. We can see the impact of the pandemic on the overall economic growth of the country. More than 25 million cases have been identified in India, with over 3 lakh people killed due to the pandemic. The basic and primary sector in the fight against pandemics is health care. The health sector receives a 137% increase in funding from India's finance ministry. Vaccines are critical in our fight against pandemics. Pharmacovigilance is a field that focuses on adverse drug reactions [1].

The rise of 'huge medical services information,' as defined by its volume, complexity, and speed, has created an intriguing and open door for research into computerized pharmacovigilance. In addition to other primary information stages, online media has thus turned into an empowering hotspot for the recognition and

C. U. Raval (✉)
Devang Patel Institute of Advanced Technology and Research (DEPSTAR), Faculty of Technology & Engineering (FTE), Charotar University of Science and Technology (CHARUSAT), Changa, India
e-mail: chintalraval.dit@charusat.ac.in

A. Makwana · D. V. Brijesh · A. H. Shah
Faculty of Technology and Engineering (FTE), Chandubhai S. Patel Institute of Technology (CSPIT), Charotar University of Science and Technology (CHARUSAT), Changa, India
e-mail: ashwinmakwana.ce@charusat.ac.in

anticipation of ADRs to advance pharmacovigilance. Many existing exploratory studies use various systems to distinguish ADRs by evaluating the relationship between medication and ADRs. However, the occurrence of ADRs can be associated with a multitude of causal variables; thus, recognizing the various factors that cause ADRs is critical.

Several countries have approved 12 different vaccines against COVID-19 for emergency use. As a result of the loss of rigorous data from long-term trials on COVID-19 vaccine protection, there is an urgent need to strengthen published advertising surveillance of unfavorable event records, particularly in low- and middle-income countries. This would necessitate continuous monitoring of vaccinated patients for potential COVID-19 vaccine adverse reactions [2].

Today, when the world desperately needs a COVID-19 vaccine, we must investigate COVID-19's adverse effects. Pharmacovigilance is a critical component of health care. In 2020, the global pharmacovigilance market is expected to be worth more than USD 6 billion, rising to USD 7 billion by 2022. The requirement for a vaccine has given machine learning a new lease on life in the field of pharmacovigilance. Vaccine Adverse Event Reporting System data is infused into our project (VAERS). This specific file contains 54 variables and 42,285 rows. Our primary goal is to forecast the death of a specific patient who has had an adverse reaction. Many ML techniques, such as regression and decision trees, are used to predict mortality [3]. We can see from the chi2 test that the history variable is a significant factor.

40.2 Materials and Methods

This study examined data, such as the medical histories of COVID-19 vaccinated patients, as well as post-vaccination outcomes and effects, and analyzed the data using statistical methods and machine learning models. After the model has been evaluated, it quantifies and ranks the importance values of the characteristics.

40.2.1 Data Collection

The researchers began with a raw dataset of vaccinated American patients, which included vaccine-related information. The first step in the machine learning pipeline is to collect statistics for training the ML model. The predictions made by ML structures are only as accurate as the statistics used to train them. Predictive models are most effective as accurate as the statistics from which they are built, so proper statistics collection practices are critical to developing high-performing models. The data was gathered from the Vaccine Adverse Event Reporting System (VAERS) of vaccinated individuals between December 2020 and March 2021, and side effects were also reported. However, for post-vaccination disease, information other than COVID-19 was excluded from this study. The total number of reports collected

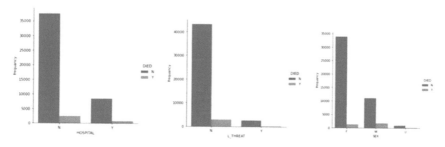

Fig. 40.1 Morbidity and mortality

was 33,797. VAERS collected data on age, gender, comorbidities, history, and post-vaccination symptoms.

40.2.2   About the Data

VAERS studies will be especially important in determining the safety of many new or significantly altered vaccines that are expected to be introduced in the near future. It is critical that the FDA/CDC continue to maintain the VAERS database and provide denominator data on the number of all vaccines distributed throughout the year, broken down by vaccine type, manufacturer, and batch, to independent researchers. This will allow for an open discussion in the scientific/medical literature on the subject of vaccines, as well as the collaboration of experts from various fields of study to help improve vaccine safety [4].

The VAERS have three individual file VAERSDATA, VAERSVAX, and VAERSSYMPTOMS. We form primary relation between that three tables and get whole dataset. The final dataset is a query of that three file (Fig. 40.1).

40.2.3   Data Pre-processing

Data pre-processing is a technique for preparing raw records for use with a machine learning model. It is the first and most important step in developing a machine learning model. It will be especially important to test larger datasets for both classification and regression problems so that we can analyze absolute volatility behaviors. We will also look into ways to combine classifier accuracy and similarity to a second classifier into a single number [5].

A real-world record typically contains noise, and missing values, and maybe in an unusable format that cannot be immediately used for ML models. Data pre-processing is required to clean the records and make them suitable for a machine learning model,

Table 40.1 Comparative analysis of different algorithms

Algorithm	Accuracy	F1-score	Cross-validation score
K-nearest neighbors	94.0682	0.8664003272105807	0.965626
Random forest classifier	96.27	0.916416	0.97626
Decision tree classifier	95.40	0.9075894647711573	0.9643
Support-vector machines	94.50%	0.8725638985386303	0.95954
Gaussian Bayes	59.55%	0.546871506451763	0.84207

which will also improve the accuracy and performance of the machine learning model [6].

Working with null values

There are several null values in the VAERS dataset. The primary purpose of null values may be to document record failure. Managing null values may be critical during the pre-processing of the dataset, as many machine learning algorithms no longer support null values. Table 40.1 shows that many categorical variables have more than 80% of null values. This is due to the VAERS guidelines. According to VAERS guidelines, such variables as DIED and L THREAT (life threatening) exist on that type of variable, and if that condition occurs, the data-field is filled; otherwise, the data-field is empty.) Encoding is used to handle null values in categorical data. We encode null values with a specific number based on that variable. Because the L THREAT (life threatening) variable has 94% null values, we encode null values with a specific number. The majority of categorical data were handled in this manner. We use the data imputation method to handle numerical data. We calculate the standard deviation and fill in the numerical value.

Feature Selection

Feature selection is a central concept in AI that has a significant impact on model presentation. The information work you use to prepare the AI model has a significant impact on the exhibition you can achieve. Include selection is a cycle in which we naturally or physically select the element that contributes the most to the indicator variable or result variable that you are interested in. For feature selection, we use two techniques.

Feature selection is a common optimization problem in which the best solution can only be found through an exhaustive search given an evaluation or search criterion. As a result, for high-dimensional problems, researchers continue to use the heuristic method with polynomial time complexity [7].

A variable that is completely useless by itself can provide a significant performance improvement when taken with others. Two variables that are useless by themselves can be useful together [8].

1. Univariate selection (for categorical variable)
2. Correlation with heatmap (for numerical variable)

Univariate Selection

The empirical results on both the real data and controlled data have shown that Chi2 is a useful and reliable tool for discretization and feature selection of numeric attributes [9].

We use chi^2 test for univariate selection. The chi square tests are used in statistics to test the independence of two events. Taking into account the data of two variables, the observed number or and the expected count β. Chi square will be diverted with each other with the number γ and the observed number of α.

$$\chi^2 = \frac{\sum (\alpha i - \beta i)^2}{\gamma i} \tag{40.1}$$

Correlation with heatmap.

The correlation explains whether one or more variables are associated with each other. These variables are the entry data functions used to predict the destination variables. The correlation, statistical methods are statistical methods that determine the movement/change of variables associated with other variables. It gives us an idea about the degree of the relationship between two variables. It is a measure of analysis of two variables that describes the relevance between the different variables. Most companies help express a topic in terms of relationships with others.

In figure of the correlation matrix, we can see that the symptom variable has a major correlation. We eliminate the symptom series variable (Fig. 40.2).

40.3 Results

In this study, we have used many different types of machine learning algorithm. The model took different feature like state, current illness, priority visit of expert and gave the prediction.

40.3.1 K-Nearest Neighbors

K-nearest neighbors is one of the maximum fundamental but crucial category algorithms in machine learning. It belongs to the supervised studying area and reveals excessive software in sample recognition, facts mining, and intrusion detection.

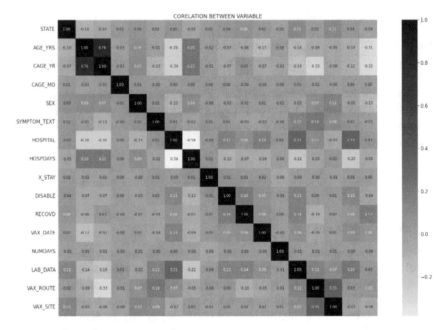

Fig. 40.2 Correlation between variables

40.3.2 Random Forest Classifier

Random forest classifiers encompass the broad umbrella of ensemble-based study skills. They are simple to set up, quick to operate, and have proven to be extremely low-risk in a wide range of sectors. The construction of multiple "simple" call trees in the coaching stage, and hence the majority vote (mode) across them in the classification stage, is a critical premise behind the random forest approach. This choose approach, among other things, has the effect of adjusting for the undesired feature of decision trees to overfit training data.

40.3.3 Decision Tree Classifier

Random forest classifiers encompass the broad umbrella of ensemble-based study skills. They are simple to set up, quick to operate, and have proven to be extremely low-risk in a wide range of sectors. The construction of multiple "simple" call trees in the coaching stage, and hence the majority vote (mode) across them in the classification stage, is a critical premise behind the random forest approach. This choose approach, among other things, has the effect of adjusting for the undesired feature of decision trees to overfit training data.

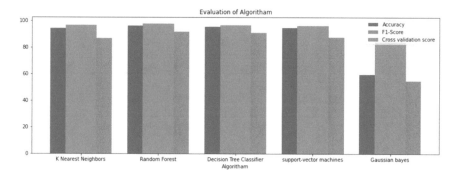

Fig. 40.3 Evaluation of algorithms

40.3.4 Support Vector Machine

The major goal is to segregate the given dataset withinside the first-class viable way. The distance among the both nearest factors is referred to as the margin. The goal is to pick out a hyperplane with the most viable margin among help vectors with inside the given dataset. Select the proper hyperplane with the most segregation from the both nearest records factors as proven withinside the proper-hand facet parent [10].

40.3.5 Naïve Bayes

Naive Bayes is a smooth technique for constructing classifiers: models that assign beauty labels to trouble instances, represented as vectors of feature values, in which the beauty labels are drawn from some finite set.

In many sensible applications, parameter estimation for naive Bayes models uses the technique of maximum likelihood; in special words, you could artwork with the naive Bayes model without accepting Bayesian possibility or using any Bayesian methods (Figs. 40.3 and 40.4).

40.4 Conclusion

In the paper, we obtained results from various algorithms, and they are quite satisfactory. However, they may be decent because we populated many null values with the functions and removed many textual variables, but processing that variable may have resulted in a more enhanced model. The VAERS database can be used for combined pharmacovigilance and machine learning research.

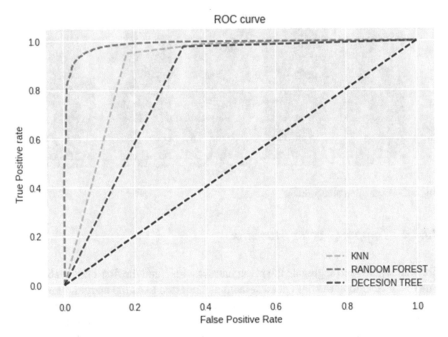

Fig. 40.4 Roc curve of best performing algorithm

References

1. Le Thanh, T., Andreadakis, Z., Kumar, A., Gómez Román, R., Tollefsen, S., Saville, M., Mayhew, S.: The COVID-19 vaccine development landscape. Nat. Rev. Drug Discovery **19**(5), 305–306 (2020)
2. Geier, D., Geier, M.: A review of the vaccine adverse event reporting system database. Expert Opin. Pharmacother. **5**(3), 691–698 (2004)
3. Lee, C., Chen, Y.: Machine learning on adverse drug reactions for pharmacovigilance. Drug Discovery Today **24**(7), 1332–1343 (2019)
4. Vaers.hhs.gov.: VAERS—data sets (2022). Available at https://vaers.hhs.gov/data/datasets. html. Accessed 31 March 2022
5. Gonzalez Zelaya, C.: Towards explaining the effects of data preprocessing on machine learning. In: 2019 IEEE 35th international conference on data engineering (ICDE) (2019)
6. Cerda, P., Varoquaux, G.: Encoding high-cardinality string categorical variables. IEEE Trans. Knowl. Data Eng. **34**(3), 1164–1176 (2022)
7. Cai, L., Wang, Y.: (2018)
8. Guyon, I., Elisseeff, A.: An introduction to variable and feature selection. J. Mach. Learn. Res. **3**, 1157–1182 (2003)
9. Liu, H., Setiono, R.: Chi2: Feature selection and discretization of numeric attributes. In: Proceedings of 7th IEEE international conference on tools with artificial intelligence (pp. 388–391). IEEE (1995)
10. Suthaharan, S.: Support Vector Machine. Machine Learning Models and Algorithms for Big Data Classification, pp. 207–235 (2016)
11. Choi, S., Oh, J., Choi, C., Kim, C.: Input variable selection for feature extraction in classification problems. Signal Process. **92**(3), 636–648 (2012)

Chapter 41
Intelligent Transportation Reminder System Using Mobile Terminal and IoT

Rui Wang, Ronaldo Juanatas, and Jasmin Niguidula

Abstract The intelligent transportation reminder system is analyzed and designed based on Android technology to solve users' complex and untimely arrival of public transportation problems. Finally, the intelligent transportation accurate reminder system is realized. Through the user's analysis of public transport demand, the system provides bus route query, transfer query, station query, regular reminder, and other functions. Based on GPS positioning, laser sensors are installed at the front and rear doors of the bus to record the number of people on the bus in real time, and upload the data to the server. When the user travels, the mobile terminal cannot only query the location information of the bus in real time, but also view the crowd on the bus, which is convenient for the user to choose the appropriate bus travel. At the same time, it can push the user's bus information regularly at the specified time, accurately remind the user to take the bus, so that the user can reasonably arrange the bus time.

41.1 Introduction

Intelligent transportation refers to the use of IoT, cloud computing, the Internet, artificial intelligence, automatic control, mobile Internet, and other technologies to establish a comprehensive, intelligent, green, and safe urban traffic ecological environment [1]. It realizes the information management of urban road traffic, reduces urban traffic congestion, and improves the intelligence and convenience of travelers.

Since the promotion of green travel by the state, a bus is an effective means to alleviate traffic congestion. It can protect the environment and improve the utilization of public resources. With the development of information technology, various public transport mobile applications are emerging endlessly. Some mobile applications can recommend bus routes according to the traveler's destination, check the attached bus stops, and weather. There are mobile applications that can find the real-time location of public transportation. Although travelers can view the real-time site of the bus,

R. Wang (✉) · R. Juanatas · J. Niguidula
Graduate Program and External Studies, Technological University of the Philippines, Manila, Philippines
e-mail: wangrui344@abc.edu.cn

© The Author(s), under exclusive license to Springer Nature Singapore Pte Ltd. 2023 461
C. So-In et al. (eds.), *Information Systems for Intelligent Systems*, Smart Innovation, Systems and Technologies 324, https://doi.org/10.1007/978-981-19-7447-2_41

sometimes they miss the bus for some reasons. There are also times when the bus is already full when they arrive at the bus station, this entails them to wait for the next bus leads to being late for work or going home late.

There are various passenger bus flow statistics, such as IC card, pressure detection, infrared detection, and video-based image detection. By using an IC card, the number of passengers on the bus can be calculated; the only problem is determining when a passenger will alight the bus. The pressure pedal method can be used to count the number of passengers and judge the number of passengers according to the number of steps and weight. However, this method cannot determine whether passengers enter or exit the bus. Moreover, the bus has a large passenger flow, the pressure pedal equipment is easy to break, and the maintenance cost is relatively high [2].

There is also a video-based image detection method, an embodiment of artificial intelligence technology. Technological setbacks like shooting angle, illumination, and shooting scene affect the quality of the target image. This occurs when there are many passengers, and the light source is not vivid, and the passengers block each other, resulting in low video quality and inaccurate statistical data [3]. In spite of these technological advancements in the transportation industry, passengers still do not want to use IC cards because they prefer cash and electronic payment [4].

Based on the encountered problems, this paper aims to design and implement an intelligent transportation application which can timely remind users of the time to go out and take the bus, so that users can arrive at the bus station in time. The application can also inform users of the exact number of passengers on the bus to determine the bus congestion. The passengers can also select a bus with few passengers.

Using big data, cloud computing, IoT, mobile application, and other information means, intelligent transportation system can store massive traffic data, and make efficient, accurate, and convenient analysis of traffic data, and provide data basis for realizing green travel so that users can master the operation status of roads in time [5, 6].

IoT mainly includes sensor network technology, RFID technology, and embedded technology [7]. The application of sensor technology in intelligent transportation systems does very well. For example, the real-time road traffic monitoring can be realized through sensor network technology [8, 9]. Mobile intelligent tracking allows users to get the latest bus routes and map-related data through the mobile service applet or APP mobile phone client docking smart transportation platform [10, 11]. Passengers can also query the travel information of the bus through the web page and WeChat official account.

The project adopts IoT technology to design a scheme for real-time statistics of the number of passengers. It can also customize and remind users to travel so that users can understand the real-time location, congestion degree, and other information of each bus. Users can timely arrange their ride plan, and solve the problems of difficult bus selection and inaccurate time control.

41.2 Methodology

The intelligent bus reminder system designed in this paper comes from the bus data of Wuhu City. It shows all bus routes of Wuhu City. In the experiment, bus route data of Wuhu City were collected, and then imported the bus route data into the server database. The service API is developed for the mobile terminal to query bus routes.

41.2.1 System Architecture Analysis

The system can be divided into three parts: data source, data storage, and interface management. The data source side is the perception layer of the IoT; it can collect and process data. The data storage part can store and manage the data obtained from the data source in the database. The interface management mainly calls the data displayed in the database in the system interface [12]. By installing positioning equipment on each bus and using a GPS positioning system can accurately detect the location information of the bus, but it cannot see the number of bus passengers. This project used the IoT technology. Dual beam-laser sensors and counters were installed at each bus front door and rear door. The laser sensor is a sensor that uses laser technology for measurement. It is composed of a laser, a laser detector and a measurement circuit. It can realize non-contact remote measurement with high speed, high precision, and large range. In the process of driving, when the bus arrives at each station, the number of people getting on at the front door and getting off at the back door at each station was counted. The difference between the number of people getting on and the number getting off is uploaded to the cloud platform public transport management service by using the 5G network, calculating the number of bus passengers at each station on the cloud platform server. The number of bus passengers is displayed on the terminal real time using the mobile app application software. Application users can monitor the congestion of the bus and choose the right time to take the bus. Figure 41.1 shows the system architecture.

41.2.2 System Model Analysis

The application built is compatible with Android mobile platform considering that the market share of the Android operating system is far ahead of other mobile terminal operating systems [13]. The system adopts MVC architecture mode, that is, "model M—view v—control C" mode [14], which separates the application program's input, processing, and output so that the same program can use different forms of expression. As shown in Fig. 41.2, in an Android platform, view plays the role of V in MVC to initiate the request. Activity and fragment play the role of C in MVC to control the flow direction of the program. Business bean was used as the model when the server

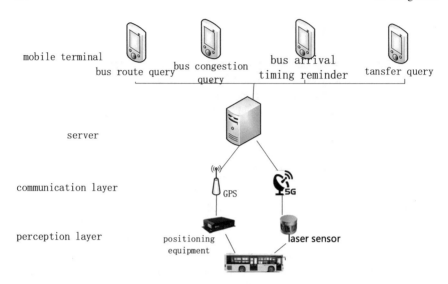

Fig. 41.1 System architecture design

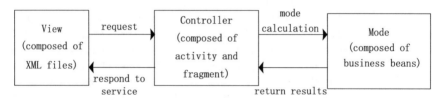

Fig. 41.2 System model diagram

returns the program processing results, and finally display them in the view. The user submits the request through the XML page. After receiving the request, the activity or fragment performs business logic processing, encapsulates the processing results with bean objects, and responds to the XML page in the activity or fragment.

41.2.3 System Interaction Analysis

The system includes a server and mobile client. The data displayed by the mobile terminal is from the server. After the client initiates, a request and submits the data in JSON format, the server gets the JSON data, parses the JSON, further processes the requested data in combination with the database, encapsulates the processing results into JSON and returns them to the client. The client parses the obtained JSON data, and displays on the page. Figure 41.3 shows system interaction diagram.

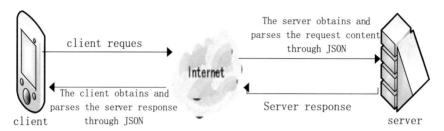

Fig. 41.3 System interaction diagram

41.2.4   Functional Analysis of the System

The system has the following functions through the user demand analysis and relevant intelligent travel function analysis [15–17].

Bus route query: in the bus route query module, enter the query interface, enter any character in the bus route, conduct a fuzzy query, and display the route information. You can view the bus route, the current vehicle location, the estimated arrival time at the next station, the number of vehicle passengers, and other information.

Transfer query: in the transfer query module, select any two stations through the drop-down box, count all the transfer schemes between the two stations, and finally display the transfer scheme on the page. On the page, you can check the information of upper station, transfer station, lower station, route station, and so on.

Site query: When the user enters any keyword in this module, the system will filter out all stations containing the keyword and display it on the page. On the details page of the station, you can see all vehicles passing through the station.

Map positioning: after the user logs in successfully, when entering the home page, it will automatically obtain the current positioning and prompt the user near the station.

Weather query: after successful login, the user will see the current temperature and weather conditions on the home page. The system provides data travel support for users' travel.

Bus arrival timing reminder: on the setting page, the user can set the number of the vehicle he often sits on and the time he needs to be reminded. The reminder contents are the location of the selected vehicle at the specified time and the expected time to arrive at the next station. After the function starts, the user will receive the arrival information reminder of the vehicle at the set specified time.

41.3   Results and Discussions

41.3.1   Development Tool Selection

Android is a free and open-source operating system based on Linux, is mainly used in mobile devices, such as smartphones and platform computers [18]. The project's mobile app adopts an Android system and uses Android studio version 3.5.3. The service side of bus cloud platform management uses MyEclipse2017-CI and apache-tomcat-8.5.65 software. Mysql database saves the data of bus cloud platform. Mysql-8.0 and Navicat 12.0.11 software was used in this project.

41.3.2   Display of the Number of Bus Passengers

After successful system login, the user interface would show the picture rotation on the top of the home page and the positioning function and weather display to obtain the current weather and geographical location in real time. The system interface will also provide the entrance of bus query, transfer query, site query, and opinion feedback. The bottom of the user interface shows the home page display, information notice, and my navigation bar. The bottom navigation bar uses the bottom navigation view tool that comes with the Android platform. Figure 41.4 shows home page of the intelligent transportation reminder system.

Select the bus query module on the home page. After inputting the initial and terminal station, the page will display the initial and terminal station, departure start and end time, detailed station of the road, current location of the vehicle, estimated arrival time at the next station, and the current number of passengers. Figure 41.5 shows the interface of bus passengers display. The perception layer collects the number of people getting on and off at each site, then saves the data in the server's database. When the user queries the bus route, the mobile terminal encapsulates the bus route number into JSON format and submits it to the server, the servlet of server. Count the current number of passengers according to the number of people getting on and getting off at each station. The passenger data and other queried bus status data are encapsulated into JSON format data and returned to the mobile terminal. At the mobile terminal, the third-party tool Gson is used to analyze the data produced by the service terminal, and the RecyclerView control technology is used to display the bus line data in the UI.

41.3.3   Bus Reminder Settings

In the "Settings" page, users can set the bus route and the reminder time. In this way, when the set reminder time reaches, the bus status of the selected bus route will

Fig. 41.4 Intelligent transportation reminder system home page

be pushed. The information will arrive every 2 min to remind users of the station to which the bus runs, so that users can plan their departure time according to the arrival time of the bus and solve the problem of uncertain time. Figure 41.6 and Fig. 41.7 show timing reminder. For information push in Android, you can select the third-party tool Aurora push, Integrate the jpush Android SDK into the application. When the message needs to be pushed to the app, call the jpush API in the program, which is a good interface provided by the third party, to push the customized data to the customer's app.

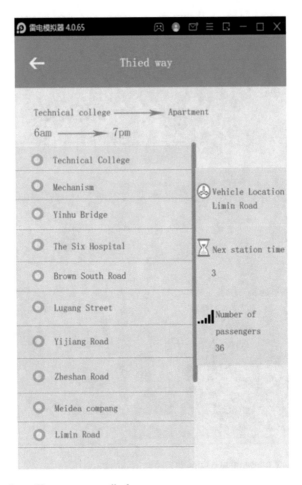

Fig. 41.5 Interface of bus passengers display

41.4 Conclusion

Based on the IoT technology and Android platform, the intelligent transportation accurate reminder system divides into the service and the mobile. This paper uses GPS technology and Internet of things technology to collect the location and number of passengers. System users can query the bus route, transfer route, bus congestion, and location information through the mobile end and regularly remind users of bus information. It is convenient for users to understand the status information of public transportation in real time, accurately plan their travel time, and solve the problems such as difficult choice and inaccurate grasp of time.

Fig. 41.6 Timing reminder setting

Fig. 41.7 Reminder push

Acknowledgements Anhui Business College. Key projects of Anhui University excellent young talents support plan. "Research on the design and application of intelligent transportation system in the context of smart thing cloud."

References

1. Sun, T.: Scheme design of urban intelligent transportation system. Modern Information Technology **12**, 76–78 (2019)
2. Intelligent transportation era: current situation and solutions of bus passenger flow monitoring industry, https://www.chinabuses.com/buses/2019/0422/article_88932.html. Accessed 22 April 2019
3. Xu, Z.: Statistical research on bus boarding and alighting passenger detection based on a video image. Beijing Jiaotong University (2021)

4. Ma, B.: Research on bus passenger flow statistics system based on IoT. Harbin Engineering University (2019)
5. Li, W.: Design and implementation of intelligent transportation big data platform system. Intelligent Building and Smart City **5**, 98–102 (2019)
6. Liu, S.: System design scheme based on intelligent transportation big data platform. Electronic Technology and Software Engineering **10**, 198 (2018)
7. Lin, X.: Analyze the application of IoT technology in intelligent transportation. Low Carbon World **1**, 167–168 (2021)
8. Hao, A.: Application of IoT technology in intelligent transportation. Times Automobile **1**, 193–194 (2022)
9. Sun, X., Wang Z.: Design and research of intelligent transportation system in Zhengzhou based on IoT technology. Automotive Practical Technology 254–256 (2018)
10. Shao, J.: Intelligent transportation service system based on mobile informatization and extensive data mining analysis. Jiangsu Communications **10**, 45–48 (2016)
11. Deng, N.: Design and research of urban intelligent bus station. Design **12**, 135–137 (2019)
12. Chen D.: Urban road intelligent traffic management system design. Xi'an Technological University (2021)
13. Yin, X.: Design of intelligent flowerpot system for the mobile terminal. Journal of Leshan Normal University **8**, 13–17 (2020)
14. Zhang, R.: Android project practice: development of intelligent agricultural mobile management system. Neusoft Electronics Press, Dalian (2015)
15. Analysis of urban public transport system. https://www.docin.com/p-2575876727.html. Accessed 10 Jan 2021
16. Jia, F.: The design of intelligent transportation extensive data systems. Digital Communication World **7**, 151–152 (2017)
17. Yang, Y., Li, X.: Research on intelligent transportation management and application based on big data technology. Journal of Chongqing University of Technology and Industry **8**, 74–79 (2019)
18. Wang, W., Ji, H., Zhang, P.: System design of intelligent travel car housekeeper platform based on android. Intelligently Networked Vehicle 39–41 (2019)

Chapter 42
Plant Leaf Diseases Detection and Classification Using Spectroscopy

Hardikkumar S. Jayswal and **Jitendra P. Chaudhari**

Abstract Detection of plant diseases is one of the important aspects in the field of agriculture. The accurate, early detection of plant diseases is very important to save the plant. Plant diseases identified by DNA (Polymerase chain reaction) and RNA-based method which is much complex and time consuming. In this paper we used spectral-based sensing with the combination of machine learning methods to detect and classify the mango disease like anthracnose, powdery mildew and sooty mold. We conducted various field experiments to acquire different vegetation reflectance spectrum profile using spectrometer at Anand Agriculture university, Gujarat, India. A total 20,000 spectral sample collected to detect and classify the diseases with training set and test set ratio is 70% and 30%, respectively. After the encoding the dependent variable training and testing data set were made for train the model. The range of wavelength used to detect a mango disease is 400 nm to 1000 nm. With the help of supervised machine learning algorithm, we achieved classifier accuracy up to 99.9980%.

Abbreviations

RS	Remote sensing
IR	Infrared
VNIR	Visible and near-infrared
QDA	Quadratic discriminant analysis
PCA	Principal Component Analysis
HLB	Huanglongbing
NB	Naive Bayes

H. S. Jayswal (✉)
Department of Information Technology, Devang Patel Institute of Advanced Technology and Research, Charotar University of Science and Technology Anand, Changa, Gujarat, India
e-mail: jayswal.hardik.kumar@gmail.com

J. P. Chaudhari
Charusat Space Research and Technology Center, Charotar University of Science and Technology Anand, Changa, Gujarat, India

RF   Random Forest
KNN   K-nearest neighbors
SVM   Support Vector Machine

42.1 Introduction

A growing population increased the demands of agricultural products [1]. A mango having rich of nutrition so it is very popular among all the fruits. Because of mango plants disease there is huge crop loss and it is effect to the mango quality also. Different disease effect to mango plant like Bacterial black spot, malformation, Shutimold, Anthracnose, Moricha disease, Root rot, Lichens, Powderymildew, Red rust, and damping off which affect the mango crop production [2, 27]. Since last many of years the crop disease detection and classification is carried out manually by the experts [3]. This manual method are tedious and time consuming. Sometimes diseases identified at late stage causes heavy losses in crop productions. Remote sensing is the procedure to fetch an information of an entity via transmitting and receiving an electromagnetic waves [16]. To detect the plant diseases there are latest techniques are available like similarity identification and classification based on deep learning models. These methods are optimum compare to the traditional methods [4, 5]. There are various methods are used to detect the plant diseases like Molecular Methods, Imaging Methods, Spectroscopic Methods [6]. Polymerase chain reaction (PCR) and quantitative polymerase chain reaction (qPCR) methods are knows as Molecular Methods which having high sensitivity. Where such techniques having a limitation like Requires DNA sequencing and design of primers. One of the popular method for plant diseases detection is imaging method, which subcategorized like Thermography, RGB and fluorescence imaging, Hyperspectral Imaging [6]. Thermography techniques detects the heat emitted by objects and it require only a thermal scan of the plant surface [7]. As Thermography some time give not proper output because of Poor specificity Requires calibration [6]. One of widely used method for plant disease detection is hyperspectral imaging. In this technique reflectance data collected over a wide spectral range, typically 350–2500 nm[7]. A Spectroscopic Methods providing accurate and timely diseases detection. A spectroscopic method are classified into three categories Visible and Infrared Spectroscopy [8], Vibrational spectroscopy [6, 9], Infrared spectroscopy (near- and mid-IR) [6, 10]. Visible and infrared (VIS/IR) spectroscopy (400–100,000 nm) is a techniques which is much useful to detect the diseases [8]. It consists of light source, light-isolating mechanisms, detector, and sampling devices [11]. There are four part of Visible and infrared namely, visible, near-infrared (NIR), mid-infrared (MIR), and far-infrared (FIR). Before to detect the diseases the wavelength must be known [12, 13]. VIS, NIR, MIR, FIR covers the wavelength range of 400–750 nm, 780–2500 nm, 2500–25,000 nm, 25,000–100,000 nm, respectively [8]. VIS spectrum has blue-green band region (400–650 nm) this band is used to identify the difference between healthy and diseases effected plant [14]. NIR consist of short-wavelength (SW-NIR) (750–1300 nm) and long wavelength (LW-NIR) (1300–2500 nm) where short and long wavelength is used for special content in the plants [8, 15]. Vibrational spectroscopy is an analytical techniques which consist of IR and RS. Both IR and RS providing detailed information about the chemical structure of analyzed samples [6, 9].

42.2 Materials and Methods

The sample was collected from Anand agriculture university, Anand, Gujarat, India. The GPS coordinates of sample collection is shown in below Fig. 42.1. Below figure shows the longitude, latitude and map where experiment carried out. An android application is used for capturing longitude and latitude. The sample of healthy and diseased leaves of mango was collected on regular interval from Anand agriculture university. To maintain the moisture and quality of samples, Each of the samples of leaves collected in air tight container and bring to Charusat space center to carried out the experiments. The GPS coordinates of experiment's carried out is shown in below Fig. 42.2. Sample collected of health and infected leaves shown in Fig. 42.3.

42.2.1 Multispectral Sensor (Pixel Sensor) Spectrometers

The Multispectral sensor (Pixel sensor) uses on-chip filtering to pack up to 8 wavelength-selective photodiodes into a compact 9×9 mm array for simple integration into optical devices. This sensor's divided the spectrum into eight separate color

Fig. 42.1 GPS coordinates of data collection At Anand Agriculture University, Gujarat

Fig. 42.2 The GPS coordinates of the experiment carried out at Charusat Space center

Fig. 42.3 Healthy and Infected mango leaves

bands with selectivity (400–1000 nm).To detect the anthracnose, powder mildew, sooty mold in mango multispectral sensors is used. Total 55 diseased sample and 55 healthy sample were measured using spectrometer. Figure 42.4 shows the experiment's carried out at university while Table 42.1 shows the detailed specification of Spectrometers [31].

Table 42.2 shows the research carried out by different researcher. Author used wavelength between VNIR (350–2500 nm) to detect the diseases. A literature review shows that different classification method used to classify the diseases among different plant. Analysis of covariance applied on wheat plant to detect and classify the yellow rust under the wavelength of VIS–NIR (460–900 nm) and achieved the 94% accuracy [26]. QDA classifier and PCA used to detect and classify the HLB

Fig. 42.4 Experiment setup using Multispectral Spectrometers

Table 42.1 Specification of multispectral sensor (pixelsensor) [30, 31]

Photodiode performance characteristics	
Dark current	2nA (Typical), 8nA (Max)
Range of spectral	400–1100 nm
Breakdown voltage	75 V
Response time	6 ns
LCC sensor	
Spectral filters	10–100 nm FWHM
Photodiodes	Si,1.0 × 0.8 mm
OEM board specifications	
Integration time	1–1024 ms
Gain reference	20–5120 nAvv
OEM board dimensions	45.72 × 21.34 mm

infected Citrus leaves under the wavelength of VNIR (350–2500 nm) and achieved more than 90% accuracy [27].

42.3   Literature Review

See Table 42.2

Table 42.2 Literature review of diseases detection and classification using spectroscopy

Plant	Diseases	Spectral range	Approach with result	References
Barley	Powdery Mildew	VNIR (400 nm–1000 nm)	Linear discriminant analysis (LDA)	[17]
Potato	Potato virus	VNIR (400 nm–1000 nm)	Deep Learning (Fully CNN)	[18]
Oilseed rape	Sclerotinia sclerotiorum	VNIR (384–1034 nm)	SVM, RBF-NN, emerging learning neural network	[19]
Apple	Apple scab	SWIR (1000–2500 nm)	Partial least square discriminant analysis	[20]
Strawberry	Anthracnose	VNIR (400–1000 nm)	least square regression and correlation measure	[21]
Spinach	Bacterial	VIS–NIR (456–950 nm)	Partial Least Squared-Discrimination Analysis (PLS-DA) 84%	[23]
Wheat	Spot disease	VIS–NIR (360–900 nm)	Nearest neighbor classifier(KNN)-91.9%	[24]
Wheat	Yellow rust	VIS–NIR (350–2500 nm)	Regression Analysis—90%	[25]
Oil palm Leaves	BSR	VIS/SW-NIR (2550–25,000 nm)	PCA, LDA, KNN, NB	[26]
Wheat	Yellow rust	VIS–NIR (460–900 nm)	Analysis of covariance (ANCOVA-94%)	[27]
Wheat	Wheat Rust	VNIR (450–1000 nm)	Statistical-based methods	[16]
Citrus	HLB infected Citrus leaves	VNIR (350–2500 nm)	QDA classifier and PCA above 90%	[29]
Salad	Salad leaf diseases detection	VNIR (350–2500 nm)	PCA and LDA 84%	[30]

Table 42.3 Comparison of NB, SVM, RF, and KNN classification methods

	Precision	Recall	F1-Score	Accuracy (%)
NB	0.71	0.75	0.72	69.372
SVM	0.96	0.93	0.94	93.095
RF	1.00	1.00	1.00	99.9934
KNN	1.00	1.00	1.00	99.9980

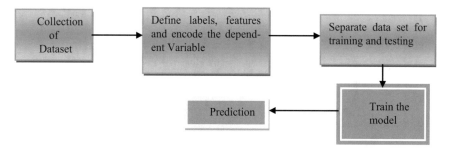

Fig. 42.5 Implementation approach to detect and classify the plant diseases

42.4   Implementation

Figure 42.5 shows the methodology to detect and classify the mango plant diseases. After the collection of dataset, the data was pre-processed and targeted, and feature variables were defined. Before train and test split target variable was encoded. After train the model prediction was made. Figure 42.6 shows the distribution of various band for disease like anthracnose, powdery mildew, sooty mold and health leaf among the eight different diode. From 7 different diode each of the diode shows the wavelength (<1000 nm) versus reception rate (0.00000–0.00150). From Figs. 42.6, 42.7 we can conclude that the sooty mold can easily differentiable then the other diseases. It shows the labeling of anthracnose, powdery mildew, sooty mold diseases. Some diseases are easily classifiable while some are highly overlap so we need complex machine learning algorithm for classification (Table 42.3).

42.5   Conclusions and Future Work

The Research results shows plants health information in real time with having an accuracy of 99%. The research focus on the applicability of spectroscopy for the detection of mango diseases like anthracnose, powdery mildew, sooty mold. A portable, multispectral sensor (pixel sensor), spectrometer used to obtain the spectral data in the range of 400–1000 nm to proceed the sample leaves. The results clearly show that the spectra of infected mango leaves can be classified from the healthy and diseases effected samples using VNIR spectroscopy. To classify the health and unhealthy plant RF, KNN, NB, SVM used. After implemented various classifier study shows KNN gives better classification among the others. Future work would involve the early disease detection in infected mango leaves.

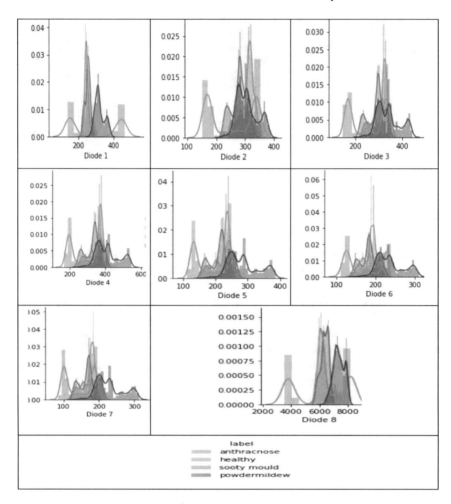

Fig. 42.6 Distribution of various band for diseases

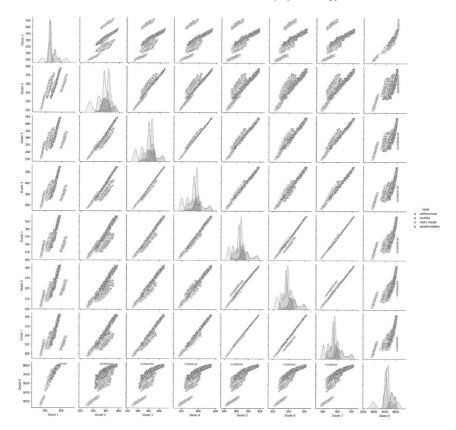

Fig. 42.7 Differentiable of diseases (mapping of each diode)

Acknowledgements We would like to thank the Charotar University of Science and Technology -CHARUSAT for funding this Research. We would also like to thanks Dr. S.H. Akbari Associate Professor, Anand Agriculture University to help the collection of data and his support during the study. We would like to express our gratitude to Prof. Rikin Nayak, Prof. Axat Patel(Charusat Space Research Center -CHARUSAT) for their continuous support during the study.

References

1. Couture, J.J., Singh, A., Charkowski, A.O., Groves, R.L., Gray, S.M., Bethke, P.C., Townsend, P.A.: Integrating spectroscopy with potato disease management. Plant Dis. **102**(11), 2233–2240 (2018)
2. Mia, M.R., Roy, S., Das, S.K., Rahman, M.A.: Mango leaf disease recognition using neural network and support vector machine. Iran Journal of Computer Science **3**(3), 185–193 (2020)
3. Mishra, P., Polder, G., Vilfan, N.: Close range spectral imaging for disease detection in plants using autonomous platforms: a review on recent studies. Current Robotics Reports **1**(2), 43–48 (2020)
4. Singh, V., Sharma, N., Singh, S.: A review of imaging techniques for plant disease detection. Artificial Intelligence in Agriculture (2020)
5. Nagaraju, M., Chawla, P.: Systematic review of deep learning techniques in plant disease detection. International Journal of System Assurance Engineering and Management **11**, 547–560 (2020)
6. Farber, C., Mahnke, M., Sanchez, L., Kurouski, D.: Advanced spectroscopic techniques for plant disease diagnostics. A review. TrAC, Trends Anal. Chem. **118**, 43–49 (2019)
7. Mahlein, A.K., Oerke, E.C., Steiner, U., Dehne, H.W.: Recent advances in sensing plant diseases for precision crop protection. Eur. J. Plant Pathol. **133**(1), 197–209 (2012)
8. Khaled, A.Y., Abd Aziz, S., Bejo, S.K., Nawi, N.M., Seman, I.A., Onwude, D.I.: Early detection of diseases in plant tissue using spectroscopy–applications and limitations. Appl. Spectrosc. Rev. **53**(1), 36–64 (2018)
9. Skolik, P., Morais, C.L., Martin, F.L., McAinsh, M.R.: Determination of developmental and ripening stages of whole tomato fruit using portable infrared spectroscopy and Chemometrics. BMC Plant Biol. **19**(1), 1–15 (2019)
10. Erukhimovitch, V., Hazanovsky, M., Huleihel, M.: Direct identification of potato's fungal phyto-pathogens by Fourier-transform infrared (FTIR) microscopy. Spectroscopy **24**(6), 609–619 (2010)
11. Qu, J.H., Liu, D., Cheng, J.H., Sun, D.W., Ma, J., Pu, H., Zeng, X.A.: Applications of near-infrared spectroscopy in food safety evaluation and control: a review of recent research advances. Crit. Rev. Food Sci. Nutr. **55**(13), 1939–1954 (2015)
12. Liaghat, S., Ehsani, R., Mansor, S., Shafri, H.Z., Meon, S., Sankaran, S., Azam, S.H.: Early detection of basal stem rot disease (Ganoderma) in oil palms based on hyperspectral reflectance data using pattern recognition algorithms. Int. J. Remote Sens. **35**(10), 3427–3439 (2014)
13. Khairunniza-Bejo, S., Vong, C.N.: Detection of basal stem rot (BSR) infected oil palm tree using laser scanning data. Agriculture and Agricultural Science Procedia **2**, 156–164 (2014)
14. Fang, Y., Ramasamy, R.P.: Current and prospective methods for plant disease detection. Biosensors **5**(3), 537–561 (2015)
15. Xu, H.R., Ying, Y.B., Fu, X.P., Zhu, S.P.: Near-infrared spectroscopy in detecting leaf miner damage on tomato leaf. Biosys. Eng. **96**(4), 447–454 (2007)
16. Maid, M.K., Deshmukh, R.R.: Statistical analysis of WLR (wheat leaf rust) disease using ASD FieldSpec4 spectroradiometer. In: 2018 3rd IEEE International Conference on Recent Trends in Electronics, Information & Communication Technology (RTEICT), pp. 1398–1402. IEEE (2018)
17. Kuska, M.T., Behmann, J., Namini, M., Oerke, E.C., Steiner, U., Mahlein, A.K.: Discovering coherency of specific gene expression and optical reflectance properties of barley genotypes differing for resistance reactions against powdery mildew. PLoS ONE **14**(3), e0213291 (2019)
18. Polder, G., Blok, P.M., de Villiers, H.A., van der Wolf, J.M., Kamp, J.: Potato virus Y detection in seed potatoes using deep learning on hyperspectral images. Front. Plant Sci. **10**, 209 (2019)
19. Kong, W., Zhang, C., Huang, W., Liu, F., He, Y.: Application of hyperspectral imaging to detect Sclerotinia sclerotiorum on oilseed rape stems. Sensors **18**(1), 123 (2018)
20. Gorretta, N., Nouri, M., Herrero, A., Gowen, A., Roger, J.M.: Early detection of the fungal disease "apple scab" using SWIR hyperspectral imaging. In: 2019 10th workshop on hyperspectral imaging and signal processing: evolution in remote sensing (WHISPERS), pp. 1–4. IEEE (2019)

21. Yeh, Y.H., Chung, W.C., Liao, J.Y., Chung, C.L., Kuo, Y.F., Lin, T.T.: Strawberry foliar anthracnose assessment by hyperspectral imaging. Comput. Electron. Agric. **122**, 1–9 (2016)
22. Teena, M., et al.: Potential of machine vision techniques for detecting fecal and microbial contamination of food products: a review. Food Bioprocess Technol. **6**(7), 1621–1634 (2013)
23. Larsolle, A., Muhammed, H.H.: Measuring crop status using multivariate analysis of hyperspectral field reflectance with application to disease severity and plant density. Precision Agric. **8**(1), 37–47 (2007)
24. Everard, C.D., Kim, M.S., Lee, H.: A comparison of hyperspectral reflectance and fluorescence imaging techniques for detection of contaminants on spinach leaves. J. Food Eng. **143**, 139–145 (2014)
25. Liaghat, S., Mansor, S., Ehsani, R., Shafri, H.Z.M., Meon, S., Sankaran, S.: Mid-infrared spectroscopy for early detection of basal stem rot disease in oil palm. Comput. Electron. Agric. **101**, 48–54 (2014)
26. Li, X.L., Ma, Z.H., Zhao, L.L., Li, J.H., Wang, H.G.: Application of near infrared spectroscopy to qualitative identification and quantitative determination of Puccinia strii formis f. sp. tritici and P. recondita f. sp. tritici. Guang pu xue yu guang pu fen xi= Guang pu **34**(3), 643–647 (2014)
27. Jayswal, H.S., Chaudhari, J.P.: Plant leaf disease detection and classification using conventional machine learning and deep learning. International Journal on Emerging Technologies **11**(3), 1094–1102 (2020)
28. Sankaran, S., Ehsani, R., Etxeberria, E.: Mid-infrared spectroscopy for detection of Huanglongbing (greening) in citrus leaves. Talanta **83**(2), 574–581 (2010)
29. Dutta, R., Smith, D., Shu, Y., Liu, Q., Doust, P., Heidrich, S.: Salad leaf disease detection using machine learning based hyper spectral sensing. In: SENSORS, 2014 IEEE, pp. 511–514. IEEE (2014)
30. https://www.idil-fibres-optiques.com/technologies-products/spectroscopy-microscopy/
31. https://www.oceaninsight.com/products/spectrometers/general-purpose-spectrometer/flame-series
32. Yeh, Y.H., Chung, W.C., Liao, J.Y., Chung, C.L., Kuo, Y.F., Lin, T.T.: Strawberry foliar anthracnose assessment by hyperspectral imaging. Comput. Electron. Agric. **122**, 1–9 (2013)

Chapter 43
Synthetic Data and Its Evaluation Metrics for Machine Learning

A. Kiran⬥ and **S. Saravana Kumar**⬥

Abstract Artificial Intelligence (AI) has become the key driving force in Industrial Automation. Machine learning (ML) and Deep Learning (DL) can be considered to be the components of AI which rely on data for model training. Data generation has increased due to the Internet, connected devices, mobile devices and social networking which in turn have also given rise to cybercrime and cyber thefts. To prevent those and preserve the identity of individuals in the public data, government and policymakers have put stringent privacy-preserving laws. The economy of data collection, quality of data in the public domain, and data bias have made data accessibility and its usage a challenge for AI/ML training for research work or industrial purposes. This has forced researchers to look into the alternative. Synthetic Data offers a promising solution to overcome the data challenges. The last few years have seen many studies conducted to verify the utility and privacy protection capability of synthetic data. However, all of these have been exploratory. This paper focuses on various methods of synthetic data generation and their validation metrics. It opens up a few questions that need further study before we conclude that synthetic data offers a universal solution for AI and ML.

43.1 Introduction

AI is one of the key technologies of fourth Industrial revolution which is known as Industry4.0. Amazon Alexa, Google DeepMind's Alpha Go, IBM Watson, Tesla's autonomous vehicle has shown that AI is the reality, and it is the future. John McCarthy first coined the work Artificial intelligence in the year 1954. The ability of machines to mimic human actions that involve thinking, analyzing and decision making is AI [1]. ML is a subset of AI and DL falls under the broader family of ML

A. Kiran (✉) · S. S. Kumar
Department of CSE, SOET, CMR University, Bangalore, India
e-mail: kiran_a@cmr.edu.in

S. S. Kumar
e-mail: sarvana.k@cmr.edu.in

[2]. The decision making ability of the machine is driven by ML and DL algorithms which in turn rely on the data for model training. Machine learning uses historic data to train and make accurate predictions. Therefore, data becomes an integral part of Machine Learning and AI. Advancement in studies, research, innovation related to machine learning is hugely dependent on access to, and availability of a large amount of data which is popularly known as big data [3]. Strict data security and privacy protection guidelines such as HIPAA, [4] California Consumer Privacy Act CCPA [5], and GDPR [4] enforce a restriction on data sharing. Along with the data privacy and protection laws, (1) Data quality and reliability [6] (2) Economy of collecting large sample data through a survey, (3) Accessibility of data that are with the government, private organizations, NGO and (4) Bias or imbalance [7] in the dataset are some of the major hurdles to the research work in ML and AI that inhibit further advancement of these technologies. These challenges have forced researchers to look through the alternative.

Synthetic Data offers a promising solution to overcome the data challenges. It is a technique where the artificial data is generated (Synthesized) from the real data such that the privacy elements in the dataset are removed without impacting the statistical properties of the dataset. Figure 43.1 shows how synthetic data is synthesized from the real data, the validation metrics are applied, and then used as a substitute to big data in research studies.

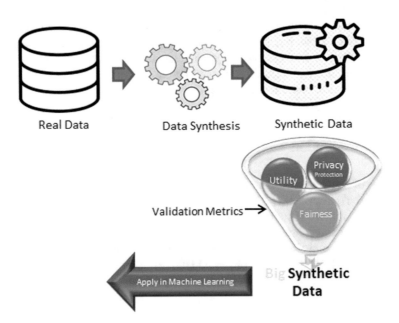

Fig. 43.1 Shows how the synthetic data is being used in machine learning as a substitute for real data

Off late, there has been a special focus on validation and applicability of synthetic data. Image processing and computer vision experiments in the field of healthcare has gained special attention from researchers, scholars, and academia. The accuracy and the usability of Synthetic data is evaluated through three metrics that is,

1. Utility which shows how close the statistical distribution of the synthetic data as compared to original data and how usable it is for statistical analysis
2. Privacy preservation shows how the synthetic data can maintain privacy in the data set
3. Fairness which is to overcome the data bias.

43.2 Artificial Intelligence, Machine Learning, and Data Challenges

Haenlein and Kaplan [8], described AI as the ability of machines to interpret the data, learn from it and adapt to perform tasks specific to goals. Established in 1950, the field has remained only of academic interest until the early part of 2000. Today AI is becoming part of every organization's transformation roadmap raising a question on how humans and machines co-exist. AI can gain human-like intelligence without the need for explicit programming. Machine Learning (ML) is a key component of AI. Das et al. [9] review how ML is related to AI and the prospective applications of ML. Alloghani et al. [10] attribute the demand for machine learning to the advent of big data, business intelligence, and their application in automation. ML algorithms can be classified into (1) Supervised Learning, (2) Unsupervised learning (3) Reinforced learning, and (4) Recommender systems. Naqa and Murphy [11] describe ML as the workhorse of modern times in the age of big data. ML is not programmed instead the algorithms are "soft coded". They are repeatedly trained with input data based on which they automatically adapt and produce the outcome for previously unseen data. The classification and predictive models in ML have found widespread application in the medical field. Kourou et al. [12] review the most recent publications in this field and present the applicability of machine learning in the early diagnosis of cancer in patients and improve the outcomes. The ability of ML to detect key features from complex datasets is also an important aspect of their study. However, access to the data has its own challenges. L'heureux et al. [13], not only call out the ability of ML to learn from big data and opportunities associated with it, but also highlight the data challenges. Jain et al. [14] studied the privacy and security concerns in big data, elaborate on what the differences between privacy and privacy requirements in big data are. Class Imbalance is the other challenge where one class has fewer samples when compared to the other. This impacts predictions in machine learning algorithms for supervised learning. Kaur et al. [15] have analyzed the impact of an Imbalanced dataset on machine learning and the corresponding solution to overcome the challenges.

43.3 Privacy Protection and Disclosure Methods

Privacy concerns have been a concern since post World War II when the camera came into existence and were used for surveillance. Internet and digital platform have raised further challenges not only due to the volume of data that is generated but they also contain personal information which the intruders can easily access that would result in identity theft. In the sub sections below, we provide a brief overview of Statistical Disclosure Control, Differential Privacy, and Privacy Enhancement Technologies.

43.3.1 Statistical Data Disclosure

Statistical Disclosure Control (SDC) or Statistical Disclosure Limitation is a technique to protect the confidential information of the individual respondents in the survey data. To meet the explosive demand for data and to overcome the legal and ethical issues of confidentiality challenges in the release of data, Rubin [16] proposed the idea of Synthetic microdata. The technique involved the artificial generation of microdata through multiple imputations [17] which was based on his own work. The potential of synthetic data is great especially when there is confidentiality concern about the release of large dataset for public usage.

43.3.2 Differential Privacy

Dwork et al. [18], introduce Differential Privacy (DP) as a set of computationally rich algorithms that provides, a meaningful, mathematically rigorous, and robust definition of privacy. The randomized algorithm is applied in such a way that the requestor will get the entire statistical feature required for the statistical analyses of the data but the individual's information in the dataset will not be compromised. DP is achieved by introducing noise to the existing data so that the probability of identifying the individual information is reduced. This is superior to other statistical disclosure control methods. However, these algorithms are computationally intensive.

43.3.3 Privacy Enhancement Technologies

Increased usage, pushing and publishing content over the internet have forced policy makers to bring stringent guidelines on privacy protection. Therefore data aggregators must ensure that privacy is protected in a rightful way. Privacy Enhancement Technologies are technologies in the field of Information and Communication Technology (ICT) that restricts or prevents unauthorized access to the personal data.

Kaaniche et al. [19] in their review create taxonomy of PETs and identify which one of these is best suited for personalized service category.

43.4   Synthetic Data and Generation Methods

The idea of disclosure control has been discussed since 1974 by Dalenius and others. Synthetic data was first introduced by Rubin in 1993 [16] in a discussion with Jabine. Synthetic data is a statistical process through which statistical information is extracted from the actual data and released for public usage. Fully synthetic data creation follows the approach introduced by Rubin whereas partially synthetic data creation was a concept suggested by Liitle in 1993. First validation of synthetic data was performed by Reiter [20] and later Raghunathan [21] used different simulations to compare synthetic data with the real data. The validation inferences of the fully synthetic data against the original data show that the results are comparable. Raghunathan [22] in his review, discussed different methods of generating and analyzing Synthetic Data, their limitations, and directions for future use. Synthetic data can be created using three approaches, Fully Synthetic which is based on the theory of multiple imputation introduced by Rubin, Partially Synthetic in which only sensitive columns are imputed, and Selective or Hybrid which uses both techniques. Figure 43.2 below gives a high-level representation of the process.

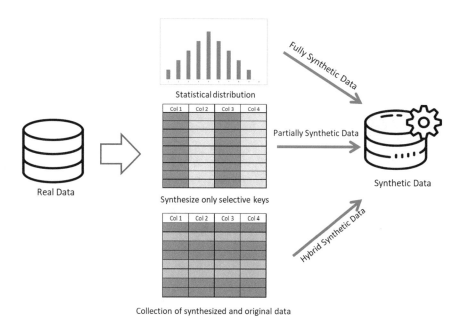

Fig. 43.2   Synthetic data type based on techniques used for generation

Emam and Mosquerea [5] provide a detailed overview of Synthetic Data, their generation techniques, and the evaluation of synthesized data for practical usage. They introduce the various data generation methods using a statistical approach as well as a machine learning approach. Reiter [23] first introduced the generation of synthetic dataset using Classification and Regression Tree (CART) which was later used by Nowok et al. [24] came up with a package in R called Synpop. Patki et al. [25] used the statistical model copula to come up with Synthetic Data Vault (SDV), which works on structural data. SDV first built a generative model of relational database and then generates the synthetic data from the relational data model. Zhang et al. [26] work has been motivated by the limitation of Differential Privacy (DP) which is inefficient with large multi-dimensional data. PrivBayes overcomes this limitation. It uses Bayesian Network to (1) to obtain the correlation between attributes and (2) approximate the distribution of a set of low-dimensional marginal. It later introduces the noise only on the low-dimensional marginal and Bayesian Network to generate approximate distribution. Finally, it samples the approximate distribution to generate Synthetic data. Inspired by this work Ping et al. [27] came up with a new synthetic data generation approach called Data Synthesizer (DS) which promises stronger privacy protection for data sharing. To overcome the challenges of classification imbalance in the dataset Chawla et al. [28], developed the Synthetic Minority Oversampling Technique (SMOTE). Generative Adversarial Network (GAN) [29, 30] and Variational Auto Encoder (VAE) are the Deep Learning techniques used in generation of Synthetic Data. Figure 43.3 shows different methods and type of synthetic data generators.

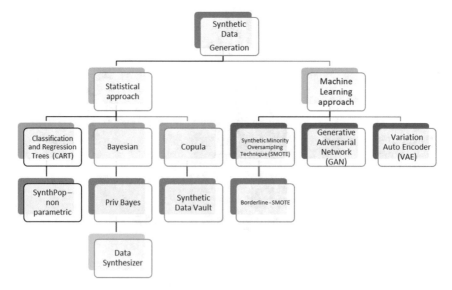

Fig. 43.3 Synthetic data generation methods

43.5   Validation Metrics

El Emam [31] summarizes seven ways of assessing the utility of synthetic data. The utility assessment must be performed before the data is shared for broader use. (1) Structural Similarity, (2) Replication of studies, (3) General Utility Metrics, (4) Subjective assessments by domain experts, (5) Bias and stability assessments, (6) Comparison with other privacy Encryption Techniques (PET), and (7) Comparison with public aggregate data are the methods suggested. These evaluations will ensure that the data scientists and researchers can comfortably use synthetic data to build models that would be used in public health, clinical trials, or any other industrial applications in the future Hittmeir et al. [32, 33] published their work on assessing the privacy and utility of synthetic dataset for both regression and classification ML tasks. They use SDV and DS for classification data and introduce one more synthetic data generation method Synthpop to evaluate regression algorithm. The data set is used in five machine learning algorithms. Output is validated by applying statistical utility on different datasets. Heyburn et al. [34] identify two specific uses of synthetic data set (1) increase the size of the minority data when the dataset is imbalanced (2) generate data when they are completely absent. Through the experiment, they explore whether synthetic data is as reliable as real data for machine learning. They use Breast cancer data as a base dataset. This is an imbalanced dataset as the number of cases of cancer patients will be very few as compared to the total population. Therefore, SMOTE and Synthpop is used in experiments on imbalanced data set for synthetic data generation. Dankar and Ibrahim [35] evaluate the different synthetic data generation methods and the utility of synthetic data by investigating (i) the effect of preprocessing of the data, (ii) parameter tuning when supervised learning is used in data generation, and (iii) sharing the machine learning results to tune the data generation models, and (iv) Propensity score to predict the accuracy of the machine learning models. DS, SDV, and Synthpop are used to generate the synthetic data sets and then their propensity score and utility metrics are compared against real data. The early results show promise but the (1) efficiency of synthesis, (2) the comparison with other PETs, the best way of data generation are to be explored further. Adoption of ML models has increased in the field of medicine creating a need for further development of privacy-preserving and fairness algorithms. Cheng et al. [36] in their work, use a Generative Adversarial Network (GAN) to generate synthetic data to improve the privacy and fairness of the data. DP is used for preserving privacy. Experimental results show that generative models do not guarantee privacy, when DP is increased, the quality of synthetic data degrades, and it also impacts the fairness of the data impacting the minority classes.

Future scope for research: There is no conclusive evidence so far that synthetic data produces consistent and reliable results. Therefore, finding a methodology for data generation and evaluation which provides the balance of (1) privacy, (2) utility, and (3) fairness could be the area of future research which is the intersection of the three circles in the below diagram (Fig. 43.4). Second area is to identify a method to assess the disclosure risk moving away from the assumption and probability of intruder's prior knowledge of data which has been used in most of the study so far.

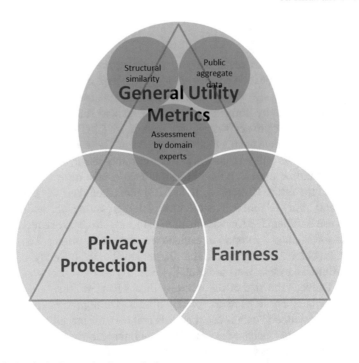

Fig. 43.4 Synthetic data evaluation methods

43.6 Conclusion

Synthetic data is gaining popularity and raising curiosity among researchers. This
could have a real potential in industrial usage provided, there are methods to prove
its consistency. The results and outcomes reviewed so far are not very conclusive.
A few questions still remain unanswered (1) Is privacy-preserving synthetic data
the best solution as compared to other privacy-preserving techniques El Emam [5],
(2) The Utility metrics are not consistent across data sets, (3) No Standardizations
techniques to check the effectiveness of synthetic data for a specific purpose is (4)
Increasing differential privacy impact the distribution of data. Differential Privacy
and PrivBayes technique created data imbalance as identified by Ganev et al. [37],
Differential Privacy also impacts the image quality in healthcare data [36]. These
open questions raise a need to explore the evaluation metrics as well as the generation
methods of synthetic data further. As of now, there is no one method that is fit for all.
This leaves open the space to explore the validation metrics specific to a field and
generalize the same which could also be an area of future research.

References

1. McCarthy, J.: Artificial intelligence, logic and formalizing common sense. Philos. Log. Artif. Intell., 161–190 (1989). https://doi.org/10.1007/978-94-009-2448-2_6
2. Ongsulee, P.: Artificial intelligence, machine learning and deep learning (2018). https://doi.org/10.1109/ICTKE.2017.8259629
3. Surya, L.: An exploratory study of DevOps and it's future in the United States. Int. J. Creat. Res. Thoughts 3(2), 2320–2882 (2016)
4. Yale, A., et al.: Generation and evaluation of privacy preserving synthetic health data. To cite this version: HAL Id: hal-03158544 (2021)
5. Emam, K., Mosquera, L., Hoptroff, R., Safari, O.M.C.: Practical Synthetic Data Generation, p. 175 (2020).
6. Liu, J., Li, J., Li, W., Wu, J.: Rethinking big data: a review on the data quality and usage issues. ISPRS J. Photogramm. Remote Sens. 115, 134–142 (2016). https://doi.org/10.1016/j.isprsjprs.2015.11.006
7. Leevy, J.L., Khoshgoftaar, T.M., Bauder, R.A., Seliya, N.: A survey on addressing high-class imbalance in big data. J. Big Data 5(1) (2018). https://doi.org/10.1186/s40537-018-0151-6
8. Haenlein, M., Kaplan, A.: A brief history of artificial intelligence: on the past, present, and future of artificial intelligence. Calif. Manage. Rev. 61(4), 5–14 (2019). https://doi.org/10.1177/0008125619864925
9. Das, S., Dey, A., Pal, A., Roy, N.: Applications of artificial intelligence in machine learning: review and prospect. Int. J. Comput. Appl. 115(9), 31–41 (2015). https://doi.org/10.5120/20182-2402
10. Alloghani, M., Al-Jumeily, D., Mustafina, J., Hussain, A., Aljaaf, A.J.: A systematic review on supervised and unsupervised machine learning algorithms for data science
11. El Naqa, I., Murphy, M.J.: Machine learning in radiation oncology. In: Machine Learning in Radiation Oncology, pp. 3–11 (2015). https://doi.org/10.1007/978-3-319-18305-3
12. Kourou, K., Exarchos, T.P., Exarchos, K.P., Karamouzis, M.V., Fotiadis, D.I.: Machine learning applications in cancer prognosis and prediction. Comput. Struct. Biotechnol. J. 13, 8–17 (2015). https://doi.org/10.1016/j.csbj.2014.11.005
13. L'Heureux, A., Grolinger, K., Elyamany, H.F., Capretz, M.A.M.: Machine learning with Big Data: challenges and approaches. IEEE Access 5, 7776–7797 (2017). https://doi.org/10.1109/ACCESS.2017.2696365
14. Jain, P., Gyanchandani, M., Khare, N.: Big data privacy: a technological perspective and review. J. Big Data 3(1) (2016). https://doi.org/10.1186/s40537-016-0059-y
15. Kaur, H., Pannu, H.S., Malhi, A.K.: A systematic review on imbalanced data challenges in machine learning: applications and solutions. ACM Comput. Surv. 52(4) (2019). https://doi.org/10.1145/3343440
16. Rubin, D.B.: Statistical disclosure limitation (SDL). J. Off. Statis., 461–468 (1993). https://doi.org/10.1007/978-0-387-39940-9_3686
17. Rubin, D.B.: An overview of multiple imputation. In: Proceedings of the Survey Research Methods Section, American Statistical Association (1988)
18. Dwork, C., Kenthapadi, K., McSherry, F., Mironov, I., Naor, M.: Our data, ourselves: privacy via distributed noise generation. Lecture Notes Computer Science (including Subseries Lecture Notes Artificial Intelligence, Lecture Notes Bioinformatics), vol. 4004 LNCS, pp. 486–503 (2006). https://doi.org/10.1007/11761679_29
19. Kaaniche, N., Laurent, M., Belguith, S.: Privacy enhancing technologies for solving the privacy-personalization paradox: taxonomy and survey. J. Netw. Comput. Appl. 171(Jan), 102807 (2020). https://doi.org/10.1016/j.jnca.2020.102807
20. Reiter, J.: Satisfying disclosure restrictions with synthetic data sets. J. Off. Stat. 18(4), 1–19 (2002) [Online]. Available: http://www.stat.duke.edu/~jerry/Papers/jos02.pdf
21. Raghunathan, T.: Multiple imputation for statistical disclosure limitation. J. Off. Stat. 19(1), 1–16 (2003) [Online]. Available: http://hbanaszak.mjr.uw.edu.pl/TempTxt/RaghunathanEtAl_2003_Multiple_Imputation_for_Statistical_Disclosure_Limitation.pdf

22. Raghunathan, T.E.: Synthetic data. Annu. Rev. Stat. Its Appl. **8**, 129–140 (2021). https://doi. org/10.1146/annurev-statistics-040720-031848
23. Reiter, J.P.: Using CART to generate partially synthetic, public use microdata. J. Off. Stat. **21**(3), 441–462 (2003) [Online]. Available: https://www.scb.se/contentassets/ca21efb41fee47d 293bbee5bf7be7fb3/using-cart-to-generate-partially-synthetic-public-use-microdata.pdf
24. Nowok, B., Raab, G.M., Dibben, C.: Synthpop: bespoke creation of synthetic data in R. J. Stat. Softw. **74**(11) (2016). https://doi.org/10.18637/jss.v074.i11
25. Patki, N., Wedge, R., Veeramachaneni, K.: The synthetic data vault. In: Proceedings—3rd IEEE International Conference on Data Science and Advanced Analytics DSAA 2016, pp. 399–410 (2016). https://doi.org/10.1109/DSAA.2016.49
26. Zhang, J., Cormode, G., Procopiuc, C.M., Srivastava, D., Xiao, X.: Priv Bayes: private data release via Bayesian networks. ACM Trans. Database Syst. **42**(4) (2017). https://doi.org/10. 1145/3134428
27. Ping, H., Stoyanovich, J., Howe, B.: Data synthesizer: privacy-preserving synthetic datasets. In: ACM International Conference Proceeding Series, vol. Part F1286 (2017). https://doi.org/ 10.1145/3085504.3091117
28. Chawla, N.V., Bowyer, K.W., Hall, L.O., Kegelmeyer, W.P.: SMOTE: synthetic minority over-sampling technique. J. Artif. Intell. Res. **16**, 321–357 (2002). https://doi.org/10.1613/jair.953
29. Goodfellow, I., Pouget-Abadie, J., Mirza, M., Xu, B., Warde-Farley, D., Ozair, S., Courville, A., Bengio, Y.: Generative adversarial nets. In: Advances in Neural Information Processing Systems, vol. 27, pp. 2672–2680 (2014) [Online]. https://proceedings.neurips.cc/paper/2014/ file/5ca3e9b122f61f8f06494c97b1afccf3-Paper.pdf
30. Frid-Adar, M., Klang, E., Amitai, M., Goldberger, J., Greenspan, H.: Synthetic data augmentation using GAN for improved liver lesion classification. In: Proceedings—IEEE International Symposium on Biomedical Imaging, vol. 2018-April, pp. 289–293 (2018). https://doi.org/10. 1109/ISBI.2018.8363576
31. El Emam, K.: Seven ways to evaluate the utility of synthetic data. IEEE Secur. Priv. **18**(4), 56–59 (2020). https://doi.org/10.1109/MSEC.2020.2992821
32. Hittmeir, M., Ekelhart, A., Mayer, R.: Utility and privacy assessments of synthetic data for regression tasks. In: Proceedings—2019 IEEE International Conference on Big Data (IEEE BigData 2019), pp. 5763–5772 (2019). https://doi.org/10.1109/BigData47090.2019.9005476
33. Hittmeir, M., Ekelhart, A., Mayer, R.: On the utility of synthetic data: an empirical evaluation on machine learning tasks. In: ACM International Conference Proceeding Series (2019).https:// doi.org/10.1145/3339252.3339281
34. Heyburn, R., et al.: Machine learning using synthetic and real data: similarity of evaluation metrics for different healthcare datasets and for different algorithms, pp. 1281–1291 (2018). https://doi.org/10.1142/9789813273238_0160
35. Dankar, F.K., Ibrahim, M.: Fake it till you make it: guidelines for effective synthetic data generation. Appl. Sci. **11**(5), 1–18 (2021). https://doi.org/10.3390/app11052158
36. Cheng, V., Suriyakumar, V.M., Dullerud, N., Joshi, S., Ghassemi, M.: Can you fake it until you make it?: Impacts of differentially private synthetic data on downstream classification fairness. In: FAccT 2021—Proceedings 2021 ACM Conference Fairness, Accountability, Transparency, pp. 149–160 (2021). https://doi.org/10.1145/3442188.3445879
37. Ganev, G., Oprisanu, B., De Cristofaro, E.: Robin Hood and Matthew effects—differential privacy has disparate impact on synthetic data (2021) [Online]. http://arxiv.org/abs/2109.11429

Chapter 44
Sacred Spaces Enduring Pro-environmental Behavior: A Case Study from Kerala Temple

M. V. Mukil, R. Athira, Tarek Rashed, and R. Bhavani Rao

Abstract In this Anthropocene era, it is relevant to understand how culture influences humans' ecological behavior. This research aims to understand the impact of the COVID-19 lockdown on temple-led eco-conservation strategies from the Sasthamcotta Shri Dharma Sastha temple, known for its natural landscapes and pro-environmental activities. A qualitative study based on semi-structured interviews was conducted virtually among the people who regularly interact with the temple. The data was collected directly from the field and is analyzed systematically based on grounded theory. The findings indicate that temples can generate eco-conservation approaches that are socio-psychologically relevant to humans, such as (a) connectedness to nature, (b) sense of place, (c) values, beliefs, and norms, and (d) general awareness. The temple also serves as a management hub for (e) pro-social activities and (f) environmental decision-making. Managerial factors such as pro-social activities and environmental decision-making were curtailed during the COVID-19 lockdown, and the strategies based on socio-psychological factors remain unchanged. According to our findings, new environmental conservation strategies should be based on socio-psychological aspects that are more in line with the mental model of the community.

44.1 Introduction

We need new strategies that are consistent in protecting the natural environment. Therefore, learning more about the culture/tradition that has played an essential role

M. V. Mukil (✉)
Amrita School for Sustainable Development, Amrita Vishwa Vidyapeetham, Amritapuri, India
e-mail: mukil.mv@ammachilabs.org
URL: https://ammachilabs.org

R. Athira
Amrita Vishwa Vidyapeetham, Amritapuri, India

T. Rashed · R. B. Rao
AMMACHI Labs, Amrita Vishwa Vidyapeetham, Amritapuri, India

© The Author(s), under exclusive license to Springer Nature Singapore Pte Ltd. 2023
C. So-In et al. (eds.), *Information Systems for Intelligent Systems*, Smart Innovation, Systems and Technologies 324, https://doi.org/10.1007/978-981-19-7447-2_44

in environmental protection for decades is necessary. The construction of Indian temples began before 800 CE. Until now, most of these temples were consistent in conserving the ecosystem. This research focuses on the temple-driven eco-conservatiofgxtreen strategies during the COVID pandemic. The COVID-19 crisis has significantly impacted the environmental, economic, and social sectors. Lockdowns, limitations on social meetings, and the distribution of COVID vaccines have all been implemented in various countries worldwide to control the outspread of COVID. As a result, the pilgrim centers and temples were only open for temple priests and authorities to perform the necessary daily rituals.

The objective of the study is to understand the impact of temple culture on psychological parameters that influence pro-environmental behavior and eco-conservation strategies. The study also investigates how these parameters change during the COVID pandemic when people's access to temples is restricted. This paper is structured by introducing the relevant Indian traditions and temples. After the introduction, theories outlining the variables influencing how people engage with environment and current literature on temple culture are focused. Based on which, the study design, Study area and Sample size is adapted followed up with Findings. Findings involves the detailed interpretation of result formulated based on the in-depth interview, and how the effect of COVID influence these factors.

44.1.1 Temples, Landscape, and Environmentally Favorable Beliefs

In India, depending upon the community and geography of the place, the temples are diverse in their architecture and functioning. Our research focuses on a temple in the Indian State of Kerala, recognized for its 'Kerala style temple' building, which features sloping roofs and more woodwork art than other architectural styles. All the temples adhere to the regional ancient Indian architecture treatises based on Vaastu Shastra (ancient Indian science on dwellings). Depending upon the architecture, Kerala temples can be classified as follows. (a) Kavu temples, (b) rock-cut or cave temples, and (c) structural temples. Towards 600 CE, evidence of tree and serpent worship has been found. Along with the major shrine of the structural temples, an area named Sarpa Kavu (Sacred grooves) would be set aside for snake worship. These sacred grooves are a pristine habitat capable of preserving biodiversity and maintaining ecological balance [9, 29]. Kerala temples also have a Theerthakkulam (Temple Pond), generally found in the temple's northeast corner. Some temples are closely associated with lakes or rivers and upkeep these water bodies. During summer, the temple ponds are used to collect and transmit water for farming. Some temples in valleys or on top of hills are believed to be the protectors of the hill and the biodiversity that exists there. There were also temple deities known as Vana Devatas (forest Deity) who dwell inside forests and guard them. A pipal tree is worshiped in almost every temple in Kerala. Similarly, many beliefs and associated rituals can be found throughout the state. Specifically in Kerala, in temples like Iringole kavu, the trees surrounding this temple are referred to as upadevatas (subordinate deities) by the locals. As a result, no trees will be cut down, including those that have already fallen. In addition to the vegetation, the water bodies surrounding the temples play an important role in numerous rituals and practices. Kerala temples such as Kulathupuzha Sastha Temple, Sasthamcotta Temple, Valamchuzhy Devi Temple, Anantha Padmanbha Swamy Temple, and many such temples are surrounded by water bodies. People consider these water bodies sacred and strive to keep them as clean as possible. The endangered Mahseer fish has been conserved at the Varadayini Mata temple in Maharashtra as these fishes are sacred to Hindus [2]. Fish is worshiped as Thirumakkal (Son of God) in some Kerala temples, and people are forbidden from fishing or polluting that area's water bodies which preserve the aquatic ecosystem [17]. Meenoottu (Fish are fed) is a ritual performed at several Kerala temples where the temple priest would feed the fishes personally during the auspicious ceremonies. Practices associated with temples have served the environment for generations. For a sustainable future, it is important to develop eco-conservation strategies which are consistent and in harmony with the mental model of the community.

44.2 Background Studies

44.2.1 Temples and Traditional Eco-Conservation Strategies Across the Globe

Environmental conservation strategies, often known as eco-conservation strategies, are management measures designed to protect the natural environment. In traditional cultures, eco-conservation strategies are linked to cultural values and beliefs. In India, there are sacred mountains, sacred water bodies, sacred forest, sacred animals, sacred flora and fauna, and so on [8, 25]. The temple's rituals and traditions motivate people to protect the environment [31]. Most temple rituals, iconographies, and festivals were developed with the environment in mind [38]. Apart from temples, the country is also home to a diverse spectrum of indigenous cultures, each with its own set of values and beliefs regarding environmental preservation [11, 37]. The eco-cultural ritual of ancient farming communities are examples of these [22]. Temples also sustain natural resources such as rivers, ponds, forests, and hills [5, 15]. Similarly, numerous communities worldwide, including the Subak village in Indonesia, practice ecologically responsible agriculture [26]. Padangtegal (Ubud) temple, an indigenous Bali heritage, has a similar temple culture. Humans and monkeys coexist there, with numerous humans providing food to the monkeys. A similar culture can also be observed in the Sasthamcotta temple [14]. The Dai villages in China safeguard the natural environment by protecting the forest cover around the hamlet and their unique rituals and attitudes toward the environment [38]. Traditional beliefs have helped the Nepal communities to establish effective water conservation methods.

The temple in the community preserves freshwater rock sprouts that conserve the community's water management system through rituals and activities [36]. Similarly, several temples in India use traditional practices to conserve the natural environment directly. Devbanis are divine forests in the Alwar area of Rajasthan. There were more Orans in the dense mountains with a large variety of species living undisturbed. These areas are dedicated to the community's native deities, Gopal Das and Shital Das. Many sacred groves and natural water resources are conserved by faith among the Bhil communities in Rajasthan. Among them, a few trees are referred to as Kalpadeva (tree deity) [20].

44.2.2 *Pro-environmental Behavior (PEB) and Eco-conservation Strategies*

In humans, the imposed environmental and personal factors have a significant impact on behavior [3]. According to the Value Belief Norm (VBN) theory, an individual's PEB can be determined by their values, beliefs, and norms. "self-interests, altruism toward other people, altruism toward other species and biosphere, conservation (traditional) values and openness to change" can predict the value orientation which leads to PEB [35]. In addition, studies show that people's interactions with nature help enhance PEB. Emotional connectedness with nature, known as the Connectedness to Nature (CN), can help increase altruistic behavior by bridging the gap between self-identity and nature. It's considered to be a predictor of both their ecological behavior and their subjective well-being. Studies also recommend using CN as an effective tool for developing eco-conservation strategies [26]. (Restall and Conrad) Another important factor that sustains ecology is having a sense of place or place or environmental identity, defined as the feeling of dwelling, subject to places/landscapes, its esthetic values, and bonds [19]. People expand themselves to include places that are meaningful to them. This mindset aids them in preserving the natural environment [27]. Pro-social behavior exerts a direct influence on PEB. Studies show interpersonal trust among tourists in place attachments and pro-social and pro-environmental behavior [30]. Pro-social behavior has also influenced people's willingness to pay for green products and help clean protected areas [37].

44.3 Research Methodology

The study aims to understand the factors that impact temple eco-conservation strategies during the COVID period. A qualitative analysis using the grounded theory method was employed in this study [6]. Based on the background study and observations from the study area, a telephonic interview was employed based on a semi-structured questionnaire. A purposive sampling method was done for the selection of respondents [13]. Initially, the factors affecting the eco-conservation strategies of the temple were analyzed from the qualitative data. Secondly, the factors influencing the eco-conservation strategies during the COVID period were identified. Analysis of the study was conducted by developing open code, axial code, and selective themes based on the data collected. From the collected data, initial codes or open codes were generated. After that, potential theories related to the open codes were identified. With the potential theories, the axial codes and selective themes for this particular scenario were found [23].

44.3.1   Data Collection

The data was collected through in-depth telephonic interviews with ten key respondents (Table 44.1). The interviewees were selected according to a purposive sampling of participants who are native and are visiting the temple regularly and actively participating in the conservation projects were chosen for the study. All these participants have been associated with the temple for at least five years. Interview recordings were used to develop codes and determine the key factors influencing the eco-conservation strategy.

Temple staff are responsible for maintenance and other administrative tasks related to the temple and its surroundings. They are employed by the Travancore Devaswom Board, a government entity that supervises several temples in Kerala's southern region. The temple's daily rituals and practices are handled by temple priests. Non-temple staff includes those who have worked for the temple in the last 10–30 years, people who reside near the temple, active participants in temple events, temple committee members, and regular visitors of the temple. These individuals and their approach to environmental conservation during this COVID pandemic guided the

Table 44.1 Demographic details of the respondents

Demographic features	Category	N (10)	Percentage
Age	18–30	3	30
	31–43	4	40
	44–56	1	10
	57–69	1	10
	70–82	1	10
Gender	Male	6	60
	Female	4	40
Occupation	Temple priest	2	20
	Temple staff	2	20
	Non-temple staff	6	60

authors to understand the temple-driven eco-conservation strategies more theoretically. The data were collected from June to August of 2021 when COVID lockdowns were at an all-time high. During that time, visitors were restricted from entering the temple.

44.3.2 Study Area

The present study mainly focused on the Hindu temple, Sasthamcotta Sree Dharma Sastha Temple (Lat: 09° 02′ 29″ N, Lon: 076° 37′ 46 ″ E) shortly Sasthamcotta temple (see Fig. 44.1) Sasthamcotta Temple is one of the Pancha Sastha temples (five major temples of the deity Sastha) in Kerala. Mandala Kala (the months from September to January) is auspicious for the deity Sastha/Ayyappa temples. The major pilgrim center of Kerala, Sabarimala, opens for people in these months. As Sasthamcotta temple is also included in the series of these Ayyappa temples, around 25,000 pilgrims visit the temple during Mandala Kala. The natural landscape of all these five temples is ecologically significant. About 200 monkeys inhabit the Sasthamcotta temple and its premises. The visitors of the temple and the people living in the community consider the monkeys as the divine retinue of the prime deity. The temple authorities have deposited funds in the bank for the monkeys' necessities, such as food and shelter. In the northwest part of the temple, there is a dining hall made for feeding the monkeys. During the auspicious days, the temple authorities also offer feasts to the monkeys. In ancient Indian scriptures, these actions can be referred to as Bhutayajna [7]. The temple is also encircled by foliage where visitors' entry is restricted (see Fig. 44.2). Among the foliage, there lies a sacred grove for snake worship. During special days dedicated to snake worship, temple priests will perform rituals in this sacred grove. In addition, the temple is encircled by Kerala's largest natural freshwater lake. A rare fish known as 'Etta' in the lake is believed to be the worshiping deity's holy retinue. The lake covers 373 ha and has a volume of 22.4 km^3. The lake of Sasthamcotta was listed as a Ramsar site in India in 2002. The Ramsar Convention establishes the Ramsar Wetland Site (an international treaty for the conservation and sustainable utilization of the wetlands). *"The ancient Sastha temple is an important pilgrimage center. Ramsar site no. 1212"* [1].

44.4 Findings

44.4.1 Factors Influencing Eco-conservation Strategies

From the detailed interviews and the observations from the place, we were able to find that temple as an entity can generate Pro-environmental behavior. Based on the background study, we can understand that all the axial codes found are important

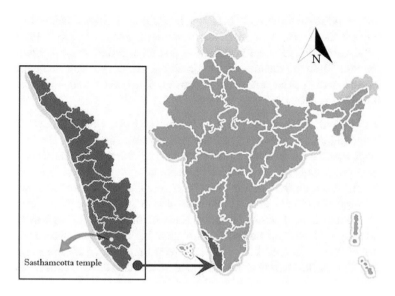

Fig. 44.1 Location of study area

Fig. 44.2 Temple landscape aerial view (Google maps, Accessed on Nov 2022)

determinants of Pro-environmental behavior. In addition, we have obtained the socio-psychological factors (a) connectedness to nature, (b) sense of place, (c) values, beliefs, and norms and (d) general awareness and managerial factors, (e) pro-social activities, and (f) environmental decision-making. Table 44.2 represents the open codes, corresponding axial codes, and selective themes. The detailed interpretation of the result is as follows.

Pro-social activities. Group activities, financial support, effects of COVID, and COVID regulations were the determined open codes for formulating the axial code of pro-social activities. In the Sasthamcotta temple, the temple 'Kara' is vital in developing pro-social activities. Kara represents the several subgroups of residents who are split based on the location of their homes. The members of Kara organize many cultural and social events inside and outside the temple, which help them engage in a wide range of pro-social and pro-environmental actions, such as planting trees, cleaning up lakes and surroundings, organizing campaigns, and so on. After the arrival of the COVID pandemic, the 'Kara-yogam' (meeting of each Kara) is not happening. During the interview, every respondent reflected on their collective efforts to protect the ecosystem. "*...we have planted many trees in the temple compound, for monkeys we have planted fruit trees... during the temple festival we used to clean the shore of lake for 'aarattu'...we have created a 'nakshtra vanam'... in future, the temple authorities should come up with more people toward nature conservation...*" Temple groups have done several eco-restoration projects. They have planted many fruit trees for monkeys, and medicinal plants, fed monkeys, and removed plastics and other contamination from the lake and nearby forest area. They may have to plant again and again or water several times due to environmental factors. Similar to the

Table 44.2 Factors affecting the temple-driven eco-conservation strategies based on the utterance

Open codes	Axial code	Selective theme
Nature and beauty Nature and happiness	Connectedness to nature	Pro-environmental behavior
Nature and place Place identity	Sense of place	
Cultural values Temple beliefs norms	Value belief and norm	
General awareness programs cleanliness COVID awareness	COVID awareness	
Group activities Financial support COVID affect COVID regulations	Pro-social activities	
Environmental ethics	Environmental decision-making	

perspectives of other indigenous traditions, their beliefs also encourage them in self-reinforced restoration [25]. In the temple, they have created 'nakshatra vanam (Star Forest)', a project to plant trees based on birth stars. There are 27 birth stars; according to Hindu astrology, each birth star is associated with a plant or a tree. Some of them are medicinal plants, and some are big trees. Which also acts as an eco-conservation strategy for them. Arattu (holy bath) will be held on the lakeshore yearly during the festival. During Arattu, the temple deity is taken to the Shasthamcotta lake for a holy bath. This lakeshore is considered to be sacred and is maintained. In Fig. 44.2 Arattu kadavu, the sacred lakeshore, is represented. Members of the Kara take extra precaution in not polluting the lakeshore and the water. Planting saplings around the lake to reduce soil erosion has been part of a project using Geotextile in recent years [32]. These temple group initiatives are important in initiating funds for all the cultural, social, and pro-environmental projects.

When the code 'COVID effect' was analyzed, it was observed that the respondents were concerned about the lack of social gatherings. Due to COVID, the temple fests and other rituals were performed with a limited population. Regular meetings of Temple groups were temporarily stopped, and no new eco-restoration initiatives were happening. Fewer visitors to the temple also limited the extra food received by the monkeys. According to social cognitive theory, the environment that one perceives can affect their personal standards [4]. Here the social norms help their self-efficacy to work for the temple by planting trees together, finding funds for these activities, arranging decorations and cleaning water bodies, serving monkeys together, and so on. They also built a dining area for monkeys to have food. The whole temple groups are showing their reverence to the deity by arranging and serving monkeys. "…*alone it is hard for me to work… TOGETHER considering each people from the kara we can achieve it…*" The respondent is aware of the importance of nature conservation, even though being in a group is more convenient for the self. The temple and the associated beliefs deliver moral courage to these temple groups, affecting their self-system toward pro-social attitude. It also encourages them toward pro-social activities [18].

Connectedness to nature and sense of place. The emotional connection between the respondent and nature is reflected in their interviews. Temple landscapes were able to generate feelings of appreciation the nature. "… *I'll stay in the temple and pray for a while, then go sit beneath a tree and watch the monkeys play while taking in the sights…*" "*… It's a peaceful space to be in…*" "*…give delight to all…*" Connectedness to nature is the attachment of the self and the natural environment. When people are exposed to nature, they feel mentally relieved. Therefore, temple landscapes improve visitors' connectedness to nature and well-being [26]. Several studies show the importance of connectedness to nature in improving pro-environmental behavior [24, 31]. All the participants much appreciate the serene beauty of Sasthamcotta temple. People visiting the temple also spend time beside the lake, offering food to monkeys and resting under the trees. A sense of place is the native people's attachment to the place [33]. The temple's landscape, such as the Sasthamcotta lake and the uniqueness of the Sasthamcotta temple offering food to monkeys, etc., invokes a sense of place in each respondent. The meaning of the name Sasthamcotta is also related to the name of the temple's deity. Sasthamcotta literally means the dwelling of Lord Sastha (deity of the temple). *"...Sasthamcotta is a place of natural beauty..."* *"When some guests visit my home. I take them to the lake and the temple."* The Sasthamcotta lake and the temple scenery are the main features of the area. Some of them stated that it was even more beautiful before. There was nothing else other than the temple and nature. The native people want the area to be undisturbed in the name of development. People's emotional attachment to the temple landscape contributes to environmental preservation. Some of them have significant personal memories (meanings) associated with the temple and its natural environment. The extended self-identity of the individual also contributes to their environmental identity, thereby protecting the protection of the natural environment [16].

Values, beliefs, and norms. The nakshatra vanam project, considering and worshiping nature as God, offering food to the monkeys, Temple Arattu at the shores of the lake, maintaining the temple premises, sacred groves, and planting fruit trees for the monkeys, are all driven by belief. *"...We all know that God is nature. Previously, we used to worship nature. Then there was idolatry. There was a time when we worshiped trees, fire, and the wind... It's not just about going to the temple to pray; it's also about conserving the forest, mountains, hills, and trees... It's not just about honoring the temple deity; it's also about worshiping the trees and the mountain, which will benefit future generations..."* According to the respondent, nature is God. They believe it is their responsibility to create safeguards for nature. Then those beliefs transformed into values and social norms. The respondent demonstrates the temple system's openness to pro-environmental initiatives and norms for environmental protection. We can also observe their belief system that aligns with Carl Jung's analytical psychology [21]. *"Everyone participates in temple activities regardless of caste or religion."* The biospheric values and altruistic values are taught through cultural values such as *"Vasudeva sarvam* (the supreme being resides in all beings), *Vasudhaiva Kutumbakam* (the family of Mother Earth), *Sarva bhuta hita* (the welfare of all beings)" [12] are extremely visible in their opinions. The value orientation among this group of people is highly oriented toward the biospheric values, which are associated with benefits to the ecosystem or biosphere [34]. Here the nonlinear philosophies around the temple became the personal moral norm and thus to eco-conservation. Along with this, constraints imposed on behalf of the faith, such as entry limits to sacred groves, preserve the forest's thickness. These constraints are social norms followed by the whole community. These norms have a significant influence on their PEB [22].

General awareness programs and environmental ethics. Along with the temple-driven projects, awareness programs have also influenced the people of Sasthamcotta to plant trees. The public awareness programs happening on July 05 as World Environment Day also have influenced their pro-environmental attitude. The social norms also have influenced them toward planting trees. Environmental ethics through "proper education" is necessary for the sustainability of nature [28]. People are aware of initiatives like planting trees, avoiding plastics, etc. Along with that, people think that they have to protect nature by making some difference by removing the grass and other *"unwanted plants for them."* But certain norms associated with the temple belief weaken such anthropocentric approaches. The temple conserves the natural environment and the existing ecology for decades through the beliefs and associated rituals of the temple, such as restrictions on visitors' entry into sacred groves. The proper awareness of environmental ethics can bring light to quality environmental decision-making.

44.4.2 Effect of COVID on Eco-conservation Strategies

COVID has negatively impacted the number of visitors and the income of the temple. Respondents have noticed a reduction in contamination around the temple environment. Before COVID, visitors used to give food to monkeys. During COVID lockdown, the number of visitors and the bonus food received by the monkeys have been reduced. Years before the COVID, the temple had opened a bank account for the basic needs of monkeys. Many people in the community and the believers have supported this initiative by offering money to the accounts. Due to this, without difficulties, monkeys are getting Temple-food service even during the COVID pandemic. Due to no social gatherings, no new Eco-restoration projects were introduced during this period. But flora and fauna around the temple are conserved with the belief.

44.5 Discussion and Conclusions

From the study, it has been found that several factors influenced the eco-conservation strategies of the temple. That includes socio-psychological factors such as the connectedness to nature, sense of place, values, beliefs, and norms. During the COVID period, community engagements and pro-social activities have decreased. Temple prevents the human-initiated anthropocentric act to protect nature. These restrictions also align with the application of 'deep ecological philosophies' [10] that improve environmental decision-making. The values and beliefs associated with the sacred spaces allow creating a sense of place and connectedness to nature that develops strong bonding with the environment.

Benefits. The study has revealed that the Indian temple-driven socio-psychological factors have an important role in achieving UN's sustainable development goal 15 (SDG-15). Directly on the target indicator SDG-15.4. Effective use of these socio-psychological factors will help architects, planners, and designers to develop frameworks that are inline with the mental model of the community. Identifying different design elements from the temple culture, can be used for improving the existing design frameworks such as 'Design for sustainable behavior' and 'Design with intend' for communities in India. Inclusion of these factors in the current public policies may also help to improve collective sustainable behavior of the community.

Limitation of the study. The current study relays on existing beliefs and values associated with certain Indian temple culture. The application of regenerating these factors can only be done on sacred spaces having existing values and beliefs. Current study is a qualitative study based on semi structured interviews that helps in understanding the various factors affecting PEB. To quantify the effect of each socio-psychological factors on PEB, studies using existing PEB scales on more sample size are required.

Future scope. The socio-psychological factors identified in the research are the determinants of pro-environmental behavior (PEB). So, to develop strategies toward eco-conservation or to improve the PEB of the people in India, the importance of socio-cultural factors like temple beliefs and worship are essential. Further studies can be done on each factors separately to understand its influence on environmental conservation. Many strategies of PEB has been discussed in ancient Indian scriptures and are left unexplored, which can be used for developing new frameworks for sustainable behavior.

Acknowledgements This project has been funded by the E4LIFE International Ph.D. Fellowship Program offered by Amrita Vishwa Vidyapeetham. I extend my gratitude to the Amrita Live-in-Labs® academic program and AMMACHI Labs for providing all the support. We would also like to express our sincere gratitude to Sri Mata Amritanandamayi Devi, Chancellor of Amrita Vishwa Vidyapeetham, and world-renowned humanitarian for guiding us and inspiring us every step of the way.

References

1. Ramsar sites information service (Aug 2002), https://rsis.ramsar.org/ris/1212. Last accessed 30 Nov 2021
2. Abumoghli, I., McCartney, E., et al.: The role of environmental and spiritual ethics in galvanizing nature based solutions (2020)
3. Bandura, A.: Human agency in social cognitive theory. Am. Psychol. **44**(9), 1175 (1989)
4. Bandura, A.: Social cognitive theory of self-regulation. Organ. Behav. Hum. Decis. Process. **50**(2), 248–287 (1991)
5. Bhagwat, S.A., Kushalappa, C.G., Williams, P.H., Brown, N.D.: The role of in-formal protected areas in maintaining biodiversity in the western ghats of india. Ecol. Soc. **10**(1) (2005)
6. Braun, V., Clarke, V.: Using thematic analysis in psychology. Qual. Res. Psychol. **3**(2), 77–101 (2006)
7. Chandrasekhar, J., Shivdas, A.: The yagna spirit–new age business dynamism. Int. J. Bus. Emerg. Mark. **6**(1), 34–53 (2014)
8. Chandrashekara, U., Sankar, S.: Ecology and management of sacred groves in kerala, India. For. Ecol. Manage. **112**(1–2), 165–177 (1998)
9. Das, D., Balasubramanian, A.: The practice of traditional rituals in naga aradhana (snake worship): a case study on aadimoolam vetticode sree nagarajaswami temple in Kerala, India. In: SHS Web of Conferences, vol. 33 (2017)
10. Devall, B., Sessions, G., et al.: Deep ecology. Pojman (1985)
11. Dudley, N., Higgins-Zogib, L., Mansourian, S.: The links between protected areas, faiths, and sacred natural sites. Conserv. Biol. **23**(3), 568–577 (2009)
12. Dwivedi, O.P.: Dharmic ecology. In: Hinduism and Ecology. pp. 3–22 (2000)

13. Etikan, I., Musa, S.A., Alkassim, R.S., et al.: Comparison of convenience sampling and purposive sampling. Am. J. Theor. Appl. Stat. **5**(1), 1–4 (2016)
14. Fuentes, A.: Naturalcultural encounters in bali: monkeys, temples, tourists, and ethnoprimatology. Cult. Anthropol. **25**(4), 600–624 (2010)
15. George, S., Nayagam, J., Mani Varghese, K.: Fungal flora of tropical riverbanks: indicator for quality running water resources. J. Hydrol. **2**, 704–707 (2018)
16. Gooden, J.L.: Cultivating identity through private land conservation. People Nat. **1**(3), 362–375 (2019)
17. Gupta, N., Kanagavel, A., Dandekar, P., Dahanukar, N., Sivakumar, K., Mathur, V.B., Raghavan, R.: God's fishes: religion, culture and freshwater fish conservation in India. Oryx **50**(2), 244–249 (2016)
18. Hannah, S.T., Avolio, B.J., Walumbwa, F.O.: Relationships between authentic leadership, moral courage, and ethical and pro-social behaviors. Bus. Ethics Q. **21**(4), 555–578 (2011)
19. Hay, R.: Sense of place in developmental context. J. Environ. Psychol. **18**(1), 5–29 (1998)
20. Jain, P.: Dharma and Ecology of Hindu Communities: Sustenance and Sustainability. Routledge (2016)
21. Jung, C.G.: Analytical psychology. In: An Introduction to Theories of Personality, pp. 53–81. Psychology Press (2014)
22. Keizer, K., Schultz, P.W.: Social norms and pro-environmental behaviour. In: Environmental Psychology: An Introduction, pp. 179–188 (2018)
23. Kendall, J.: Axial coding and the grounded theory controversy. West. J. Nurs. Res. **21**(6), 743–757 (1999)
24. Klaniecki, K., Leventon, J., Abson, D.J.: Human–nature connectedness as a 'treatment' for pro-environmental behavior: making the case for spatial considerations. Sustain. Sci. **13**(5), 1375–1388 (2018)
25. Long, J.W., Lake, F.K., Goode, R.W., Burnette, B.M.: How traditional tribal perspectives influence ecosystem restoration. Ecopsychology **12**(2), 71–82 (2020)
26. Mayer, F.S., Frantz, C.M., Bruehlman-Senecal, E., Dolliver, K.: Why is nature beneficial? the role of connectedness to nature. Environ. Behav. **41**(5), 607–643 (2009)
27. Naiman, S.M., Allred, S.B., Stedman, R.C.: Protecting place, protecting nature: predicting place-protective behaviors among nature preserve visitors. J. Environ. Stud. Sci. **11**(4), 610–622 (2021)
28. Norton, B.G.: Environmental ethics and weak anthropocentrism. In: The Ethics of the Environment, pp. 333–350. Routledge (2017)
29. Notermans, C., Nugteren, A., Sunny, S.: The changing landscape of sacred groves in Kerala (India): a critical view on the role of religion in nature conservation. Religions **7**(4), 38 (2016)
30. Ramkissoon, H.: Perceived social impacts of tourism and quality-of-life: a new conceptual model. J. Sustain. Tour., 1–17 (2020)
31. Rosa, C.D., Profice, C.C., Collado, S.: Nature experiences and adults' self-reported pro-environmental behaviors: the role of connectedness to nature and childhood nature experiences. Front. Psychol. **9**, 1055 (2018)
32. Sitharam, T., Kolathayar, S., Yang, S., Krishnan, A.: Concept of a geotechnical solution to address the issues of sea water intrusion in ashtamudi lake, Kerala. In: Civil Infrastructures Confronting Severe Weathers and Climate Changes Conference, pp. 238–246. Springer (2018)
33. Stedman, R.C.: Is it really just a social construction?: the contribution of the physical environment to sense of place. Soc. Nat. Resour. **16**(8), 671–685 (2003)
34. Stern, P.C., Dietz, T.: The value basis of environmental concern. J. Soc. Issues **50**(3), 65–84 (1994)
35. Stern, P.C., Dietz, T., Abel, T., Guagnano, G.A., Kalof, L.: A value-belief-norm theory of support for social movements: the case of environmentalism. Hum. Ecol. Rev., 81–97 (1999)
36. Tripathi, M., Hughey, K.F., Rennie, H.: The role of sociocultural beliefs in sustainable resource management: a case study of traditional water harvesting systems in Kathmandu valley, Nepal (2018)

37. Trivedi, R.H., Patel, J.D., Savalia, J.R.: Pro-environmental behaviour, locus of control and willingness to pay for environmental friendly products. Mark. Intell. Plann. (2015)
38. Wang, H.F., Chiou, S.C.: Study on the sustainable development of human settlement space environment in traditional villages. Sustainability **11**(15), 4186 (2019)

Chapter 45
A Hybrid Convolutional Neural Network–Random Forest Model for Plant Disease Diagnosis

Lavika Goel and Jyoti Nagpal

Abstract One of the primary factors influencing agriculture crop yield production is the occurrence of plant diseases. Recent advancement in deep learning techniques has found their utilization in crop disease recognition which provides high accuracy. In this context, this paper represents a hybrid approach for an automatic plants disease classification model which utilizes a transfer learning neural network ensemble with a machine learning classifier. Transfer learning is an approved technique of deep learning where pre-trained architectures solve natural language processing problems or computer vision classification problems. This research work has performed a multi-class classification of distinct plant diseases on the plant Village dataset using VGG16 and ResNet50 (pre-trained deep learning models) ensembled with Random Forest Classifier. In this Hybrid approach, the last layers of the pre-trained deep learning neural networks are replaced by the Random Forest classifier, and a comparative analysis is performed with state-of-art algorithms, which depicts that Resnet50 ensembled with Random Forest provides an accuracy of 0.980. Also, comparing it with Traditional machine learning models, the accuracy for Random Forest on the same dataset is 0.927.

45.1 Introduction

For the sustainable development of any country, agriculture is considered one of the primary factors since it has a significant impact on society and the economy. Farmers suffer from the problem that they cannot detect the plant diseases at the correct time or the correct disease to treat the crop diseases on time; as a result, farmers suffer major economic losses. To aid decision-making in agriculture, self-activating computerized systems built on artificial intelligence approaches have been adduced in the literature

L. Goel (✉) · J. Nagpal
Malaviya National Institute of Technology, Jaipur, India
e-mail: lavika.cse@mnit.ac.in

J. Nagpal
e-mail: 2020rcp9583@mnit.ac.in

© The Author(s), under exclusive license to Springer Nature Singapore Pte Ltd. 2023
C. So-In et al. (eds.), *Information Systems for Intelligent Systems*, Smart Innovation, Systems and Technologies 324, https://doi.org/10.1007/978-981-19-7447-2_45

509

[1]. Specific image preprocessing tasks such as image enhancement, color or shape transformation, and image segmentation are required prior to extracting an efficient feature set [3]. Convolutional neural networks (CNN) are a genre of neural networks that is the basic building block of Deep learning. In contrast to its former models, the supreme feature of CNN is that it detects the features automatically without enumerating human supervision.

Various CNN models [9], such as LeNet, AlexNet, ZFNet, GoogLeNet, Visual Geometry Group16 (VGG16), VGG19, Residual Network (ResNet), XceptionNet, DenseNet, SqueezeNet have been used in literature for specific agricultural-based applications as described in Table 45.1. Hassan et al. [7] worked on transfer learning using a pre-trained model of ShallowVGG16 on Corn, Potato, and tomato leaf images. A Hybrid technique using shallow VGG with Random Forest and Shallow VGG with Xgboost is proposed to detect and classify plant diseases. The results depict that the hybrid of ShallowVGG with Xgboost provides better results while measuring on different parameters. The resultant accuracy of shallow VGG and Xgboost is 95.70%, while shallow VGG with Random Forest provides the accuracy of 91.68% in 10 epochs. Further in [13], the authors represented a hybrid approach of deep CNN with nature-inspired optimization. This work uses transfer learning pre-trained models, MobileNetv2 and DenseNet201, to obtain the feature vectors. After combining the feature vectors together from both the pre-trained models and optimizing using Whale Optimization Algorithm (WOA), it gives an accuracy of 95.7%, which is better while taking the independent feature from each model. In [10], authors applied a hybrid approach using a pre-trained VGG16 model with Multiclass SVM, which provides a classification accuracy of 99.4%. The innovative idea of this research is that the corresponding diseased crop Eggplant dataset was unavailable, so the dataset was created in both field and laboratory conditions for five major diseases. The main challenge of CNN is that it suffers from a wide range of hyperparameters in particular architectures. The authors in [8] utilize the orthogonal learning particle swarm optimization algorithm to optimize the parameters generated by VGG16 and VGG19 models, which provides better accuracy than InceptionV3 and Xception models.

45.2 Proposed Approach

Research shows that despite using a very large network with millions of parameters, shallow CNN can extract the feature information precisely. So, in this proposed approach, the Hybrid of CNN with machine learning classifier is applied to reduce the complexity and reduce the number of parameters. The proposed architecture is customized in two sections, first is the CNN external network for extracting the features, and the second is a conventional random forest classifier. This research work represents the fusion of deep learning, and machine learning approaches VGG16 + Random Forest, ResNet50 + Random Forest. This model's required input image size is 256 * 256 * 3. Figure 45.1 demonstrates the framework of the proposed

Table 45.1 Literature survey on the applications of the conventional deep learning models in plant disease diagnosis

Paper	Specific crop	Dataset taken	Deep learning (pre-trained) model used	Classification accuracy
[6]	14 crop-species	Plantvillage dataset	AlexNet, GoogleNet	Alexnet = 93.88% GoogleNet = 95.47%
[7]	Potato, corn, and tomato	Plantvillage and from the field	Shallow VGG as a Feature Extractor and Random Forest, and Xgboost as a Classifier	Shallow VGG + RF = 91.68% Shallow VGG + Xgboost = 95.70%
[8]	Maize leaf	Plantvillage dataset	Feature fusion VGG16 + VGG19 which are optimized by (Orthogonal learning PSO)	VGG16 + VGG19 + OLPSO = 98.2%
[10]	Eggplant	Own dataset (created In laboratory and field conditions)	VGG16 used as a feature extractor and multiclass SVM as a classifier	VGG16 + MSVM = 94%
[11]	Six different crops	Laboratory and field dataset	Dense convolutional neural network	Dense CNN = 99.19%
[13]	Citrus fruit	Own dataset acquired from fields	Feature vectors of DenseNet20, MobileNetv2 are optimized by Whale Optimization Algorithm (WOA)	DenseNet20 + MobileNetv2 + WOA = 95.7%

model. Initially, the input image dataset is preprocessed, including image transformation, enhancement, noise removal, etc. In the proposed work, the images after preprocessing are fed distinctly to the VGG16 and ResNet50 models, and feature sets obtained from both models are then fed to the Random Forest (machine learning classifier). The output of the preprocessed image is provided as an input to the CNN architecture. The CNN architectures extract the feature information using convolutional and pooling layers further, the flattened feature vector of the last convolutional block is provided as an input to the Random Forest classifier. Therefore, the overall cost of computation of this model is reduced compared to the Deep learning architectures. In this work, the implementation is performed on three types of crops: Apple, Corn, and Potatoes. The dataset of each crop is available online at the Plantvillage website [12]. The Apple dataset contains 1073 images of three categories, named apple_blackrot, apple_rust, apple_scab, apple_healthy. The dataset of corn consists of 1406 images of four kinds of diseases images namely, corn_grayleaf spot, corn_commonrust, corn_leafblight, corn_healthy, and Potato leaf dataset containing 790 images of 3

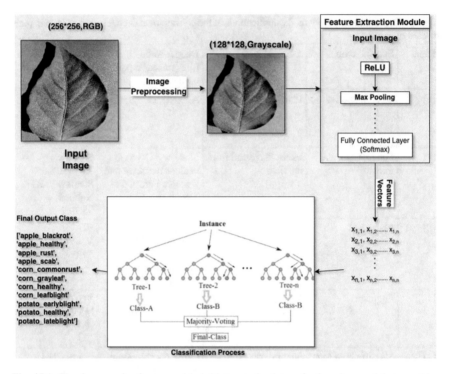

Fig. 45.1 The framework of proposed hybrid (pre-trained transfer learning model + machine learning) approach for plant diseases classification

categories which is potato_late blight, potato_early blight, and potato_healthy. Next, the pseudocode for the proposed model is described below.

45.2.1   Step by Step Procedure of Implementing Hybrid Model 1 [VGG-16 + Random Forest]

1. Loading the dataset.
2. Splitting datasets into Training [80% of Each Crop] and Testing [20% of Each Crop].
3. Train the VGG16 model by giving the input image dataset, which includes the following procedure.

   - The initial two convolutional layers of VGG16 are comprised of a total of 64 filters having size 3 * 3, which changes the output dimension by 224 * 224 * 64.
   - The resultant output is then fed to a max-pooling layer with filter size 2 * 2 and stride is 2, reducing the output dimensions to 112 * 112 * 64.

- The subsequent two convolutional layers comprise 124 filters having size 3 * 3, followed by a max pool layer of the same filter dimension as in the previous max-pooling layer, which outturn the output 56 * 56 * 128.
- The fifth to seventh convolutional layers uses 256 kernels, which is followed by the max pool layer, and the output dimensions become 28 * 28 * 256.
- The successive eight to thirteen layers are two distinct sets of convolutional layers having filter size 3 * 3, having 512 kernels, and followed by a max-pooling layer with the stride of 1.
- The output is then flattened, which becomes 7 * 7 * 512.

4. The final flatten output of VGG16 last convolutional block is fed as an input to the Random Forest Classifier.
5. Evaluate the performance measure based on the Confusion Matrix.

45.2.2   Step by Step Procedure of Implementing Hybrid Model 2 [ResNet50 + Random Forest]

1. Loading the dataset.
2. Splitting datasets into Training [80% of Each Crop], Testing [20% of Each Crop].
3. Train the ResNet50 model by giving the input image dataset, which includes the following procedure.

- Initially, it contains the Convolutional block having 64 filters with a filter size of 7 * 7, and a stride of 2. The output of the first layer is given its input to the next max-pooling layer having a stride of 2.
- In the next convolution block, there are 64 filters of 1 * 1, followed by 64, 3 * 3 filters, and the last 256, 1 * 1 filter. These three layers are replicated three times, given as nine layers.
- In the next convolutional block, the 128 filters of 1 * 1, followed by 128 filters of 3 * 3, and 512 filters of size 1 * 1, repeated four times, giving the output as 12 layers in this part.
- The next block consists of 256 filters of 1 * 1, two more 3 * 3 256 filters, and another 1 * 1, 1024 filters are repeated six times, providing 18 layers.
- In the last block, 512 filters of 1 * 1, two more 512 3 * 3 filters, and 2048, 1 * 1 repeated three times, provides a total of 9 layers that average pooling is performed.

4. The final flatten output feature vector of ResNet50 last convolutional block is fed as an input to the Random Forest Classifier.
5. Evaluate the performance measure based on the Confusion Matrix.

45.3   Results and Discussion

The experimental implementation from acquiring images to preprocessing and feature extraction using transfer learning networks is executed in Anaconda-python 3.6 using OpenCV and Keras Library. The experimental results including the comparison analysis of the traditional machine learning algorithms and detailed results of the proposed techniques are described in the following subsections.

45.3.1   Comparison of Machine Learning Classifier in Past Literature

The proposed methodology used a tenfold cross-validation technique to compare different machine learning models. Cross-validation is a technique to evaluate predictive models by dividing the global sample into training and testing separately. In tenfold cross-validation, the original sample is randomly divided into ten equal-sized subsamples. Out of the ten subsamples, one subsample is chosen as a validation set for testing, while the remaining nine sets are considered as train sets. This process is repeated ten times; each set becomes a validation set precisely once. Ten results are obtained from these ten processes, which are then averaged to produce a single result. Random forest performs best among all the machine learning classifiers with an accuracy of 0.899, followed by Logistic regression (0.797) and K-nearest neighbor (0.794). Figure 45.2 describes the comparative accuracy analysis of Logistic regression (LR), K-nearest neighbor (KNN), Naive Bayes (NB), Linear discriminant analysis (LDA), Classification and Regression Trees (CART), Random Forest (RF), and Support Vector Machine (SVM). The objective behind selecting the random forest classifier is that random forest performs best compared with the state-of-the-art machine learning classifiers. Figure 45.3 shows the precision, recall, and F1- score values of the Random Forest classifier for each iteration of 10 cross-validation.

45.3.2   Performance Evaluation of Proposed Hybrid Methodology

Figure 45.4 illustrates the graph between the proposed techniques and their corresponding validation accuracy, which shows that the proposed hybrid approaches ResNet50 + RF and VGG16 + RF provide an accuracy of 0.980 and 0.97, which is highest among the rest of the conventional machine learning and deep learning methods.

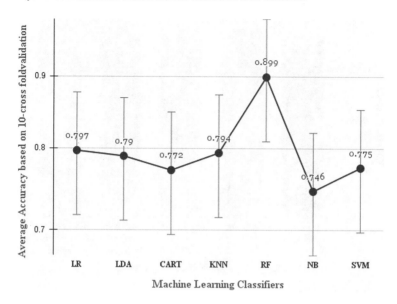

Fig. 45.2 Comparative analysis of machine learning classifier along with the average accuracy based on 10-cross fold validation

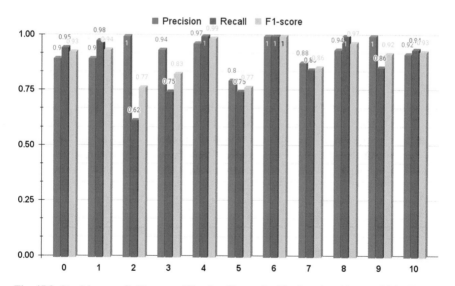

Fig. 45.3 Precision, recall, F1-score of Random Forest classifier based on 10-cross fold validation

45.4 Conclusion and Future Scope

This paper represents a plant leaf disease classification model based on the fusion of convolutional neural network and Random Forest classifier. The motivation behind

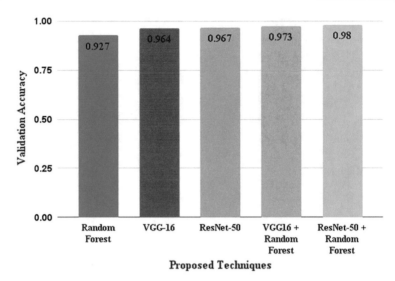

Fig. 45.4 Proposed techniques along with its validation accuracy

the fusion of deep learning and machine learning algorithms is that the deep learning algorithms outturn a large number of parameters, increasing the model complexity. In this hybrid approach, the number of parameters is reduced by replacing the last layer of the pre-trained convolutional neural network with the random forest classifier. A comparative analysis is performed on the machine learning approaches and the proposed Hybrid approaches. The experimental implementations shows that both hybrid models perform better than traditional machine learning algorithms. Among all the classifiers, the fusion of ResNet50 and RF provides the highest accuracy of 0.980 on the validation dataset. In the future, the approach can be improved by applying parameter selection algorithms to improve model performance. The proposed system can also be integrated with cloud-based systems in order to provide more efficient results. Another future aspect of this hybrid model is to test the model on a large dataset with high feature dimensionality, varying classes, and data sizes on real-world conditions.

References

1. Wani, J.A., Sharma, S., Muzamil, M., Ahmed, S., Sharma, S., Singh, S.: Machine learning and deep learning based computational techniques in automatic agricultural diseases detection: methodologies, applications, and challenges. Arch. Comput. Meth. Eng. **29**, 641–677 (2021)
2. Vishnoi, V.K., Kumar, K., Kumar, B.M: Plant disease detection using computational intelligence and image processing. J. Plant Dis. Protect. **128**, 19–53 (2020)
3. He, K., Zhang, X., Ren, S., Sun, J.: Deep residual learning for image recognition. In: 2016 IEEE Conference on Computer Vision and Pattern Recognition (CVPR), pp. 770–778 (2016)

4. Fieres, J., Schemmel, J., Meier, K.: Training convolutional networks of threshold neurons suited for low-power hardware implementation. In: 2006 IEEE International Joint Conference on Neural Network Proceedings, pp. 21–28 (2006)
5. Siyah, B.: Plant disease using Siamese network—Keras [Online]. Available: https://www.kaggle.com/bulentsiyah/plant-disease-using-siamese-network-keras (2020)
6. Mohanty, S.P., Hughes D.P., Salathé M.: Using deep learning for image-based plant disease detection. Front. Plant Sci. **7** (2016)
7. Hassan, S.K., Jasiński, M., Leonowicz, Z., et al.: Plant disease identification using shallow convolutional neural network. Agronomy **11**(12), 2388 (2021)
8. Darwish, A., Ezzat, D., Hassanien, A.: An optimized model based on convolutional neural networks and orthogonal learning particle swarm optimization algorithm for plant diseases diagnosis. Swarm Evol. Comput. **52** (2020)
9. Goncharov, P., Ososkov, G.A., Nechaevskiy, A., Uzhinskiy, A., Nestsiarenia, I.: Disease detection on the plant leaves by deep learning. In: Advances in Neural Computation, Machine Learning, and Cognitive Research II (2018)
10. Krishnaswamy Rangarajan, A., Raja, P.: Disease classification in eggplant using pre-trained VGG16 and MSVM. Sci. Rep. **10**(1), 2322 (2020)
11. Tiwari, V., Joshi, R.C., Dutta, M.K.: Dense convolutional neural networks based multiclass plant disease detection and classification using leaf images. Ecol. Inform. **63** (2021)
12. https://www.kaggle.com/emmarex/plantdisease
13. Rehman, Z., Ahmed, F., Khan, M., et al.: Classification of citrus plant diseases using deep transfer learning. Comput. Mater. Continua. **70**(1), 1401–1417 (2022)
14. Padol, P.B., Yadav, A.A.: SVM classifier based grape leaf disease detection. In: 2016 Conference on Advances in Signal Processing (CASP), pp. 175–179 (2016)
15. Joshi A.A., Jadhav, B.D.: Monitoring and controlling rice diseases using image processing techniques. In: 2016 International Conference on Computing, Analytics and Security Trends (CAST), pp. 471–476 (2016)
16. Vallabhajosyula, S., Sistla, V., Kolli, V.K.: Transfer learning-based deep ensemble neural network for plant leaf disease detection. J. Plant Dis. Protect., 1–14 (2021)
17. Patel, B., Sharaff, A.: Rice crop disease prediction using machine learning technique. Int. J. Agri. Environ. Inform. Syst. **12**(4), 1–15 (2021)
18. Sethy, P.K., Barpanda, N.K., Rath, A., Behera, S.: Deep feature-based rice leaf disease identification using support vector machine. Comput. Electron. Agric. **175**(1), 105527 (2020)

Chapter 46
Voice Data-Mining on Audio from Audio and Video Clips

A. Sai Tharun, K. Dhivakar, and R. Nair Prashant

Abstract The lockdown due to Covid-19 has resulted in us relying heavily on technology to communicate and see each other. That, in turn, has led to an explosion of applications that provide audio and video conferencing solutions; which, in turn, has resulted in the generation of vast volumes of visual and auditory data. This ocean of data is mainly untapped and has vast potential to help us build applications that can one day hope to, truly, assist people. All this data, particularly auditory data, can be used to build applications that can support us by communicating with us and understanding what we say, making it comfortable and closer to humans. This data can also be processed to give us more insights into the behaviour of humans and what makes us truly us. All this and many more insights, which are waiting to be tapped, can be garnered from this goldmine of data, thereby allowing us to bridge the huge gap between man and machine.

46.1 Introduction

We are being overflown with a huge amount of data every day and given the current lifestyle, even basic human interactions happen via a digital medium through video/audio conferencing applications like Teams, Zoom, etc. This audio data is something a computer cannot natively understand and hence difficult to gather insightful information that can be used in a lot of other places to make this new normal even better. One such example where it can be used is for missed meetings and sessions, going through every one of those missed sessions is time-consuming

A. Sai Tharun (✉) · K. Dhivakar · R. Nair Prashant
Department of Computer Science and Engineering Amrita School of Engineering,
Amrita Vishwa Vidyapeetham, Coimbatore, India
e-mail: cb.en.u4cse18250@cb.students.amrita.edu

K. Dhivakar
e-mail: cb.en.u4cse18314@cb.students.amrita.edu

R. Nair Prashant
e-mail: prashant@amrita.edu

and would help if we can get some key insights on what transpired without having to watch the entire meeting again.

The gathering of this information can be achieved by using a technique known as Voice data mining, which is the process of extracting useful information from voice recordings. This has many benefits, such as, it can help organizations to better understand their customers, and employees, and improve their products and services. With the increasing use of voice-enabled devices and applications, there is a growing need for efficient methods of extracting information from voice.

Thus, we intend to build a comprehensive solution consisting of a pipeline with various ML models, trained with the help of data from various data sets and processes of various algorithms, which can be used to mine data from audio, where the audio can be from a video or can just be an audio sample, thereby being able to provide useful and workable insights, which can then be shown to the user or used to assist the users in various other frontiers.

46.2   Architecture Diagram

See Fig. 46.1.

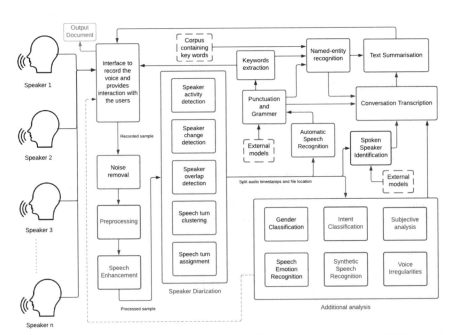

Fig. 46.1 Architecture diagram depicting the process flow

46.3 Proposed System

46.3.1 Noise Removal

This module attempts to separate noise from the given audio sample. It is capable of reducing noise in the audio sample by sampling the noise and filtering out the noise component per sample, using a technique known as spectral grating. This is seen as the most effective method to reduce noise and has been the preferred method of noise reduction in the audio industry [1]. The algorithm is as in [2].

46.3.2 Voice Activity, Speaker Change, Speech Overlap Detection and Embedding Extraction

(VAD), also known as speech detection, aims to detect the presence or absence of speech and differentiates a speech section in the given audio from the non-speech sections. Used in a variety of speech-based applications, especially in speech recognition and speech coding.

In Speaker Change Detection (SAD) we find different speakers by finding the cosine distance between waveforms and in Speech Overlap Detection we find if conversations overlap by checking how the mean audio waveform is different from each other in the neighbourhood.

Audio signals are one of the trickiest data for any machine learning method to process. Representing an audio signal in a way that can be readily processed by a machine learning algorithm is challenging. Usually, this is done through high dimension data in the vector space.

So to effectively process them in the simplest way possible we need to convert this data into embeddings. We choose d-vector embeddings [3, 4] as they seemed right for our use case (Fig. 46.2).

46.3.3 Speaker Diarization

Speaker diarization is the process of automatically identifying who spoke when in an audio recording [5, 6]. It is a challenging problem because speakers can overlap in conversation, and there can be background noise that makes it difficult to hear individual voices. It can be used for a variety of applications, such as meeting transcription, call centre monitoring, and automatic identification of speakers in video recordings.

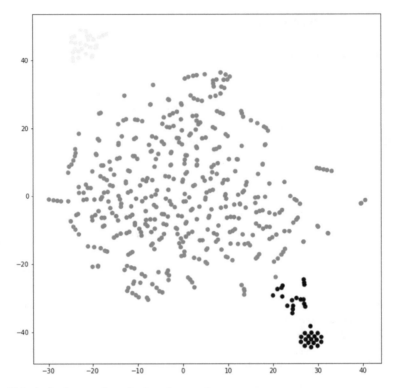

Fig. 46.2 Audio features clustering based on speaker

The main approaches are *bottom-up* and *top-down* [7]. We opted for a top-down approach called speaker diarization with (LSTM) [8]. (LSTM) speaker diarization is the process of using an LSTM recurrent neural network to identify speakers in a given audio recording. The accuracy of the diarization depends on many factors, including the quality of the audio signal, the number of speakers in the recording, and the duration of the recording.

The input to the network is typically a sequence of (MFCC)s extracted from the audio signal. The network is trained on a labelled dataset of speech utterances, where each utterance is associated with a speaker ID. The output of the network is a sequence of speaker labels, one for each time frame in the input (MFCC) sequence.

The system first extracts a set of acoustic features from the audio signal that is representative of the speaker's voice. Using (SAD) and SCD we get the intersection of these points, and we get the accurate point where speech turns occur. Using the embedding (D Vector representations) from these segments we can perform clustering of the data. These clusters then correspond to each segment of the speaker.

46.3.4 Voice Gender Identification Module

This module is capable of identifying the gender of a speaker, given we can provide the trained voice model for both the male and female speakers beforehand, with reasonable accuracy [9, 10]. The module performs exceptionally well in identifying female speakers while taking a slight hit on the performance in identifying male speakers using both the pre-trained gender models.

The module works on the principle of (GMM), a type of machine learning algorithm used to classify data into many different categories based on the probability distribution. To create the (GMM) we extract the (MFCC)s of the audio file and pass it on as the input for the (GMM). (MFCC)s are some transformed values of the signal in the cepstral domain. Speech is produced by humans by applying a filter, by our vocal tract, and so the properties of the vocal tract are what are responsible for giving shape to the spectrum of the signal. These properties are what govern the type of sound produced and MFCCs features can be used to best represent this shape.

This algorithm when trained on a dataset consisting of 5 distinct male speakers and 5 distinct female speakers with 1 min of speech per speaker (about 10 mins of speech data) can give us an accuracy of about 84% (Fig. 46.3). The accuracy in detecting female speech is about 95 % while for the male counterpart it drops to about 73% (Tables 46.1 and 46.2).

```
-0CAdy06NRo.wav
        detected as -  male
        scores:female  -19.82327929396574 ,male  -19.504943102943518

-0DO0ulATPY.wav
        detected as -  female
        scores:female  -20.235419774177824 ,male  -20.702589387109185

-0kDcUEDfmY.wav
        detected as -  female
        scores:female  -19.127272835057735 ,male  -19.13024382455876

-0l8Gjpj4Fs.wav
        detected as -  male
        scores:female  -19.831754648246886 ,male  -19.468401142807398
```

Fig. 46.3 Gender identification—model training for male

Table 46.1 Gender identification—classification report

	Male	Female	Total
Male	398	148	546
Female	27	531	558

Table 46.2 Gender identification—results

Male model accuracy	72.89%
Female model accuracy	95.16%
Overall accuracy	84.03%

46.3.5 Speech Emotion Recognition

(SER) is the task of classifying the emotion conveyed by a spoken utterance into one of a set of predefined categories [11]. The most common set of categories used in (SER) research is the six basic emotions [12], namely happiness, sadness, anger, fear, surprise, and disgust.

(SER) could be used to provide feedback to callers in customer service, monitor employees' emotional states, automatically analyse the emotions conveyed in speeches or provide input to affective computing applications such as virtual agents and intelligent tutoring systems. It is a challenging task.

1. Acoustic speech signal contains a lot of information not related to the emotions being conveyed (e.g. words spoken, intonation, speaking rate, etc.).
2. The same emotion can be conveyed in different ways by different people, and even by the same person at different times. For example, the same anger can be expressed with different degrees of intensity and can be accompanied by different facial expressions, body language, and vocal qualities.

The first step in any SER system is the extraction of low-level and high-level features from the speech signal (Table 46.3).

Once the features are extracted, they can be used to train an ML model for (SER) (Table 46.4). The most common ML approach to (SER) is to use a (SVM).

Table 46.3 Low-level and high-level features

Low-level features	High-level features
Typically hand-crafted by experts and designed to capture information about the acoustic properties of the speech signal relevant for emotion recognition	Typically extracted from the speech signal using (ASR) or (NLP) techniques
For example, spectral features such as the (MFCC)s capture the spectral envelope of the speech signal, while prosodic features such as pitch and speaking rate capture information about the intonation of the speech [13]	For example, lexical features such as the number of words and the number of different words can be extracted using (ASR), while syntactic features such as the number of clauses and the number of different clause types using NLP

Table 46.4 Speech emotion recognition results

Male model training accuracy	99.79%
Male model testing accuracy	89.06%
Female model training accuracy	99.98%
Female model testing accuracy	94.51%
Overall training accuracy	97.87%
Overall testing accuracy	88.06%

Other machine learning algorithms that have been used for (SER) include decision trees, k-nearest neighbours (KNN), Convolutional Neural Network (CNN) [14] and neural networks (NNs) [15].

46.3.6   Speech to Text—Automatic Speech Recognition (ASR)

Speech to Text conversion is the bedrock of this project. Not only did it require a fast and simple method to convert speech-to-text but also required an authentic conversion. Initially, we explored many methods, ranging from (HMM) models to deep feed-forward and recurrent neural networks.

We used End-to-end Automatic Speech Recognition(ASR). Unlike traditional methods which mostly were phonetic-based where it mandated the need for different unique modules and models which each focussed on pronunciation, acoustic representation, and language model.

The (ASR) system we picked up was from Facebook called wav2vec2. This approach has several advantages over the more traditional methods that are based on the analysis of spectrograms.

1. It can directly learn from the raw waveforms, thereby does not require any hand-crafted features or preprocessing. This is important because the quality of the features that can be extracted from the waveforms is often limited by the signal-to-noise ratio.
2. It can capture a wide range of temporal information. This is because the deep learning models that are used can learn from long sequences of waveforms.
3. It is scalable, efficient, robust, and resistant to overfitting while also generalizing well to new data.

46.3.7   Spoken Speaker Identification Module

This module attempts to match the given audio sample with the speaker, who spoke the audio. It works on the principle of (GMM)s and (MFCC)s and is capable of

```
anthonyschaller-20071221-\wav\a0494.wav
anthonyschaller-20071221-\wav\a0495.wav
modeling completed for speaker: anthonyschaller.gmm with data point = (1595, 40)
Apple_Eater-20091026-yul\wav\a0052.wav
Apple_Eater-20091026-yul\wav\a0053.wav
Apple_Eater-20091026-yul\wav\a0054.wav
Apple_Eater-20091026-yul\wav\a0055.wav
Apple_Eater-20091026-yul\wav\a0056.wav
modeling completed for speaker: Apple_Eater.gmm with data point = (2077, 40)
Ara-20091020-vni\wav\a0517.wav
Ara-20091020-vni\wav\a0518.wav
Ara-20091020-vni\wav\a0519.wav
Ara-20091020-vni\wav\a0520.wav
Ara-20091020-vni\wav\a0521.wav
modeling completed for speaker: Ara.gmm with data point = (3644, 40)
```

Fig. 46.4 Spoken speaker training

Fig. 46.5 Performing
spoken speaker identification

```
anthonyschaller-20071221-\wav\a0500.wav
-23.607237149278646
                detected as -  anthonyschaller
Apple_Eater-20091026-yul\wav\a0057.wav
-25.180983346663556
Speaker undetected
Apple_Eater-20091026-yul\wav\a0058.wav
-22.558332218193616
                detected as -  Apple Eater
Ara-20091020-vni\wav\a0522.wav
-20.44295855644294
                detected as -  Ara
```

Table 46.5 Spoken speaker identification—results

Undetected	8.82%
No. of wrong predictions	2%
Error	1.18%
Accuracy	98.82%

identifying the speaker, given we can provide the trained voice model of the speaker beforehand, with reasonable accuracy.

To build this module we have extracted 40-dimensional features from the audio of a speaker, of these 20 are extracted (MFCC) features while the other 20 are (MFCC) features derived from the first 20 extracted ones (Figs. 46.4 and 46.5).

This algorithm when trained on a dataset consisting of 34 distinct speakers with 5 audio clips per speaker (about 510 s (34 * 5 * 3 secs per recording (average)) of speech data) can give us an accuracy of about 90% (Table 46.5).

46.3.8 Punctuation and Grammar

One of the requirements, for an (ASR) System, is Punctuation Restoration and Grammatical Correction. The reason for Punctuation Restoration is straightforward, the speech data does not contain the intricacies of punctuation, people invariably speak in phrases rather than sentences and also miss use the use of complex sentences when talking. Text without punctuation might be meaningless. This means the text generated from (ASR) is fine for short sentences but will not be legible as the number of sentences increases.

Punctuation can not be restored from audio and so we restored it by using a linguistic approach, i.e. from its text equivalent (unsegmented textual content (Fig. 46.6)). To do so, we employed an (RNN) model, which is a bidirectional (LSTM), to restore inter-word missing punctuations (Fig. 46.7).

Grammatical correction is paramount for (ASR), but only because it acts as a nice guard to correct spelling and grammar. It also adds in punctuation, typically the ones that punctuation restoration can not solve like a colon, as well.

46.3.9 Keyword Extraction

It is a process of automatically identifying the most important keywords in a document. This is typically done by analysing the document's text and extracting the most relevant and important terms. We explored some popular options like Textrank [16] and KeyBert [17]. We used KeyBert to get the potential keywords and then used POS tagging to get noun-based keywords (Fig. 46.8).

46.3.10 Named Entity Recognition

This module is capable of extracting/highlighting, in the given textual content, all the named entities like names of persons, organizations, locations, expressions of

```
PACT IN U S EDUCATION AND GLOBAL HEALTH WITH GLOBAL HEALTH BECAUSE OF SOME NEW VACCINESA THAT WE'VE BEEN INVOLVED WITH AND TH
OSE E'VE REALLY GOTTEN OUT WE INOW SWUNG FOR THE FENCES THERE AND ACTUALLY A IT SUCCEEDED EDUCATION'S BEEN A LOT TOUGHER WHERE
A ALTHOUGH WE'VE DONE SOME GOOD CURRICULUM AND DONE SOME THINGS FOR TUTORING AND KIDS GETTING FEED BACK YOU KNOW WE DON'T HAVE
A THE DROP OUT RAT GRADUATION PREPAREDNESS STILL YOU KNOW THE SYSTEM IS NOT DRAMATICALLY IMPROVED AND NOWHERE NEAR AS GOOD AS W
E WANT TO BE I FEL LIKE YOU ALWAYS SEEM TO SWING FOR THE FENCES IN GENERAL IS JU SORT OF A NEW WAY OF FRAMING IT BUT I HAVE TO
```

Fig. 46.6 Before models being applied

```
"Pact in u, s, education and global health with global health, because of some new vaccinesa that we've been involved with. And
those e've really gotten out. We inow swung for the fences there and actually a it succeeded. Education's been a lot tougher, w
here a although we've done some good curriculum and done some things for tutoring and kids getting feed back. You know, we do
n't have a the drop out rat graduation preparedness. Still, you know, the system is not dramatically improved and nowhere near
as good as we want to be. I fel, like you always seem to swing for the fences in general is ju sort of a new way of framing it.
```

Fig. 46.7 After models being applied

```
Das : Hi and welcome to the a16z podcast. I'm Das, and in this episode, I talk SaaS
go-to-market with David Ulevitch and our newest enterprise general partner Kristina Shen. The
first half of the podcast looks at how remote work impacts the SaaS go-to-market and what the
smartest founders are doing to survive the current crisis. The second half covers pricing
approaches and strategy, including how to think about free versus paid trials and navigating
the transition to larger accounts. But we start with why it's easier to move upmarket than
down… and the advantage that gives a SaaS startup against incumbents.David : If you have a
cohort of customers that are paying you $10,000 a year for your product, you're going to find
```

Fig. 46.8 Keywords generated from KeyBert

```
[(European, 'B', 'NORP'), (authorities, 'O', ''), (fined, 'O', ''), (Google, 'B', 'ORG'), (a, 'B', 'MONEY'), (record, 'I', 'MON
EY'), ($, 'I', 'MONEY'), (5.1, 'I', 'MONEY'), (billion, 'I', 'MONEY'), (on, 'O', ''), (Wednesday, 'B', 'DATE'), (for, 'O', ''),
(abusing, 'O', ''), (its, 'O', ''), (power, 'O', ''), (in, 'O', ''), (the, 'O', ''), (mobile, 'O', ''), (phone, 'O', ''), (mark
et, 'O', ''), (and, 'O', ''), (ordered, 'O', ''), (the, 'O', ''), (company, 'O', ''), (to, 'O', ''), (alter, 'O', ''), (its,
'O', ''), (practices, 'O', ''))]
```

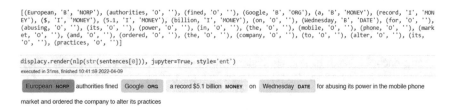

Fig. 46.9 Sample named entities

times, etc. Beyond regular named entities we can also extract/highlight words of our choice by providing our corpus. The algorithm for this process is

1. Apply word tokenization and part-of-speech tagging to the sentence/paragraph.
2. Apply verbatim, extract part-of-speech and lemmatize this sentence.
3. Extract or highlight the necessary tokens for further analysis.
4. Visualize the chosen tokens and how they are built up (optional) (Fig. 46.9).

46.3.11   Conversation Transcription

Conversation transcription is the process of converting long audio sequences into written text. This can be done in real-time, or after the fact. A transcription is a valuable tool, as it allows them to capture the content of meetings and other verbal interactions. It can also be used for things like interviews, focus groups, and phone calls.

Our approach for this particular module is to aggregate the previous models' results and try to present the text simply and legibly. To do so, we opted to have a script-inspired format to present the speech logs with the timestamp, along with its classification results. This is combined with a simple downloading mechanism which cuts the bigger audio clip into its components and also prepares a PDF document with similar details as in the interface.

Summary: 1

Martin Luther King's dream was that people would not be judged by the color of their skin, but by the content of their charact
er. John Peterson believes that identity politics is predicated on the idea that the appropriate way to classify people is by
their group identity. The issue is whether the individual identity is primary and the group identity is secondary. Jordan asks
if there's evidence of structural inequality and oppression, why not have a conversation about having women represented in all
professions at all levels?

Fig. 46.10 Summary text

46.3.12   Text Summarization

Text summarization [18] is the process of reducing a text document to create a summary that retains the most important points of the original document. This is done by extracting key sentences and creating a shorter version of the text that still conveys the most important information [19, 20].

There are two main types of summarization: *extractive* [17] and *abstractive* [21]. BART [22] is a statistical model, based on the idea of Bayesian inference (a method of statistical inference that uses Bayes' theorem to update the probabilities of events) that can be used for both regression and classification.

We used a Bart model optimized for Summarization [23], which had a limitation from a transcription perspective. The model was optimized for a general document summarization rather than summarization for transcription in a script-like format. We enhanced the model because we felt that it was sufficient to summarize (Fig. 46.10) the content, the only thing it wasn't aware of was the possibility of more than one speaker and the difference between first-person and third-person. This was also because other multi-speaker-aware Summarization techniques reduced the content a lot and gave an incomplete summary.

We used the SAMSUM dataset which contains audio/text transcripts along with message logs. Our Bart model was enhanced by training with it, and was optimized to give better results by the target summary in the dataset (Fig. 46.11).

46.4   Results and Discussion

1. The Open-source APIs for speaker diarization were not accurate enough for our analysis and abstract of the entire process through which the diarization is occurring, due to which we built our own custom pipeline.
2. For the speech-to-text model we have used an existing model (accurate one available) named Wav2vec since the model we tried to build did not give a good result due to insufficient training. The results of these models have been verified with the means of (WER). Out of the available models, we choose Wav2Vec2-large-960h-self due to the balance between speed and efficiency.
3. In Speech turn clustering, we optimize speaker count by providing approximate speaker count since it helps us to compute a distance threshold from the dendrogram and support the Agglomerative clustering process. Classifying overlapped

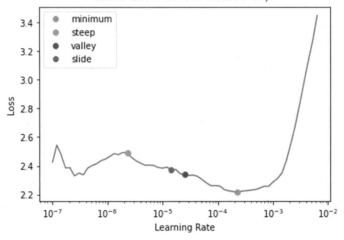

Fig. 46.11 Noise removal in model training

speakers into their own speech clusters, so that we would still get the information out of the conversation in overlapped speech.

4. The outputs that come out of the box from (ASR) require punctuation restoration to make them more readable. However, the model we used still does not identify more obscure punctuation like colons, semicolons, etc. We further employed grammar and spelling correction, which in turn added the extra punctuation we were looking for.

5. Restored punctuation was not indicative of the way the speaker meant it to sound, i.e. sometimes even questions get terminated with a period, which implies various prosodic features could have been employed within the context of the model to get better results.

6. The accuracy of the gender identification module is higher for females at 93% when compared to males at 73%. The major reason for this could be the bass, given that a male voice has a higher bass than its female counterpart and is not being recorded properly by the recording devices [24].

7. Gender identification module categorizes speech into two binary categories [25] rather than diverse categories in the gender spectrum. Just as how humans can not make a distinction between these categories, it's very hard for a system to make the same. So we used new terms such as Sounds-Like-Male(SLM) and Sounds-Like-Female(SLF).

8. With keyword generation, we realized that at times the word returned as a keyword is a probabilistic measure, therefore we used (POS) tagging to identify the nouns that happen to be the keywords we need.

```
[ 00:00:00.010 --> 00:00:16.630]

[ Sounds Like Male ] [ fear ]
Speaker 2 : Martin luther king's dream was that there would come a time when people would not be judged by the color of their
skin, but by the content of their character. How is today's identity politics consistent with that vision,
```

    ▶ 0:00 / 0:15 ──────── ◀) ⋮

```
[Pause] 1 second(s)

[ 00:00:19.159 --> 00:02:13.992]

[ Sounds Like Male ] [ happy ]
Speaker 1 : I don't think it's consistent with that vision. At all I mean the problem I have with identity politics, as as a m
ode of philosophical apprehension, is that it's predicated on the idea that the appropriate way to classify people is by their
group identity in whatever fragmentary formulation that might take in the multiplicity of ways that people can be divided into
groups and tha, the classical post, modern, and, I would also say, mark's way of viewing the world, even though those two thing
s shouldn't be allowed together, they tend to be is that group identity takes priority over individual identity and I think tha
t's precisely the opposite of what martin luther king was hoping for and working for, and I think it's unbelievably dangerous,
because, partly because when you, when you assume that people should primarily be identified by their group, then you can also
attribute group guilt to them by their group, and then things go downhill very very rapidly, and we've had no shortage of evide
nce of that sort of thing happening, say throughout the twentieth century. The particular groups that you are more concerned ab
out than others, for example the liberal party as terry battlesadila- is a group other group that you think odd it is. There is
n't a problem with groups. The problem is assuming that the fundamental way that you should categorize people is with their gro
up identity. Obviously we all belong to groups. The issue is whether not the individual identity is primary and the group ident
ity is secondary or the group identity is primary and the individual identity is secondary. If you're a proponent, for example,
of equality, of outcome of quotas, then you de facto accept the proposition that it's the group identity. That is primary and t
here's all sorts of dangers that are associated with that. That far outweighs whatever good you're likely to do.
```

    ▶ 0:00 / 1:54 ──────── ◀) ⋮

```
[Pause] 5 second(s)

[ 00:04:02.782 --> 00:04:03.732]

[ Sounds Like Female ] [ angry ]
[NOISE/Overlapping Conversations]
```

    ▶ 0:00 / 0:00 ──────── ◀) ⋮

    Download Files

    -

```
Duration of Audio:              253.77092970521542
Time Taken for Complete Process: 512.0224027633667
```

Fig. 46.12 Transcription report

9. To identify a given speaker, their pre-trained voice model is required, which establishes a baseline to compare with. Regardless, training does not require huge audio data, a dataset ranging from 3 to 5 recordings (2–3 mins per recording) is enough to train a model with an accuracy of about 80%.

10. Rather than finding emotion for the source audio in (SER), a more effective method seemed to be uncovering the emotions for its augmented forms and then finding the common ones. However, where they disregard emotions as noise since they do not enhance the speech-to-text process and do not further any knowledge or understanding [26].

11. Summarization performs well enough if a proper context is given. With larger clips and more conversations, it accurately picks summaries from key points and thereby giving multi-speaker-aware summaries.

12. The transcribed sentences can then be printed as a PDF with proper formatting for the convenience of the user (Fig. 46.12).

Table 46.6 Time taken for end to end processing

Duration of audio (in s)	120.0	254	402	604	650	724
Time taken (in s)	190.8	512	1540.3	1037.8	1052.3	1122.9

46.5 Conclusion

By creating a pipeline, and defining the process, we were able to mine valuable insights from a given audio data. Various insights such as 'who spoke what and when?', 'Name and gender of the speaker', 'The emotional state under which the said content is spoken', 'Summary of the conversation', etc., have been extracted and presented to the user in the form of a neat PDF transcription document.

From testing (Table 46.6) we were able to conclude that the costliest operation is the (ASR) 46.3.6 which has a complexity of **O(m*n)** where m is the number of audio segments and n is the length of the longest audio segment in seconds. For the 402 secs audio the time taken was huge since it had an audio segment which was about 2 mins long, while the rest had all their audio segments below 45 secs.

The solution is far from perfect and further improvements can be done. A few places where the solution becomes questionable is

1. Quality of the device being used to record plays a huge role in determining the effectiveness of the analysis.
2. Punctuation restoration model we build is not so accurate in identifying more obscure punctuation, which can in turn impact the processes that depend on its output.
3. The solution capable of running locally, is still computationally intensive, due to which it does not run as expected in a low-resource device.
4. To identify the spoken speaker we would need the trained model of the speaker, obtaining which has privacy constraints.
5. Noise removal works with consistent sounds, and if the noise captured is inconsistent we will need to further pre-process it, which introduces an additional overhang on the system.

The entire code base with the process is available in the repository [27].

46.6 Future Enhancement

1. Preprocessing and speech enhancement steps defined in the modules are hard to implement as they are very dependent on current and future studies. Moreover, the current open-source publications go for resource-intensive methods which could not be used in the real world.

2. In (SER) a new dataset can be prepared with accurate labels for its corresponding audio clips along with audio of varying duration. In many places, the emotion is mislabeled or can also be an altogether new emotion.

3. A better RNN punctuator can be made by including contextual and prosodic information into its scope of work.

4. Synthetic speech recognition [28] is still a new field of study and would require huge exploration and additional studies to build an effective dataset and some fundamental ideas can be formalized like its preprocessing steps.

5. For the sentiment [29] and intent classification, and subjective analysis modules, we had to prepare a dataset with the labels we were looking for. An extensive literature survey should have been done that could have been its own body of study.

6. Future enhancements can also include employing federated learning [30].

References

1. Li, B.: A principal component analysis approach to noise removal for speech denoising. In: 2018 International Conference on Virtual Reality and Intelligent Systems (ICVRIS) (2018). https://doi.org/10.1109/icvris.2018.00111

2. Sainburg, T.: Noise reduction using spectral gating in python. https://timsainburg.com/noise-reduction-python.html

3. Wang, S., Qian, Y., Yu, K.: What does the speaker embedding encode? In: Interspeech, pp. 1497–1501 (2017)

4. Wang, Q., Downey, C., Wan, L., Mansfield, P.A., Moreno, I.L.: Speaker diarization with lSTM. In: 2018 IEEE International Conference on Acoustics, Speech and Signal Processing (ICASSP), pp. 5239–5243. IEEE (2018)

5. Wang, C., Tang, Y., Ma, X., Wu, A., Okhonko, D., Pino, J.: fairseq s2t: fast speech-to-text modeling with fairseq. arXiv preprint arXiv:2010.05171 (2020)

6. Gonina, E., Friedland, G., Cook, H., Keutzer, K.: Fast speaker diarization using a high-level scripting language. In: 2011 IEEE Workshop on Automatic Speech Recognition & Understanding, pp. 553–558. IEEE (2011)

7. Evans, N., Bozonnet, S., Wang, D., Fredouille, C., Troncy, R.: A comparative study of bottom-up and top-down approaches to speaker diarization. IEEE Trans. Audio Speech Lang. Process. **20**(2), 382–392 (2012)

8. Kumar, A.G., Sindhu, M., Kumar, S.S.: Deep neural network based hierarchical control of residential microgrid using lSTM. In: TENCON 2019—2019 IEEE Region 10 Conference (TENCON) (2019). https://doi.org/10.1109/tencon.2019.8929525

9. Kumar, S., Gornale, S.S., Siddalingappa, R., Mane, A.: Gender classification based on online signature features using machine learning techniques. Int. J. Intell. Syst. Appl. Eng. **10**(2), 260–268 (2022). https://ijisae.org/index.php/IJISAE/article/view/2020

10. Buyukyilmaz, M., Cibikdiken, A.O.: Voice gender recognition using deep learning. In: Proceedings of 2016 International Conference on Modeling, Simulation and Optimization Technologies and Applications (MSOTA2016) (2016). https://doi.org/10.2991/msota-16.2016.90

11. Sezgin, M.C., Gunsel, B., Kurt, G.K.: Perceptual audio features for emotion detection. EURASIP J. Audio Speech Music Process. **2012**(1), 1–21 (2012). https://doi.org/10.1186/1687-4722-2012-16

12. Ekman, P., Friesen, W.V.: Constants across cultures in the face and emotion. J. Pers. Soc. Psychol. **17**(2), 124–129 (1971). https://doi.org/10.1037/h0030377

13. Kumaran, U., Radha Rammohan, S., Nagarajan, S.M., Prathik, A.: Fusion of mel and gamma-tone frequency cepstral coefficients for speech emotion recognition using deep C-RNN. Int. J. Speech Technol. **24**(2), 303–314 (2021). https://doi.org/10.1007/s10772-020-09792-x

14. Trigeorgis, G., Ringeval, F., Brueckner, R., Marchi, E., Nicolaou, M.A., Schuller, B., Zafeiriou, S.: Adieu features? end-to-end speech emotion recognition using a deep convolutional recurrent network. In: 2016 IEEE International Conference on Acoustics, Speech and Signal Processing (ICASSP) (2016). https://doi.org/10.1109/icassp.2016.7472669

15. Karthiga, M., Sountharrajan, S., Suganya, E., Sankarananth, S.: Sentence semantic similarity model using convolutional neural networks. EAI Endorsed Trans. Energy Web **8**(35), e8 (2021)

16. Bird, S., Klein, E., Loper, E.: Natural language processing with Python: analyzing text with the natural language toolkit. O'Reilly (2009)

17. Liu, Y.: Fine-tune BERT for extractive summarization. arXiv preprint arXiv:1903.10318 (2019)

18. Lalithamani, N.: Text summarization. J. Adv. Res. Dyn. Control Syst. **10**(3), 1368–1372 (2018)

19. Raj, D., Geetha, M.: A trigraph based centrality approach towards text summarization. In: 2018 International Conference on Communication and Signal Processing (ICCSP) (2018). https://doi.org/10.1109/iccsp.2018.8524528

20. Mohan, G.B., Kumar, R.P.: A comprehensive survey on topic modeling in text summarization. In: Micro-Electronics and Telecommunication Engineering, pp. 231–240 (2022). https://doi.org/10.1007/978-981-16-8721-1_22

21. Gupta, S., Gupta, S.K.: Abstractive summarization: an overview of the state of the art. Expert Syst. Appl. **121**, 49–65 (2019)

22. Chipman, H.A., George, E.I., McCulloch, R.E.: BART: Bayesian additive regression trees. Ann. Appl. Stat. **4**(1), 266–298 (2010)

23. Lewis, M., Liu, Y., Goyal, N., Ghazvininejad, M., Mohamed, A., Levy, O., Stoyanov, V., Zettlemoyer, L.: BART: denoising sequence-to-sequence pre-training for natural language generation, translation, and comprehension. CoRR abs/1910.13461 (2019). arXiv:1910.13461

24. Shankara, R.S., Raghaveni, J., Pravallika, R., Sravya, Y.V.: Classification of gender by voice recognition using machine learning algorithms. Int. J. Adv. Sci. Technol. **29**(06), 8083–8098 (2020). http://sersc.org/journals/index.php/IJAST/article/view/25200

25. Pondhu, L.N., Kummari, G.: Performance analysis of machine learning algorithms for gender classification. In: 2018 Second International Conference on Inventive Communication and Computational Technologies (ICICCT) (2018). https://doi.org/10.1109/icicct.2018.8473192

26. Catania, F., Crovari, P., Spitale, M., Garzotto, F.: Automatic speech recognition: Do emotions matter? In: 2019 IEEE International Conference on Conversational Data & Knowledge Engineering (CDKE) (2019). https://doi.org/10.1109/cdke46621.2019.00009

27. Dhivakar, K., Tharun, A.S.: Voice-data-mining. https://github.com/DhivakarK-git/Voice-Data-Mining

28. Singh, A.K., Singh, P.: Detection of AI-synthesized speech using cepstral & bispectral statistics. In: 2021 IEEE 4th International Conference on Multimedia Information Processing and Retrieval (MIPR) (2021). https://doi.org/10.1109/mipr51284.2021.00076

29. Sreevidya, P., Murthy, O.R., Veni, S.: Sentiment analysis by deep learning approaches. TELKOMNIKA (Telecommun. Comput. Electron. Control) **18**(2), 752 (2020). https://doi.org/10.12928/telkomnika.v18i2.13912

30. Konečný, J., McMahan, H.B., Yu, F.X., Richtárik, P., Suresh, A.T., Bacon, D.: Federated learning: strategies for improving communication efficiency. arXiv preprint arXiv:1610.05492 (2016)

Chapter 47
Using Optimization Algorithm to Improve the Accuracy of the CNN Model on the Rice Leaf Disease Dataset

Luyl-Da Quach⑩, **Anh Nguyen Quynh**⑩, **Khang Nguyen Quoc**⑩, and **Nghe Nguyen Thai**⑩

Abstract Rice plays an essential role in daily meals. Therefore, planting and tending to play a significant role, however, the disease is an issue that needs attention and monitoring. In this work, we propose an approach to improve the accuracy of the prediction model using CNN algorithm on rice leaf dataset with 7532 samples with 5 different diseases such as bacterial blight, blast, red strip, tungro, and brown spot. This dataset uses data augmentation methods with rotations, width range shift 0.2, height shift 0.2, vertical flip, and horizontal flip. Finally, with the application of optimization models such as Adaptive Gradient Algorithm (Adagrad), Root Mean Square Propagation (RMSProp), and Adaptive Moment Estimation (Adam), the Adam optimal algorithm results in stability and accuracy. 98.06%, higher than the other 2 algorithms 72.70 and 96.86%.

47.1 Introduction

In Asian countries, rice is a food that plays an essential role in daily meals, and it contributes significantly to the GDP of a huge countries, including Vietnam [1]. Vietnam is the world's fifth-largest rice exporter in volume and third in export. Rice cultivation plays an important role not only in the context of food security but also in reducing greenhouse gas emissions. Rice production and export affect people globally, as rice supplies half of the population that is constantly growing,

L.-D. Quach (✉) · A. N. Quynh · K. N. Quoc
FPT University, Can Tho, Vietnam
e-mail: luyldaquach@gmail.com

A. N. Quynh
e-mail: AnhNQCE160515@fpt.edu.vn

K. N. Quoc
e-mail: KhangNQCE160710@fpt.edu.vn

N. N. Thai
Can Tho University, Can Tho, Vietnam
e-mail: ntnghe@cit.ctu.edu.vn

especially in the developing regions of Asia, Latin America, and Africa. Realizing the increased demand for rice, many studies have been conducted to identify rice diseases monitored in vials or in the laboratory. Visual inspection is often time-consuming and requires a professional to perform. Therefore, there have been many studies on rice disease diagnosis by machine learning and image processing.

The use of image processing techniques yields positive signals from a variety of studies and techniques. In the study [2], the authors identified undamaged, healthy, and sick plants at 3 weeks of age. Flatbed scanners were used to record images of infected seedlings and controls to measure their morphological and color characteristics. Using the support vector classifier model was used with the basic properties and optimal model parameters for the SVM classifier. Research [3, 4] uses K-means and multi-SVM clustering models with image processing techniques on Matlab to detect leaf diseases and natural products to obtain fast results. The study [5] identified brown planthopper disease on rice stalks, and the phone image processing technique reduced the population thickness. The primary processing techniques include segmentation, image enhancement, median filtering, and K-means clustering. Research [6] develops an automated system to accurately identify and classify different diseases by characteristics such as color and shape combined with the Minimum Distance Classifier and k-Nearest Neighbor classifier, which are relatively positive. Research [7] using image processing techniques has dealt with the different classifications of plant diseases. Research [8] uses EfficientNet to classify 4 types of rice leaves to run on Android mobile devices. The use of imaging is more effective for identifying diseases caused by organisms, bacteria, or viruses in crops.

Besides using methods in image processing, processing methods in machine learning are also interesting. Research [9] uses an SVM classification model combined with deep CNN on a 1080 images of early ripening disease datasets with an accuracy of about 97.5%. Research [10] using a neural network based on deep learning CNN has significantly improved the ability to identify diseases on rice pest images with models VGG16, InceptionV3, MobileNet, NasNet Mobile, and SqueezeNet with a result of 93%. Research [11] uses a Bayesian optimization algorithm combined with a neural network for depth analysis to detect and classify rice diseases from rice leaf images. Based on the MobileNet structure and reinforcement learning mechanism, the study applied to classifying 4 rice diseases with a result of 94.56%. Study [12] used transfer learning to improve predictive learning by using a complex neural network to recognize the Grain discoloration disease of rice with an accuracy of about 88.2%. The research [13] classified rice data on 5 diseases based on contour, quality, and color with the Naïve Bayes classification model.

In this research, a deep CNN-based system has classified rice data on 5 diseases, including the data of the collection study and the data set of the research [14]. The article's contribution is the evaluation of 3 optimization algorithms, Adagrad, RMSprop, Adam, combined with a deep CNN algorithm on a dataset of 7532 disease samples on rice leaves with 5 different diseases.

Fig. 47.1 Sample image of disease on rice leaves, from left: Bacterial blight, Blast, Red strip, Tungro, Brown spot

47.2 Proposed Method

The study detailing the data set and proposed implementation methods in this section will be discussed in the appropriate subheadings.

47.2.1 Dataset

The dataset contains 7,032 leaf pathogens, including bacterial blight, blast, brown spot, tungro, and red stripe. The initial data set with 5 different layers, including Bacterial Blight, Blast, Tungro, Brown Spot and Red Stripe, was taken at the Mekong Delta Rice Institute, Vietnam, with a Samsung A72 phone. This image has a resolution of 235×426, 351×277, etc. From there, they were combined with the data set of the study [13] consisting of 5932 images using a Nikon DSLR-D5600 with an 18-lens lens—55 mm high resolution. The images of diseased rice leaf samples are shown in Fig. 47.1.

To enhance the data, the study performed simple image rotation and flip operations at different angles such as 45°, width range shift 0.2, height shift 0.2, vertical flip, and horizontal flip. Since there are more enhancement images, the opportunity to evaluate the algorithms used for testing is better. Table 47.1 lists the names and number of images used to test the algorithm. The samples are randomly divided according to the 70:30 ratio components to train the deep CNN.

47.2.2 Optimization Algorithm

Adagrad [14] is an advanced machine learning technique that performs gradient descent by varying the learning rate. Adagrad is further improved by giving correct learning weights based on the input before it to adjust the learning rate in the most optimal direction instead of with a single learning rate for all nodes.

Table 47.1 Statistical table of the number of rice leaf diseases used in the assessment

Leaf diseases	Number of original images	Number of images used of augmentation	Number of images used for training and validation
Bacterial blight	1884	1318	6590
Blast	1740	1218	6090
Red stripe	300	210	1050
Tungro	1608	1125	5625
Brown spot	1500	1050	5250
Total	7032	4922	24,610

$$W_{t+1} = W_t - \frac{\eta}{\sqrt{G_t + \epsilon}} \cdot g_t \qquad (47.1)$$

With η is learning rate, ϵ is guaranteed parameter, g_t is gradient at time t, G_t is the diagonal matrix containing the square of the derivative of the parameter vector at t.

Root Mean Square Propagation (RMSProp) [15] uses the mean square of the gradient to normalize the gradient. That has the effect of balancing the step size—reducing the step for large gradients to avoid Exploding Gradient, and increasing the step for small gradients to avoid Vanishing Gradient. RMSProp automatically adjusts the learning rate, and chooses a different learning rate for each parameter (47.2) and (47.3).

$$E\left[g^2\right]_t = 0.9E\left[g^2\right]_{t-1} + 0.1g_t^2 \qquad (47.2)$$

$$W_{t+1} = W_t - \frac{\eta}{\sqrt{E\left[g^2\right]_t + \epsilon}} \cdot g_t \qquad (47.3)$$

Which $E\left[g^2\right]_t = :$ running average at time step t.

Adam (Adaptive Moment Estimation) [16] is an adaptive learning rate method that calculates individual learning rates for different parameters. Adam uses estimates of the first and second intervals of the slope to adjust the learning rate for each weight of the neural network. Adam algorithm is described:

$$m_t = \beta_1 m_{t-1} + (1 - \beta_1)g_t \qquad (47.4)$$

$$v_t = \beta_2 v_{t-1} + (1 - \beta_2)g_t^2 \qquad (47.5)$$

With m_t, v_t is estimates of the first moment (the mean) and the second moment (the uncentered variance) of the gradients, respectively; β_1, β_2 is speed of movement. In the initial time steps, and when the decay rate is small, m_t and v_t are initialized as vectors of 0, Adam's authors find that they are biased toward 0. They compute bias-corrected first and second time estimates:

$$\hat{m}_t = \frac{m_t}{1 - \beta_1^t} \tag{47.6}$$

$$\hat{v}_t = \frac{v_t}{1 - \beta_2^t} \tag{47.7}$$

Then they yield Adam update rule, like we have seen in RMSProp:

$$W_{t+1} = W_t - \frac{\eta}{\sqrt{\hat{v}_t + \epsilon}} \cdot \hat{m}_t \tag{47.8}$$

47.2.3 Rice Leaf Disease Identification by Convolutional Neural Network (CNN)

CNN is used to extract from a specific class, and the result is a feature vector [17]. Features are put into a fully connected layer to identify rice leaf diseases using the softmax function for each layer generated by the CNN network with many layers, including feature class and feature vector. The use of optimization algorithms with weights such as weights—w and the bias—b library is more suitable for architectures consisting of Conv and Pooling classes, respectively, such as Fig. 47.2. In this research, some popular adaptive gradient methods will be evaluated including RMSProp, Adagrad, Adam. The evaluation is based on the criteria of loss of value and correct recognition rate. Based on the results to get the impact assessment of the optimal algorithm on the network model applied in the classification of rice leaf diseases.

First layer has a convolution layer of 32 filter size and filter size (3, 3), then a max pool layer of stride (2, 2). Then there are 3 sets similar to the first layer built with filter increments of 64, 128, 256 respectively. This image is passed to the stack of the convolution layer. In the convolution and max-pooling layers, the filters we use are 3 * 3 (like VGG16) instead of 11 * 11 in AlexNet and 7 * 7 in ZF-Net.

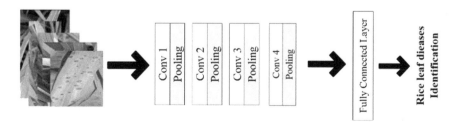

Fig. 47.2 Rice leaf disease identification by convolutional neural network (CNN)

After the stack of convolution and max-pooling layers, we have a feature map (2, 2, 256). We flatten this output to make it a feature vector (1, 1024). Then there are 3 fully connected layers (like VGG16), the first one takes input from the last feature vector and outputs a vector (1, 4096), the second also outputs a vector of size (1, 4096) but the third layer outputs 5 channels for 5 classes (5 types of diseases distinguish), i.e., the 3rd fully connected layer is used to implement softmax function to classify 5 classes. All hidden layers use ReLU as their activation function.

47.3 Evaluating Optimal Algorithms for Convolutional Neural Network

47.3.1 Model Evaluation

The model is evaluated using accuracy, and the loss function is: cross-entropy. Accuracy is an instance that is precisely determined from the total number of data sets. In other words, it is the correct predictions made by the model relative to the total predictions. The following equation:

$$\text{Accuracy} = \frac{\text{Correct Prediction Count}}{\text{Total Predictions}} \qquad (47.9)$$

The loss function evaluates the effectiveness of the 'learning' algorithm for the model on the data set used. It returns a non-negative actual number representing the difference between two quantities: the predicted label, and y, the correct label. This is a reward-punishment mechanism, the model will have to pay the penalty each time the prediction is wrong, and the penalty is proportional to the size of the error. The goal of any supervised math problem includes reducing the total penalty payable. In the ideal case $a = y$, the loss function should return a minimum of 0. Cross-entropy (CE) is considered a measure of the classifier's performance.

$$\text{Cross Entropy} = -\sum_{i}^{c} a_i \log(p_i) \qquad (47.10)$$

where

c is the number of classes,
a_i the actual value,
p_i is the predicted value.

Table 47.2 Accuracy comparison table of 3 algorithms

Epoch	Adagrad	RMSProp	Adam
200	72.70%	96.82%	99.34%

47.3.2 Results

With the installation of 3 optimal algorithms to evaluate the data set, the results are obtained in Table 47.2. The first column contains the number of epochs to compare with the 3 optimal algorithms. Which, dominating with high accuracy is the Adam algorithm with an accuracy of 99.34%, followed by Adagrad and RMSProp with an accuracy of 72.7 and 96.82%.

From the graph (Fig. 47.3) it can be seen that the Adam algorithm has the lowest amplitude, hardly changing too much after only the first 50 epochs. Meanwhile, RMSprop, and Adagrad are the two algorithms with the larger variation during epochs. It can be seen that the correct recognition rate of the model is influenced by the optimization algorithm. Specifically, for an algorithm with a high loss rate like Adagrad, the accuracy rate is low, about 72.70% on the Testing set. Meanwhile, low-loss algorithms such as RMSProp and Adam give better accuracy rates, with 96.82% and 99.34% results, respectively, when used on the same network architecture model.

47.4 Conclusion

The research found that the Adam algorithm combines the 'gradient descent with momentum' algorithm and the 'RMSP' algorithm, so it inherits the strengths of the two methods to create the optimal gradient descent. This result is because the Adam algorithm controls the rate of gradient descent in such a way that there is minimal oscillation when it reaches the global minimum while taking large enough steps to overcome the local minimum barriers along the way. Therefore, combine the features of the above methods to reach the global minimum effectively. Building on the previous model's strengths, the Adam Optimizer gives much better performance than the previously used models. It outperforms them by a large margin to generate optimized gradient descent.

Through the research, the performance evaluation of the proposed model and the accuracy of the model with 3 optimal algorithms: Adagrad, RMSProp, and Adam, are relatively satisfactory. Initially, the deep feature of the fully connected layer is extracted and classified the image into 5 layers corresponding to 5 diseases: Bacterial Blight, Blast, Brown Spot, Tungro, and Red Stripe. At the fully connected layer, data classification will be performed by activating the softmax function to calculate the probability at the output layer. Moreover, the research has used the optimization algorithm to increase the model's accuracy. The results from the above study have evaluated the impact of the optimization algorithm on the correct classification of results in the image recognition problem. The study also provides more understanding

Fig. 47.3 Accuracy rate of
optimization algorithms on
training and testing sets, top
to bottom is Adagrad,
RMSprop and Adam
algorithm

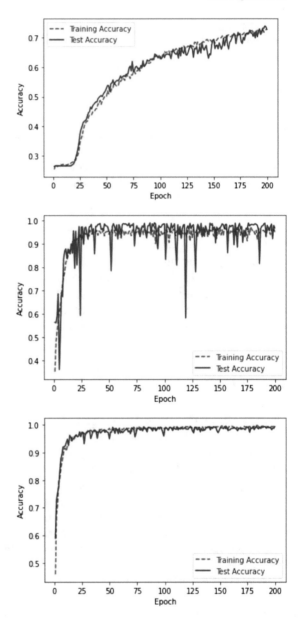

of the optimization algorithm through the results from the evaluation experiment,
thereby helping us make a reasonable choice of algorithms in building, training, and
evaluating network models.

In addition, further research develops an integrated application to match other
devices. For future works, we will consider integrating and deploying unmanned
aircraft systems and real-time systems. All for data collection and analysis, early

detection of diseases in rice, timely limitation, and prevention, preventing the disease from spreading. Moreover, the research will develop a map generation system and epidemic map application. Finally, we will combine the location data obtained from the drone for mapping with the disease detection results on the generated map and evaluate the accuracy of the whole system.

References

1. Clauss, K., Ottinger, M., Leinenkugel, P., Kuenzer, C.: Estimating rice production in the Mekong Delta, Vietnam, utilizing time series of Sentinel-1 SAR data. Int. J. Appl. Earth Obs. Geoinform. **73**, 574–585 (2018). https://doi.org/10.1016/j.jag.2018.07.022
2. Chung, C.-L., Huang, K.-J., Chen, S.-Y., Lai, M.-H., Chen, Y.-C., Kuo, Y.-F.: Detecting Bakanae disease in rice seedlings by machine vision. Comp. Electron. Agri. **121**, 404–411 (2016). https://doi.org/10.1016/j.compag.2016.01.008
3. Raut, S., Fulsunge, A.: Plant Disease Detection in Image Processing Using MATLAB, 6, 9 (2007)
4. Krishnan, H., Priyadharshini, K., Gowsic, M., Mahesh, N., Vijayananth, S., Sudhakar, P.: Plant disease analysis using image processing in MATLAB. In: 2019 IEEE International Conference on System, Computation, Automation and Networking (ICSCAN), pp. 1–3. IEEE, Pondicherry, India (2019). https://doi.org/10.1109/ICSCAN.2019.8878753
5. Sethy, P.K., Dash, S., Barpanda, N.K., Rath, A.K.: A novel approach for quantification of population density of rice brown plant hopper (RBPH) using on-field images based on image processing, 6, 5 (2019)
6. Joshi, A.A., Jadhav, B.D.: Monitoring and controlling rice diseases using image processing techniques. In: 2016 International Conference on Computing, Analytics and Security Trends (CAST), pp. 471–476. IEEE, Pune, India (2016). https://doi.org/10.1109/CAST.2016.7915015
7. Tichkule, S.K., Gawali, Dhanashri.H.: Plant diseases detection using image processing techniques. In: 2016 Online International Conference on Green Engineering and Technologies (IC-GET), pp. 1–6. IEEE, Coimbatore, India (2016). https://doi.org/10.1109/GET.2016.7916653
8. Thai-Nghe, N., Tri, N.T., Hoa, N.H.: Deep learning for rice leaf disease detection in smart agriculture. In: Dang, N.H.T., Zhang, Y.-D., Tavares, J.M.R.S., Chen, B.-H. (eds.) Artificial Intelligence in Data and Big Data Processing, pp. 659–670. Springer International Publishing, Cham (2022). https://doi.org/10.1007/978-3-030-97610-1_52
9. Hasan, Md.J., Mahbub, S., Alom, Md.S., Abu Nasim, Md.: Rice disease identification and classification by integrating support vector machine with deep convolutional neural network. In: 2019 1st International Conference on Advances in Science, Engineering and Robotics Technology (ICASERT), pp. 1–6. IEEE, Dhaka, Bangladesh (2019). https://doi.org/10.1109/ICASERT.2019.8934568
10. Rahman, C.R., Arko, P.S., Ali, M.E., Iqbal Khan, M.A., Apon, S.H., Nowrin, F., Wasif, A.: Identification and recognition of rice diseases and pests using convolutional neural networks. Biosys. Eng. **194**, 112–120 (2020). https://doi.org/10.1016/j.biosystemseng.2020.03.020
11. Wang, Y., Wang, H., Peng, Z.: Rice diseases detection and classification using attention based neural network and Bayesian optimization. Expert Syst. Appl. **178**, 114770 (2021). https://doi.org/10.1016/j.eswa.2021.114770
12. Duong-Trung, N., Quach, L.-D., Nguyen, M.-H., Nguyen, C.-N.: Classification of grain discoloration via transfer learning and convolutional neural networks. In: Proceedings of the 3rd International Conference on Machine Learning and Soft Computing—ICMLSC 2019, pp. 27–32. ACM Press, Da Lat, Viet Nam (2019). https://doi.org/10.1145/3310986.3310997

13. Mohapatra, D., Tripathy, J., Patra, T.K.: Rice disease detection and monitoring using CNN and Naive Bayes classification. In: Borah, S., Pradhan, R., Dey, N., Gupta, P. (eds.) Soft Computing Techniques and Applications, pp. 11–29. Springer Singapore, Singapore (2021). https://doi.org/10.1007/978-981-15-7394-1_2

14. Duchi, J., Hazan, E., Singer, Y.: Adaptive subgradient methods for online learning and stochastic optimization, 39

15. Hinton, G., Srivastava, N., Swersky, K.: Neural networks for machine learning lecture 6a overview of mini-batch gradient descent

16. Kingma, D.P., Ba, J.: Adam: A Method for Stochastic Optimization, http://arxiv.org/abs/1412.6980 (2017)

17. Lopes, U.K., Valiati, J.F.: Pre-trained convolutional neural networks as feature extractors for tuberculosis detection. Comput. Biol. Med. **89**, 135–143 (2017). https://doi.org/10.1016/j.compbiomed.2017.08.001

Chapter 48
Evaluation and Techniques of Automatic Text Summarization

Deepali Vaijinath Sawane, Sanjay Azade, Shabeena Naaz Khan,
Sarfaraz Pathan, Sonali Gaikwad, and Jyotsna Gaikwad

Abstract A progression of script summarization of compress original manuscript
and produces a summary containing significant sentences without changing the
meaning of source manuscript and includes all significant vital information from
the source manuscript. Now-a-days, text summarization has become very popular
and applicable for search engine, business analysis, market review, medical records,
news article summary, government officials, and researchers. The most indispensable concepts, while viewed commencing the summary conclusion, refer the term
taking out script sorting along with conceptual script sorting. An extract the scripts
with highly rank statement review is designation in addition to enlargement as well
as at this moment investigate in addition with transmitted collaborate with conceptual summarization and instantaneous summarization. Even though nearby enclose
been accordingly an assortment of accomplishment within the attainment of data
sources, mechanism, and approaches available, in attendance be nix several credentials with the purpose of be capable of endow with an expansive representation of the
contemporary circumstances of investigate pasture. This article makes accessible a
bulky and competent assessment of investigate in the pasture of text summarization
available from preceding ten existence as the consequence of extraction of investigate studies for the acknowledgment and investigation to portray investigate focus or
topical trends, data sources, preprocessing, features, investigation studies, methods,
costing, lacunas, and future scope.

D. V. Sawane (✉) · S. Azade · S. N. Khan · S. Pathan
MGM University, Aurangabad, Maharashtra, India
e-mail: dsawane@mgmu.ac.in

S. Azade
e-mail: sazade@mgmu.ac.in

S. Gaikwad
Shri Shivaji College, Chikhli, Buldhana, Maharashtra, India

J. Gaikwad
Deogiri College, Aurangabad, Maharashtra, India

48.1 Introduction

The mechanized manuscript summarization is the procedure apropos sinking the text manuscript through a workstation curriculum in organizes to construct a summary that retains the majority imperative spot of the inventive manuscript. As the difficulty of information excess has full-fledged, and as the extent of data has increased, so has interest in automatic summarization. Technologies that can construct logical summary take into description variables such as length, script approach, and verbal communication regulations. The summarization classifies between two categories extractive and abstractive. Extractive summarization methods make things easier the trouble of summarization into the problem of recognizing a representative separation of the sentences in the source manuscript [1]. Abstractive summarization may invent novel sentences, hidden in the source document. The purpose of this research is hub on sentence-based extractive manuscript summarization. The extractive summarization methodology is naturally based on techniques for sentence taking out and intend to wrap the set of sentences that are most essential for the generally perceptive of a known manuscript [2]. The huge growing and straightforward simplicity of use of in sequence on the World Wide Web have newly resulted in combing up the traditional linguistics problem—the compression of data from source documents. This assignment is fundamentally a statistics compression procedure. The objective of mechanized manuscript depiction exists precipitate preeminent origin of wellspring manuscript hooked on a minuscule description conserving via origin of statistics origin contented and in general significance [3]. Manuscript depiction is the terminology of mechanically designing a compacted description of a given manuscript preserving its data content. Automatic article summarization is a considerable research part in ordinary lexical mechanism procedure. The skill of mechanized script depiction is raising furthermore may perhaps make available a resolution in preeminent direction of via origin source excess trouble [4]. An executive summary, occasionally known as a supervision summary, is a small manuscript or part of a manuscript, created for business purposes, that summarizes a larger description or application or an assembly of connected information in such an approach that readers can quickly become familiar by means of a bulky corpse of objects devoid of having to understand it all [5]. It regularly contains a concise declaration of the difficulty or proposal enclosed in the major article(s), background data, brief study, and most significant results. It is proposed as an assist to decision-making by managers and has been described as perhaps the majority essential part of a business strategy [6]. The text summarization study tendency has also undergone a minor modify in the history of some past years where novel possessions have emerged that are trends that are most important to optimization, how to optimize text summarization presentation in arrange to obtain elevated correctness[7].

48.2 Functioning of Text Summarization

The working apropos conceptual script depiction along with uprooting manuscript depiction such as: conceptual script depiction exist preeminent assignment of producing a small and brief abstract summary that captures the outstanding and novel concepts of the resource manuscript. The generated summaries potentially contain new phrases and sentences that may not appear in the source text. Abstractive summarizers are supposed since they do not perform decide on sentences beginning the originally given manuscript passage to make the summary. As a substitute, they construct a paraphrasing of the most significant contents of the specified manuscript, using a terminology position dissimilar from the innovative manuscript. Extractive text summarizations focuses at recognizing the most significant data that are then extracted and set mutually to form a brief summary. Abstractive summary creation rewrites the complete manuscript by structure inner semantic demonstration, and then, a summary is shaped by means of natural language processing (Fig. 48.1).

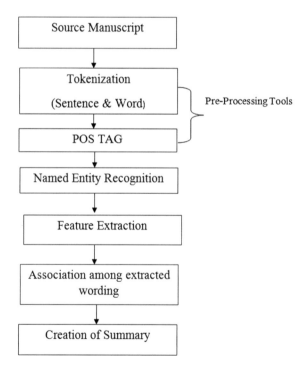

Fig. 48.1 Functioning methodology of text summarization

48.2.1 Features of Manuscript Summarization

Manuscript summarizers make out and extract foremost sentences commencing the spring manuscript and focal point on them to form a diminutive summary. An inventory of features as defined underneath can be referred for recognize the significant sentences of the source manuscript.

a. **Term Frequency**—a large amount essential terms contributed as a consequence of records as well as facts take place established resting on expression rate of recurrence; thus, essential sentences are those words that ensue frequently. The habitually happening statement enlarged keep count of statement. The largest part ordinary determine extensively worn to estimate the word frequency is TF–IDF [8].

b. **Location**—this expression is collaborated on the perception that foremost sentences are situated at prejudiced situation in manuscript or in editorial, such start or end of an article. Initial and previous sentence of article has superior possibility to be integrated in summary [8, 9].

c. **Cue Method**—result of constructive or disapproval of word on the sentence mass to point toward importance or significant proposal such as cues: "in summary," "in result," "the manuscript description and explanation" [10].

d. **Title/Headline statement**—the expressions in the bearing and designate of a script that acquire situate in statement are absolutely recognized [3].

e. **Sentence length**—keeps in sight the amount of summary. In general, especially extended as well as exceptionally diminutive pronouncements come about besides not apposite for précis [5].

f. **Similarity**—the comparison canister be designed by resources of lexical statistics along with records. It represents comparison among preeminent statement as well label apropos preeminent manuscript, along with correspondence linking the statement and lingering statement apropos the manuscript [11, 12].

g. **Proper noun**—for manuscript depiction, statement having appropriate nouns come about imperative resembling person's forename, location, and position [13].

h. **Proximity**—the space flanked by manuscript module wherever attributes happen exists a formative thing intended meant intended refers to designing associations among attributes [14].

48.3 Literature Review

See Table 48.1.

Table 48.1 Review on text summarization

Author and year	Techniques	Drawback/disadvantage
Moratanch and Chitrakala [8]	Rule-based method	As evaluate to particular manuscript summarization, multidocument summarization has more likelihood of uncertainty. That outcome in assignment of higher gain values to some terminology unacceptably by the sentence scoring algorithm. Lexical chain may help to solve this problem
Bui et al. [5]	Neural network method	This research work is not simultaneously work with single document and multidocument
Widyassari et al. [15]	Query-based method	The inadequacy is summaries controlled by the input chart fashioned by unusual clustering methods are identified for the performance, and therefore, it is not provided appropriate result
Bhatia and Jaiswal [16]	Knowledge-based methods	Constraint of the knowledge-content perspective exist with the intention of the assembly of preeminent topic knowledge does not acquire into description the observed substance structure of the documents to be summarized
Wang et al. [17]	Machine learning and text mining	The limitation of this performance makes user to go through sometimes—irrelevant credentials also therefore its not designed appropriate summary
Barzilay and McKeown [18]	Feature-based machine learning method	The limitation is, it virtually infeasible (given memory limitations of current hardware) to prepare a model which encodes all of them into vectors and afterward generates a summary
Kikuchi et al. [19]	Unsupervised method	The limitation is evaluations have low standard scores for summaries despite the consequences of the scheme
Hirao et al. [20]	Feature selection method	In this paper, assignment must be effectively quantifying dealings with client

(continued)

Table 48.1 (continued)

Author and year	Techniques	Drawback/disadvantage
Harabagiu and Lacatusu [21]	Extractive text summarization	This system does not need a manuscript summary corpus, but only user criticism in order to gain knowledge, and this is an key concern ever while the enlargement of summaries is a protracted and tiresome assignment, and that there are only a few credentials summary corpus

48.4   Motivation

The manuscript summarization is the procedure of recognizing the majority of significant data in a manuscript of associated manuscript and assigning it in fewer spaces and typically by a factor of five to ten than the original text [19]. A quantity of the reasons intended for motivation of manuscript summarization is as discussed:

- To keep up with the world associations by listening to news bulletin and people base investment decisions on stock market updates [20].
- Community even goes to cinema mostly on the sources of reviews they have seen [20, 21].
- Among created summaries, populace can create valuable decisions in stipulated time [22].
- The motivation is also to construct such implement which is technically convenient and appropriate summaries create successfully [23].

48.4.1   Problem Domain

Automatic script summarization has built up hooked on an indispensable constituent of everyday life apposite to the effortlessness of exploit of enormous quantity of records that demand to survive summarized preordained for researchers consequently to facilitate they be capable of construe noteworthy filling in not as much of time. It has been understood meant for decades that more and more data are appropriate accessible and that tools are desirable to handle it [24]. Only, in recent times, however, does it appear that an enough amount of the data is automatically accessible to create an extensive necessitate for automatic summarization [25]. It is significant to extract critical data or compress data in order to minimize quantity of time invested to evaluation this enormous data [26].

48.4.2 Problem Definition

There are lengthy sentences in extract than standard length. Therefore, occasionally, even the component of sentences which is not essential is also integrated, which outcome in space expenditure [27]. The significant or appropriate data are typically extending transversely statements, as well as withdrawal depiction incapable near imprison that is unless the narrative exist extended as much as necessary to hold all those sentences. Originally, the sentence extraction is frequently accompanied to tribulations in general consistency of the summary generation process. The recurrent problems concern "dangling" anaphora, and the sentences are frequently enclose demonstrative, which be defeated their instances apropos when dropping away of perspective [28]. The poorer quality so far, edging mutually decontextualized extracts possibly will go ahead to an ambiguous explanation of anaphors comparable crisis survive among chronological terminology [29]. These tribulations happen to further crucial in the enormous scrip depiction summarization process, from the time when withdraw come about strained commencing various origin of resources manuscript. A typical come within reach up to focusing these tribulations included situation of mechanism withdrawn of records.

48.5 Solution Methodologies for Manuscript Depiction

Manuscript depiction is generally separated into conceptual as well as withdrawal. The concise explanation regarding apiece perspective is as follows.

48.5.1 Abstractive Summarization Approach

Abstractive Summarization Approach summarizations by means of abstractive methodologies are generally classified hooked on two credentials: structured-based approach and semantic-based approach.

A. **Structured-Based Approach**
Structured-based approach designed chiefly indispensable original sources from the editorial from beginning to end cognitive schemes such as standard format, mining convention, and other designed format such as tree, ontology, and guide and corpse phrase construction [30]. The concise abstract of all the methodology under structured-based approach is explained as follows:

- Tree Based Method—this process is used by the author, Barzilay and McKeown [16], Kikuchi et al. [17], Hirao et al. [18]. It refers an enslavement tree construction to indicate the text of a commentary. It refers moreover a language generator or an algorithm for production of summary. The advantage of this method is, it walks on units of the given manuscript study and easy to summary. The constraint is, it lacunas an inclusive sculpt which would encompass a conceptual manifestation for comfortable assortment.

- Template-Based Method—this performance is used by the author, Harabagiu and Lacatusu [19], Oya et al. [20]. It uses a template to signify a whole article. Linguistic and lexical methodology or extraction convention is synchronized to be acquainted with text scraps that determination be calculated into stencil slots. The advantages ofthis method is, it generates summary is extremely coherent since it relies on appropriate data recognized by information extraction method. Limitation is, it requires designing of templates, and generalization of template is too difficult.
- Ontology-Based Method—This method is worn by the author Lee and Jian [21], Viswanath (2006), Ramezani et al. (2015), Ragunath and Sivaranjani [22] make use of ontology, or it also known as knowledge base method to get better the progression of summarization. It exploits fuzzy ontology to grasp uncertain sources that uncomplicated sphere ontology cannot drawing relation is easy appropriate to ontology. It refers ambiguity at sensible quantity. This approach is incomplete to Chinese news only. It used for generate rule-based methodfor handling uncertainty is a composite task.
- Lead and Body Phrase Method—this method is used by the author Tanaka et al. [23]. This scheme is referred scheduled the functions of phrases together with incorporation and replacement that have indistinguishable template pinnacle hunk in the escort and corpse statement in organize to rephrase the lead statement. The advantages, it is good for semantically suitable revisions for revising guide statements. The constraint is parsing errors degrade sentential entirety such as grammaticality and replication. Its aim is on rewriting system and lacks an absolute representation which would comprise an abstract demonstration for comfortable collection.
- Rule-Based Method—Genest and Lapalme [24]—these procedure scripts to be make synopsis are performed in requisites of classified and averification of circumstances. It has a possible for generating summaries with superior key point's compactness than present situation of skill. The disadvantage of this method is that entire regulations and pattern are physically on paper, which is boring and time-consuming task.

B. Semantic-Based Approach

In semantic-based budge in the direction of, semantic manifestation of script is worn to endow with for natural language generation (NLG) format. These procedures deliberate on be acquainted with noun turn of phrase and verb phrase by production elsewhere linguistic sources of editorial. Concise theoretical of all the methodology under semantic-based move toward is providing as follows:

- Multimodal semantic model—Greenbacker [25]—a semantic representation, which captures approach and involvement involving concepts, is building to indicate the contents of multimodal research papers. A considerable assistance of this configuration is with the intention of it methodology intangible summary, whose revelation is stupendous from the occasion when it collaborates pertinent textual and graphical satisfied from the comprehensive script. The restriction of this constitution is that it is in the flesh evaluated by humans.

- Information Item-Based Method—Genest and Lapalme [26], Mallett et al. [27]—the filling of summary is formed from intangible manifestation of reserve script, relatively than from statement of resource editorial. The intangible manifestation is statistics article, which is the ostensible component of cogent statistics in a source script. The most important potency of this method is that it generates small, rational, significant data with key points, high quality, and fewer redundant summaries. It avoided owing to the involvedness of constructing substantial and grammatical statement commencing them. Linguistic eminence of summaries is enormously petite owing to imprecise parses.
- Semantic Graph-Based Method—Moawad and Aref [28], Ganesan et al. [29], Plaza et al. (2011) [30], Subramaniam and Dalal (2015) [31]. This performance is worn to abridge a script by preparing a semantic graph defined as rich semantic graph (RSG) for the creative manuscript, compress the created semantic graph. It produces concise, coherent, and less significant amount redundant and grammatically accurate article. This technique is incomplete to solo or single manuscript abstractive summarization.

48.5.2 Extractive Summarization Techniques

An extractive summarization performance collaboration of specifying considerable statement of lines, paragraphs, etc., in view of the fact that the pioneering script and merged them into a small amount form. The significance of sentences is determined based on statistical and linguistic features of sentences. The concise abstract of techniques of extractive-based approach as follows:

- Term Frequency–Inverse Document Frequency Method—Fachrurrozi et al. (2013) [32], Pimpalshende (2013) [33]. Statement promptness is divergent as the enlarge quantity of statement in the script that encircle that expression. Subsequently, this condemnation vectors are measured by similitude to the vagueness, and the greatest ranked statement is preferred to be component of the summary.
- Cluster Based method—Twinandilla et al. [34], Bazghandi et al. (2012) [35], Deshpande and Lobo (2013) [36]. It is perceptive to believe that summaries should deal with dissimilar themes appearing in the article. If the article compilation for which summary is being created is of entirely dissimilar topics, article clustering becomes approximately necessary to produce a consequential summary. The drawback is summary is not created appropriately with using clustering technique and geospatial methods.

- Graph-Based Method—Mihalcea et al. (2013) [37], Sarda and Kulkarni [30], Chen and Kit [32], Verma and Verma [38]. The chart theoretic manifestation of passages arranged performance of acknowledgment of themes. Afterward than the average preprocessing ladder, specifically, stemming and prevent word amputation; statement in the script are performed as nodes in an undirected diagram.
- Machine Learning Method—Sarkar et al. (2011) [33], Bazrfkan and Radmanesh [39]. The summarization course of action is constructed as a cataloguing difficulty: Statements are distinguished as summary statement, and summarized sentences depend on the features that they acquire.
- Deep Learning method—Awasthi et al. (2011) [40], Lozanova et al. [41], Ahuja et al. [37]. This system represents a wider variety of manuscript summary, and consequences are powerfully benefited. The drawback is necessity large dataset to execute summarization procedure.
- Neural Network Method—Sarda and Kulkarni [30], Sinha et al. [42], Lata et al. [38]. This performance included grounding the neural networks to learning the several categorizations of statement that supposed to be integrated in the summary. It refers three crusted provide for frontward neural network.
- Fuzzy Logic—Patil and Mane (2014) [43], Lloret et al. (2008) [44]. This performance is flexible for every attribute of a script such as assessment to caption, sentence coverage and association to explanation statement, etc., as the contribution of the fuzzy arrangement. The drawback is it complex commission for the conclusion consumer to go throughout huge quantity of appropriate retrieved manuscript, access, and arrange—the necessary essential data.
- Query-Based Method—Imam et al. [45]. In query-based manuscript summarization coordination, the statements in a given script are measured on the basis of promptness ranked of circumstances. It refers vector space imitation representation.
- LSA Method—Patil and Mane [43], Ozsoy et al. [46]. This technique accesses the name LSA because SVD useful to script expression matrices, cluster script that are semantically coupled to apiece supplementary, motionless while they do not contribute to ordinary terminology.

48.5.3 Feature-Based Method

Roth and Small [40], Liu (2020) [39], Lee and Landgrebe, Fellow [40], Uddin et al. [41]. This technique recognizes the most significant sentences from the source manuscript and extracted on the basis of interactive feature space construction protocol for extraction process, and it positively included in the summary.

48.6 Conclusion

This review paper discusses a small amount of the abstractive techniques and extractive techniques of text summarization. Abstractive summarization is the procedure of creating a summary of a source manuscript from its innovative ideas, not through repetition precisely the majority significant sentences from original article. These are an imperative and confront assignment in natural language processing. An extractive summary is an assortment of significant sentences or essential key points from the original or source text manuscript that from the bottom explained superiorly innovative and source text manuscript. A variety of techniques of extractive approach have emerged in the past ten years. But, it is difficult to judge that how a large amount of better interpretive complexity, at judgment or manuscript stage, contributes to great presentation. Devoid the use of natural language processing, the created summaries may not be a good deal precise in terms of semantics. If the contribution of source manuscript is wrapping multiple topics, it becomes complicated to produce an objective and meaningful summary without changing the meaning of source manuscript. For this reason, deciding appropriate weights of individual features is significant as excellence of absolute summary depend on the source manuscript.

References

1. King, M., Zhu, B., Tang, S.: Optimal path planning. Mob. Robots **8**(2), 520–531 (2001)
2. Allahyari, M., Pouriyeh, S., Assefi, M., Safaei, S., Trippe, E.D., Gutierrez, J.B., Kochut, K.: Text summarization techniques: a brief survey (2017). arXiv preprint arXiv:1707.02268
3. Abualigah, L., Bashabsheh, M.Q., Alabool, H., Shehab, M.: Text summarization: a brief review. In: Recent Advances in NLP: The Case of Arabic Language, pp. 1–15 (2020)
4. Mishra, R., Bian, J., Fiszman, M., Weir, C.R., Jonnalagadda, S., Mostafa, J., Del Fiol, G.: Text summarization in the biomedical domain: a systematic review of recent research. J. Biomed. Inform. **52**, 457–467 (2014)
5. Widyassari, A.P., Rustad, S., Shidik, G.F., Noersasongko, E., Syukur, A., Affandy, A.: Review of automatic text summarization techniques and methods. J. King Saud Univ. Comp. Inf. Sci. (2020)
6. Thapa, S., Adhikari, S., Mishra, S.: Review of text summarization in Indian regional languages. In: Proceedings of 3rd International Conference on Computing Informatics and Networks, pp. 23–32. Springer, Singapore (2021)
7. Indu, M., Kavitha, K.V.: Review on text summarization evaluation methods. In: 2016 International Conference on Research Advances in Integrated Navigation Systems (RAINS), pp. 1–4. IEEE (2016, May)
8. Bui, D.D.A., Del Fiol, G., Hurdle, J.F., Jonnalagadda, S.: Extractive text summarization system to aid data extraction from full text in systematic review development. J. Biomed. Inform. **64**, 265–272 (2016)
9. Modi, S., Oza, R.: Review on abstractive text summarization techniques (ATST) for single and multi documents. In: 2018 International Conference on Computing, Power and Communication Technologies (GUCON), pp. 1173–1176. IEEE (2018, September)
10. Masum, A.K.M., Abujar, S., Talukder, M.A.I., Rabby, A.S.A., Hossain, S.A.: Abstractive method of text summarization with sequence to sequence RNNs. In: 2019 10th International

Conference on Computing, Communication and Networking Technologies (ICCCNT), pp. 1–5. IEEE (2019, July)

11. Suleiman, D., Awajan, A.: Deep learning based abstractive text summarization: Approaches, datasets, evaluation measures, and challenges. Math. Probl. Eng. **2020** (2020)

12. Syed, A.A., Gaol, F.L., Matsuo, T.: A survey of the state-of-the-art models in neural abstractive text summarization. IEEE Access **9**, 13248–13265 (2021)

13. Alomari, A., Idris, N., Sabri, A.Q.M., Alsmadi, I.: Deep reinforcement and transfer learning for abstractive text summarization: a review. Comput. Speech Lang. **71**, 101276 (2022)

14. Moratanch, N., Chitrakala, S.: A survey on extractive text summarization. In: 2017 International Conference on Computer, Communication and Signal Processing (ICCCSP), pp. 1–6. IEEE (2017, January)

15. Bhatia, N., Jaiswal, A.: Automatic text summarization and it's methods-a review. In: 2016 6th International Conference-Cloud System and Big Data Engineering (Confluence), pp. 65–72. IEEE (2016, January)

16. Wang, M., Wang, M., Yu, F., Yang, Y., Walker, J., Mostafa, J.: A systematic review of automatic text summarization for biomedical literature and EHRs. J. Am. Med. Inform. Assoc. **28**(10), 2287–2297 (2021)

17. Barzilay, R., McKeown, K.R.: Sentence fusion for multidocument news summarization. Comput. Linguist. **31**(3), 297–328 (2005)

18. Kikuchi, Y., Hirao, T., Takamura, H., Okumura, M., Nagata, M.: Single document summarization based on nested tree structure. In: Proceedings of the 52nd Annual Meeting of the Association for Computational Linguistics (Volume 2: Short Papers), pp. 315–320 (2014, June)

19. Hirao, T., Nishino, M., Yoshida, Y., Suzuki, J., Yasuda, N., Nagata, M.: Summarizing a document by trimming the discourse tree. IEEE/ACM Trans. Audio, Speech, Lang. Process. **23**(11), 2081–2092 (2015)

20. Harabagiu, S.M., Lacatusu, F.: Generating single and multi- document summaries with gistexter. In: Document Understanding Conferences, pp. 11–12 (2002, July)

21. Oya, T., Mehdad, Y., Carenini, G., Ng, R.: A template- based abstractive meeting summarization: leveraging summary and source text relationships. In: Proceedings of the 8th International Natural Language Generation Conference (INLG), pp. 45–53 (2014, June)

22. Lee, C.S., Jian, Z.W., Huang, L.K.: A fuzzy ontology and its application to news summarization. IEEE Trans. Syst. Man Cybern. Part B (Cybernetics) **35**(5), 859–880 (2005)

23. Ragunath, R., Sivaranjani, N.: Ontology based text document summarization system using concept terms. ARPN J. Eng. Appl. Sci. **10**, 2638–2642 (2015)

24. Tanaka, H., Kinoshita, A., Kobayakawa, T., Kumano, T., Kato, N.: Syntax-driven sentence revision for broadcast news summarization. In: Proceedings of the 2009 Workshop on Language Generation and Summarisation (UCNLG + Sum 2009), pp. 39–47 (2009, August)

25. Genest, P.E., Lapalme, G.: Fully abstractive approach to guided summarization. In: Proceedings of the 50th Annual Meeting of the Association for Computational Linguistics (Volume 2: Short Papers), pp. 354–358 (2012, July)

26. Greenbacker, C.: Towards a framework for abstractive summarization of multimodal documents. In: Proceedings of the ACL 2011 Student Session, pp. 75–80 (2011, June)

27. Genest, P.E., Lapalme, G.: Framework for abstractive summarization using text-to-text generation. In: Proceedings of the Workshop on Monolingual Text-to-Text Generation, pp. 64–73 (2011, June)

28. Mallett, D., Elding, J., Nascimento, M.A.: Information- content based sentence extraction for text summarization. In: International Conference on Information Technology: Coding and Computing, 2004. Proceedings. ITCC 2004, vol. 2, pp. 214–218. IEEE (2004, April)

29. Moawad, I.F., Aref, M.: Semantic graph reduction approach for abstractive text summarization. In: 2012 Seventh International Conference on Computer Engineering and Systems (ICCES), pp. 132–138. IEEE (2012, November)

30. Ganesan, K., Zhai, C., Han, J.: Opinosis: a graph based approach to abstractive summarization of highly redundant opinions (2010)

31. Sarda, A.T., Kulkarni, A.R.: Text summarization using neural networks and rhetorical structure theory. Int. J. Adv. Res. Comp. Commun. Eng. **4**(6), 49–52 (2015)
32. Chen, X., Kit, C.: Higher-order constituent parsing and parser combination. In: Proceedings of the 50th Annual Meeting of the Association for Computational Linguistics (Volume 2: Short Papers), pp. 1–5 (2012, July)
33. Verma, P., Verma, A.: A review on text summarization techniques. J. Sci. Res. **64**(1) (2020)
34. Sarkar, K., Nasipuri, M., Ghose, S.: Using machine learning for medical document summarization. Int. J. Database Theory Appl. **4**(1), 31–48 (2011)
35. Bazrfkan, M., Radmanesh, M.: Using machine learning methods to summarize Persian texts. Indian J. Sci. Res **7**(1), 1325–1333 (2014)
36. Awasthi, I., Gupta, K., Bhogal, P.S., Anand, S.S., Soni, P.K.: Natural language processing (NLP) based text summarization-a survey. In: 2021 6th International Conference on Inventive Computation Technologies (ICICT), pp. 1310–1317. IEEE (2021, January)
37. Lozanova, S., Stoyanova, I., Leseva, S., Koeva, S., Savtchev, B.: Text modification for Bulgarian sign language users. In: Proceedings of the Second Workshop on Predicting and Improving Text Readability for Target Reader Populations, pp. 39–48 (2013, August)
38. Sinha, A., Yadav, A., Gahlot, A.: Extractive text summarization using neural networks (2018). arXiv preprint arXiv:1802.10137
39. Lata, K., Singh, P., Dutta, K.: A comprehensive review on feature set used for anaphora resolution. Artif. Intell. Rev. **54**(4), 2917–3006 (2021)
40. Roth, D., Small, K.: Interactive feature space construction using semantic information. In: Proceedings of the Thirteenth Conference on Computational Natural Language Learning (CoNLL-2009), pp. 66–74 (2009, June)
41. Lee, C., Landgrebe, D.A.: Feature extraction based on decision boundaries. IEEE Trans. Pattern Anal. Mach. Intell. **15**(4), 388–400 (1993)
42. Ahuja, N., Bansal, R., Ingram, W.A., Jude, P.M., Kahu, S., Wang, X., Fox, E.A.: Big Data text summarization (2018)
43. Froud, H., Lachkar, A., Ouatik, S.A.: Arabic text summarization based on latent semantic analysis to enhance Arabic documents clustering (2013). arXiv preprint arXiv:1302.1612
44. Imam, I., Nounou, N., Hamouda, A., Khalek, H.A.A.: An ontology-based summarization system for Arabic documents (OSSAD). Int. J. Comp. Appl. **74**(17), 38–43 (2013)
45. Ozsoy, M., Cicekli, I., Alpaslan, F.: Text summarization of turkish texts using latent semantic analysis. In: Proceedings of the 23rd İnternational Conference on Computational Linguistics (Coling 2010), pp. 869–876 (2010, August)
46. Pallavi, D.P., Mane, P.M.: A comprehensive review on fuzzy logic & latent semantic analysis techniques for improving the performance of text summarization. Int. J. Adv. Res. Comp. Sci. Manage. Studies (IJARCSMS) **2**(11) (2014)
47. Uddin, J., Islam, R., Kim, J.M.: Texture feature extraction techniques for fault diagnosis of induction motors. JoC **5**(2), 15–20 (2014)

Chapter 49
Photo Classification Using Machine Learning to Understand the Interests of Tourists

Suguru Tsujioka, Kojiro Watanabe, and Akihiro Tsukamoto

Abstract In this study, we attempt to cluster photos posted on Tripadvisor. Many tourists post photos on travel service platforms, such as Tripadvisor, and these photos represent the interests of the tourists. However, these photos are not often analyzed due to the time and effort involved. With the proposed method, features are extracted, and dimensionality is reduced using a machine learning method. The photos are then automatically classified using an unsupervised learning method. The clustering results showed that similar photos were classified into the same cluster. The proposed method can reduce the cost of interpreting tourists' interests.

49.1 Introduction

In this study, we propose an automatic classification method for a set of photos taken from user-generated content on Tripadvisor (https://www.tripadvisor.com/), which is an online travel service platform. Online travel service platforms contain a large amount of user content, such as reviews and ratings posted by people who have visited tourist spots. For example, the Fushimi Inari Shrine section of Tripadvisor contains 20,664 reviews (including 11,233 in English) from foreign visitors (as of January 24, 2022). Many photos have been included in those postings in recent years. These photos represent the interests of visitors. Therefore, an analysis of these photos can reveal the interest tourists have in tourist attractions.

Tourist attraction operators search and analyze reviews related to their own operations on social networking and travel service platforms. They are aware that visitors' postings contain information about the attractions and the drawbacks of tourist destinations. However, the volume of postings is enormous, and the operators do not have the energy or the time to adequately analyze the images.

S. Tsujioka (✉)
Shikoku University, Tokushima, Japan
e-mail: tsujioka-s@shikoku-u.ac.jp

K. Watanabe · A. Tsukamoto
Tokushima University, Tokushima, Japan

The proposed method automatically classifies photos using machine learning. This method can be used to classify a large number of photos on online travel platforms. The proposed classification method can be used to reduce the cost of labor for photo analysis, and it can be used by tourist attraction operators to understand, quickly and easily, the strengths and weaknesses of their own tourist operations.

49.2 Literature Review

Studies have been conducted to interpret the interests of tourists using web content. Many of these studies have used social networking services (SNSs) data for the analysis. A large number of postings are made on SNS by tourists; these posts are thought to express the interests of the tourists. For this reason, many studies have attempted to analyze SNS posts with the aim of discovering tourists' interests at low cost. Paolanti et al. collected a set of tweets related to Cilento, a famous tourist destination in Southern Italy, and applied deep learning methods to the tweets to understand the interests of tourists [1]. Each tourist attraction was evaluated by classifying the tweets posted about Cilento into three categories: positive, negative, and neutral. Padilla et al. collected tweets from tourists visiting the city of Chicago and analyzed these data [2]. The results revealed the impact of weather and other external factors on tourist interests.

As mentioned above, most of the studies that aim to interpret the impressions of tourist destinations focus on the analysis of posted text. However, only a few studies have analyzed posted photos. An analysis of submitted photos is considered useful for planning measures to attract tourists. However, the technical difficulties of data collection and image analysis and the high costs in terms of time and effort are obstacles to the analysis.

49.3 Analysis Data

In this study, photos posted by tourists on Tripadvisor are classified in an attempt to understand tourist interests. The posted content on travel platform services contains less noise than do generic SNS, such as Twitter. Generic social networking sites such as Twitter often include posts for marketing purposes by stakeholders, and posts unrelated to tourist attractions. These posts can become noise, and classification costs may be excessive.

Fushimi Inari Shrine and Kenrokuen Garden were selected as the tourist attractions for classification. The top thirty sightseeing attractions in Japan according to Tripadvisor [3] are listed in Table 49.1. Our previous study showed that many foreign visitors perceive temples, shrines, and pagodas as similar religious structures [4]. For this reason, they were combined as one type. As can be seen in Table 49.1, religious structures and nature gardens account for more than 75% of the Top 30. Therefore,

Table 49.1   Top thirty sightseeing attractions in Japan according to Tripadvisor [3]

Types of tourist attractions	Tourist attractions (The popularity rankings of tourist attractions are shown in parentheses.)
Religious structures	Fushimi Inari Shrine (1st), Itsukushima and Itsukushima Shrine (3rd), Todaiji Temple (4th), Sanjusangen-do (7th), Mt. Koya Okunoin (8th), Kinkakuji Temple (10th), Naritasan Shinshoji Temple (12th), Hasedera Temple (13th), Toshogu Shrine (15th), Daishoin Head Temple (17th), Shoshazan Engyoji Temple (20th), Sensoji Temple (21th), Meiji Shrine (22th), Otagi Nenbutsuji Temple (24th), Chureito Pagoda (25th), Byodoin Temple Phoenix Hall (27th), Eikando Temple (28th)
Nature garden	Shinjuku Gyoen (6th), Kenrokuen Garden (11th), Nara Park (14th), Shukkeien Garden (16th), Shiratani Unsuikyo Ravine (23th), Ritsurin Garden (29th)
Museum	Hiroshima Peace Memorial Museum (2nd), The Hakone Open-Air Museum (5th)
Others	Himeji Castle (9th), Lake Kawaguchiko (18th), Shirakawago Thatched Roof Village (19th), Kyoto Station Building (26th), Tokyo Disney Sea (30th)

religious buildings and gardens are the subject of analysis, and Fushimi Inari Shrine and Kenrokuen Garden are employed as the representatives of these types.

Photos obtained from user reviews of web postings about these two attractions on Tripadvisor were analyzed. The 500 most recent photos submitted by July 1, 2022, were obtained. These photos were not at the original full size but were thumbnails that appeared in the list of posts. This is because high resolution was not necessary for our proposed classification method.

49.4   Research Method

The proposed method is intended to cluster a large number of photos without using training data. Figure 49.1 shows the processing procedure of the proposed method.

The details and significance of each step are as follows.

Feature Extraction using VGG16. A convolutional neural network (CNN) is used to extract the features (edges, colors, etc.) of each photo. In this study, VGG16 [5], a CNN model that has been trained on a large data set, is employed. By using the 13 layers, excluding the coupling layers, 8192 dimensional features can be obtained.

Dimensionality reduction and visualization using t-SNE. t-distributed stochastic neighbor embedding (t-SNE) is used to reduce the dimensionality of the features. The purpose of the dimensionality reduction is to make the features visible and understandable to the analyst. The hyper parameters of the clustering method are determined through trial and error by the analyst. t-SNE is a manifold learning

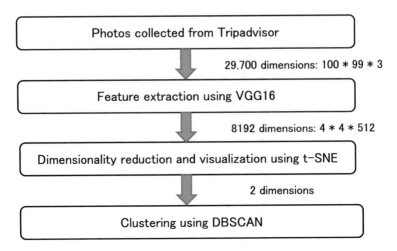

Fig. 49.1 Processing procedure of the proposed analytical method

method that can perform dimensionality reduction while preserving the positional relationships between elements in higher dimensions [6]. In this study, we reduce the dimensionality to two dimensions.

Clustering using DBSCAN. Density-based spatial clustering of applications with noise (DBSCAN) is used to cluster a large number of photos arranged in two-dimensional space. DBSCAN has the following advantages: (1) it does not require a certain number of clusters, and (2) it does not assume that the clusters are spherical in shape. On the other hand, it requires a neighborhood distance and a minimum number of components as hyper parameters, and it has the disadvantage that the clustering results change dramatically depending on the values of these parameters [7]. Therefore, in this study, appropriate hyper parameters are set based on the configurations visualized in two-dimensional space by t-SNE.

49.5 Findings and Discussion

49.5.1 Clustering Results

Figure 49.2 shows the results of dimensionality reduction and visualization using t-SNE on the collected set of photos. The hyper parameters of t-SNE in this study are perplexity: 20, and number of iterations: 5000. The boundaries between clusters are ambiguous, as shown in Fig. 49.2. Tourists take pictures of various compositions according to their individual interests. The results of the visualization show this diversity.

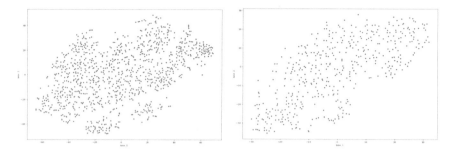

Fig. 49.2 Visualization results using t-SNE (Kenrokuen garden on the right and Fushimi Inari Shrine on the left)

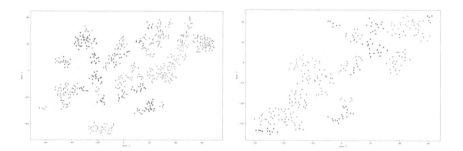

Fig. 49.3 Clustering results with DBSCAN (Kenrokuen garden on the right and Fushimi Inari Shrine on the left)

Next, clustering by DBSCAN is performed on the group of elements reduced to two dimensions by t-SNE. As a result, 42 clusters are obtained for Fushimi Inari Shrine and 22 clusters for Kenrokuen Garden. The cluster distribution is shown in Fig. 49.3. Note that the color of the clusters in Fig. 49.3 is limited to 9 colors for visibility, but there is no relationship between the clusters and their color. The hyper parameters of DBSCAN in this study are density-reachability: 3, and density-connectivity: 6. These values are set so as not to miss any clusters that can be seen in Fig. 49.2.

49.5.2 Evaluation

To verify the validity of these clustering results, all photos comprising each cluster are visually checked. When there is a common subject that accounts for more than half of the photos comprising each cluster, this is designated as the *main subject*. The clustering results with their evaluations of precision, recall, and *F*-values are shown in Table 49.2; precision indicates the percentage of agreement with the main subject in each cluster. Recall indicates the probability that each photo is classified as

Table 49.2 Evaluation of the clustering results

	Number of clusters	Precision	Recall	F-value
Kenrokuen garden	22	0.431	0.736	0.544
Fushimi Inari shrine	42	0.691	0.540	0.606

belonging to an appropriate cluster. The F-value is calculated as the harmonic mean of these two values.

Fushimi Inari Shrine achieves greater than 50% in both recall and precision, and by analyzing the photos in each cluster, we can determine tourist interest at low cost. On the other hand, the precision of Kenrokuen is much lower than that of Fushimi Inari-taisha Shrine. In a natural park such as Kenrokuen, most of the scenes consist of trees, which makes it difficult to distinguish between compositions.

49.6   Conclusion

In this study, photos posted on Tripadvisor were clustered to discover tourist interests. The analyses for the two targeted tourist attractions resulted in F-values greater than 0.5. The group of photos included in the clusters shows the interest of a large number of tourists. These photos capture many of the attractions that attraction operators wish to promote. On the other hand, there are also many compositions in these clusters that do not appear in guidebooks or official web pages. These compositions still hold appeal for tourists, and the proposed method is useful for discovering these unknown tourism resources.

Acknowledgements   This work was supported by JSPS KAKENHI Grant Number JP 19K12569.

References

1. Paolanti, M., Mancini, A., Frontoni, E., Felicetti, A., Marinelli, L., Marcheggiani, E., Pierdicca, R.: Tourism destination management using interest analysis and geo-location information: a deep learning approach. Inf. Technol. Tour. **23**(2), 241–264 (2021)
2. Padilla, J.J., Kavak, H., Lynch, C.J., Gore, R.J., Diallo, S.Y.: Temporal and spatiotemporal investigation of tourist attraction visit interest on Twitter. PLoS ONE **13**(6), e0198857 (2018)
3. tsunagu Japan. https://www.tsunagujapan.com/top-30-attractions-in-japan-for-international-tra velers/. Last accessed 9 July 2022
4. Tsujioka, S., Watanabe, K., Tsukamoto, A.: Tourism analysis using user-generated content: a case study of foreign tourists visiting Japan on TripAdvisor. Tour. Sustain. Dev. Rev. **1**(1), 57–64 (2020)
5. Simonyan, K., Zisserman, A.: Very deep convolutional networks for large-scale image recognition. arXiv preprint arXiv, 1409–1556 (2014)

6. Van der Maaten, L., Hinton, G.: Visualizing data using t-SNE. J. Mach. Learn. Res. 9(11) (2008)
7. Ester, M., Kriegel, H.P., Sander, J., Xu, X.: A density-based algorithm for discovering clusters in large spatial databases with noise. Kdd **96**(34), 226–231 (1996)

Chapter 50
Combining Contrast Limited Adaptive Histogram Equalization and Canny's Algorithm for the Problem of Counting Seeds on Rice

Luyl-Da Quach⬤, **Phuc Nguyen Trong**⬤, **Khang Nguyen Hoang**⬤, and **Ngon Nguyen Chi**⬤

Abstract Counting the number of rice seeds is essential in assessing the quality of rice varieties such as yield, rice diseases (sloppy rice), etc. Besides the machine learning approach, there are disadvantages. Contrary to machine learning, the counting approach utilizes image processing. This research contributes to counting the number of rice grains per panicle by combining contrast limited adaptive histogram equalization and Candy algorithm on a data set of 150 rice samples. The method is conducted through the steps of denoising, converting RGB color channels to LAB, image segmentation and contouring, and finally counting. The study's results were evaluated by comparing counting by hand and by algorithm; the error result was ~0.902%.

50.1 Introduction

In an image, counting the number of instances of an object is approached from two different aspects: (1) counting from the trained object detector and (2) training the object counter. For the first method, training the object requires an extensive data set of objects and is given the object's label, appearance, and the result of a classifier. With the second method, the approach is quite simple when just providing the object image to count the number of objects with the image processing method [1].

L.-D. Quach (✉) · P. N. Trong · K. N. Hoang
FPT University, Can Tho, Vietnam
e-mail: luyldaquach@gmail.com

P. N. Trong
e-mail: phucntce160315@fpt.edu.vn

K. N. Hoang
e-mail: KhangNHCE171197@fpt.edu.vn

N. N. Chi
Can Tho University, Can Tho, Vietnam
e-mail: ngonnc@ctu.edu.vn

C. So-In et al. (eds.), *Information Systems for Intelligent Systems*, Smart Innovation, Systems and Technologies 324, https://doi.org/10.1007/978-981-19-7447-2_50

The object detector-based approach has had many studies applying machine learning methods to implement. In particular, research [2] has been performed to determine the total number of wheat grains on an actual photograph in the field by applying the K-mean clustering algorithm to extract the morphological and color features of the plant wheat with accuracy up to 98.5%. Research [3, 4] uses machine learning, and Yolo shows relatively high accuracy on the data set. Research [5, 6] uses a deep learning neural network to count the number of cereals in a tray at the laboratory. However, as mentioned, the use of this method will be limited because of the need for large data sets.

Contrary to the above approach, the approach using counters by image processing overcomes the limitations of the data set and shows positive results. The research [7–9] used computer vision techniques to count shrimp larvae in the laboratory. The study [10] applied the Canny contour extraction technique to counting the number of wheat seeds and tomatoes with an accuracy of up to 94.34%. Research [11] uses RGB images and thermal images to assess whether the quality of rice grains is full or unsaturated, which will affect rice yield. Research [12] uses segmentation and touch algorithm to perform counting of rice grains on a tray.

In general, the studies show that the effectiveness of the object counting algorithm with the primary purpose of affecting the quality of the objects will affect the productivity from which the evaluation can be made quality, predicting the yield of the respective object. However, no studies have counted the number of wet rice grains per cotton. This also provides similar solutions for evaluating the yield and quality of rice grains, such as saturated rice or unsaturated rice flat or not, using unmanned equipment to assess rice yield. For that reason, in this study, the research team counted the number of rice grains by the RGB color image recognition method and applied the Canny contour extraction technique under the condition that the rice was still on cotton in the laboratory. Experiment with steps as shown in Fig. 50.1.

- Step 1: The wet rice data set was obtained from the Mekong Delta Rice Institute (10.1229° N, 105.5860° E). The image of wet rice was taken with Samsung Galaxy Note 10 with camera model SM-N770F, and the image size is 1816×4032 pixels.
- Step 2: Resize the image for consistency between photos.
- Step 3: Label the image with manual grain counting to perform the comparison.
- Step 4: Denoise the image.
- Step 5: Segment the image and extract the contour.
- Step 6: Export the results of the number of seeds and evaluate the results.

Fig. 50.1 Process of data processing to perform counting of rice grains per cotton

50.2   Enhancement of Images

Photos are taken in a laboratory, so noise affects the image such as lights, camera stability. These factors affect the image quality and will cause some noise in the image. Therefore, it is necessary to enhance and blur to limit noise in the image before going into object contour detection. The image enhancement and noise processing are handled as follows.

50.2.1   Converting Images from RGB Color to LAB Color

Before entering contour decomposition, image denoising is required for higher image quality and is a stepping stone for more accurate contour decomposition. The RGB color space is an associative color space where all colors are obtained by a linear combination of red, green, and blue values. The three channels are correlated by the amount of light hitting the surface experimentally [13]. Besides, the LAB color space will have three components L: brightness (intensity), a: color components from green to magenta, b: color components from blue to yellow.

In the RGB color space, color information is separated into three channels, but the same three channels also encode luminance information. Whereas LAB color space, L channel is independent of color information and encodes only luminance. The other two channels are color-coded. From the characteristics of the LAB color channel, we will apply L channel to help reduce noise and improve image quality.

```
lab = cv.cvtColor(resizeImg, cv.COLOR_BGR2LAB)
```

With resizeImg, the image has been resized with the size of 300×700 pixels based on the original image.

50.2.2   Separation of Images into Different Color Channels

Split the above-converted image to different channels like Fig. 50.2 using split statements like below.

```
l, a, b = cv.split(lab)
```

50.2.3   Contrast Limited Adaptive Histogram Equalization

Contrast limited adaptive histogram equalization (CLAHE) application for adaptive histogram equalization has little contrast. Histogram equalization is the adjustment

Fig. 50.2 Image illustrating the process of image denoising and enhancement, from left to right and top-down is after separating color channels corresponding to L channel, a channel, b channel; image after merge image clahe and channel a, b; denoised image from the original image, denoised image

of the histogram to the equilibrium state, making the distribution of pixel values not clustered at a narrow interval but "stretched". The camera is often affected by lighting conditions. That causes many photos to be dark or too bright. Histogram equalization is a compelling image pre/post-processing method. Especially in many problems in computing vision, this image preprocessing method gives very high data quality, significantly improving the quality of deep learning models [14].

The steps to calculate the histogram balance in the adjustment are performed according to the following steps:

- Step 1: Call the transformation function to be determined as K(i) with i in the interval [0,255].
- Step 2: Count the number of pixels for each light level, the result is a histogram H(i)
- Step 3: Calculate the "cumulative function" Z for each light level according to the formula:

$$Z(i) = \sum_{j=0}^{i} H(i) \tag{50.1}$$

In which, Z(i) is the total number of pixels with value $\leqslant$ i.

- Step 4: Using the transform function K at a brightness level i is calculated as follows:

$$K(i) = \frac{Z(i) - \min(Z)}{\max(Z) - \min(Z)} * 255 \tag{50.2}$$

However, the feature of histogram equalization will make the background more visible, but the object will become blurred because its histogram is not limited to a specific area. To solve this problem, adaptive histogram equalization (CLAHE) is used. The image is divided into small blocks called "tiles" (Size is 8×8 by default in OpenCV). Then, each of these blocks is histogram equalized as usual. So in a small area, the histogram will be confined to a small area (unless there is noise). If there is noise, it will be amplified. To avoid this, a contrast limit is applied. If any histogram pane exceeds the specified contrast limit (40 by default in OpenCV), those pixels will be clipped and uniformly distributed to the other panes before applying histogram equalization. After equalization, to remove artifacts in cell borders, apply bilinear interpolation.

The specific application of adaptive histogram equalization is used below the code, the reflection limit (clip limit) is specified as 1, and the tile size is 8×8.

```
clahe = cv.createCLAHE(clipLimit=1.0, tileGridSize=(8,8)) cl =
clahe.apply(l)
```

50.2.4 Combining CLAHE with l-channel Enhanced in Step 3 with Channels a and B

This combining step will help the image improve image quality and eliminate noise. At this time, the image of the rice grain can be seen clearly and clearly. The result is when in the LAB color channel has a blue-violet background.

```
limg = cv.merge((cl, a, b))
```

50.2.5 Converting Images from LAB Color to RGB Color and Enhancement of Image Quality

Since the picture is in the LAB color channel, the picture will not be the same as the real thing. In this step, research will convert the color channel of the image back to RGB and get a denoised and enhanced image.

```
final = cv.cvtColor(limg, cv.COLOR_LAB2BGR)
```

Image enhancement increases the brightness of the wet rice grain in the image, which makes the contrast between the wet rice grain and the background of the stem and.

leaves more transparent, which is beneficial for the extraction of contour features of the rice grain. Wet rice grain is more accurate.

50.3 Image Segmentation and Wheat Ear Contour Extraction

After image enhancement, there will be a clear difference between the wet rice grain's color and the stem's background color and leaves. The area containing the wet rice seeds will be lightened against the background. In the subsequent step, research will go to the border counting process, calculating the total number of rice grains.

The process of determining the contour and calculating the number of rice grains is done according to the steps below:

50.3.1 Converting Image to Grayscale

In image processing, converting color images to grayscale images is extremely common. A color image is just a collection of numeric matrices of the same size. When we want to process the information on an image, it is easier to process the data on a single numeric matrix instead of multiple numeric matrices. The original image is saved as RGB. This means there will be three corresponding gray matrices for red, green, and blue. Now, we need to find a way to combine these three gray matrices into a single matrix. The matrix, after being aggregated, will be called grayscale converting. One of the popular recipes for doing that is

$$Y = 0.2126R + 0.7152G + 0.0722B \tag{50.3}$$

where Y is the gray matrix to be searched, R is the red-gray matrix of the image, G is the gray-green matrix of the image, and B is the gray-blue matrix of the image.

This conversion is supported by OpenCV using the cvtColor function with the COLOR_BGR2GRAY parameter.

```
src_gray = cv.cvtColor(final, cv.COLOR_BGR2GRAY)
```

50.3.2 Blur Image

Image blur is used a lot in image processing and has many important roles. The blur effect will help reduce noise in the image, smoothing the image. Smoothing the image will reduce edge sharpness, instead, the smooth area will spread out.

The window size of the filters is usually a positive integer (3, 5, 7, 9, ...). Because of the odd size, we will only have 1 pixel in the center of the kernel. Kernels are usually designed to be square (width = height). Many popular noise reduction filters exist box filter, Gaussian Filter, and median filter. In this study, box filter will be applied to reduce noise.

The box filter is designed by setting each value on the filter to: $\frac{1}{W*H}$.

That is if the width and height of the filter are equal to 3, the value on the convolve window will be: 1/9.

```
src_gray = cv.blur(src_gray, (3,3))
```

50.3.3 Converting the Image to Canny to Reveal the Contours

There are often components such as smooth areas, corners/edges, and noise in images. Edges in an image have essential features, usually belonging to the object in the image. Therefore, to detect edges in the image, we will apply the Canny algorithm [15] to process.

The Canny algorithm has four main steps: noise reduction, gradient calculation and direction, non-maximum suppression, and threshold filtering. OpenCV supports implementing the Canny algorithm through the Canny() function.

```
canny = cv.Canny(src_gray, 220, 255)
```

50.3.4 Finding Contours

Contour is a closed curve and is represented by a list of points. This set of points has the same color or intensity of light. Contours have practical applications in image processing, especially in object zoning and recognition.

In OpenCV, contour works on single-channel images but is most effective on binary images. To find the correct contour, we need to binarize the image. Specifically, the Canny edge decimation method applied above is one of the binary methods. When we binarize the image, we will get an image with a black background and a white object.

Finding contours in OpenCV, we will use findContours() function. This function outputs a list of contours found; each contour is a list of point coordinates. The contours query parameter used in the study is CV_RETR_TREE, and the method to estimate the number of vertices of each contour CV_CHAIN_APPROX_NONE helps to list all the vertices of the contour polygon.

```
(cnt, heirarchy) = cv.findContours(canny.copy(),
cv.RETR_EXTERNAL, cv.CHAIN_APPROX_NONE)
```

50.3.5   Contour Estimation and Approximation

Using the Ramer–Douglas–Peucker (RDP) algorithm [16], the contour approximation aims to simplify a polyline by reducing its vertices to a threshold value. In other
words, we take a curve and reduce its number of vertices while keeping its shape.

The application of contour approximation will help estimate the shape of the
contour and thereby help us to illustrate the contour lines corresponding to each
grain of rice.

```
contours_poly = [None] * len(cnt) boundRect = [None] * len(cnt)
centers = [None] * len(cnt) radius = [None] * len(cnt)
for i, c in enumerate(cnt):
   contours_poly[i] = cv.approxPolyDP(c, 3, True)
   boundRect[i] = cv.boundingRect(contours_poly[i])
   centers[i], radius[i] =
cv.minEnclosingCircle(contours_poly[i])
```

50.3.6   Creating a Box for the Border to Draw and Finish
Counting the Number of Rice Grains

In the last step, after determining the shape of each border, we will create a box for
each border. Here, each box will correspond to the number of rice grains, which is
also the research paper's result (Fig. 50.3).

```
drawing = np.zeros((thresh.shape[0], thresh.shape[1], 3),
dtype=np.uint8)
```

Fig. 50.3 Illustrating the results of the image obtained after the process of determining the contour
and counting the number of rice grains, from the left is the result of the image obtained when
converting to grayscale, after blurring, the image after applying the Canny algorithm; the image of
the number of contours is shown after calculating the approximation, the image is obtained after
boxing the contour

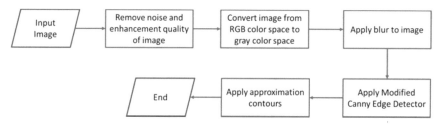

Fig. 50.4 Schematic estimation of the number of wet rice seeds based on edge detection using Canny algorithm

```
for i in range(len(cnt)):
    color = (rng.randint(0,256), rng.randint(0,256),
rng.randint(0,256))
    cv.drawContours(drawing, contours_poly, i, color)
    cv.rectangle(drawing, (int(boundRect[i][0]),
int(boundRect[i][1])),
        (int(boundRect[i][0]+boundRect[i][2]),
int(boundRect[i] +boundRect[i])), color, 2)
print('Number of Rice: ', len(cnt))
```

Figure 50.4 shows a diagram of estimating the total number of wet rice grains using the Canny edge detection technique. The diagram applies additional noise reduction and image enhancement techniques. In the paper, after applying the Canny edge detection technique, we also apply the contour approximation technique to work how the Canny algorithm applies, thereby helping to consider the accuracy of the application. Using the Canny edge detection technique in calculating total grain on wet rice.

50.4   Statistical Analysis

To determine the algorithm's accuracy, we manually counted 150 rice samples. The error of each grain is determined by subtracting the grain of the corresponding sample counted by the algorithm. If the error is negative, the corresponding sample counted by the algorithm has more particles than the sample counted manually and vice versa. When considering the algorithm results, it can be seen that the error based on each sample is quite high ($20 \rightarrow 35\%$). However, in general, the error based on manual counting compared to counting by an algorithm is not much different; below is the calculation of the error based on the population in Table 50.1.

Table 50.1 Statistical results of the number of samples and the error between hand counting and algorithm

Hand counting	Algorithm counting	Wrong number (rice)	Wrong number (%)
9.862	9.773	89	~0.902%

50.5 Conclusion

Through the table of analytical data, we can see that, if considering each sample, the algorithm still has a significant error rate (20 → 35%). Considering a large set of samples, the algorithm gives results close to the total number of manually counted samples with an error rate (~0.902%), thereby showing the algorithm's ability to apply in reality. The research has improved the algorithm compared to previous studies, including improving the quality of images before inclusion to determine the contour of the grain. Adjust the contour level to suit the image characteristics of the samples for more accurate counting. However, there are some limitations of the algorithm. In calculating the number of contours, the algorithm still cannot clearly distinguish between the stem's and the seed's contours. But in terms of generality on a large data set, the algorithm still gives results with a low error rate.

References

1. Lempitsky, V., Zisserman, A.: Learning to count objects in images, 9
2. Xu, X., Li, H., Yin, F., Xi, L., Qiao, H., Ma, Z., Shen, S., Jiang, B., Ma, X.: Wheat ear counting using K-means clustering segmentation and convolutional neural network. Plant Methods **16**, 106 (2020). https://doi.org/10.1186/s13007-020-00648-8
3. Armalivia, S., Zainuddin, Z., Achmad, A., Wicaksono, M.A.: Automatic counting shrimp larvae based you only look once (YOLO). In: 2021 International Conference on Artificial Intelligence and Mechatronics Systems (AIMS), pp. 1–4. IEEE, Bandung, Indonesia (2021). https://doi.org/10.1109/AIMS52415.2021.9466058
4. Yeh, C.-T., Ling, M.-S.: Portable device for ornamental shrimp counting using unsupervised machine learning. Sens. Mater. **33**, 3027 (2021). https://doi.org/10.18494/SAM.2021.3240
5. Liu, T., Chen, W., Wang, Y., Wu, W., Sun, C., Ding, J., Guo, W.: Rice and wheat grain counting method and software development based on Android system. Comput. Electron. Agric. **141**, 302–309 (2017). https://doi.org/10.1016/j.compag.2017.08.011
6. Deng, R., Tao, M., Huang, X., Bangura, K., Jiang, Q., Jiang, Y., Qi, L.: Automated counting grains on the rice panicle based on deep learning method. Sensors **21**, 281 (2021). https://doi.org/10.3390/s21010281
7. Morimoto, T., Zin, T.T., Itami, T.: A study on abnormal behavior detection of infected shrimp. In: 2018 IEEE 7th Global Conference on Consumer Electronics (GCCE), pp. 291–292. IEEE, Nara (2018). https://doi.org/10.1109/GCCE.2018.8574860
8. Khantuwan, W., Khiripet, N.: Live shrimp larvae counting method using co- occurrence color histogram. In: 2012 9th International Conference on Electrical Engineering/Electronics, Computer, Telecommunications and Information Technology, pp. 1–4. IEEE, Phetchaburi, Thailand (2012). https://doi.org/10.1109/ECTICon.2012.6254280
9. Thai, T.T.N., Nguyen, T.S., Pham, V.C.: Computer vision based estimation of shrimp population density and size. In: 2021 International Symposium on Electrical and Electronics Engineering (ISEE), pp. 145–148. IEEE, Ho Chi Minh, Vietnam (2021). https://doi.org/10.1109/ISEE51682.2021.9418638
10. Rozikin, C., Suriansyah, M.I., Suharso, A., Nugroho, M.F.E., Sanjaya, N.: The detection and counting of object bottles in the boxes based on image processing using watershed algorithm, Vol. 10 (2021)
11. Kumar, A., Taparia, M., Madapu, A., Rajalakshmi, P., Marathi, B., Desai, U.B.: Discrimination of filled and unfilled grains of rice panicles using thermal and RGB images. J. Cereal Sci. **95**, 103037 (2020). https://doi.org/10.1016/j.jcs.2020.103037

12. Tan, S., Ma, X., Mai, Z., Qi, L., Wang, Y.: Segmentation and counting algorithm for touching hybrid rice grains. Comput. Electron. Agric. **162**, 493–504 (2019). https://doi.org/10.1016/j.compag.2019.04.030
13. Loesdau, M., Chabrier, S., Gabillon, A.: Hue and saturation in the RGB color space. In: Elmoataz, A., Lezoray, O., Nouboud, F., Mammass, D. (eds.) Image and Signal Processing, pp. 203–212. Springer International Publishing, Cham (2014). https://doi.org/10.1007/978-3-319-07998-1_23
14. Zuiderveld, K.: Contrast limited adaptive histogram equalization. Graphics Gems, 474–485
15. Rong, W., Li, Z., Zhang, W., Sun, L.: An improved Canny edge detection algorithm. In: 2014 IEEE International Conference on Mechatronics and Automation, pp. 577–582. IEEE, Tianjin, China (2014). https://doi.org/10.1109/ICMA.2014.6885761
16. Zhao, L., Shi, G.: A method for simplifying ship trajectory based on improved Douglas-Peucker algorithm. Ocean Eng. **166**, 37–46 (2018). https://doi.org/10.1016/j.oceaneng.2018.08.005

Chapter 51
Remote Health Monitoring System Using Microcontroller—Suitable for Rural and Elderly Patients

Md. Khurshed Alam, Imran Chowdhury⬤, Al Imtiaz⬤, and Md. Khalid Mahbub Khan⬤

Abstract Amid and post-COVID-19 pandemic, the matter of being in touch with patients to monitor their health matrices became somewhat challenging, especially in the rural areas of countries like Bangladesh and for elderlies. To address this issue, a patient health monitoring system is developed using a Programmable Intelligent Computer (PIC) microcontroller and Global System for Mobile Communications (GSM) protocol with the help of a pulse sensor, IR sensor, photodiodes, temperature sensor, etc., to measure 3 (three) crucial health matrices such as heartbeat/pulse, oxygen saturation level, and body temperature from a fingertip of the patient in 20 s remotely. Whenever the system measures the health matrices, it sends a short message service (SMS) report to a personal caretaker over GSM automatically. If the system finds any anomaly based on predefined threshold levels for each health parameter, it sends a SMS alert report to the designated doctor automatically as well. A prototype of the developed system is made, verified, and tested to be working perfectly as designed and programmed. In the experiment with the developed system, heart rate ranged from 61 to 105 bmp, body temperature ranged from 95.3 to 99.1 °F, and oxygen saturation was minimum at 97%. According to the set threshold levels, which led to an automatic SMS alert to the caretaker's mobile phone.

51.1 Introduction

The expanded utilization of versatile innovations and gadgets in the well-being zone has expedited unprecedented impact on the diagnostics and health monitoring systems. But, because of the lack of sufficient health infrastructure in rural areas, and considering the pandemic like COVID-19, it may not be conceivable for a nurturer

Md. K. Alam (✉) · A. Imtiaz · Md. K. M. Khan
Department of Computer Science and Engineering (CSE), University of Information Technology and Sciences (UITS), Dhaka 1212, Bangladesh
e-mail: khurshedfs@gmail.com

I. Chowdhury
Department of Electrical and Electronic Engineering (EEE), University of Information Technology and Sciences (UITS), Dhaka 1212, Bangladesh

© The Author(s), under exclusive license to Springer Nature Singapore Pte Ltd. 2023
C. So-In et al. (eds.), *Information Systems for Intelligent Systems*, Smart Innovation, Systems and Technologies 324, https://doi.org/10.1007/978-981-19-7447-2_51

or a specialist to talk physically from one patient to another in a hospital for a regular well-being checkup persistently. As a consequence, the basic health matrices may not be found timely and effectively. Moreover, when therapeutic crises happen to the patient, they are regularly oblivious and incapable to press a crisis caution button [1]. Besides, amid the COVID-19 pandemic, keeping many patients together for a long duration or even getting in touch with them became an uncertain matter. Moreover, currently, in most cases, individuals living in rural areas cannot get benefited from preventive well-being administrations because of the absence of proper technology-oriented infrastructure, which leads to unfortunate and severe medical outcomes for those individuals. Furthermore, as the general population becomes older, the need for continuous support for the elderly grows with the disparity in family structure [2].

Ceaseless checking of an individual's well-being through alterable biomedical gadgets is currently conceivable with numerous alterable well-being units, especially in urban areas. These alterable gadgets consistently measure the patient's health matrices and when a side effect or anomaly occurs, they may send data about the patient's well-being condition to the relatives and the specialist [2, 3]. Gadgets that give ceaseless observing of these patients are over the top expensive, especially for rural and elderly patients, and require a prepared workforce to utilize them [4].

Numerous studies on similar topics of interest are done in recent years [1–15]. In [4], the authors tried to solve the mentioned issues by designing a system using Arduino and Bluetooth technology for remotely monitoring heart rate, blood pressure, and body temperature. In [11], the authors used Raspberry Pi and GSM technology to monitor the same health matrices remotely, and in [13], the authors used NodeMcu and Wi-Fi technology to measure and monitor heartbeat remotely. According to almost all the mentioned studies, taking advantage of readily available, reliable, smart, and connected gadgets could solve or at least ease all those issues discussed above. Platforms of smart gadgets are a reality now in the technologically advanced modern era that is changing our lives already. In the coming years, the gap between urban and rural people will become shorter and shorter and they will be able to keep pace with other places and incorporate more technological benefits into their life [10]. One of those beneficiary areas could be the well-being sector. And, one of such smart and connected gadgets is the microcontroller, which is readily available, relatively cheap but reliable. Microcontrollers are also known as embedded controllers since they and their supporting circuitry are most of the time embedded or constructed into the device they are programmed to control [16, 17].

Keeping the discussed issues in mind, in the developed health monitoring system presented in this paper, an alterable gadget or device is intended to quantify indispensable qualities, for example, heartbeat, blood saturation, and body temperature, which legitimately concern well-being. These quantities or health matrices are collected by sensors from the tip of the patient's finger and smartly processed by a microcontroller to check for anomalies based on predefined threshold levels for each quantity or health parameter. The after-effects of this checking are programmed to be transferred to GSM-based specific mobile phones with discernible visual alarm making the device or system a connected one. The aim of the presented work is to make a contribution to bringing comfort in medical works and closing the physical gap between the doctors or well-being specialists and patients by adopting the following benefits: (i) ease of use, (ii) patient convenience, (iii) remote detection and alert, (iv) affordable medical cost, and (v) better life for rural and elderly people.

51.2   Methods and Materials

51.2.1   Components and Peripheral Devices

The developed health monitoring system incorporates 4 (four) major components to do its function: a microcontroller as the brain of the system; a heartbeat, blood saturation, and temperature sensor to measure or collect the health matrices; a GSM module to transfer the health matrices remotely; and an LCD display to show the measured health matrices on demand. The total list of required components and peripheral devices is provided in Table 51.1.

51.2.2   Microcontroller (PIC16F876A)

The main component used to develop the system is a microcontroller, which is the brain for the system's function. A microcontroller (sometimes called μC or MCU) can be referred to as a little computer that is a programmable chip containing a processor center, memory, and programmable input/output (I/O) peripherals. Microcontrollers have ended up common and exceptionally well known in numerous regions, and can be found in numerous categories of items like tech gears, microwave broilers, refrigerators, automobiles, etc. Even tall tech robots contain microcontrollers [17, 18]. The microcontroller used in the presented work is PIC16F876A (Fig. 51.1), which is from the PIC family. Microchip technology is the company that makes PIC (commonly pronounced "pick") family of microcontrollers. This family is basically based on the PIC1650 originally developed by the Microelectronics Division of General Instrument. Although initially, the term PIC was referred to as "Peripheral Interface Controller", currently it stands for "Programmable Intelligent Computer" [19].

51.2.3   System Model and Block Diagram

The developed system is comprised of both equipment and computer programs. On the equipment side, 3 (three) types of sensors are utilized to measure the health matrices. Subsequently, a microcontroller coordinates with the GSM module to

Table 51.1 System's components and peripheral devices

Components/devices	Specifications/values
Veroboard	2 × 8 cm, dotted, single side
Transformer (center tapped)	Primary: 220 V at 50 Hz, Secondary: 12–0–12 V, 1 A
Crystal oscillator	4–16 MHz
Buck converter module (LM2596, DC–DC step-down)	Input voltage: 3–40 V, Output voltage: 1.2–37 V, Output current: 3A (max), 150 kHz
IR LED (infrared transmitter)	700 nm–1 mm wavelength, 30°–40° viewing angle
Photodiode (PD333-3B/H0/L2)	5 mm round, V_R: 32 V
Bipolar transistor (8050)	NPN, Pc = 1 W, Vce = 25 V, Veb = 6 V, Ic = 1.5 A, 100 MHz
ON/OFF push button	6 × 6 × 4.3 mm
LCD display (1602A)	16 × 2, 5 V
Op-Amp (LM324)	Quad Op-Amp, 1.3 MHz bandwidth, 100 dB voltage gain, 3–30 V
GSM module (SIM800)	850 MHz GSM, GPRS multi-slot class 12, 4.5–12 V
Resistor (RES)	½ W, 220 Ω, 2.2 kΩ, 10 kΩ, 33 kΩ, 1 MΩ
Variable resistor (POT)	201, 5 kΩ
LED (red)	5 mm round, V_R: 5 V, V_F: 3–3.6 V
Voltage regulator (LM7805)	Input voltage: 7–25 V, Output voltage: 4.8–5.2 V, Operating current: 5 mA, Output current: 1.5 A (max)
Temperature sensor (LM35)	− 55 to 150 °C Range, Linear 10 mV/°C scale factor, 4–30 V
Microcontroller (PIC16F876A)	28-pin package, 22 I/O pins, 20 MHz clock input, 14 KB program memory, 368 bytes RAM, 256 bytes EEPROM, 10-bit ADC 5 channels, two analog comparators, 2–5.5 V
Non-polarized capacitor (ceramic)	100 V (100 μF)
Polarized capacitor (electrolytic)	16 V (1000 μF), 25 V (100 μF), 35 V (1000 μF), 50 V (0.1 μF, 10 μF)

Fig. 51.1 PIC16F876A microcontroller [19]

collect, process, and transfer the data. In addition, LCD shows the pulse result as well. On the programming side, a set of computer codes is written to perform the desired functions. A block diagram is presented in Fig. 51.2 to provide an overall easy visualization of the equipment side of the entire system including all of its peripheral devices. From the diagram, a discrete idea of all the incorporated modules/devices and their responsibilities can be achieved on a macro level. On the input side of the microcontroller, there are 3 (three) sensors as mentioned: heartbeat, blood saturation, and temperature sensor. On the output side of the microcontroller, there is an LCD display and a GSM module. There are 2 (two) GSM receivers considered in this system: one for the doctor and another for the personal caretaker of the patient. As the diagram indicates, the system is running on a 12 V DC supply converted from the 220 V AC using a center-tapped transformer. The LCD is fed by the 12 V DC whereas the microcontroller and GSM module is fed by a 5 V DC, which is converted from the 12 V DC using a fixed voltage regulator. The sensor modules are fed by a 3.3 V DC which is also converted from the 12 V DC using a DC–DC step-down converter.

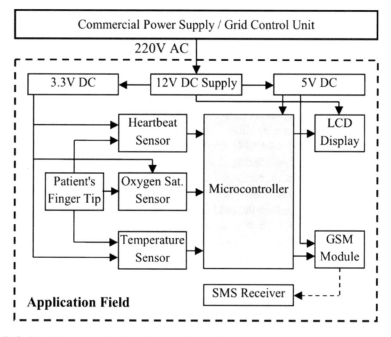

Fig. 51.2 Block diagram of the developed health monitoring system

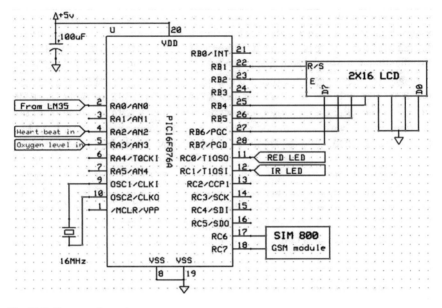

Fig. 51.3 System schematic

51.3   Electronic Circuit/Hardware Interfacing

The microcontroller is the main component around which the whole system is developed. PIC16F876A microcontroller is used in this case to make decisions such as generating and sending alert SMS based on the sensor inputs. Figure 51.3 shows the schematic of circuit interfacing of the system, where temperature, heartbeat, and blood saturation sensors are connected to pin 2, 4, and 5, respectively. The heartbeat and blood saturation sensors are made with a combination of IR LED, photodiode, and regular LED (red), where the red LED and IR LED are connected to pin 11 and 12, respectively. The LCD display is connected to pin 22–23, 25–28, and the GSM module is connected to pin 17–18. Other connections are maintained accordingly.

51.4   Microcontroller Programming

A microcontroller can be called a platform for programming electronic devices. The microcontroller used in the presented work is PIC16F876A from the PIC family. Therefore, the functionality of the microcontroller depends on the programming that follows the general attributes of PIC programming [18, 20]. The programming codes that are written for the system in the presented work are described using the flowchart in Fig. 51.4. According to the flowchart, every time the health matrices are checked an SMS is generated and gets sent to the personal caretaker of the patient. But whenever

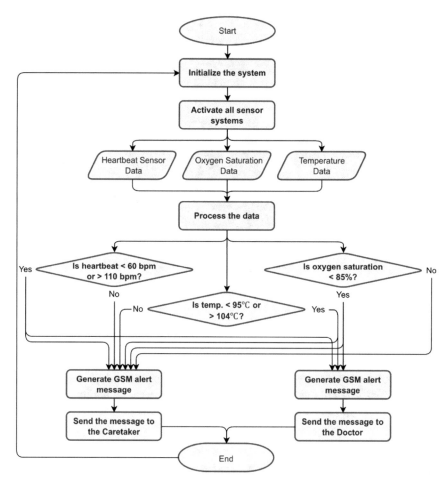

Fig. 51.4 Programming flowchart of the developed health monitoring system

any of the 3 (three) health matrices cross the defined threshold levels the SMS gets sent to both the personal caretaker and the doctor.

51.5 Results and Discussion

51.5.1 Prototype Implementation

The developed health monitoring system is practically implemented based on the block diagram and circuit design discussed above using the components and peripheral devices mentioned in Table 51.1. The written programming codes based on the

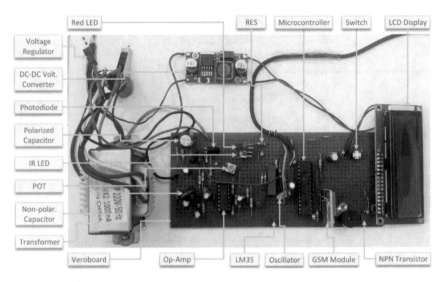

Fig. 51.5 Implemented prototype of the developed system with labeling

flowchart discussed above are burnt on the microcontroller to achieve its functionality. The implemented prototype is created by soldering most of the electronic components onto a 2 × 8 cm single-sided Veroboard. The transformer, voltage regulator, and DC–DC step-down converter are kept separate from the Veroboard, which are later glued to a white plastic board together with the Veroboard for easier mobility. Figure 51.5 shows the prototype of the practical device with the labeling of its components.

51.5.2 Results from Prototype Testing

For verification of the implemented prototype's operation according to the design and burnt program, a demonstration and its outcomes are shown in this section where 2 (two) dummy patients were invited to test their health matrices. The outcomes are justified according to the microcontroller programming written for the system and their logical workflow, which satisfied the desired results. Figures 51.6, 51.7 and 51.8 show some of the snapshots of the demonstration and its results, which correspond to the previously mentioned process steps of the programming flowchart. Table 51.2 shows the accumulated demo test results.

Fig. 51.6 Dummy patient-1 placed his fingertip in the sensor module and the microcontroller is processing the data

Fig. 51.7 Health matrices of dummy patient-1 are displayed on the LCD after the sensor data is being processed

Fig. 51.8 Generated message containing the health matrices of dummy patient-2 along with a location identifier is being sent to the caretaker's mobile phone number

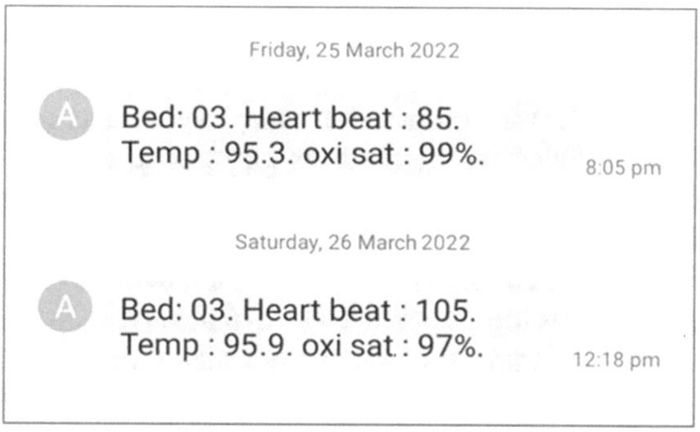

Friday, 25 March 2022

A   Bed: 03. Heart beat : 85.
Temp : 95.3. oxi sat : 99%.
     8:05 pm

Saturday, 26 March 2022

A   Bed: 03. Heart beat : 105.
Temp : 95.9. oxi sat : 97%.
     12:18 pm

Table 51.2 Accumulated test results (health matrices) for both dummy patients	Number of demo test	Heart beat (BPM)	Body temperature (°F)	Oxygen saturation (%)
	Patient-1, test-1	61	99.1	99
	Patient-2, test-1	85	95.3	99
	Patient-3, test-2	105	95.9	97

51.5.3 How the Demonstration Worked

After the prototype device is turned ON, the tip of one of the fingers (index finger in this case) of the dummy patient is placed in the sensor module, and a push-button (ON–OFF switch) is pressed so that the sensor module starts working. The sensor module is mainly comprised of 4 (four) components: IR LED, red LED, photodiode, and LM35. The IR LED is used to get the heartbeat/pulse data, the combination of a red LED and a photodiode is used to get the blood saturation data, and the combination of IR LED and LM35 is used to get the body temperature data. The procedure for getting the data is stated below step by step.

After the patient's finger is placed in the sensor module and the push-button is pressed, the IR LED flickers every 1000 ms to get the heartbeat or pulse data from the bloodstream. Then, the IR LED is turned OFF automatically and the red LED and photodiode turns ON to get the data for blood saturation based on light throughput from the red LED. After that, the red LED and the photodiode are turned OFF automatically and IR LED turns ON again to get the body temperature data with the help of LM35. All these happen in 20 s while the patient's fingertip is placed in the sensor module, and then the measured data get transferred to the microcontroller and appear in the LCD.

51.5.4 Discussion

In the "Results from Prototype Testing" subsection, the generation and sending of SMS alerts is demonstrated for one phone number only that is assigned to a caretaker. The SMS alert for the doctor is supposed to be generated and sent when any of the health matrices go below or beyond the threshold levels, which did not occur in this case for any of the dummy patients as per Table 51.2 and Fig. 51.4. The SMS is sent with an identifier "Bed: 03" to confirm the location or identification of the patient. This identifier could be anything else such as the apartment/street number if the patient is not staying at the hospital or the name/ID of the patient. The dummy patients were considered to be ones with the ability to lean and move their fingers to use the device with or without anyone's assistance. But if the patient is not able to (completely or partially) move his or her finger, then the sensor module needs to be separated to make a strap that can be attached to the patient's finger, which is under consideration now. The data collected using the device are not stored anywhere currently that can be used to make a progress curve, which is also under consideration.

The work presented in this paper aimed to bring contribution to bridging the gap between a doctor or specialist and a patient. This will potentially be helping, at least partially, in automating the workflow of regular medical checkups and minimizing the required human resources, which is essential in rural areas. It will also help in cutting down the requirements for doctors or specialists to make physical visits, which is challenging when the patients are not in the hospital and living in a rural area. The system will also help in reducing hospital stays, re-admissions, and waiting in a long queue for regular checkups as well, which is a big concern for elderly patients.

As mentioned in the literature review in the Introduction section, recently developed health monitoring system mostly relies on modern but a little more expensive electronic devices such as Raspberry Pi, Arduino, NodeMcu, etc. [1–15]. But the work presented in this paper employed a traditional microcontroller to cut down the cost of development since it is intended for application in rural areas of countries like Bangladesh and elderlies. Because, affording health monitoring devices or any medical gadgets for those people is almost a luxury that in most cases they do not have, especially for elderlies who do not have any earning sources or health insurance and are dependent on others.

51.6 Conclusion

In the post-COVID-19 era, because of the financial setback, issues of social distancing, and already existing lack of proper health infrastructures, going to a hospital for regular health checkups or affording costly health monitoring gadgets became an inconvenience and somewhat challenging for the people living in rural areas, especially for the elderly, in countries like Bangladesh. To address and solve this issue, the presented remote health monitoring system is developed. The system is practically implemented by creating a prototype, where most of the electronics are soldered onto a 2 × 8 cm Veroboard. The system takes 12 V DC supply converted from 220 V AC using a center-tapped transformer. The LCD is fed by the 12 V DC, the microcontroller and GSM module are fed by a 5 V DC, and the sensor modules are fed by a 3.3 V DC using a voltage regulator and a buck converter (DC–DC step-down converter). In the experiment with the developed system, the minimum and maximum heart rate were found at 61 and 105bmp, respectively, the minimum and maximum body temperature were found at 95.3 and 99.1 °F, respectively, and the minimum oxygen saturation was found at 97%. According to the set threshold levels (Fig. 51.4), this led to an automatic SMS alert to the caretaker's mobile phone. Future scopes of the presented work may involve an extended health monitoring system based on the Internet of Things (IoT) and designing a web user interface to perform several control activities and show real-time graphs and history, etc.

References

1. Jebane, P., Anusuya, P., Suganya, M., Meena, S., Diana, M.: IoT based health monitoring and analysing system using Thingspeak Cloud & Arduino. Int. J. Trendy Res. Eng. Technol. **5**, 1–6 (2021)
2. Roman Richard, A.A, Sadman, M.F., Mim, U.H., Rahman, I., Rahman Zishan, M.S.: Health monitoring system for elderly and disabled people. In: 2019 International Conference on Robotics, Electrical and Signal Processing Techniques (ICREST), pp. 677–681. IEEE (2019)
3. Joyia, G.J., Liaqat, R.M., Farooq, A., Rehman, S.: Internet of Medical Things (IOMT): applications, benefits and future challenges in healthcare domain. J. Commun. **12**, 240–247 (2017). https://doi.org/10.12720/jcm.12.4.240-247
4. Majumder, S., Rahman, M.A., Islam, M.S., Ghosh, D.: Design and implementation of a wireless health monitoring system for remotely located patients. In: 2018 4th International Conference on Electrical Engineering and Information & Communication Technology (iCEEiCT), pp. 86–91. IEEE (2018)
5. Wijaya, N.H., Fauzi, F.A., Helmy, T.E., Nguyen, P.T., Atmoko, R.A.: The design of heart rate detector and body temperature measurement device using ATMega16. J. Robot. Control. **1**, 40–43 (2020). https://doi.org/10.18196/jrc.1209
6. Tunggal, T.P., Juliani, S.A., Widodo, H.A., Atmoko, R.A., Nguyen, P.T.: The design of digital heart rate meter using microcontroller. J. Robot. Control. **1**, 141–144 (2020). https://doi.org/10.18196/jrc.1529
7. Devis, Y., Irawan, Y., Junadhi, Zoromi, F., Herianto, Amartha, M.R.: Monitoring system of heart rate, temperature and infusion in patients based on microcontroller (Arduino Uno). J. Phys. Conf. Ser. **1845**, 012069 (2021). https://doi.org/10.1088/1742-6596/1845/1/012069
8. Viswavardhan Reddy, K., Kumar, N.: A framework for remote health monitoring. In: Advances in Intelligent Systems and Computing, pp. 101–112. Springer, Singapore (2021)
9. Srivastava, R.N., Padole, D.: Design of IoT based remote patient health care monitoring system. In: 2021 International Conference on Computational Intelligence and Computing Applications (ICCICA), pp. 1–6. IEEE (2021)
10. Kirtana, R.N., Lokeswari, Y.V.: An IoT based remote HRV monitoring system for hypertensive patients. In: 2017 International Conference on Computer, Communication and Signal Processing (ICCCSP), pp. 1–6. IEEE (2017)
11. Chowdary, K.C., Lokesh Krishna, K., Prasad, K.L., Thejesh, K.: An efficient wireless health monitoring system. In: 2018 2nd International Conference on I-SMAC (IoT in Social, Mobile, Analytics and Cloud), pp. 373–377. IEEE (2018)
12. Fakhrulddin, S.S., Gharghan, S.K.: An autonomous wireless health monitoring system based on heartbeat and accelerometer sensors. J. Sens. Actuator Netw. **8**, 39 (2019). https://doi.org/10.3390/jsan8030039
13. Abdullah, S.K.: Remote heart rate monitor system using NodeMcu microcontroller and easy pulse sensor v1.1. IOP Conf. Ser. Mater. Sci. Eng. **518**, 052016 (2019). https://doi.org/10.1088/1757-899X/518/5/052016
14. Nowshin, N., Mazumder, P., Soikot, M.A., Probal, M., Qadir, M.U.: Designing and implementation of microcontroller based non-invasive health monitoring system. In: 2019 International Conference on Robotics, Electrical and Signal Processing Techniques (ICREST), pp. 134–139. IEEE (2019)
15. Nath, A., Ashraf-Ur-Rahman, M., Kader, M.A., Muttaqui, K.M.H., Alam, M.E., Arafat, Y.: Development of a wireless automotive health monitoring system for doctor's chamber. In: 2019

10th International Conference on Computing, Communication and Networking Technologies (ICCCNT), pp. 1–5. IEEE (2019)

16. Hussain, A., Hammad, M., Hafeez, K., Zainab, T.: Programming a microcontroller. Int. J. Comput. Appl. **155**, 21–26 (2016). https://doi.org/10.5120/ijca2016912310

17. Chowdhury, I., Ahmed, T.: Design and prototyping of sensor-based anti-theft security system using microcontroller. Int. J. Eng. Res. Technol. **V10**, 58–66 (2021). https://doi.org/10.17577/ijertv10is030019

18. Ahmed, T., Sheik Md. Kazi Nazrul Islam, Chowdhury, I., Binzaid, S.: Sustainable powered microcontroller-based intelligent security system for local and remote area applications. In: 2012 International Conference on Informatics, Electronics & Vision (ICIEV), pp. 276–280. IEEE (2012)

19. PIC16F876A | Microchip Technology. https://www.microchip.com/en-us/product/PIC16F876A. Last accessed 25 April 2022

20. Ahmed, T., Chowdhury, I.: Design of an automatic high precision solar tracking system with an integrated solar sensor. In: 2019 5th International Conference on Advances in Electrical Engineering (ICAEE), pp. 235–239. IEEE (2019)

Chapter 52
A Review on Twitter Data Sentiment Analysis Related to COVID-19

Tasleema Noor and Rakesh Kumar Godi

Abstract On March 11, 2020, Dr. Tedros Adhanom Ghebreyesus, Director-General of the WHO, pronounced the outbreak a pandemic. The term "pandemic" refers to a disease that spreads rapidly and engulfs an entire geographic region. Coronavirus is a brand-new viral disease named after the year it first appeared. There is a scarcity of academic research on the subject to help researchers. Social media content analysis can reveal a lot concerning the general temperament and mood of the human race. In the field of sentiment analysis, deep learning models have been widely used. Sentiment analysis is a set of techniques, tools, and methods for detecting and extracting information. People have been using social networking sites like Twitter to voice their opinions, report realities, and provide a point of view on what is happening in the world today. Folks have always used Twitter to share data about the COVID-19 pandemic. People randomly share data visualizations from news revealed by organizations and the government. The numerous studies surveyed are selected based on a similarity. Every paper which is supervised performs sentiment analysis of Twitter data. Various studies have made used a fusion of diverse word embedding's with either machine learning classifiers or deep learning classifiers. Albeit the interpretation of single classifiers is satisfactory, the studies those proposed hybrid models have shown outstanding performance. On top of that transformer based models demonstrated quality results. It is concluded that using hybrid classifiers on Twitter data for sentiment analysis can surpass the achievements of the single classifiers.

52.1 Introduction

When WHO announced the coronavirus eruption it caused havoc and infected masses of people worldwide. In a number of studies conducted around the world, COVID-19 has been shown to have a negative impact on people's psychological well-being. Social media can be a valuable mode for learning about people's lives and opinions [1]. The domain of NLP and its application to social media analysis has grown at a

T. Noor (✉) · R. K. Godi
CVR College of Engineering, Ibrahimpatnam, Hyderabad, Telangana, India
e-mail: tasleemanoor5@gmail.com

© The Author(s), under exclusive license to Springer Nature Singapore Pte Ltd. 2023
C. So-In et al. (eds.), *Information Systems for Intelligent Systems*, Smart Innovation, Systems and Technologies 324, https://doi.org/10.1007/978-981-19-7447-2_52

breakneck pace. However, using NLP-methods to deduce a text's intrinsic meaning remains a difficult task. Sentiment analysis comes into play here, with the goal of developing efficient processes for extracting the writer's sentiment from the text automatically. In this study, Twitter data is used as a source for textual analysis. It is used to gauge public opinion and, in particular, to follow the impressions of the advancement of fright related to COVID-19. Context-aware proposes to leverage ambiguity are absent or ineffective in the lack of tone of the text and the expressions of the user. Social media can provide information about current sentiment and its network effects. This data covers a wide range of social phenomena, which are frequently discussed on social media [2]. DNN can be significantly more productive in the matter of parameters and computation for the same level of accuracy. Every approach of deep learning has exclusive set of benefits and drawbacks. Instead of trying to solve all problems with integrated models for the same task using a single machine learning approach, a combination of hybrid system and collaborative functions can express potential pitfalls of a single system better. Some approaches may struggle to perform well on various types of data, with poor performance and accuracy in sentiment analysis [3].

In the fourth quarter of 2020, approximately 1.8 billion active Facebook users were found. Seventy-five per cent of the whole percentage of Internet users are currently active on at least any one of the social media. Apple's iMessage, Tencent's WeChat, and Google's YouTube have all surpassed one billion monthly active users. At the phrase, sentence, or document level, sentiment analysis aims to categorize and determine the polarity of a subjective thought. It has several applications in various sectors. However, it is fraught with a slew of issues and difficulties in the area of NLP [4]. In terms of user-generated content, Web 2.0 is primarily defined by engagement on social media. Sentiment analysis using online reviews is gaining a lot of traction as the amount of messaging data and social media information grows exponentially. Online reviews provide many businesses a chance to grow by examining customer reviews [5]. It is expected that social media posts will direct people to reliable information. However, in the vast majority of cases, the information led to incorrect decisions. People's minds had already been disturbed by the coronavirus; now, opinions and tweets about COVID-19 are alarming [6].

COVID-19 has had a significant impact on a number of aspects of our lives. COVID-19 appears to have triggered anxiety, depression, loneliness, violence, and substance abuse, just as the SARS was linked to patients and medical personnel suffering from PTSD, stress, and psychological distress [7]. Twitter has grown in importance as a tool for gathering public opinion and is increasingly being used in public health research. It provides reflections on a wide range of viewpoints in various locations, attitudes, and public debates [8]. The COVID-19 pandemic has resulted in a rise in the usage of social media as a forum for discussing different pandemic-related topics. To achieve some level of herd immunity, it is estimated that minimum 70% of the population needs to be vaccinated. The public's support for vaccination is critical to achieving this goal. Previous research has associated social media activity with anti-vaccine movements and vaccine hesitancy [9].

52.2 Related Works

Research work in the field of natural language processing is expanding in various languages and applications. Researchers are experimenting with various combinations of classifiers to achieve better results. Hence, this review is conducted on the sentiment analysis of Twitter data. As a part of research work, articles analysing tweets of different languages in distinct fields and much more are studied. In [10], study is conducted on NepCOV19Tweets dataset which is a collection of Nepali tweets freely available on Kaggle. The evaluation metrics used in both [11] and [10] are accuracy, F1-score, recall, and precision. Machine learning algorithms used are SVM+Linear, SVM+RBF, XGBoost, ANN, RF, NB, LR, and KNN. For the proposed CNN models and state-of-the-art ML algorithms, comparison is carried out using three different feature extraction methods. The feature extraction methods used are domain-specific probability-based, fastText-based, and domain-agnostic probability-based. Results conclude that fastText performs better than other methods and gives highest accuracy with SVM+RBF model in terms of precision and XGBoost in terms of F1-score, recall, and accuracy. The accuracy of the proposed CNN model significantly outperforms the other models with 68.7% accuracy. By overcoming the limitations mentioned in the article, there is a scope of improving accuracy.

Study in [11] is conducted on the same dataset as [10] with different classifiers. After pre-processing the data, comparison between TF-IDF, fastText, and a hybrid feature extraction technique (TF-IDF + fastText) is analysed with state-of-the-art models like logistic regression (LR), KNN, Naïve Bayes, decision tree, random forest, extra tree classifiers (ETC), AdaBoost, MLP-ANN, SVM + LR, and SVM + RBF. Results show that the hybrid model SVM + RBF with hybrid feature method achieves the highest accuracy of 72.1% among the others. Though, it has achieved highest accuracy the scope of attaining a better score is possible using various word embeddings and classifiers.

Alouffi et al. [12] introduced an optimized model for detecting misleading information on COVID-19. Three datasets are collected from Internet. The first dataset consists of Facebook posts. The second dataset is classified as 0 (fake news) and 1 (true news). The third dataset is classified as (1) false, (2) true, and (3) partially false news. The datasets are pre-processed and cleaned. The TF-IDF word embedding is used for DL and ML classifiers. The proposed model uses TF-IDF with CNN and LSTM as classifier using hyperparameter optimization. Performance is calculated using the cross-validation and test performance. For Dataset1, the proposed model achieved 93.24% of accuracy in testing performance which is highest among all the other models. For Dataset2, the proposed model achieved 97.7% accuracy in testing performance. And for Dataset3, the proposed model achieved 77.24% accuracy in testing performance.

In study conducted by Rodrigues et al. [13] live tweets are collected using an official Twitter API called Tweepy. Collected tweets are used as input data and filtered, tokenized, lemmatized/stemmed, and stopwords are removed under pre-processing process. Two various types of feature extraction methods, i.e. TF-IDF and bag of words (BOW) are used. This study is for spam detection and sentiment analysis on static dataset (spam or ham) and live dataset (positive, negative, and neutral), respectively. The classifiers tested for spam detection are logistic regression, multinomial Naïve Bayes, Bernoulli Naïve Bayes, random forest, decision tree, and support vector machine. The classifiers tested for sentiment analysis are SVC, stochastic gradient descent (SGD), RF, Naïve Bayes (NB), LR, recurrent neural network (RNN), long-short term memory, bidirectional LSTM, and convolutional neural network. MNB and LSTM achieved 97.78 and 98.74% accuracy for spam classification. SVM and LSTM achieve 70.56 and 73.81% accuracy for sentiment analysis. Detecting spam tweets gave an accurate sentiment analysis. Correlating the spam tweets with the accounts might help in further study.

The study in [14] aims to bridge the gap addressed in the research works. The gap referred is, the studies conducted on Arabic languages are few. Hence, the proposed study focusses on the Arabic tweets collected from Tweepy. The data is pre-processed. Unrelated and duplicate tweets are taken off to clean the data. After building the dataset, the tweets were annotated manually by experts. The dataset was found imbalanced and to overcome this problem SMOTENC is amalgamated. Word2Vec is used for feature extraction. The experiment is conducted on single classifiers and ensemble classifiers on two CBOW pre-trained models named as Arabic.news and CC.AR.300. Results reveal that the ensemble classifiers performed better than the single classifiers. Single classifiers without SMOTENC achieved an average F1-score of 78.59% for CC.AR.300 model and 77.93% for Arabic.news model. Likewise single classifiers with SMOTENC performed better compared to without SMOTENC. Ensemble classifiers without SMOTENC accomplished better results than single classifiers. More specifically, the ensemble classifiers with SMOTENC attain the highest average F1-score of 83.81% for CC.AR.300 model and 84.61% for Arabic.news model. For future enhancement, different datasets with various ensemble classifiers and feature extraction methods can be explored.

Chintalapudi et al. [15] aim to perform sentiment analysis on the Twitter data using DL models. The dataset is gathered from github.com. Pre-trained BERT model is compared to LSTM, SVM, and LR. The BERT model if fine-tuned to aim for better results than the other three classifiers. LR, LSTM, and SVM give 75%, 65%, and 74.75% accuracy, respectively. While the fine-tuned model beats the other three classifier giving 89% accuracy.

The article by Samuel et al. [16] focuses on the sentiment insights of the public using the tweets from Twitter. The data is collected using the Twitter API. The R software is utilized. Authors conducted textual analytics and Twitter analytics. Classification models like linear regression, Naïve Bayes, logistic regression, and KNN. Geo-tagged analytics and sentiment analytics are also performed to gain insights.

The study in [17] indicates towards the critical issue of mental health among Indians caused by anxiety, stress, and trauma. This study discloses that expressing feelings on social media platforms is a way to cope up with mental stress caused during the pandemic. A total of 840,000 Tweets are collected using Tweepy API with keywords like (COVID and stress), (COVID and trauma), and (COVID and anxiety) from April to October 2020 (120,000 tweets per month). This study concludes that

5.23%, 34.4%, and 60.3% of the tweets collected are positive, negative, and neutral, respectively. The second part of the study focuses on LDA. LDA topic modelling is used to understand the major concerns caused by pandemic (COVID-19) among the public. As LDA works best with huge data, it is suitable for this study. Few of the topics listed in the paper are death, lockdown, increase in COVID-19 cases, travelling, and vaccine. For further study, different cultures, different biographies, and other modelling methods can be surveyed.

The authors, El Azzaoui et al. [18] came up with SNS big data framework to predict the pandemic outbreaks. Twitter case study is taken to analyse on the framework. The framework is clearly explained in the article. The framework consists of four layers namely user layer, fog layer, cloud layer, and application layer. Data is collected over a month from USA users. After collecting the data, it is then pre-processed and tweets are selected using the key hashtag. Features are extracted using user's location, subjectivity, polarity, and symptoms. After comparison, if the tweets are not similar they are placed in post-processing list. It is then classified as SIR model. SIR refers to susceptible, infected, and removed tweets. The predicted cases are compared to the confirmed cases. A total of nine models are compared out of which the highest prediction is made by logistic. As a conclusion, SNS analysis achieved an accuracy of 98.9%. There are limitations of this study which can be rectified for future works.

The study in [19] compares the public sentiments of nine megacities during COVID-19 pandemic using the Twitter application from 2 March to 21 May 2020. The listed cities are Singapore, Boston (US), Washington (US), New York Chicago (US), City (US), Seattle (US), Los Angeles (US), Rome (Italy), and London (UK). The nine megacities are selected based on their coronavirus epidemic time, confirmed case growth rate, size of population, and geographical position. About forty keywords with a significant relationship with coronavirus are employed to preserve as filters. Just the tweets pertinent to the topic, such as "outbreak", "COVID", "quarantine", and "cough". The sentiments are classified either as positive (1) or negative (0). The entire dataset used is an amalgamation of "Sentiment140" and "Twitter Sentiment Corpus" datasets. Basic pre-processing is performed. The authors noticed that the number of tweets in English speaking countries are more than the number of tweets from countries where other languages are spoken. As the tweets are small, the number of bigrams and unigrams is almost the same. Firstly, machine learning classifiers like logistic regression and Naïve Bayes are tested with BOW. Then, deep learning classifiers like CNN and LSTM are used for analysis. To understand the relation between tweets, Spearman's rank correlation is calculated. The process followed to rank the sentiment score is clearly explained by the authors in the article. Amidst all the classifiers tested, LSTM gave the best precision. The findings of this study indicate that quarantine measures had a positive impact on public sentiment amid pandemic, and it also urges for fast and effective quarantine orders to be issued at the onset of contagious diseases in densely populated areas. The list of limitations mentioned in the paper can be eliminated to allow more research work.

The study in [20] focuses on the effects of the pandemic on the airlines business. This paper aims to acknowledge the seriousness of coronavirus and the proposed approach introduces a new method for extracting pertinent data from people's social media activity on Internet in order to develop usable predictors for stock return volatility forecasting. This study is very helpful in many ways to understand and predict the stock market. To understand the stock market changes, authors have collected data from three big airlines of USA namely United Airlines, American Airlines, and Delta Airlines which is referred as UA, AA, and DA in the paper. Closing rates of airlines is collected from Yahoo Finance since 1 of January, 2015 to 30 of April 2020. Authors have compared the flu occurred in 2017-2018 and COVID-19 based on the total cases, deaths, fatality rate, and start date of the outbreak. An assumption made by the authors is that the behaviour and the interests of a common man can help in understanding the unstable state of the airlines stock prices. Around 30,000 tweets are collected from 10 Twitter handles like CNBC, CNN, NBC, MSNBC, etc. Authors have selected particularly these airlines because they fit the aspects of the study. ARMA and GARCH models are used to forecast the values. A combination of baseline, MCASES, MCOVID, and MICP models with mean model (ARMA) and variance model (GARCH) is analysed to forecast the returns and unsteadiness in the UA, AA, and DA airlines stock. The analysis reveals that the pandemic has caused a lot of damage to the aviation industry. As the COVID cases are increasing in USA and throughout the world it is mandatory for the government to step up to curb the spread of virus. During April, small scale airlines like Trans States Airlines and Compass Airlines suffered losses just like the big companies have suffered. But the smaller ones had a great impact. It is the same situation with the aviation industry all over the world.

Trajkova et al. [21] crawled 5409 visualizations from Twitter from 14 April 2020 to 9 May 2020. The ultimate goal is to gain a better understanding of the COVID-19-related Twitter visualizations that people have created, as well as to look into the problems that come with trying to interpret such posts. Although a part of authors analysis is quantitative, the majority of authors work is qualitative, which is not a novel method of gaining information in visualization research. The bottom-up inductive approach is embedded in the Twitter dataset. To collect the data, keywords like coronavirus, SARS-CoV-2, COVID-19, and SARS-CoV-2. And to obtain the visualizations keywords like graph, chart and visualization is used. Along with collecting the relevant data, authors have collected some additional data which is later concluded as not necessary and useful for the analysis. Analysis is performed on the 540 random posts. A four step iterative process of inductive coding is performed to answer all the research questions. There are four research questions which the authors have acknowledged: "What are people posting?", "Do people retweet more the visualizations that are created by individuals or organizations?", "What is the relationship between variables $1/x$ and $2/y$ and the number of retweets?", and "What challenges may arise from these casual visualizations?". The first research question is answered by: comparing the posts from individuals and organizations, exploring the different types of visualizations in the dataset using the Type-by-Task Taxonomy (TTT), portraying the sources of data which was used to create the visualizations (results show that majority of the sources are not specified), categorizing the data. The answer to second research question is: To see if there were any contrasts in retweets among generation types, a Kruskal-Wallis H test is used with two levels of categories:

individual and organization. According to a visual examination of a boxplot, the distributions of retweet scores were parallel for all groups. The difference in median retweet count between the two levels of visualization generation is statistically significant. Answer to third research question is: People are more active with visualizations when 1/x is location or geography and 2/y is deaths, as per the statistical analysis. This supports the previous theory that people at the time were more concerned with the death rate than with the demography. Apparently, people may be more concerned about the impact of the outbreak in places they reside or in places they need or intend to visit. The forth research question is answered as: The five challenges are listed are mistrust, proportional reasoning, temporal reasoning, and misunderstanding about virus and cognitive bias. The limitation of this study mentioned is that the research is limited to only one social media platform. For future research, data from other social platforms can be considered. Study provides general guidelines for people who create and interpret casual data visualizations. Finally, the research results can aid the news outlets and government entities in gaining a clear picture of the types of data that people are most interested in, as well as the challenges that citizens might face while doing so analysing pandemic visualizations.

Divyapushpalakshmi et al. [22] utilized the updated edition of the SemEVAL2017 dataset consisting of around 20,000 tweets. The data is pre-processed using regular techniques like converting to lower case, tokenization, remove usernames and hashtags, remove stopwords, remove non-alphabetical words, and lemmatization. The authors have proposed a hybrid model named as HDL-SA which means a hybrid deep learning model introduced for sentiment analysis. Sentiment analysis is carried out by a scale classification of five points. Sentiment analysis is performed using artificial neural network. For weight optimization, modified salp swarm algorithm (MSSA) is applied. This algorithm is inspired from the nature of salps of swarming while they scavenge and explore the seas. MSSA is a modified version of salp swarm algorithm (SSA) which is much more robust. Evaluation of the hybrid model is calculated using the F-score, precision, recall, and accuracy metrics. The results of the proposed model are best when compared to the other existing models. The author describes the ways of extending his work for the future researchers (Table 52.1).

52.3   Sentiment Analysis

Service providers can use sentiment analysis to determine where they fall short and which aspects of their service satisfy customers. Sentiment analysis classifies positive and negative emotions in text data, which can be difficult to discern in subtle wordplays. Every day, a massive amount of data is generated by online reviews, which must be utilized and meaning extracted from text [5].

Sentiment analysis is the process of analysing text using machine learning to determine the text's polarity. Twitter is a popular source for sentiment analysis because it serves as a platform for scientists, politicians, athletes, celebrities, and field experts

Table 52.1 Tabular description of recent works

Works by	Purpose	Dataset	Model	Results
Qaid et al [23]	To predict COVID-19	The datasets consists of X-ray images. Two datasets are used out of which one is collected from Kaggle and the other one from a hospital in Saudi Arabia	Four models are used: CNN, VVG16/VVG19, hybrid model: CNN + ML, and hybrid model: VVG16/VVG19 + ML	Proposed hybrid models perform better than all the other models
Zain et al. [24]	To forecast pandemic	The data is collected from the dashboard of WHO COVID-19 site	Proposed model: CNN + LSTM baseline models, statistical models, linear models, ensemble models, and ML models	Using the metrics MAPE, RMSE, and RRMSE the accuracy of the proposed model is calculated. The forecasting performance of the proposed model is efficient
Alam et al. [25]	Analyse the vaccine responses	All COVID-19 vaccines tweets from Kaggle	LSTM and Bi-LSTM	90.59% and 90.83% accuracy is achieved by LSTM and Bi-LSTM, respectively
Godi Rakesh et al. [26]	Detect seasonal outbreaks	The author's previous works dataset is analysed	KNN, SVM, DT and RF	Random forest attains highest accuracy
Lamsal [27]	Provides design and analysis of large-scale COVID-19 tweets dataset	COV19Tweets dataset	Filtering tweets using geo-tagged tweets	A new dataset named as GeoCOV19Tweets is generated
Ridhwan et al. [28]	Understand public sentiment in Singapore on COVID-19	Tweets are gathered using snscrape with keywords	VADER and RNN. topic modelling using LDA, GSDMM, and pyLDAvis	35 topics are distributed out of 33 are selected using GSDMM model. Analysis shows that the public of Singapore support the government measures
Das et al. [29]	Sentiment evaluation and predictive analysis	Tweets related to Indian PM Modi's speech and Donald Trump's press briefing	Proposed model: deepCNN	90.67% deep CNN with softmax performs better than deep CNN with ReLU and other models

to express their views on a given topic. Through social media, users are able to create content ranging from daily events to far-reaching incidents [6].

It is a crucial task in NLP that uses rule-based, supervised, or unsupervised machine learning or deep learning techniques to categorize the sentiment expressed in a text. Positive emotions, negative emotions, and neutral emotions are the most common classifications [8].

52.4 Data Collection

COVIDSENTI is a two-month archive of tweets with 90,000 unique tweets from 70,000 users who met the selection criteria out of 2.1 million tweets collected between February and March 2020. Only English tweets are used. The Twitter API library Tweepy was used to collect the dataset [2].

In [3], various public datasets are used for a specific application domain. The target of the experiment is to recognize whether the models give steady accurate results irrespective of the dataset type and size. Having a balanced class distribution is important to ensure that prior probabilities are not biased for training models and doing classification. A total of eight datasets were used for this study. Music reviews, Sentiment140, Cornell movie reviews, Tweets Semeval, Tweets Airline, Book reviews, IMDb movie reviews (2), and IMDb movie reviews (1).

The dataset, Sentiment140 is collected from Kaggle consisting of six features. The Twitter search API was used to extract 1.6 million tweets. Because of the volume, manually annotating the tweets would have been a time-consuming and labour-intensive task. The authors of the dataset annotated the tweets by taking the emoticons' noise into account when determining whether the tweet was positive or negative [4].

The first dataset, Airline Sentiment Data is collected from an online platform with 82,842 reviews. The data includes reviews from travellers from 85 different countries who have used 186 different airlines. The second dataset, Twitter Airline Sentiment data is provided by CrowdFlower's library. It consists of 15 columns with 12,195 negative and positive reviews [5].

The dataset is obtained from the IEEE data port. The sentiment scores and IDs of tweets about the COVID-19 pandemic on Twitter [6].

The Twitter data used in this study came from an open source, publicly accessible IEEE website. This is due to the Twitter API's restriction on accessing data older than one week. From March to April 2020, hydrated tweets were downloaded in a CSV file. The final data included tweets with Canada in the user location attribute and social distancing-related keywords. [1]

The data is gathered via API from the Twitter. The data is collected using an application programme interface. It is a user interface that connects the user to the source. The source is a Twitter website that contains users' tweets. The airline dataset is used has attributes such as tweet sentiment, tweet ID, comments, tweet count, tweet created, negative reason, time zone, and location [30].

The 11,974 tweets are searched in Philippines in "English" and "Tagalog" languages using the RM Search Twitter operator in the month of March 2021 [7].

Snscrape package was used to obtain tweet IDs and Tweepy to gather the data. A combination of different hashtags and keywords was used to fetch 2,678,372 English tweets between November and January 2021 [8].

A total of 4,552,652 tweets were collected using snscrape tool posted between January 2020 and January 2021 [9].

52.5 Data Pre-processing

Data pre-processing is one of the most crucial and first step in sentiment analysis is the data pre-processing.

- The raw data collected from online platforms is unstructured, noisy, and informal. In most cases, hashtags are not necessary for sentiment, but they can have an impact on performance.
- The text is cleaned up by removing any unnecessary hashtags.
- The text is case-folded. All capitalized letters are folded to lower case to avoid misreading the same word as a different word due to capitalization.
- Word segmentation is performed to segregate the hashtags like "coronavirus" to "corona virus".
- Stop words are commonly removed from textual data to reduce noise. Stop words have no effect on understanding the sentiment valence of a sentence.
- Lemmatization processes morphological analysis of words, then returns them to their dictionary form.
- The goal of the textual data analysis was to remove stopwords, punctuation, white space, hyperlinks, and @mentions. Special numbers, punctuation, and characters are removed from the dataset because they do not aid in the detection of sentiment.
- Retweets are either excluded while retrieving the data or eliminated after collecting the dataset.
- Duplicate tweets are filtered out.

52.6 Labelling Data

After retrieving the data, tweets are labelled using different tools. Data is labelled based on the polarity score of the tweets. In [3, 5], tweets are labelled as positive and negative tweets. In [2, 4, 7, 30], tweets are labelled as positive, negative, and neutral.

In [6], tweets are labelled are positive, weakly positive, strongly positive, neutral, negative, weakly negative, and strongly negative.

In [9], tweets are labelled as pro-vaccine, anti-vaccine, and vaccine hesitant (Table 52.2).

Table 52.2 Various methods used by authors for annotation

Labelled as	Annotation	Method
Positive, negative, and neutral	Annotated by authors	Shahi et al. Sitaula et al.
Positive and negative	Annotated by experts	Al-Hashedi et al.
Positive, negative, and neutral	TextBlob	Naseem et al. Praveen et al.
Positive and negative	TextBlob	El Azzaoui et al.
Highly positive, positive, neutral, negative, and highly negative	Annotated using CrowdFlower	Divyapushpalakshmi et al.

52.6.1 TextBlob

TextBlob is a NLP Python library. Natural Language ToolKit was used extensively by TextBlob to complete its tasks. It allows users to work with tasks like classification by providing easy access to a large number of lexical resources. It is a simple library that allows for complex operations.

52.6.2 VADER

VADER is an open source which stands for Valence Aware Dictionary and Sentiment Reasoner with MIT License. It is a high-performance computing lexicon and rule-based tool. VADER has proven to be quite effective. This is because VADER not only displays the positivity and negativity scores, but also the degree to which a sentiment is positive or negative [8].

52.6.3 SentiStrength

SentiStrength is used to detect sentiment polarity because it was created using a lexicon-driven method. It has the most average human-level accuracy when compared with other Twitter analysis tools [31]. It SentiStrength is a popular standalone tool that works without any additional hardware or software. Emotions, negating words, idioms, booster words, emoticons slang, and booster words are also detected by the tool. For each text, the SentiStrength tool generates two scores. Positive emotion is represented by a single score ranging from 1 to 5, with 1 indicating not positive and 5 indicating extremely positive. The negative emotion score ranges from 1 to 5, with 1 indicating no negative emotion and 5 indicating extremely negative emotion [1]

52.7   Feature Extraction

52.7.1   Tf-idf

Term frequency inverse document frequency of records is the process of determining how similar a word in a corpus is to the given text. The meaning of a word grows in proportion to how many times it appears in the text, but this is offset by the corpus's word frequency. It is represented mathematically as

$$\mathrm{idf(word)} = \ln\left(\frac{n_{\text{documents in analysis}}}{n_{\text{documents word appear in}}} \right)$$

52.7.2   Word2Vec

It generates word vectors, which are distributed numerical depictions of the word features. Word features are words that represent the context of single words in the vocabulary. Through the created vectors, word embeddings eventually assist in establishing the association between words with similar meanings. Tomas Mikolov of Google [62] published Word2vec in 2013. A large corpus of datasets was used to train an unsupervised learning model. With a matrix $D \times W$, here, D is denoted as the aggregate of documents and W is the dimension of word embedding, the dimension of one-hot encoding is much more than the dimension of Word2Vec. The CBOW and skip-gram models are present in Word2vec. The unavailability of support for out-of-vocabulary words is, however, one of the major disadvantages of using Word2vec. It uses the distinct token for terms or words that are never encountered in the dictionary [3].

52.7.3   GloVe

"Global Vectors" is the abbreviation for GloVe. As previously mentioned, GloVe captures both local and global statistics of a corpus in order to generate word vectors.

52.7.4   fastText

The Facebook Research Team created fastText as a library for learning sentence classification and word representations easily.

52.8   Models

The authors set a benchmark dataset named as COVIDSenti [2]. The whole dataset is divided into three sub-datasets naming them as COVIDSenti-A, COVIDSenti-B, and COVIDSenti-C. A combination of machine learning classifiers with word embeddings like TF-IDF, Word2Vec, GloVe, and fastText are compared. Similarly, deep learning classifiers are also compared. Hybrid models like IWV and HyRank are considered for comparison. Attention models or transformer models like distilBERT, BERT, XLNET, and ALBERT are analysed.

The proposed models [3] are four hybrid deep learning models. CNN + LSTM with ReLU as the classifier and BERT/Word2vec as feature extraction method and CNN + LSTM with SVM as the classifier and BERT/Word2vec as feature extraction method are the two hybrid models. LSTM + CNN with ReLU as the classifier and BERT/Word2vec as feature extraction method and LSTM + CNN with SVM as the classifier and BERT/Word2vec as feature extraction method are the other two models. To show that hybrid models works better, single models like SVM, LSTM, and CNN are also compared.

Gaye et al. [4] conducted a sentiment classification on the Twitter data using the proposed hybrid model. Three feature extraction methods are used. They are TF-IDF, bag of words (BOW), and feature union. The third method refers to the combination of TF-IDF and BOW. Logistic regression, random forest, and AdaBoost machine learning classifiers and LSTM deep learning classifier are considered. The proposed model is the mix of LR and LSTM. The tweets are re-annotated with TextBlob to compare with actual annotations.

The proposed model is a hybrid model containing CNN and LSTM. The architecture is clearly explained in the paper [5] with a diagram. Proposed model is compared to the single models.

The authors named their proposed model as hybrid heterogeneous SVM model (H-SVM) [6]. It is multi-kernel-based SVM. It makes use of H-EOM for evaluating the space of the ith nominal feature. Analysis of 20 tweets, 50 tweets, 250 tweets, and 600 tweets along with recurrent neural network and support vector machine. The list of the most frequently used words are visualized in a graph.

The proposed method is sequential minimal based optimization decision tree (SMODT). This optimization method [30] is used to extract features and classic ML classifiers for sentiment classification.

A study of Filipinos' attitudes towards COVID-19 vaccines was conducted on the social networking site Twitter. The findings of this study can assist the Philippine government in making informed verdict about funding, vaccination rollout strategy, and vaccination provision. It automatically labels the polarity of tweets in both English language and Filipino language to categorize them as per their polarity, which may be put to use in closely related studies. Naïve Bayes is used as the classifier [7].

52.9 Results

Experimental results in [2] show that SVM with TF-IDF gives 84.5 accuracy, Random forest with Word2Vec gave 76.9% accuracy, random forest with GloVe gave 72.6%, and random forest with fastText gave 84.5% accuracy. DCNN-(GloVe + CNN) deep learning classifier gave 86.9% accuracy. IWV and HyRank gave 77.1% and 88.1% accuracy, respectively. All the transformer language models achieve above 92% accuracy, BERT outperforms other models with 94.8%.

Accuracy, precision, recall, F-score, and AUC achieved from the single models and hybrid models with BERT and Word2Vec for all the eight sets of data are compared in the paper [3]. The results reveal that when the pre-trained model (BERT) as feature extraction method and hybrid classifiers together performs better than the single models and Word2Vec + hybrid model. Though the pre-trained models performed well, the computation complexity and time complexity is high. This limitation can eradicated by hyper tuning the pre-trained model.

The findings reveal that the highest accuracy obtained for Airlinequality by ML and DL models is 86.0% and 86.3% accuracy, respectively. For the Twitter data, ML and DL achieved 88.0% and 86.2% accuracy, respectively. The proposed model [5] attains an accuracy of 87.6% and 87.5% for the two datasets. For further enhancement, optimization techniques can be applied for good accuracy.

The performance is evaluated to the positive, negative, and neutral tweets obtained from the sentiment polarity [1]. Dataset is distributed in 80:20 ratio. Initially, the test is conducted on 10% of the dataset, then 20% of the dataset and finally on the complete dataset. The accuracy achieved for the 10% data is 87% using the linear kernel on negative and positive sentiments, for 20% data is 81% using the RBF kernel with SVM on negative and positive sentiments only. For the complete dataset, 71% accuracy is achieved using the RBF kernel for all the sentiments.

The analysis results show that KNN, SVM, KNN + SVM, SMO, DT, and SMO + DT reached to an accuracy of 67%, 68%, 76%, 86.23%, 80%, and 89.47% [30], respectively.

The 993 leftover tweets were analysed out of which 83.38% of the tweets have positive nature, 8.26% have negative nature and 8.36% have the neutral nature. Using these annotations Naïve Bayes classifier [7] gives 81.77% accuracy score.

When the sentiment analysis was performed on 2,678,372 tweets, 30.3% and 42.8% tweets were classified as negative and positive tweets. VADER classifies a tweet sentiment with an F1 of 0.96. When Pfizer [8] was introduced there was a rise in number of tweets which was later stabilized as neutral sentiment.

The tweets in [9] are distributed as positive, negative, and neutral of 34%, 25%, and 41% of the whole dataset, respectively. Sentiments of the public towards the vaccines and various vaccine brands is shown in a graphical presentation. The negative posts towards Pfizer has increased over the time. Results in US shows that initially tweets suggesting pro-vaccine were high then they have reduced and vaccine hesitant tweets are increased. In Australia, pro-vaccine tweets were decreased and hesitant tweets

Table 52.3 Results shown by Gaye et al. [4]

Annotations	Models	Method	Best classifier	Metric	Result
Original	ML classifier	TF-IDF	LR	Accuracy	0.75
Original	ML classifier	BOW	LR	Accuracy	0.74
Original	ML classifier	Feature union	LR	Accuracy	0.78
Original	Proposed model	–	LR-LSTM	Accuracy	0.80
TextBlob	ML classifier	TF-IDF	LR	Accuracy	0.95
TextBlob	ML classifier	BOW	LR	Accuracy	0.97
TextBlob	ML classifier	Feature union	LR	Accuracy	0.98

are increased. The same situation is shown in India, UK, and Canada. In Ireland, anti-vaccine tweets are high in number by the end.

The analysis is performed on both the machine learning and proposed models with actual annotations and TextBlob annotations. The proposed model [4] attains best accuracy results (Table 52.3).

52.10   Conclusion

The study particularly focuses on the sentiment analysis of people across the world during pandemic times. Twitter data is deemed as one of the most utilized source of data collection for such type of analysis. This study is not constricted to any confines but rather focused on understanding the sentiments of people. This study involves diversified works of several authors. In conclusion, all the papers surveyed the hybrid models or stacked models conquers the highest accuracy score when compared to the single classifiers for sentiment analysis. The results achieved from hybrid classifiers are outstanding. The usage of combination of various classifiers with different word embeddings has proved to be a good course of action for exploratory analysis.

52.11   Future Work

From the observations made from the existing studies on sentiment analysis of tweets, it is culminated that sentiment analysis on Twitter data best fits with hybrid classifiers. Hence, in future works implementing of hybrid models will be exercised for acquiring progressing results.

References

1. Shofiya, C., Abidi, S.: Sentiment analysis on COVID-19-Related social distancing in Canada using Twitter data. Int. J. Environ. Res. Public Health **18**(11), 5993 (2021). https://doi.org/10.3390/ijerph18115993

2. Naseem, U., Razzak, I., Khushi, M., Eklund, P.W., Kim, J.: COVIDSenti: a large-scale benchmark Twitter data set for COVID-19 sentiment analysis. IEEE Trans. Comput. Social Syst. **8**(4), 1003–1015 (2021). https://doi.org/10.1109/tcss.2021.3051189

3. Dang, C.N., Moreno-García, M.N., De la Prieta, F.: Hybrid deep learning models for sentiment analysis. Complexity **2021**, 1–16 (2021). https://doi.org/10.1155/2021/9986920

4. Gaye, B., Zhang, D., Wulamu, A.: A tweet sentiment classification approach using a hybrid stacked ensemble technique. Information **12**(9), 374 (2021). https://doi.org/10.3390/info12090374

5. Jain, P.K., Saravanan, V., Pamula, R.: A hybrid CNN-LSTM: a deep learning approach for consumer sentiment analysis using qualitative user-generated contents. ACM Trans. Asian Low-Resour. Lang. Inf. Process. **20**(5), 1–15 (2021). https://doi.org/10.1145/3457206

6. Kaur, H., Ahsaan, S.U., Alankar, B., Chang, V.: A proposed sentiment analysis deep learning algorithm for analyzing COVID-19 tweets. Inf. Syst. Front. **23**(6), 1417–1429 (2021). https://doi.org/10.1007/s10796-021-10135-7

7. Villavicencio, C., Macrohon, J.J., Inbaraj, X.A., Jeng, J., Hsieh, J.: Twitter sentiment analysis towards COVID-19 vaccines in the Philippines using Naive Bayes. Information **12**(5), 204 (2021). https://doi.org/10.3390/info12050204

8. Liu, S., Liu, J.: Public attitudes toward COVID-19 vaccines on English-language Twitter: a sentiment analysis. Vaccine **39**(39), 5499–5505 (2021). https://doi.org/10.1016/j.vaccine.2021.08.058

9. Yousefinaghani, S., Dara, R., Mubareka, S., Papadopoulos, A., Sharif, S.: An analysis of COVID-19 vaccine sentiments and opinions on Twitter. Int. J. Infect. Dis. **108**, 256–262 (2021). https://doi.org/10.1016/j.ijid.2021.05.059

10. Sitaula, C., Basnet, A., Mainali, A., Shahi, T.B.: Deep learning-based methods for sentiment analysis on Nepali COVID-19-Related tweets. Comput. Intell. Neurosci. **2021**, 1–11 (2021). https://doi.org/10.1155/2021/2158184

11. Shahi, T., Sitaula, C., Paudel, N.: A hybrid feature extraction method for Nepali COVID-19-related tweets classification. Comput. Intell. Neurosci. **2022**, 1–11 (2022). https://doi.org/10.1155/2022/5681574

12. Alouffi, B., Alharbi, A., Sahal, R., Saleh, H.: An optimized hybrid deep learning model to detect COVID-19 misleading information. Comput. Intell. Neurosci. **2021**, 1–15 (2021). https://doi.org/10.1155/2021/9615034

13. Rodrigues, A.P., Fernandes, R., Shetty, A., Lakshmanna, K., Shafi, R.M.: Real-time Twitter spam detection and sentiment analysis using machine learning and deep learning techniques. Comput. Intell. Neurosci., 1–14 (2022).https://doi.org/10.1155/2022/5211949

14. Al-Hashedi, A., Al-Fuhaidi, B., Mohsen, A.M., Ali, Y., Gamal Al-Kaf, H.A., Al-Sorori, W., Maqtary, N.: Ensemble classifiers for Arabic sentiment analysis of social network (Twitter data) towards COVID-19-Related conspiracy theories. Appl. Comput. Intell. Soft Comput. **2022**, 1–10 (2022). https://doi.org/10.1155/2022/6614730

15. Chintalapudi, N., Battineni, G., Amenta, F.: Sentimental analysis of COVID-19 tweets using deep learning models. Infect. Dis. Reports **13**(2), 329–339 (2021). https://doi.org/10.3390/idr13020032

16. Samuel, J., Rahman, M.M., Ali, G., Esawi, E., Samuel, Y.: COVID-19 public sentiment insights and machine learning for tweets classification (2020). https://doi.org/10.31234/osf.io/sw2dn

17. Praveen, S., Ittamalla, R., Deepak, G.: Analyzing Indian general public's perspective on anxiety, stress and trauma during COVID-19—a machine learning study of 840,000 tweets. Diabetes Metab. Syndr. **15**(3), 667–671 (2021). https://doi.org/10.1016/j.dsx.2021.03.016

18. EL Azzaoui, A., Singh, S.K., Park, J.H.: SNS big data analysis framework for COVID-19 outbreak prediction in smart healthy city. Sustain. Cities Soc. **71**, 102993 (2021). https://doi.org/10.1016/j.scs.2021.102993

19. Yao, Z., Yang, J., Liu, J., Keith, M., Guan, C.: Comparing tweet sentiments in megacities using machine learning techniques: in the midst of COVID-19. Cities **116**, 103273 (2021). https://doi.org/10.1016/j.cities.2021.103273

20. Deb, S.: Analyzing airlines stock price volatility during COVID-19 pandemic through internet search data. Int. J. Financ. Econ. (2021). https://doi.org/10.1002/ijfe.2490

21. Trajkova, M., Alhakamy, A., Cafaro, F., Vedak, S., Mallappa, R., Kankara, S.R.: Exploring casual COVID-19 data visualizations on Twitter: topics and challenges. Informatics **7**(3), 35 (2020). https://doi.org/10.3390/informatics7030035

22. Divyapushpalakshmi, M., Ramalakshmi, R.: An efficient sentimental analysis using hybrid deep learning and optimization technique for Twitter using parts of speech (POS) tagging. Int. J. Speech Technol. **24**(2), 329–339 (2021). https://doi.org/10.1007/s10772-021-09801-7

23. Qaid, T.S., Mazaar, H., Al-Shamri, M.Y., Alqahtani, M.S., Raweh, A.A., Alakwaa, W.: Hybrid deep-learning and machine-learning models for predicting COVID-19. Comput. Intell. Neurosci. **2021**, 1–11 (2021). https://doi.org/10.1155/2021/9996737

24. Zain, Z.M., Alturki, N.M.: COVID-19 pandemic forecasting using CNN-LSTM: a hybrid approach. J. Control Sci. Eng. **2021**, 1–23 (2021). https://doi.org/10.1155/2021/8785636

25. Alam, K.N., Khan, M.S., Dhruba, A.R., Khan, M.M., Al-Amri, J.F., Masud, M., Rawashdeh, M.: Deep learning-based sentiment analysis of COVID-19 vaccination responses from Twitter data. Comput. Math. Methods Med. **2021**, 1–15 (2021). https://doi.org/10.1155/2021/4321131

26. Godi, R. K., Balaji, G. N., Vaidehi, K.: A study of physiological homeostasis and its analysis related to cancer disease based on regulation of pH values using computer-aided techniques. In: Advances in Intelligent Systems and Computing, pp. 725-734 (2020). https://doi.org/10.1007/978-981-15-1097-7_61

27. Lamsal, R.: Design and analysis of a large-scale COVID-19 tweets dataset. Appl. Intell. **51**(5), 2790–2804 (2020). https://doi.org/10.1007/s10489-020-02029-z

28. Mohamed Ridhwan, K., Hargreaves, C.A.: Leveraging Twitter data to understand public sentiment for the COVID-19 outbreak in Singapore. Int. J. Inf. Manage. Data Insights **1**(2), 100021 (2021). https://doi.org/10.1016/j.jjimei.2021.100021

29. Das, S., Kolya, A.K.: Predicting the pandemic: Sentiment evaluation and predictive analysis from large-scale tweets on COVID-19 by deep convolutional neural network. Evol. Intel. (2021). https://doi.org/10.1007/s12065-021-00598-7

30. Naresh, A., Venkata Krishna, P.: An efficient approach for sentiment analysis using machine learning algorithm. Evol. Intel. **14**(2), 725–731 (2020). https://doi.org/10.1007/s12065-020-00429-1

Chapter 53
Risk Reduction and Emergency Response Utilizing Beacon and Mobile Technology

Joferson Bombasi, Ronaldo Juanatas, and Jasmin Niguidula

Abstract Disasters like earthquake and fire causes panic when people are contained in establishment where power can be cutoff for a longer period due to occurrence of disaster. This paper aims to utilize beacon and mobile technology to help in the risk reduction and emergency response of building occupants in a disaster or prolonged power outage. A system like risk reduction and emergency response using beacon and mobile technology could help in the emergency response and risk mitigation of the building officials involve in the disaster and rescue operation of a establishment. The system design has two major components: the beacon and the mobile device. The beacons can be strategically attached to walls or ceilings to create a path to locate the nearest emergency exit inside the building. An API such as TextToSpeech function of the Android or AVSpeechSynthesizer instance of IOS are voice technology or the text to speech API that converts the distance data into a natural-sounding voice. The mobile device must be programmed to continuously scan and detect beacons, select the closest beacon, reads distance information, and convert it to voice data. A survey was conducted on the study respondents such as mall and hotel managers, facility officers, civil engineers, NDRRMC/BFP and DRRM personnel, building occupants, including PWD and senior citizens. All respondents had a strong agreement that the risk reduction and emergency response utilizing beacon and mobile technology as an effective technology for disaster response. The system design may be an input to elevate the building management system through the use of available technology like the beacon and mobile devices.

J. Bombasi (✉)
Far Eastern University, Alabang, Philippines
e-mail: jlbombasi@feualabang.edu.ph

R. Juanatas · J. Niguidula
Technological University of the Philippines, Manila, Philippines
e-mail: ronaldo_juanatas@tup.edu.ph

J. Niguidula
e-mail: jasmin_niguidula@tup.edu.ph

53.1 Introduction

53.1.1 A Subsection Sample

In the event of a strong earthquake or fire, the electrical power inside a building normally goes out to mitigate the risk of spreading fire. Power generators may not be enough to power the light of an entire building because power loss takes longer. There are cases where emergency lights installed in the building are too dim or would not work due to a lack of periodic maintenance. In the case of fire, visibility is a concern due to heavy smoke. It is difficult to find the nearest fire or emergency exit inside the building amidst the clouds of smoke. Furthermore, the smoke coming from fire affects the visibility of emergency lights, signage, and maps in the wall, thus unable to serve their purpose [1]. CNN Philippines posted an article last 2015 states that 72 people were killed by fire in a footwear company in Valenzuela City. The dead bodies were trapped on the second floor of the building [2]. Rappler also published an article last December 2017 where 37 people were reportedly killed by fire in NCC shopping mall in Davao City. People were trapped on the top floor [3]. Another disaster incident posted by CBSNEWS last April 2019, stating that many people got trapped inside a collapsed building in Pampanga amidst a power interruption brought by a 6.1 magnitude earthquake. The earthquake toppled electrical posts and knocked out power in large parts of Pampanga [4]. These were unfortunate events that could have been avoided or, at least, reduced the severity of the loss of lives and properties if only the building emergency plans worked successfully as planned.

During calamities, different agencies' emergency responses are being undertaken to mitigate risk to save lives and properties. Emergency response activities involve responding to an incident, crisis, or disaster and managing that incident at the scene. Should an incident escalate to the crisis or disaster stage, a crisis management team (CMT) should take over managing the crisis to its conclusion. Suppose the crisis or disaster does cause damage to a company building, facility or operation. In that case, the CMT should hand over to a business continuity team the responsibility of recovery and resumption [5].

The Incident Command System (ICS) of the Philippine National Disaster Risk Reduction and Management Council is an example of emergency command that ensures disaster risks will be prevented. Incident management teams are prepared to address the needs of the affected population when a disaster occurs. The ICS is a tool to organize on-scene operations for a broad spectrum of emergencies from minor to complex incidents with priority objectives of saving lives and ensuring the safety of responders. The ICS, established under RA 10,121, enables effective and efficient incident management by integrating facilities, equipment, personnel, procedures, and communications within a typical organizational structure.

A study using mobile phone technologies for disaster risk management indicated that the reach of mobile networks has expanded significantly over the last decade, with approximately 93% of the global population being covered by a mobile broadband network. The ownership of mobile phones in Southeast Asia rising six-fold in just five years. The study also revealed that at the end of 2019, more than 3.7 billion people were connected to the mobile internet. An increased mobile phone coverage, ownership, and use have improved communications access to more people and vulnerable communities than ever before. Such access presents new opportunities for reducing risks from disasters [6].

This paper aims to utilize mobile technology to help in the risk reduction and emergency response of building occupants in a disaster or prolonged power outage. The system design is aimed to consider assistive technology (AT) concept to consider utilization of people with disabilities. Assistive technology (AT) refers to products, equipment, and systems that enhance learning, working, and daily living for persons with disabilities [7].

53.2 Methodology

The study utilized the developmental method to systematically examine products, tools, processes, and models to provide reliable, usable information to practitioners and theorists [8]. The agile software development was applied to develop the system. Agile methods or agile processes generally promote a disciplined project management process that encourages frequent inspection and adaptation, a leadership philosophy that encourages teamwork, self-organization and accountability, a set of engineering best practices intended to allow for rapid delivery of high-quality software, and a business approach that aligns development with customer needs and company goals [9].

53.2.1 Respondents of the Study

See Table 53.1.

Table 53.1 Respondents of the study

Respondents	Number
Mall manager	2
Hotel manager	2
Facility officer	6
Civil engineer	3
NDRRMC/BFP	8
LGU	3
Building occupant	2
PWD	2
Senior citizen	2

53.2.2 Beacon Technology

A beacon is a small device that constantly sends out radio signals to nearby devices such as smartphones and tablets, containing a small amount of data. To achieve the desired signal coverage, the signal strength and time between each signal can be configured. The beacons are commonly attached on walls or ceilings. The signals emitted by the beacon can trigger an action to the mobile phones. Beacons work using Bluetooth low energy (BLE), a version of the more common Bluetooth protocol, designed to use very little power and send less data—typically 1–20% of standard Bluetooth power and at 15–50% of the speed. The beacons can be powered by a battery and battery life can be anywhere from 1 month to 2–3 years. There is also an available variety of beacons that uses USB/mains power or solar panels [8].

53.2.3 Mobile Application Specifications

Beacon technology can communicate on both IOS and Android platforms. The mobile devices must be programmed to continuously scan and detect beacons, select the closest beacon, read distance information, and convert it to voice data. The device program includes automatic activation of the device Bluetooth to enable communication with available beacons in the area automatically. The TextToSpeech function of the Android or AVSpeechSynthesizer instance of IOS allows converting distance data into voice data. The voice data is vital to guide the building occupants to the nearest building exit.

53.3 Results and Discussion

The integration of technology to risk reduction and emergency response can enhance the building's emergency plan and response in case of emergencies like fire, earthquake, floods, and other catastrophe. The system design of the risk reduction and emergency response using beacon and mobile technology is shown in Fig. 53.1.

The risk reduction and emergency response using beacon and mobile technology system design have two major components: the beacon and the mobile device. The beacon component is a wireless technology that can send a signal to mobile devices. The beacons are strategically attached to walls or ceilings to create a path to locate the nearest emergency exit inside the building. Almost all smartphones nowadays are Bluetooth enabled, making them ready to communicate with any beacon available. An API such as TextToSpeech function of the Android or AVSpeechSynthesizer instance of IOS is voice technology or the text to speech API that converts the distance data into a natural-sounding voice. The mobile device must be programmed

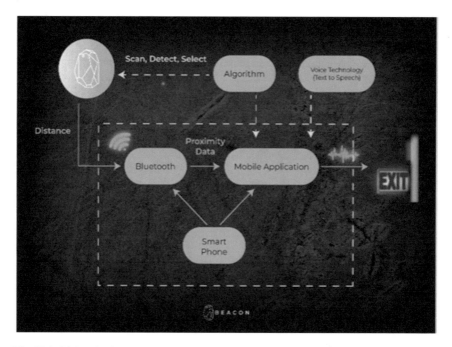

Fig. 53.1 Risk reduction and emergency response using beacon and mobile technology system design

to continuously scan and detect beacons, select the closest beacon, reads distance information, and convert it to voice data (Fig. 53.2).

The map shows the sample distance of the beacons, using 70 m range. The building length is 56.5 m, so beacons are placed to each end of the building. If the building is compliant with both building and fire codes, then the distance from exits at any point in the building shall not exceed 61 m. From 60- to 70-m range BLE proximity beacon can cover this distance. The number of beacons needed to install in a building is equal to the total number of emergency or fire exits in the floor plan of the building.

Beacon count = Total number of emergency/fire exit in the floor plan.

Example: 10 emergency exits in the floor plan = 10 BLE proximity beacons.

Additional beacons are required in large hotels or shopping malls, where a stairway or a balcony can serve as an alternative exit or passageway. Suppose the length or line of travel from a particular area or floor going to an alternative exit or passageway exceeds 70 m. In that case, an additional beacon must be placed for every 70-m point.

Beacon count = Total number of emergency/fire exit in the floor plan + additional one beacon for every 70-meter point per alternative exit or passageway.

Example A: 10 emergency exits in the floor plan + 5 alternative exits (within 70 meters) = 15 BLE proximity beacons

Example B: 10 emergency exits in the floor plan + 5 alternative exits (within 100 meters) = 20 BLE proximity beacons

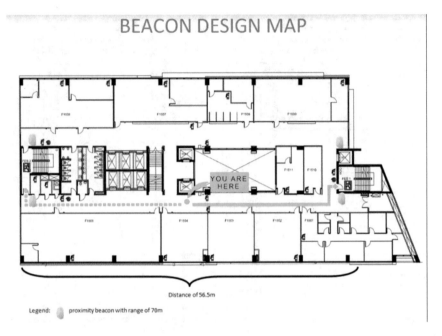

Fig. 53.2 Beacon mapping with 70 m range

The mapping of beacons could identify the locations of the beacons inside the building. The floor plan can be used to validate the beacon count. One beacon must be placed on each exit and alternative passageway. The beacons must be installed ideally 3–4 m from the ground (above crowds or objects) or in the ceiling and must establish a line of sight to any person at any point in the area.

Since the signal transmitted by the proximity beacon is in the 2.4 GHz radio frequency, the signal can be distorted by interferences. Therefore, beacons are best fixed in low interference potential materials like wood, synthetic materials, or glass. The dot in the beacon must be facing upward if installed in the walls (Fig. 53.3).

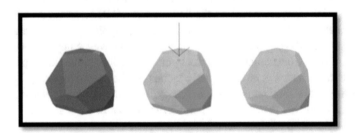

Fig. 53.3 Sample beacon orientation

53.3.1   Mobile Device Specifications

All mobile devices with Bluetooth 4.0 or newer are compatible with the beacon component. These mobile devices have BLE support and are designed to lower power consumption in communicating proximity data from beacon sensors. iPhone 4S or newer are BLE-compatible devices. Android 4.3 and higher have built-in platform support for BLE. The mobile application needs to be programmed to continuously scan and detect beacons, select the closest beacon, reads distance information, and convert it to voice data. When the closest beacon has been selected, which is the nearest emergency exit from the point of location of the user, the beacon will send distance data every short time interval. The mobile device will continuously convert it to voice data that will serve as the user's guide to reach the nearest emergency exit for a safe escape.

Figure 53.4 shows the sample screen of the mobile device when the beacon signal is detected. There are two exits detected, the immediate exit with 0.46 m and the nearest exit with a distance of 2.41 m. Aside from being shown on the device screen, the proximity of the exit is also being relayed through speech just in case the building occupant cannot read the screen due to disability or because of an emergency situation. A voice guide will provide the distance every short time until the building occupant reaches the nearest emergency exit. This makes the application assistive and can be used by people with eye impairment.

A survey was conducted on the study respondents such as mall and hotel managers, facility officers, civil engineers, NDRRMC/BFP and DRRM personnel, building occupants, including PWD and senior citizens. They strongly agreed that the mobile

Fig. 53.4 Sample mobile device screen when beacon detected

application is an essential tool for an emergency response since considering that it can support mitigating risks during a disaster. A 5-point Likers scale was used, and a mean rating of 4.93 was given by the respondents who strongly agreed that the mobile application could enhance emergency response. The perceived usefulness of the technology, or the extent to which respondents believe it may improve emergency response and minimize or eliminate disaster risks, is rated 4.88, which is commensurate to strongly agree. The perceived ease of use of the technology, or the degree to which respondents believe it does not require excessive effort to use, is rated 4.96, which is commensurate to strongly agree. The complexity of the technology, or the extent to which respondents believe the technology is simple to grasp, apply, and utilize, is rated 4.96, which is commensurate to strongly agree. The NDRRMC and BFP personnel's perceptions are critical in this study because the two government agencies are in charge of the country's disaster management program. The NDRRMC and BFP respondents agreed that technology could improve emergency response and minimize or eliminate disaster risks, scoring 4.78 and 4.94, respectively. The survey results show that the respondents strongly agree that the beacon and mobile technology are a means to improve emergency response and reduce disaster risks.

53.4 Conclusion

Integrating technology into disaster management would improve risk reduction and emergency response. The beacon and mobile technology system design for risk reduction and emergency response is developed to provide an assistive response to people during emergencies inside the building. This study utilized mobile technology to help in the risk reduction and emergency response of building occupants in the event of a disaster or prolonged power outage. The design will make everyone in the building an emergency responder during an emergency. It will provide voice assistance to help them escape from the building and save their lives. This study will make everyone in the building the first line of response in saving their own lives and not waiting anymore for external responders. This includes senior citizens and persons with disability. The system was rated in terms of emergency response, perceived usefulness of the technology, ease of use, and simplicity of the system. All respondents had a strong agreement that the risk reduction and emergency response utilizing beacon and mobile technology is an effective technology for disaster response. The system design may be an input to elevate the building management system through the use of available technology like the beacon and mobile devices.

References

1. Jones, H.: Disaster management: risk reduction, response strategy and recovery. In: Callisto Reference (2018); Author, F., Author, S.: Title of a proceedings paper. In: Editor, F., Editor, S. (eds.) Conference 2016, LNCS, vol. 9999, pp. 1–13. Springer, Heidelberg (2016)
2. Mullen, J.: 72 killed in fire at footwear factory in Philippines. CNN, 14 May 2015. [Online]. Available: https://edition.cnn.com/2015/05/14/asia/philippines-factory-fire/index.html. Accessed 2021. Author, F.: Contribution title. In: 9th International Proceedings on Proceedings, pp. 1–2. Publisher, Location (2010)
3. Agence France-Presse, 37 feared dead in Davao City mall fire, 24 December 2017. [Online]. Available: https://www.rappler.com/nation/192131-davao-city-nccc-mall-fiare-death-toll/. Accessed 2021.
4. Earthquake in Philippines collapses building, damages Clark International Airport, 22 April 2019. [Online]. Available: https://www.cbsnews.com/news/earthquake-in-philippines-injured-trapped-collapsed-building-porac-clark-airport/. Accessed 2021
5. Fischer, R.J., Halibozek, E.P., Walters, D.C.: Contingency planning emergency response and safety. In: Elsevier Public Health Emergency Collection, pp. 249–268 (2019)
6. Using mobile phone technologies for disaster risk management: reflections from SHEAR, June 2021. Reliefweb (2021). [Online]. Available: https://reliefweb.int/report/world/using-mobile-phone-technologies-disaster-risk-management-reflections-shear-june-2021. Accessed 2021
7. Earthquake "What is AT?" ATiA Assistive Technology Industry Association (2022). [Online]. Available: https://www.atia.org/home/at-resources/what-is-at/
8. Rita Richey, E.J.K.: Developmental research methods: creating knowledge from instructional design and development practice. J. Comput. High. Educ. **16**(2), 23–38 (2005)
9. What is Agile? What is Scrum?. cprime An Alten Company (2022). [Online]. Available: https://www.cprime.com/resources/what-is-agile-what-is-scrum/

Chapter 54
Arduino and NodeMCU-Based Smart Soil Moisture Balancer with IoT Integration

Mubarak K. Kankara, Al Imtiaz◎, Imran Chowdhury◎, Md. Khalid Mahbub Khan◎, and Taslim Ahmed◎

Abstract Without proper moisture in the soil, the process of agriculture can fall in danger, which can lead to even an economic collapse for a country. However, over-irrigation, under irrigation, or improper water distribution can result in crop damage and reduced productivity, which leads to waste of valuable resources including water. To contribute to addressing this issue, a smart soil moisture balancer is developed based on Internet of Things (IoT), with the help of a soil moisture sensor, water pump control, water flow meter, water level indicator, Arduino Uno, and NodeMCU with built-in Wi-Fi (IEEE 802.11b Direct Sequence) module. The developed system intelligently controls the irrigation pump's switching based on the data collected from a soil moisture sensor. The water level indicator provides data on water availability in the storage, and the water flow meter provides data on water flow rate, which gets transmitted to the ThingSpeak IoT server that stores the data and generates graphs to help with the analysis and making future decisions. A prototype of the developed system is made, verified, and tested to be working perfectly as designed and programmed. In the experiment with the prototype, it is found that the system saves 36.17% of water in case of sandy soil, 37.08% and 32.90% in case of clay soil and loamy soil, respectively. On average, the system saves 35.38% of the water, which in turn can save other intertwined resources like time and energy, keeping the efficiency of the irrigation system.

M. K. Kankara (✉) · A. Imtiaz · Md. K. M. Khan
Department of Computer Science and Engineering (CSE), University of Information Technology and Sciences (UITS), Dhaka 1212, Bangladesh
e-mail: mubarakabeer2015@gmail.com

I. Chowdhury
Department of Electrical and Electronic Engineering (EEE), University of Information Technology and Sciences (UITS), Dhaka 1212, Bangladesh

T. Ahmed
Department of Electrical and Electronic Engineering (EEE), Rajshahi Science and Technology University (RSTU), Natore 6400, Bangladesh

54.1   Introduction

As a result of both population growth and rising earnings, food demand is expected to keep on rising as well. As per the United Nations' (UN) World Population Prospects: the 2017 Revision, the total world population will grow from 7.8 to 9.8 billion in 2050 [1], resulting in more mouths to feed. Developing countries will account for the vast majority of population growth. Because of this, the required amount of food is expected to touch nearly 3 billion tons by 2050, according to the UN. This increased demand for food required increased and optimal usage of every process in agriculture, one of which is irrigation. Current irrigation systems are mostly manual that causes waste of water, and energy, and are not ideal for optimal yields. Agriculture uses 85% of available freshwater resources worldwide, according to World Bank statistics, and this percentage will continue to grow as a result of population growth and increased food demand. As a result, there is a pressing need to make water management systems reliant on science and innovation, including technological, agronomic, managerial, and institutional advances. We can handle water waste and maximize scientific techniques in irrigation systems by applying technology and innovation, which will significantly improve water usage and efficiency. One of such technologies is the IoT which is booming currently in the agriculture and farming sector in optimizing every step of the process, including irrigation [2–4].

IoT enables us to capture data from various devices called "things" which can be sensors, computers, smartphones, household appliances, or other objects. This information can then be stored in the cloud or web saver and can be retrieved later to improve decision-making. This technology plays an outstanding role in so many fields, and agriculture is not left behind. The IoT framework comprises web-enabled smart devices that use embedded technologies like processors, sensors, and communication hardware to store, transmit, and respond to data collected from their surroundings. Sensor data is exchanged between IoT sensors via linking to an IoT gateway or other edge node, where it is either sent to a server for storage or processed locally. These devices often communicate with one another and take action based on the information they share. The gadgets carry out a fair amount of work without human intervention, but humans may use them to set them up, transmit commands, and retrieve data [5, 6]. The major components of the IoT are shown in Fig. 54.1.

The rapid rise of IoT-based technologies is upgrading virtually every industry, shifting the industry away from statistical to quantitative techniques. In current times, farmers have been utilizing mostly manual irrigation systems through manual control in which the irrigation is performed at a regular interval which leads to improper utilization of water and sacrificing productivity. Through automation, IoT has the ability to make agricultural industry measures more productive by minimizing human interference, which effectively can be called smart agriculture.

A comprehensive study has been carried out in recent years where some of the efficient and effective IoT-based technologies were recommended in the topic of interest [2, 3, 7–22]. In [2], the authors have tried to solve the mentioned issues by developing an IoT-based smart irrigation system using Arduino Uno and Bluetooth technology

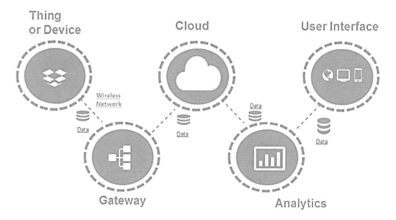

Fig. 54.1 Major components of IoT (https://www.rfpage.com/)

by monitoring the soil moisture level. In [4, 9], the authors used NodeMCU and its Wi-Fi module for developing the IoT-based smart irrigation system by monitoring the soil moisture, temperature, humidity, etc. In [16], the authors used an AVR micro-controller and ESP8266 Wi-Fi module for developing the IoT-based smart irrigation system by monitoring not only the soil moisture but also the humidity, light intensity, and temperature. This smart agricultural market is estimated to grow to $11.23 billion in US (United States) dollars by 2022, as per 2017's Research and Markets Forecast. With an annual growth rate of 20% continuously, the global market size of smart agriculture is forecasted to triple by 2025 to $15.3 billion (particularly in comparison with just around $5 billion in 2016). In response to that, the agriculture industry and farmers are already into IoT-based solutions that allow farmers to minimize waste and increase efficiency from the number of fertilizers used to the amount of water made available by the farmer to his crops efficiently, saving the resources like water, energy, etc.

Keeping the discussed issues in mind, the developed soil moisture balancer presented in this paper is intended to overcome the unnecessary water flow into the agricultural lands by alerting the pump control to either turn ON or OFF with respect to the soil dryness or wetness, based on measured soil moisture content and the amount of water usage. The central processing unit of the system also includes a communication gateway such as a Wi-Fi module, to send data to an IoT server in real time, and relay the information to the user's device such as a computer or hand-held devices like a smartphone or tablet for analysis.

Table 54.1 System's
components and peripheral
devices

Components/devices	ID/remarks
Arduino Uno R3	ATmega328P based
NodeMCU	ESP8266-12E
Water level indicator	P35, floating
Water flow meter	YF-S201, hall-effect
Soil moisture sensor	FC-28
Organic light-emitting diodes (OLED)	0.96″ 12C
Liquid–crystal display (LCD)	16 × 2 LCD
Relay	LM393, single channel voltage comparator
Pump control	Mini submersible
Breadboard	MB-102, full
Connecting wires	Jumper, MM MF FF

54.2 Methods and Materials

54.2.1 Components and Peripheral Devices

The developed soil moisture balancer system incorporates various electronic devices
and components, e.g., an Arduino and a NodeMCU board as the brains of the system,
sensors to measure soil moisture, water level, and water flow, a controller to control
equipment like the pump, displays to present information, etc., to do its intended
function. The total list of required components and tools is provided in Table 54.1.

54.2.2 System Model and Block Diagram

The developed system is comprised of both equipment and computer programs. On
the equipment side, 3 (three) types of sensors are utilized to measure soil moisture,
water level, and water flow. Subsequently, an Arduino Uno and a NodeMCU are
used as the IoT design platforms, where the NodeMCU coordinates with its built-
in Wi-Fi module to transfer the data to an IoT server (ThingSpeak). In addition, 2
(two) displays are used to show the results as well. On the programming side, a set
of computer codes is written to perform the desired functions. A block diagram is
presented in Fig. 54.2 to provide an easy visualization of the entire system. From the
diagram, a discrete idea of all the incorporated modules/devices and their responsibilities can be achieved on a macro level. On the input side of the Arduino Uno,
there are 2 (two) sensors: soil moisture and water level; and on the output side, there
is an LCD display and a relay module. On the input side of the NodeMCU, there is

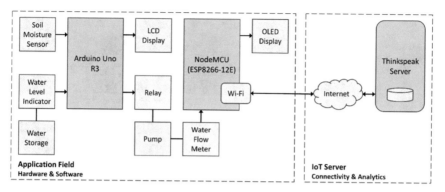

Fig. 54.2 Block diagram of the developed system

a water flow meter; and on the output side, there is an OLED display. The built-in Wi-Fi module of the NodeMCU is used to connect the system to the ThingSpeak server.

54.3 Electronic Circuit/Hardware Interfacing

Arduino Uno R3 and NodeMCU ESP8266-12E are used to make decisions such as turning ON the water pump and sending results to the cloud based on data from sensors. Figure 54.3 shows the schematic of circuit interfacing of the developed system, where the soil moisture and water level sensors are interfaced to the Arduino's A0 and pin-2, respectively, and the relay and LCD display are controlled by pin-3 and pin-4, 5, respectively. The water flow meter is interfaced to NodeMCU's D4, and the OLED display is controlled by D1 and D2, respectively. The complete interfacing is better depicted in Tables 54.2 and 54.3.

54.4 Software Programming and IoT Server Integration

54.4.1 Programming Flowchart

Both Arduino and NodeMCU are microcontroller-based devices. The microcontroller used in an Arduino is ATmega328P from Atmel. The ESP8266 in a NodeMCU is a low-cost Wi-Fi chip that has a microcontroller capability. Therefore, the functionality of an Arduino and NodeMCU depends on the programming that follows the general attributes of Atmega and ESP8266 programming. The Arduino IDE software is used to program both Arduino Uno and NodeMCU. The programming codes that are written for the system in the presented work are described using the flowchart

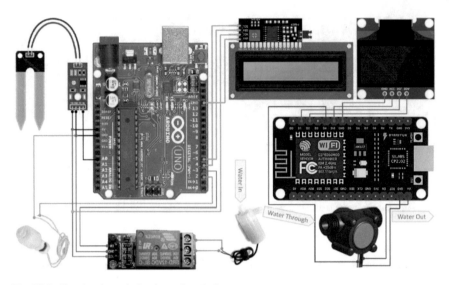

Fig. 54.3 Circuit schematic/hardware interfacing

Table 54.2 Interfacing between Arduino Uno and its components (pin-to-pin)

Arduino Uno	Soil moisture sensor	Water level indicator	LCD display	Relay
GND	GND	GND (−)	GND	GND
VCC (5 V)	VCC		VCC	VCC
Pin-5			SDA	
Pin-4			SCL	
A0	AO (anode)			
Pin-2		VCC (+)		
Pin-3				IN

Table 54.3 Interfacing between NodeMCU and its components (pin-to-pin)

NodeMCU	Water flow meter	OLED display
Vin	Red (VCC)	
GND	Black (GND)	GND
D4	Yellow (hall effect)	
3V3		VCC
D1		SCL
D2		SDA

in Fig. 54.4. According to the flowchart, every time the soil moisture sensor senses dryness the system checks for water availability through the water level indicator and decides whether the water pump should be turned ON or OFF. The system keeps the record of the rate and volume of water flow using the water flow meter and sends the data to the ThingSpeak IoT server.

54.4.2   Sensors and Parameter Setup

Soil Moisture Sensor. The level of the soil moisture sensor changes based on the soil's resistance [23, 24]. The driver LM393 voltage comparator relay is a double differential measuring stick that compares the sensor's tension to a 5 V voltage level. The sensor values range from 0 to 1023; 0 being the wettest state and 1023 being the driest state. Based on the soil attribute the calibration of the soil moisture sensor can be changed in programming, which is done in the following way:

```
if (sensorValue <= 500){ //Soil Value Level Reached
digitalWrite(PumpMotor, HIGH); //Pump OFF
lcd.setCursor(8,1);
lcd.print("PUMP OFF")
```

Water Level Indicator. The water level indicator includes a reed-magnetic switch with floating magnets that leads when water is available. When the Arduino Uno reads the status of the soil moisture using the soil moisture sensor and the soil happens to be dry, then it checks the availability of water in the water storage using the water level sensor. If the water is available then the system notifies with the text "WATER OK" on the LCD screen, then the pump turns ON and automatically turns OFF when an adequate amount of water is supplied. The pump control is driven by a relay circuit. However, when water is unavailable, then the system notifies with a text "NO WATER" on the LCD screen. For any other condition, the pump remains OFF and the status of the moisture and pump will be displayed on the LCD screen.

Water Flow Meter. When the pump control is turned ON, the water passes through the water flow meter. Every revolution of the meter produces an electrical pulse from an inbuilt magnetic hall-effect sensor. Counting the pulses from the sensor's output can be used to calculate the water flow rate. Each pulse contains about 2.25 ml. Though this sensor is the least expensive and one of the best, it is not the most accurate one as the value of water volume fluctuates slightly depending on sensor orientation, fluid pressure, and flow rate. A significant amount of calibration is required to achieve a precision of more than 10%. But for the proof of concept and making the prototype, this sensor is used as it is one of the least expensive ones. Because the pulse signal is a simple square wave, logging it and converting it to liters per minute using the formula below makes it simple [25].

$$\frac{F}{7.5} = Q(\text{L/min}) \tag{54.1}$$

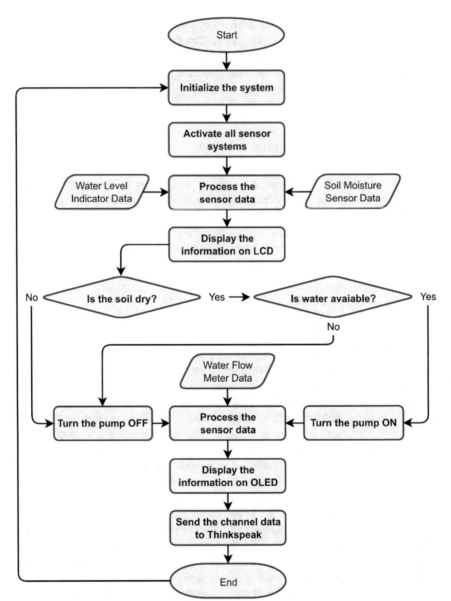

Fig. 54.4 Programming flowchart of the developed system

where F is the pulse frequency in hertz (Hz), and Q is the discharge or water flow rate. The pulse frequency depends on the water speed, and water speed depends on the pressure that drives the water through the pipelines. There is a known and constant cross-sectional area of the pipe, and if the water velocity is known, the water flow rate can be calculated as

$$Q = A \times V \left(\mathrm{m}^3/\mathrm{s} \right) \tag{54.2}$$

where A is the cross-sectional area of the pipe and V is the water velocity. From the previous equation of water flow rate, the volume of water can be calculated as [25]

$$\text{Water volume} = Q \times t(\mathrm{s}) \times \frac{1}{60(\mathrm{s})} (\mathrm{L}) \tag{54.3}$$

$$\text{Water volume} = \frac{F(\text{pulses}/\mathrm{s})}{7.5} \times t(\mathrm{s}) \times \frac{1}{60(\mathrm{s})} (\mathrm{L}) \tag{54.4}$$

$$\text{Water volume} = \frac{\text{pulses}}{7.5 \times 60(\mathrm{s})} (\mathrm{L}) \tag{54.5}$$

where t is the time elapsed for water flow in seconds.

54.4.3 Setting Up IoT Server (ThingSpeak)

In order to be called IoT, the end-point systems or devices need to be connected to a cloud server to be able to store data and make analytical decisions. For the work presented in this paper, the popular IoT platform ThingSpeak is chosen. There are some steps to integrate the end device with ThingSpeak. A gist of which is creating channels for different types of data, generating API Keys for each type of data, and incorporating the API Keys in the written code for the end device.

After generating the API keys the incorporation in the written code is done in the following manner. For the proof of concept and demonstration purpose, only one channel creation for the water flow meter is shown (Fig. 54.5).

```
String apiKey = "KBD1JSZTUKCXJ15V";
const char *ssid = "mubarak";
const char *pass = "005kkr";"
```

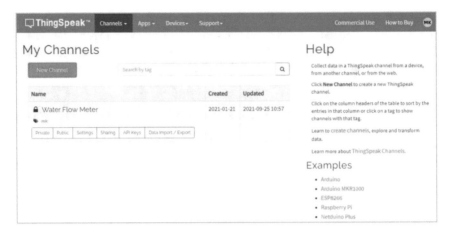

Fig. 54.5 List of the created channel(s)

54.5 Results and Discussion

54.5.1 Prototype Implementation

The developed soil moisture balancer system presented in this paper is practically implemented based on the block diagram and circuit design discussed above using the components and peripheral devices mentioned in Table 54.1. The written programming code based on the flowchart discussed above is burnt on the Arduino Uno and NodeMCU to achieve its functionality. The implemented prototype is created by interfacing all the electronic components using full-size MB-102 breadboards. Almost all the components in the system are running on a 5 V DC supply, except for the OLED screen which takes 3.3 V DC. Figure 54.6 shows the prototype of the practical device with the labeling of its components.

54.5.2 Results from Prototype Testing

In order to test the prototype, the soil moisture sensor is buried inside some dry soil surface (Fig. 7c) carefully keeping the fact in mind that the sensor wirings are not waterproof. For precision sensing, it is recommended to position the sensor near the roots of the plants. The water level indicator is placed in the water storage (tank), and the water flow meter is already connected through the output pipeline. After the power is turned ON, the system worked as designed and programmed by delivering water to the soil. The LCD screen showed the soil moisture sensor, water availability, and pump status (Fig. 7a); and the OLED screen showed the water flow rate and water

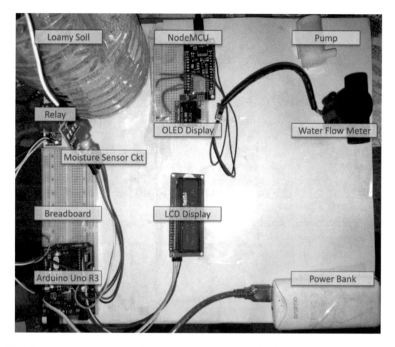

Fig. 54.6 Implemented prototype of the developed system with labeling

volume (Fig. 7b). The IoT server status is also checked and Fig. 54.8 shows that the created channel data (water flow meter) is transferred to the ThingSpeak server.

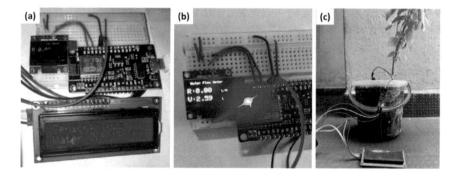

Fig. 54.7 Prototype testing. **a** Soil moisture sensor, water availability, and pump status. **b** Water flow rate and water volume. **c** Calibrating soil moisture sensor for clay soil

Fig. 54.8 Sensor data is being transferred to the ThingSpeak server

54.5.3 Experimental Results Based on Soil Types

The level of soil moisture varies for different types of soil. So, an experiment is done taking the 3 (three) types of soil (clay, loamy, and sandy) into account by calibrating the soil moisture sensor threshold as per the sensitivity of each soil type to observe how the water consumption by soil varies, with and without using the soil moisture sensor. The obtained experimental results are shown in Tables 54.4 and 54.5.

Table 54.4 Results for variation in water consumption using soil moisture sensor

Soil type	Time	Moisture level (dry condition)	Moisture level (wet condition)	Water volume in liters (L)
Sandy	08:00:03 a.m.	1023	450	1.80
	10:04:08 a.m.	1000	446	1.79
	12:03:00 p.m.	1021	448	1.81
Clay	08:30:28 a.m.	1022	288	1.49
	11:30:08 a.m.	1001	296	1.52
	03:00:06 p.m.	1014	290	1.53
Loamy	09:00:26 a.m.	1021	459	1.70
	11:04:33 a.m.	1010	457	1.67
	01:09:50 p.m.	1020	460	1.71

Table 54.5 Results for variation in water consumption without using soil moisture sensor

Soil type	Time	Water volume in liters (L)
Sandy	08:00:46 a.m.	2.71
	10:08:18 a.m.	2.95
	12:01:11 p.m.	2.80
Clay	08:39:12 a.m.	2.30
	12:39:09 a.m.	2.50
	03:00:18 p.m.	2.41
Loamy	09:00:39 a.m.	2.72
	11:04:10 a.m.	2.21
	01:05:59 p.m.	2.63

54.5.4 Discussion

Averaging the data from Tables 54.4 and 54.5, the differences in water consumption between the cases when the soil moisture sensor is used and not used are calculated and shown in Fig. 54.9. Also, the percentage of water consumption for the case when the soil moisture sensor is used with respect to the case when the sensor is not used are calculated and presented in Table 54.6. According to Table 54.6, using the soil moisture sensor 36.17% (100 − 63.83) of water is saved in the case of sandy soil, 37.08% (100 − 62.92) saved in the case of clay soil, and 32.90% (100 − 67.10) in case of loamy soil.

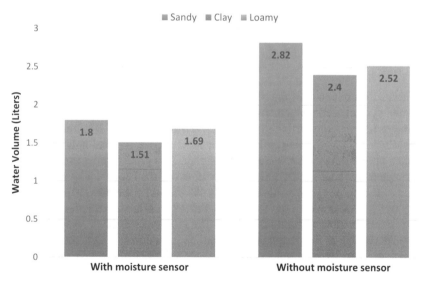

Fig. 54.9 Comparison of water consumption between using soil moisture sensor and without using the sensor

Table 54.6 Percentage of water consumption for using soil moisture sensor with respect to not using the sensor

Soil type	Avg. water volume with sensor (in L)	Avg. water volume without sensor (in L)	% Water consumption with sensor
Sandy	1.80	2.82	63.83
Clay	1.51	2.40	62.92
Loamy	1.69	2.52	67.10

For the proof of concept in the case of incorporating IoT in the developed system, only one sensor's (water flow meter) data is being channeled to the server. But channeling the soil moisture sensor's data would provide a diverse analytical advantage for better decision-making, which is under consideration in the next improvement. In case of testing the developed system for water consumption by 3 (three) types of soil, a 2-L water bottle is used to contain loamy soil, a 2-L and 550-ml plastic buckets are used to contain clay soil and sandy soil, respectively. The resultant data may vary if the sample sizes change.

The work refers to the fact that the methodology used to track the moisture content of the soil here allows agriculturalists to use moisture calculation and automatic irrigation with the potential to eliminate unnecessary irrigation cycles and save a huge amount of water. The developed system is feasible and cost-effective for optimizing the irrigation system in small-scale farms by embedding multiple soil moisture sensors, and probably multiple water pumps depending on the size of the field.

54.6 Conclusion

To address the efficiency of the irrigation system, and save precious resources, the presented system is developed. According to the system, the water releases to the field only when it is required based on the soil moisture level. This is accomplished by sensing the soil moisture using the FC-28 moisture sensor and accordingly controlling the water pump using the LM393 relay controller. This process proceeds only when the storage has enough water, which is determined by the P35 water level indicator. Also, the amount of water being released is calculated by the YF-S201 water flow meter, which gets stored in the IoT server ThingSpeak in real time for producing analytics and fine-tuning future decisions. The prototype of the developed system is tested according to the design and programming. In doing so, it is found that the system saves 36.17% of water in case of sandy soil, 37.08% in case of clay soil, and 32.90% in case of loamy soil. Overall, the system saves 35.38% of the water on average, which in turn can save other intertwined resources like time and energy, increasing the profit margin for farmers. Future scopes of the presented work may involve (i) incorporating humidity, temperature, and other sensors, (ii) introducing cloud controlling, (iii) providing moisture adjustment mechanisms based on crop type, etc.

References

1. World Population Prospects: The 2017 Revision | United Nations, https://www.un.org/en/desa/world-population-projected-reach-98-billion-2050-and-112-billion-2100
2. Ashwini, B.V.: A study on smart irrigation system using IoT for surveillance of crop-field. Int. J. Eng. Technol. **7**, 370–373 (2018). https://doi.org/10.14419/ijet.v7i4.5.20109
3. Madushanki, A.A.R., N, M., A., W., Syed, A.: Adoption of the Internet of Things (IoT) in agriculture and smart farming towards urban greening: a review. Int. J. Adv. Comput. Sci. Appl. **10**, 11–28 (2019). https://doi.org/10.14569/IJACSA.2019.0100402
4. Vijay, Saini, A.K., Banerjee, S., Nigam, H.: An IoT instrumented smart agricultural monitoring and irrigation system. In: 2020 International Conference on Artificial Intelligence and Signal Processing (AISP), pp. 1–4. IEEE (2020)
5. Farooq, H., Rehman, H.U.R., Javed, A., Shoukat, M., Dudely, S.: A review on smart IoT based farming. Ann. Emerg. Technol. Comput. **4**, 17–28 (2020). https://doi.org/10.33166/AETiC.2020.03.003
6. Antony, A.P., Leith, K., Jolley, C., Lu, J., Sweeney, D.J.: A review of practice and implementation of the Internet of Things (IoT) for smallholder agriculture. Sustainability **12**, 3750 (2020). https://doi.org/10.3390/su12093750
7. Glória, A., Dionisio, C., Simões, G., Cardoso, J., Sebastião, P.: Water management for sustainable irrigation systems using Internet-of-Things. Sensors **20**, 1402 (2020). https://doi.org/10.3390/s20051402
8. Hsu, W.-L., Wang, W.-K., Fan, W.-H., Shiau, Y.-C., Yang, M.-L., Lopez, D.J.D.: Application of Internet of Things in smart farm watering system. Sensors Mater. **33**, 269 (2021). https://doi.org/10.18494/SAM.2021.3164
9. Sai, T., Proeung, B., Tep, S., Chhorn, S., Pec, R., Nall, V., Ket, P., Oeurng, C., Hel, C.: Prototyping of smart irrigation system using IoT technology. In: 2021 7th International Conference on Electrical, Electronics and Information Engineering (ICEEIE), pp. 1–5. IEEE (2021)
10. Nagaraj, P., Muneeswaran, V., Pallikonda Rajasekaran, M., Muthamil Sudar, K., Sumithra, M.: Implementation of automatic soil moisture dearth test and data exertion using Internet of Things. In: Emerging Technologies in Data Mining and Information Security, pp. 511–517. Springer, Singapore (2021)
11. Sharma, B.B., Kumar, N.: IoT-based intelligent irrigation system for paddy crop using an internet-controlled water pump. Int. J. Agric. Environ. Inf. Syst. **12**, 21–36 (2021). https://doi.org/10.4018/IJAEIS.20210101.oa2
12. Bentabet, D.: Cloud-IoT platform for smart irrigation solution based on NodeMCU. In: Lecture Notes in Networks and Systems, pp. 689–696. Springer, Cham (2021)
13. Nadhori, I.U., Udin, M., Al Rasyid, H., Ahsan, A.S., Mauludi, B.R.: Internet of Things based intelligent water management system for plants. In: 2nd International Conference on Smart and Innovative Agriculture (ICoSIA 2021), pp. 331–336. Atlantis Press (2022)
14. Drakshayani, S., LaksmiManjusha, Y., Ramadevi, P., Madhuravani, V., Suguna, K.R.: Smart plant monitoring system using NodeMCU. In: 2022 International Conference on Electronics and Renewable Systems (ICEARS), pp. 525–530. IEEE (2022)
15. Khoa, T.A., Trong, N.M., Bao Nhu, L.M., Phuc, C.H., Nguyen, V., Dang, D.M.: Design of a soil moisture sensor for application in a smart watering system. In: 2021 IEEE Sensors Applications Symposium (SAS), pp. 1–6. IEEE (2021)
16. Subashini, M.M., Das, S., Heble, S., Raj, U. and Karthik, R.: Internet of Things based wireless plant sensor for smart farming. Indones. J. Electr. Eng. Comput. Sci. **10**, 456 (2018). https://doi.org/10.11591/ijeecs.v10.i2.pp456-468

17. Srilakshmi, A., Rakkini, J., Sekar, K.R., Manikandan, R.: A comparative study on Internet Of Things (IoT) and its applications in smart agriculture. Pharmacogn. J. **10**, 260–264 (2018). https://doi.org/10.5530/pj.2018.2.46
18. Kiani, F., Seyyedabbasi, A.: Wireless sensor network and Internet of Things in precision agriculture. Int. J. Adv. Comput. Sci. Appl. **9**, 99–103 (2018). https://doi.org/10.14569/IJACSA.2018.090614
19. Muley, R.J., Bhonge, V.N.: Internet of Things for irrigation monitoring and controlling. In: Advances in Intelligent Systems and Computing, pp. 165–174. Springer, Singapore (2019)
20. G. S. Campos, N., Rocha, A.R., Gondim, R., Coelho da Silva, T.L., Gomes, D.G.: Smart and green: an Internet-of-Things framework for smart irrigation. Sensors **20**, 190 (2019). https://doi.org/10.3390/s20010190
21. Kyu Park, J., Kim, J.: Agricultural environment monitoring system to maintain soil moisture using IoT. J. Internet Things Converg. **6**, 45–52 (2020). https://doi.org/10.20465/KIOTS.2020.6.3.045
22. Sanjeevi, P., Prasanna, S., Siva Kumar, B., Gunasekaran, G., Alagiri, I., Vijay Anand, R.: Precision agriculture and farming using Internet of Things based on wireless sensor network. Trans. Emerg. Telecommun. Technol. **31**, e3978 (2020). https://doi.org/10.1002/ett.3978
23. Chowdhury, I., Binzaid, S.: AgriTronX—a solar powered off-grid automated cultivation system applicable also in Piper Betle (paan) growth. In: 2012 International Conference on Informatics, Electronics & Vision (ICIEV), pp. 320–325. IEEE (2012)
24. Binzaid, S., Chowdhury, I.: Soil erosion prevention by sustainable phytoremediation process using solar irrigation and fertilization system. Int. J. Sci. Res. Publ. **4**, 1–13 (2014)
25. Admin: IoT Water Flow Meter using ESP8266 & Water Flow Sensor. https://how2electronics.com/iot-water-flow-meter-using-esp8266-water-flow-sensor/

Chapter 55
Privacy Preserving Keyword Search Over Hash Encrypted Data in Distributed Environment

K. S. Sampada⑩ **and N. P. Kavya**

Abstract In trusted public cloud, allowing data consumers to recover the information with full integrity. The most of research efforts were focused on a single data owner for safe cloud-based encryption of encrypted data. Additionally, searchable encryption makes it possible for a data user to locate a specific encrypted document from a cloud of encrypted data by searching for keywords (KS). This study isn't useful for retrieving relevant results from keyword searches. Such issues may be overcome using the secure and privacy-preserving keyword search retrieval (SPKSR) technology, which allows users to recover hashed encrypted files from cloud data that is also hashed encrypted. Three distinct entities are included in the proposed system: (a) cloud server (CS), (b) data owner (DO), and (c) data users (DU). It is the CS that generates a searchable index tree for the owner's hashed encrypted files. The CS keeps the index tree structure and the hashed encrypted document set in its own storage. The encrypted data is hashed and DU executes the "search" over it. Analyses and comparisons of the experimental findings of the suggested system with the other existing system demonstrate its superiority.

55.1 Introduction

In general, CC is a cloud computing utility that allows customers to save their data in the cloud from a distance. The on-demand high-quality apps and services provided by a shared pool of customizable computing resources assist them in this way [1]. Such cloud resources are often distributed and dynamically re-allocated by various users based on user requirements. The maintenance as well as storage of data on the user's end are reduced by putting it on the cloud [2]. The data are always encrypted before they are saved in the cloud to ensure their security [3]. As a result, it is vital that DO provide the public CS access to their data while allowing DU access [4]. Modern studies' system design focuses only on DO, indicating that DO and DU may easily interchange and transmit hidden information in their solutions. When a

K. S. Sampada (✉) · N. P. Kavya
Computer Science and Engineering, RNS Institute of Technology, Bangalore, India
e-mail: k.s.sampada@gmail.com

© The Author(s), under exclusive license to Springer Nature Singapore Pte Ltd. 2023   637
C. So-In et al. (eds.), *Information Systems for Intelligent Systems*, Smart Innovation, Systems and Technologies 324, https://doi.org/10.1007/978-981-19-7447-2_55

system has a large number of DOs, transmitting confidential information results in significant communication overhead [5–7]. Multiple searchable encryption methods have recently been developed to protect data searches inside encoded files and documents [8]. Additionally, the new keyword searching technique is used to get data in a safe and privacy-preserving method and to search data effectively [9]. Searchable encryption (SE), multi-keyword SE, single-keyword SE, fuzzy keyword SE, ranked KS, plain-text fuzzy KS, as well as Boolean KS, are the search techniques [10, 11]. Selective file retrieval is possible with such KS approaches [12]. In addition to supporting multi-keyword ranked search (MKRS), fuzzy KS, and similarity search, safe search over encrypted data significantly saves storage and computation costs. These schemes all fall within the single-owner prototype category [13]. Using any relevant terms, a user may search for or retrieve relevant information [14]. Search engine (SE) techniques are still lacking in their capacity to provide customized, private, and flexible search options [15, 16]. The following is a summary of the paper's original drafting structure. There are three sections in this paper. For this strategy, a review of relevant literature is provided in Sect. 55.2. Section 55.3, a review of the suggested technique, Sect. 55.4, and a brief conclusion in Sect. 55.5.

55.2   Related Work

See Table 55.1.

55.3   Secure Keyword Search Retrieval Over Encrypted Cloud Data

Better data outsourcing and data services are made possible by the emergence of the CC methodology. However, there are also privacy risks with storing sensitive data on the cloud. Users often encrypt their data before uploading it to the cloud to ensure confidentiality, which presents a major obstacle to effective data use. There are several challenges to traditional data use, thus making it easier to explore encrypted cloud data is critical. Using hashed encrypted cloud data, this study presents an effective SPKSR method for retrieving hashed encrypted documents. Figure 55.1 illustrates the suggested framework. The DO, DU, and CS are three of the suggested system's components. The CS receives the hashed encrypted document collection as well as the produced searchable index tree from DO. It is also up to the DO to make sure that the documentation is kept up to date. The DO creates the update data locally and then submits it to CS during an update. Index tree structure as well as encrypted document collection are stored in the CS.

Table 55.1 Summary of literature review on PPKS

Author	Technique	Methodology	Remarks
Peng et al. [17]	Tree-based ranked multi-KS strategy for multiple DO	DO can encrypt their individual-based index, and the cloud can be allowed to combine indexes' unknowing index data to effectively integrate indexes	Not very efficient
Zhang et al. [18]	Trustworthy KS scheme on encrypted cloud storage	Honest CSs were shielded from the malicious Dos by the characteristics like user-side verifiability, encryption keys, fair payment, and server-side verifiability, as well as no TTP	Still a need for more security [19]
Hahn et al [20]	Effective and easy multi-user similarity search algorithms for cloud storage	Untrusted CS may now compare the similarity of multidimensional searchable indexes and queries using this approach. The similarity search in this method used adaptive semantic security	This design necessitated additional security measures
Xiang et al. [19, 21]	Privacy-preserving query strategy for effectual similarity query in encrypted multidimensional data	p-stable distribution-based locality-sensitive hashing similarity search approach (pLSH) was used	The scheme used only experimental data and was found to be secure and capable of protecting both the confidentiality of the data and the privacy of the users
Zhangjie et al. [21, 22]	Central keyword semantic extension ranked	Strategy employs two safe searchable encryption methods to satisfy different confidentiality requirements under two different threat models	Experiments on a real-world data set showed that this strategy worked as promised and was completely secure and efficient

(continued)

Table 55.1 (continued)

Author	Technique	Methodology	Remarks
Wang et al. and Jie et al. [22, 23]	Ranked multi-keywords fuzzy search strategy supporting ranges query	Stable LSH functions with a BF ("Bloom filter") should be used in the first index layer to provide multi-keyword fuzzy matching and to generate a score table for each document to achieve ranked search results. Normal BF was used for the secondary index layer to get range query targets by changing numerical values into specific characters and infusing the m-bit array with them	This system's accuracy and efficiency can be further improved
Wan et al. and Yinbin et al. [24, 25]	Verifiable KS on encrypted data in multiple owners	Multi-signature methods were used. A random prototype was used in the scheme's security evaluation, and the results showed that it was protected against IND-CKA	There was no correlation between DO count and verification time, and this experience result across real-world datasets demonstrates this approach's practical usefulness and viability

Suggested scheme searches for the hashed encrypted information. Using the shared nonce number as well as hashed decryption key provided by the DO, the DU may decode the matching hashed encrypted documents. There are three entities that constitute this suggested system: DO, DU, and CS.

55.3.1 Input Details

The input for this project is taken from "Electronic Medical Records" (EMPs). EMR refers to a digital record of a patient's medical history that may be generated, collected, maintained, and reviewed by qualified physicians and other staff in a healthcare facility. At this time, the patient number is quickly growing across the world, and healthcare administrators are using EMR to keep track of their patient's information. The patient details are stored in the EMR including "pname," "pid," "pgender," "pillness", "plocation", and "paddress". Healthcare organizations around

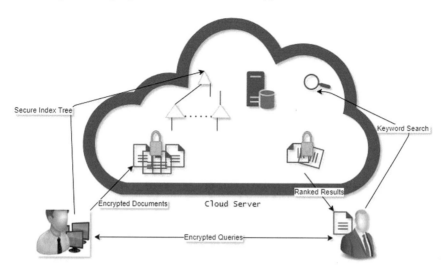

Fig. 55.1 Architecture of PPKS over encrypted data in cloud

the world work together to keep track of these records. The data collected is kept in the cloud, where it may be accessed more quickly and efficiently. Attributes are encrypted using KS to improve security in the cloud. The encrypted data is maintained by the EMR to ensure the safety of patient information. The healthcare industry and the data owner participate together in this part. DU and the data owner are referred to as "patients" and "Doctors", respectively. To decrypt EMR data, DO uses a nonce-valued hash key to deliver the encrypted EMR data to CS; DU decrypts the data using this key. Data may be retrieved using one of three search methods: single key KS, multi-keyword rank, or multi-keyword Boolean. **Sample Heading (Third Level).** Only two levels of headings should be numbered. Lower level headings remain unnumbered; they are formatted as run-in headings.

55.3.2 Data Owner Entity

A searchable index tree and hashed encrypted documents are outsourced to CS to maintain security. An entropy-based index tree creation method is used to construct hashed encrypted files, and the searchable index tree is created using this method as well. Each one is highlighted in the list that follows.

Modified Hashed Encrypted Documents

This approach uses "Hashed modified elliptic curve cryptography" (HMECC) and generates hash codes for encrypted files written using algorithm SHA 512. Using ECC method, data files are encrypted and can only be decoded by a specific person who has been granted access to them. This ECC algorithm was chosen because of this security concern. Using the ECC algorithm ensures that files are safe from access by anybody except the intended recipient. A sort of public-key cryptography implementation is on the basis of ECC algorithm. Using a prime number function, a curve with defined base points is the foundation of this approach. The maximum value may be set using this function. Here is a mathematical representation of the ECC.

$$b2 = a^3 + ra + s; 4r^3 + 27s \neq 0 \tag{55.1}$$

where a and b are coordinate values, whereas 's' and 'r' are integers. Encryption strength only relies on the method used to generate the key in a cryptographic operation. All three sorts of keys are required in the suggested scheme. The initial action is to encrypt the server's public key and generate it. Generate a private key on server-side and decode message using this key. Generating a secret key from the public, private, and curve-based keys is the third stage. The secret key serves as a safety measure. All users have a public key, making it easy for unauthorized people to access the contents. Secret keys are produced and only provided to authorized users for this reason. The SHA 512 method is used to generate hash codes for the hashed public and private keys, which are then used in the encryption and decryption processes. Taking into account the given equation,

$$R_G = \prod (N_G, P_G) \tag{55.2}$$

The public key is represented by R_G, the private key is marked by N_G, which is an arbitrary integer, and point on curve is given by P_G. Secret key: By adding the R_G, P_G, and N_G, the secret key is generated:

$$S_G = \sum (R_G, N_G, P_G) \tag{55.3}$$

S_G represent secret key, and its calculation is defined by Eq. (55.3). Encryption: The values obtained are encrypted once S_G is generated. There are two ciphertexts in the encrypted data, which are mathematically defined as shown below.

$$C1 = ((K : 1, 2, ...(n-1)) * P_G) + S_G \tag{55.4}$$

$$C2 = M + ((K : 1, 2, ...(n-1)) * H(R_G)) + S_G \tag{55.5}$$

Here, the two cypher messages are depicted by $C1$ and $C2$, and the random number produced in the range of $(1$ to $n-1)$ is K, the original message is given by M, and a hash value of public key represented by $H(R_G)$. Decryption: S_G is encrypted with two ciphertexts in this case

$$M = (((C_2 - H(N_G)) * C_1) - S_G) \tag{55.6}$$

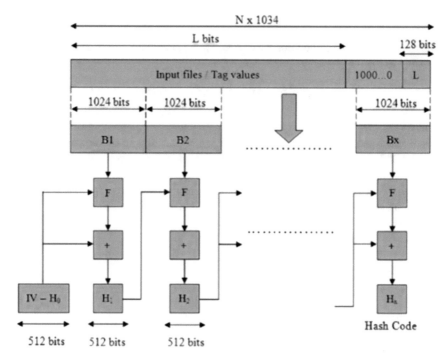

Fig. 55.2 SHA 512 algorithm

In this case, M stands for the original message, while $H(N_G)$ represents the private key's hashed value. S_G is deducted from the normal decryption equation during the decryption process.

SHA 512 algorithm: Several computations are carried out in this process. Figure 55.2 demonstrates each stage in detail. A multiple of 1024 bits is added to the message bits as a padding step in the encryption process. Then, each of these 1024-bit chunks is split into two smaller ones. The hash code for the first block is formed by combining the initializing vector with the first block. The previously produced hash codes are used to link the subsequent blocks. Figure 55.2 depicts the SHA 512 algorithm in action. The CS outsources the hashing and encryption of these encrypted documents.

55.3.2.1 Index Tree Generation

Entropy-based index tree generation is used to construct the index tree structure. Pre-processing, entropy computation, and entropy-based index tree formation are all expressly included in index tree generation.

Pre-processing

Suppose that the set of documents is ΔD, with t documents in it. It is, therefore, possible to display the appropriate data set as

$$\Delta D = D1; D2; D3 : \ldots Dt \tag{55.7}$$

where $D1, D2, D3. \ldots Dt$ represents the data set's documents. Splitting, itemset, stop words removal, CF_I as well as frequent itemset, are all part of the pre-processing process. It is the process of dividing a large amount of data into smaller subsets. Big data input is broken down into smaller chunks in the pre-processing step, making it easier to extract the data's characteristics. The data set contains a significant amount of data, making it difficult to find specific items using a keyword search. To overcome this challenge, the data set has been split into a series of documents. Assume that the data set $\Delta D = D1, D2, D3. \ldots Dt$ is the input. In this method, each document is broken down into equal portions $R1, R2, \ldots Rt$. For example, document D1's spilt function may be written as

$$S1 = D1R1; R2; : \ldots Rt \tag{55.8}$$

assuming that t is length of the words included in file. All of the dataset $D1, D2, D3. \ldots Dt$ undergoes the splitting operation indicated by the common split array

$$S[i] = S1 + S2 + S3+ : \ldots St \tag{55.9}$$

where $S_t = t$th document's Splitted array (Dt).

Stop words removal: After the splitting process is complete, the undesirable words are removed using a technique known as "Stop-word removal". Stop words and stemming are two indexing techniques that are often used together. Stop words are those appear often but have low semantic weight, making them ineffective in assisting in the retrieval process. A frequent word (like "the") is omitted from the sentence. In IR, it's common practice to remove them from the index. Itemset A the term itemset refers to a collection of items. The remaining words are shown as itemset after the removal of stop words. Frequent itemset Frequency is defined as the number of times a certain item appears in a given set. Another term used for this is "support count", "frequency", or "count" of elements in the set. "Closed frequent itemset" (CFI): It is the frequent itemset item with the greatest number of occurrences. Entropy calculation Pre-processing CFI values are used to estimate entropy in entropy estimation. It is a measure of the average amount of information. The entropy of every class is an arbitrary variable whose predicted value or average equals the total amount of information in each class. To generate the index tree, the entropy calculation is taken into consideration. The entropy formula is as follows:

$$\text{Entropy} = -\sum_{b=1}^{M} P(\text{CFI}_i) \log_2(P(\text{CFI})) \tag{55.10}$$

$P(\text{CFI})$ is the probability of obtaining ith CFI from set D. Entropy values gathered during the training stage are used to generate the index tree.

Entropy-based index tree (EIT) formation: The CFI's entropy values are used as input in this algorithm. Entropy-based index trees are returned as a result of the algorithm (EIT). The first step is to get the CFI's entropy values. The header node is selected based on a manually defined threshold value. As the entropy tree increases, so do its leaf nodes, which are based on the header node. Whenever the data's entropy value exceeds the header node's entropy value, the data is stored in the appropriate portion of the tree. Data whose entropy value is less than the header node's entropy are positioned to the left of the tree as a result. The CS handles these EITs.

55.3.3   Cloud Server Entity

The data owners can upload as well as retrieve files in the cloud server entity. This cloud server's security is critical for both storing and retrieving data. Documents have been hashed and encrypted, and the CS is where they are stored with their respective index trees. Consequently, the CS performs the update after getting the information from the DO. As soon as the CS receives a request for a search, it begins searching for results that match the user's specified query.

55.3.4   Data User Entity

DUs are those who have the ability to search encrypted data. The DU requests a specific document from the CS using a search request in the form of a keyword. The CS now asks for the hashed decryption key and the nonce number for the specified document from the DO entity. Every time a document is decrypted, a unique nonce number is generated, enhancing the system's overall security. The DO delivers the hashed decryption key to DU through the CS, along with a nonce number in 0 to 1 range.

$$D_K = H(D_K) + N_c \tag{55.11}$$

It's D_K and N_c that constitute the hashed decryption key. Decryption is possible when the DU gets the nonce number as well as the hashed decryption key.

$$O_{D_K} = H(D_K) - N_C \tag{55.12}$$

55.4   Result and Discussion

SPKSR, the proposed privacy-preserving keyword search over hashed encrypted data, has been implemented. The results of experiments were assessed using a variety of criteria, including encryption time, decryption time, index construction time, computational time, "Key generation time" (KGT), security level, and "Query retrieval time" (QRT). With the current method, these values were calculated and compared.

55.4.1   Encryption Time

It represents the length of time necessary to encrypt the plain text. "Elliptic-curve cryptography" (ECC) and "Rivest-Shamir-Adleman" (RSA) are used to compare the suggested HMECC's encryption time. The following graph demonstrates how the time it takes to encrypt cloud storage data depends on the number of records it contains. Figure 55.3 shows that the encryption time increases as the number of attributes increases. The present RSA technique performs the worst here, requiring 35,443 ms to encrypt 1000 entries. In comparison with the RSA method, the ECC technique consumes 33,244 ms to encrypt 1000 records. Compared to conventional RSA and ECC, the proposed HMECC takes a little less time to encrypt data. Therefore, it is proven that the newly designed model consumes less time to encrypt the input document.

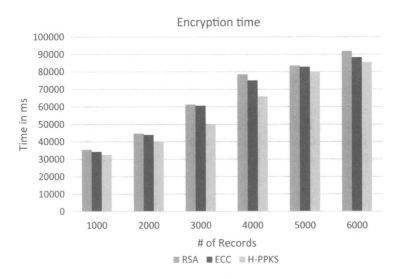

Fig. 55.3 Encryption time

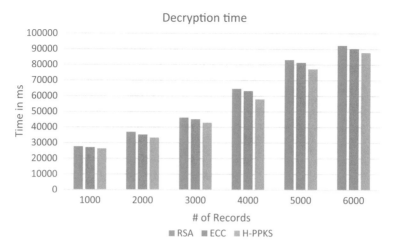

Fig. 55.4 Decryption time

55.4.2 Decryption Time

In computing, the DT is the amount of time it takes to get the original information or plain text from the network. For the present ECC and RSA and the planned HMECC, this data is evaluated. Figure 55.4 shows that the proposed HMECC method has a lower DT than the current RSA and ECC algorithms. The suggested model takes 87,755 ms to decode a 6000-record document, while RSA and ECC require 92,323 and 90,343 ms, correspondingly, to decrypt the document. This finding demonstrates that the suggested HMECC outperforms the industry's best approaches.

55.4.3 Index Construction Time

For the entropy-based index tree, this is the time it takes to construct it. Figure 55.5 shows how tree construction time varies depending on the amount of data. Milliseconds are used to measure time. The suggested SPKSR only constructs the tree after computing the entropy performance, saving time. A comparison of the current system's encryption time fluctuation is shown in this graph. The graph exhibits that the current MCKSMAT method takes longer than the suggested SPKSR system. A comparison of the SPKSR and existing methods shows that SPKSR is a superior option.

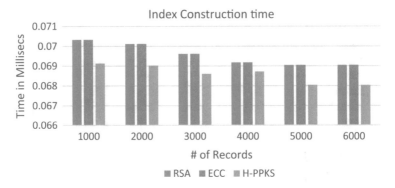

Fig. 55.5 Index construction time

55.4.4 Computational Time

It is a measure of the amount of time needed to carry out a computational activity. An evaluation of the new SPKSR approach is made in terms of Query retrieval time, key generation time, and security level.

Query retrieval time shows the overall amount of time it took to get the data. The suggested SPKSR takes 39 ms to retrieve a query containing 40,000 records, while the conventional MCKS MAT takes 41 ms. As a result, the suggested SPKSR system was found to be superior to the best available approaches.

Security level: Cloud storage security is of the utmost importance. The degree of security varies according to the number of records. The suggested HMECC algorithm has a security level of 15; however, the current ECC method, which also uses RSA, has levels of 13 and 7, which is less than the suggested HMECC system. The suggested method is more secure than the current RSA and ECC algorithms for all records. Because of this, the proposed HMECC algorithm is clearly more secure than the RSA and ECC algorithms.

Time taken to create a key is known as key Generation Time (KGT). Based on the total amount of records stored, a different KGT is computed for each new record. The performance of the cryptographic method improves if the KGT is reduced. As a result, the KGT for the suggested HMECC is low. This comparison shows that the proposed HMECC generated the key in a smaller duration of time.

55.5 Conclusion

This paper proposes a method for securely and privately retrieving hashed, encrypted cloud data to do keyword searches. Metrics were used for evaluating the efficiency of the developed system. The suggested KS over encrypted data system performed very well in terms of QRT and computing time, according to the results of the performance study. The suggested KS over encrypted data system performed very well in terms of QRT and computing time, according to the results of the performance study. The suggested work features low encryption time, low computing time, low DT, as well as low index creation time, and excellent security. MCKSMAT and the suggested work are superior in terms of efficiency as well as security, as shown by the comparative results. Thus, the suggested method is more secure than the already existing approach, and it consumes less time performing the same function. In further study, improved encryption methods utilizing distributed keys and cloud-based secure keyword search retrieval may greatly boost the system's performance.

References

1. Ratankumar, A.: Privacy preserving multi-keyword ranked search over encrypted cloud data. Int. J. Adv. Electron. Comp. Sci. 3(9), 222–233 (2016)
2. Shekokar, N., Sampat, K., Chandawalla, C., Shah, J.: Implementation of fuzzy keyword search over encrypted data in cloud computing. Proc. Comput. Sci. 45, 499–505 (2015)
3. Shen, Z., Shu, J., Xue, W.: Keyword search with access control over encrypted data in cloud computing. In: 2014 IEEE 22nd International Symposium of Quality of Service (IWQoS), pp. 87–92. IEEE (2014)
4. Pandiaraja, P., Vijayakumar, P.: Efficient multi-keyword search over encrypted data in untrusted cloud environment. In: 2017 Second International Conference on Recent Trends and Challenges in Computational Models (ICRTCCM), pp. 251–256. IEEE (2017)
5. Ambika Ganguli, M., Maddodi, P.S., Saralaya, S.: Multi keyword search over encrypted cloud. Adv. Comput. 7(2), 58–61 (2017)
6. BalaAnand, M., Karthikeyan, N., Karthik, S., Sivaparthipan, C.B.: A survey on BigData with various V's on comparison of apache Hadoop and apache spark. Adv. Nat. Appl. Sci. 11(4), 362–370 (2017)
7. Maram, B., Gnanasekar, J.M., Manogaran, G., Balaanand, M.: Intelligent security algorithm for UNICODE data privacy and security in IOT. SOCA. 13(1), 3–15 (2018). https://doi.org/10.1007/s11761-018-0249-x
8. Li, J., Wang, Q., Wang, C., Cao, N., Ren, K., Lou, W.: Fuzzy keyword search over encrypted data in cloud computing. In: Infocom, 2010 proceedings IEEE, pp. 1–5. IEEE (2010)
9. Naga Aswani, P., Shekar, K.C.: Fuzzy keyword search over encrypted data using symbol-based Trie-traverse search scheme in cloud computing (2012)
10. Akkalkot, J., Shanmug, P.S.: A survey on keyword-based search mechanism for data stored in cloud. Int. J. Comput. Sci. Mob. Comput. 5(5), 235–240 (2016)
11. Miao, Y., Ma, J., Liu, X., Zhang, J., Liu, Z.: VKSE-MO: verifiable keyword search over encrypted data in multi-owner settings. Sci. China Inform. Sci. 60(12), 122105:1–122105:15 (2017)
12. Meharwade, A., Patil, G.A.: Efficient keyword search over encrypted cloud data. Procedia Comp. Sci. 78, 139–145 (2016)
13. Sonawane, K.: Multi-keyword ranked search over encrypted cloud data with multiple data owners. IJEDR. 5(3), 1–15 (2017)
14. Li, J., Chen, X.: Efficient multi-user keyword search over encrypted data in cloud computing. Comput. Inform. 32(4), 723–738 (2013)
15. Shen, Z., Shu, J., Xue, W.: Preferred keyword search over encrypted data in cloud computing. In: 2013 IEEE/ACM 21st International Symposium on Quality of Service (IWQoS), pp. 1–6. IEEE (2013)

16. BalaAnand, M., Karthikeyan, N., Karthik, S., Varatharajan, R., Manogaran, G., Sivaparthipan, C.B.: An enhanced graph-based semi-supervised learning algorithm to detect fake users on Twitter. J. Supercomput. **75**(9), 6085–6105 (2019). https://doi.org/10.1007/s11227-019-029 48-w

17. Peng, T., Lin, Y., Yao, X., Zhang, W.: An efficient ranked multi-keyword search for multiple data owners over encrypted cloud data. IEEE Acc. **6**, 21924–21933 (2018)

18. Zhang, Y., Deng, R.H., Shu, J., Yang, K., Dong, Z.: TKSE: trustworthy keyword search over encrypted data with two-side verifiability via blockchain. IEEE Acc. **6**, 31077–31087 (2018)

19. Xiang, G., Lin, X., Wang, H., Li, B.: Privacy preserving query over encrypted multidimensional massive data in cloud storage. Wuhan Univ. J. Nat. Sci. **23**(2), 163–170 (2018)

20. Hahn C.H., Shin, H.J., Kwon, H., Hur, J.: Efficient multi-user similarity search over encrypted data in cloud storage. Wirel. Pers. Commun., 1–17 (2018)

21. Zhangjie, F., Wu, X., Wang, Q., Ren, K.: Enabling central keyword-based semantic extension search over encrypted outsourced data. IEEE Trans. Inf. Forensics Secur. **12**(12), 2986–2997 (2017)

22. Wang, J., Yu, X., Zhao, M.: Privacy-preserving ranked multi-keyword fuzzy search on cloud encrypted data supporting range query. Arab. J. Sci. Eng. **40**(8), 2375–2388 (2015)

23. Bala Anand, M., Karthikeyan, N., Karthik, S.: Designing a framework for communal software: based on the assessment using relation modelling. Int. J. Parallel Prog., 1–15 (2018). https://doi.org/10.1007/s10766-018-0598-2

24. Wan, Z., Deng, R.H.: VPSearch: achieving verifiability for privacy-preserving multi-keyword search over encrypted cloud data. IEEE Trans. Dependable Secure Comput. **15**(6), 1083–1095 (2016)

25. Sivaparthipan, C.B., Karthikeyan, N., Karthik, S.: Designing statistical assessment healthcare information system for diabetics analysis using big data. Multi. Tools Appl., 1–4 (2018)

Chapter 56
Promoting Network Connectivity in a Textile Showroom

A. Balasubramanian, V. Sanmathi, J. Kavya, and R. Amarnath

Abstract Social network analysis is an important tool for constructing, analyzing and understanding networks within an organization and using this information to improve communication and productivity of the organization. Some communication issues, hierarchical problems within the sales department and communication gap between the employers and customers prevented smooth functioning in one of the textile showrooms in Tamil Nadu, India. To address these issues the social network analysis was carried out. The outcome of this analysis, suggestions, recommendations, implementation of the recommendations and its positive impact on the overall functioning of the showroom are reported in this paper.

56.1 Introduction

Social network analysis (SNA) is an invaluable and effective tool for visually analyzing the patterns of relationships, which reveals a number of interesting and actionable points [1]. People who are highly central in networks can be identified, and this information is useful in reallocating informational domains, so that the group as a whole is more effective. Also identifying who is peripheral in a network and finding ways to engage these people will ensure that expertise resident in a given network is effectively used. SNA is also a very effective tool for promoting collaboration and knowledge sharing within important groups in an organization such as research and development section, core functions and strategic business units [2, 3]. This concept has now been well developed, and there are several books on this topic such as Cross and Parker [4], Andrews and Knoke [5], Baker [6], Monge and Eisenberg [7] and Scott [8].

A. Balasubramanian (✉) · V. Sanmathi · J. Kavya · R. Amarnath
Department of Mass Communication, Amrita Vishwa Vidyapeetham, Coimbatore, India
e-mail: balasubramanian@cb.amrita.edu

Presence of isolated vertices in any network is a serious problem which needs to be addressed. Finding the reason for an actor to be an isolated vertex and steps to be taken for making the person to interact with other members of the network are essential. The problem of dealing with workplace isolation has been reported in Mulki et al. [9] and Orhan et al. [10]. The way the workers respond when faced with personal problems at workplace has been explored in Holgate et al. [11].

SNA can also be used to identify bottleneck elements who are responsible for disturbing the smooth functioning of an organization. A case study of such a situation has been presented in Cross and Parker [4, pp. 31–35]. It is a social network analysis in a consulting firm with a highly skilled group of employees who had advanced degrees or extensive industry experience. By integrating such a highly specialized expertise, the management hoped to differentiate the firm from competitors. However, the social network analysis resulted in the identification of one employee who appeared to play a critical role in holding the group together, but in reality he was a bottleneck, fragmenting the group. After nine months from the date of removal of the bottleneck element the two groups of the organization became a well-integrated group and started functioning more effectively. This has also been reported in Borgatti and Cross [12].

We encountered a similar situation in our study of the functioning of a texting showroom in Tamil Nadu, India. In this paper SNA in this textile showroom, identification of bottleneck elements, recommendations and the effects of implementing the recommendations are reported.

56.2   Social Network Analysis in a Textile Showroom

A textile showroom in Tamil Nadu, India, which was established in 2007 deals with readymade garments for men, women, kids and newborns. The organization is known for quality service. Readymade apparels of high quality in different materials are made available at an affordable cost. The textile showroom has six branches, and all of them are having a good online rating with a score of four and above in the five-point scale. This organization has set up a strong footing with a tremendous base of clients, which keeps on developing constantly.

In one of their showrooms some communication issues, hierarchical problems within the sales department and communication gap between the employers and customers were brought to the notice of the proprietors. Hence, the management contacted the researchers to investigate this problem and provide suggestions for improving the overall performance in the showroom. Since the way in which works get done in an organization depends on the invisible network of communications among the employees, we decided to adopt social network analysis as a tool to address this problem.

56.2.1 About the Organization

The branch of the showroom where we have undertaken this study was started in 2018. It has three managers and twenty-seven salespersons. For the purpose of this study, the three managers are denoted by m_1, m_2 and m_3. The twenty-seven salespersons are denoted by $s_1, s_2, \ldots, s_{27}$ and among them s_{24}, s_{25}, s_{26} and s_{27} are godown workers. The godown incharge is s_{24} and s_{25}, s_{26}, and s_{27} are working under s_{24}.

The twenty-seven salespersons are divided and put into fifteen different sections in the showroom. The designation of m_1 is the product incharge, but he does the work of showroom incharge. The designation of m_2 is floor incharge, but he works as a supervisor or assistant showroom incharge. The overall showroom incharge is m_3, and he is a newly appointed employee. Godown is also under the supervision of m_2. All other sections except billing section come under m_2. The billing section is taken control by m_1, and sometimes when there is too much rush he also works there.

Even though the organization has an efficient marketing strategy, brand image and profitable sales throughout the year, there are some communication issues and hierarchical problems within the sales department. In this branch both offline and online reviews about the sales and the service are not satisfactory. Lots of complaints about the employees were emerging in this showroom.

56.2.2 Data Collection

Data was collected through questionnaires and personal interview. The responses of the employees for the following four questions were collected.

Q1: If you face any problem while you are carrying out your work in the shop, from whom will you seek help?
Q2: When you have new ideas or suggestions for improving the performance of the shop, with whom will you share and discuss it further?
Q3: If someone working with you goes on leave, you may find it difficult to carry out your regular work. Give the list of such persons.
Q4: Give the list of persons in your shop with whom you have regular communication.

During the collection of data, the dominance and controlling nature of the managers m_1 and m_2 were seen. With the responses collected from the employees for these questions, we constructed four networks. For the construction of the networks, Pajek was used. This software was developed by Mrvar and Batagelj [13] for analyzing large networks. In each of these networks the set of thirty employees of the showroom form the vertex set, and the edge between two employees indicates the existence of communication between them.

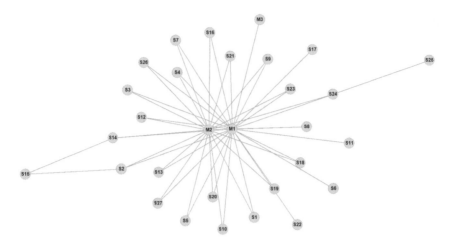

Fig. 56.1 Graph G_1 for question Q_1

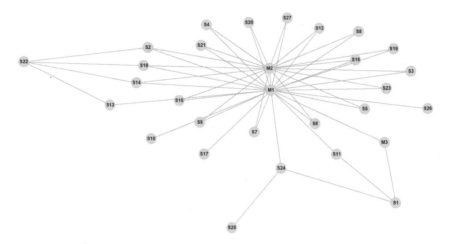

Fig. 56.2 Graph G_2 for question Q_2

56.2.3 Analysis of Graphs

The graph for the question Q_1 is given in Fig. 56.1.

Of the twenty-seven salespersons, eighteen of them are connected only with m_1 and m_2. Three salespersons s_2, s_{14}, s_{24} are also connected to m_1 and m_2 and to one more salesperson. Four salespersons s_6, s_{11}, s_{17} and s_{22} are connected to exactly one of the vertices m_1 or m_2. Thus, among the twenty-seven salespersons only two of them are not directly connected to either m_1 or m_2. The manager m_3 is connected only with m_1. Thus, it is clearly seen that m_1 and m_2 are bottleneck elements who prevent communication among the salespersons (Fig. 56.2).

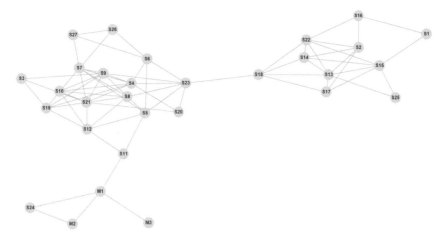

Fig. 56.3 Graph G_3 for question Q_3

The graph G_2 is almost similar to G_1. In the graph G_2 out of the twenty-seven salespersons only five of them are not connected to m_1 and m_2. Fourteen of them are connected to only m_1 and m_2, four of them are connected to m_1, m_2 and another salesperson, five of them are connected to either m_1 or m_2 but not both. Thus the dominant nature of m_1 and m_2 is seen clearly from G_1 and G_2.

The graph G_3 for Q_3 is given in Fig. 56.3.

The graph G_3 has three clusters, namely cluster 1 consisting of all female workers, cluster 2 consisting of male workers and cluster 3 consisting of the three managers m_1, m_2, m_3, and the godown incharge s_{24}. Based on personal interviews we concluded that the formation of cluster 1 and cluster 2 is mainly due to the dominance of m_1 and his instruction to the members of the two clusters not to communicate with other groups. Hence, there is communication only within the cluster. It is also seen that m_1 is blocking the communication of m_2, m_3 and s_{24} with other employees. In fact m_1, m_2 and s_{24} are contacting each other but m_3 is contacting only m_1. Thus m_1 turns out to be bottleneck element blocking the overall communication in the entire group.

The graph G_4 corresponding to Q_4 is given in Fig. 56.4. This is a disconnected graph having seven isolated vertices and five more components. In the regular communication process the manager m_3 is also isolated. This graph shows that there is regular communication only between the employees of the same section and as a consequence the salespersons do not have any knowledge about the products of the other sections. Hence, if a salesperson is shifted to another section he or she is not able to fulfill the customer needs.

One of the complaints from the proprietors of the showroom is that, employees are taking leave more often and they quit their jobs frequently without prior information. This is due to the dominance of m_1 and m_2.

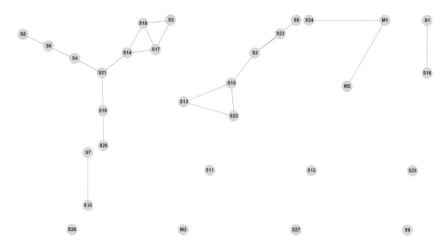

Fig. 56.4 Graph G_4 for question Q_4

56.2.4 Recommendations

1. The manager m_1 is the bottleneck element in the showroom, and he is blocking the overall communication. The manager m_2 has joined hands with m_1 and is also supporting the processes of blocking communication. Hence, it is recommended that m_1 should be warned or removed and m_2 should be under continuous observation. If m_1 is removed, then m_3 should be given the overall responsibility of supervising the employees.
2. The proprietors of the company should regularly visit the showroom and monitor the incharge and also communicate with the employees.
3. Employees should be encouraged to have interpersonal communication with the opposite gender, which will develop positive work culture.
4. Training sessions and one-day camps should be conducted at regular intervals to educate the employees about all the products in the showroom.
5. Shuffling of the workers regularly between the sections in the organization should be done periodically. Employees should be allowed to have conversations and discussions regarding the products and their prices when they do not have any customers to attend. This will ensure that the salespersons will become familiar with products of all the sections.

56.2.5 Implementation

When the above recommendations were provided to the proprietors of the showroom, they initially felt that m_1 and m_2 are in the organization for a longer period and were hesitant to take action against them. However, they started monitoring their activities closely and were convinced that the recommendations must be implemented. Accordingly, m_1 was removed from the showroom, and m_3 was appointed as overall incharge. A warning was given to m_2, and his activities were monitored on a regular basis. The proprietors started visiting the showroom and interacted with the employees in an informal manner. After three months the work culture of the showroom and feedback from both customers and employees have given more positive results. Buyer and seller satisfaction is achieved. The sales rate has also increased.

A letter of appreciation to the researchers was handed over to them by the management.

56.3 Conclusion and Scope

This study indicates how social network analysis can be successfully carried out in organizations for improving the overall functioning and ensuring smooth communication between employees. This tool may be more effective and useful in organizations which involve research and development units.

References

1. Vidhya, V., Ambily, A.S.: A study on the impact of social networking on micro, small and medium enterprise with reference to Ernakulam district. Am. Int. J. Res. Humanit. Arts Soc. Sci. **22**(1), 158–165 (2018)
2. Cross, R., Borgatti, S., Parker, A.: Making invisible work visible: using social network analysis to support strategic collaboration. Calif. Manage. Rev. **44**(2), 25–46 (2002)
3. Nikhil, K.S., Ambika, B., Judy, M.V.: Iterative parallel genetic algorithm for detecting communities in social networks. ARPN J. Eng. Appl. Sci. **12**(23), 6944–6948 (2017)
4. Cross, R., Parker, A.: The Hidden Power of Social Networks. Harvard University Press, Boston (2004)
5. Andrews, S.B., Knoke, D.: Networks in and Around Organizations: 16 (Research in the Sociology of Organizations). JAI Press, Stanford (1999)
6. Baker, W.: Networking Smart: How to Build Relationships for Personal and Organizational Success. McGraw-Hill, New York (1994)
7. Monge, P.R., Eisenberg, E.M.: Emergent Communication Networks. In: Jablin, F., Putnam, L., Roberts, K., Porter, L. (eds.) Handbook of Organizational Communication. Sage Publications, Newbury Park (1987)
8. Scott, J.: Social Network Analysis, 2nd edn. Sage Publications, Thousand Oaks (2000)
9. Mulki, J.P., Locander, W.B., Marshall, G.W., Harris, E.G., Hensel, J.: Workplace isolation, salesperson commitment, and job performance. J. Pers. Selling Sales Manag. **28**(1), 67–78 (2008)
10. Orhan, M.A., Rijsman, J.B., van Dijk, G.M.: Invisible, therefore isolated: Comparative effects of team virtuality with task virtuality on workplace isolation and work outcomes. J. Work Organ. Psychol. **32**(2), 109–122 (2016)
11. Holgate, J., Keles, J., Pollert, A., Kumarappen, L.: Workplace problems among kurdish workers in London: experiences of an 'invisible' community and the role of community organisations as support networks. J. Ethn. Migr. Stud. **38**(4), 595–612 (2012)

12. Borgatti, S., Cross, R.: A social network view of organizational learning: relational and structural dimensions of 'know who'. Manage. Sci. **49**, 432–445 (2003)
13. Mrvar, A., Batagelj, V.: Analysis and visualization of large networks with program package Pajek. Complex Adapt. Syst. Model. **4**, Article number: 6 (2016)

Chapter 57
Classroom Beyond Walls: Perspectives of Students and Teachers on Pedagogy and Features of MOOCs

Vijay Makwana and Pranav Desai

Abstract There is no exaggeration in stating that massive open online courses (MOOCs) have become an integral part of education today. After the pandemic and emergence of the new normal, sudden rise of online learning is witnessed. MOOC, which was an idea, has become an industry now. MOOC, which emerged around 2010s as a movement to democratize learning, has become the next big thing in education now. However, there are very few studies or published data regarding MOOC learners and their experience of learning. The purpose of the present study is to conduct comparative analysis of MOOC platforms in terms of features, content, and user satisfaction. The researchers in the present study conducted an online survey of 250 students and interviews of 30 students of Charotar University of Science and Technology (Gujarat, India) who enrolled for MOOCs offered on five different platforms—Coursera, EdX, Future Learn, Udemy, and Alison. In addition, 10 teachers who integrated MOOC courses in their course transaction were interviewed. It was found that, the chief reasons for undertaking a MOOC are academic requirement, professional development, and interest in the subject matter. The esthetic and simplified presentation of content and ease of access are the major reasons for the popularity of a few platforms. Also, it was found that, MOOC has abysmal completion rates, and a heavy proportion of MOOC attendees does not complete the courses they enrolled for, and the reasons found for the same are lack of motivation and consistency. This study provides actionable insights for online education in general, MOOC in particular.

V. Makwana · P. Desai (✉)
Faculty of Management Studies, Indukaka Ipcowala Institute of Management, Charotar University of Science and Technology (CHARUSAT), Changa, India
e-mail: pranavdesai.mba@charusat.ac.in

V. Makwana
e-mail: vijaymakwana.cs@charusat.ac.in

57.1   Introduction

If you aspire to learn brand management without attending a B-School or study social media marketing or even Greek literature, MOOCs have opened the gates of learning for masses. With the MOOC revolution, it has become possible for anyone to acquire new skills or enhance the knowledge one already has. Since MOOCs are offered virtually, it merely requires Internet connectivity, some allocation of time and interest in the subject matter, and the learning is just a tab or a click away. Online learning which was seen as additional track is a part of mainstream education now. A decade back, MOOCs, which began as a movement of free and open education, are delivering top-grade content from best schools and teachers. MOOCs have upend everything in higher education.

57.1.1   The Rise of MOOC Phenomenon

> High-quality education provided by MOOCs can be a significant factor in opening doors to opportunity—even among the college-educated.
> —Daphne Koller.

This quote by Daphne Koller, a Stanford professor and founder of Coursera aptly captures the cardinal role MOOCs (Massive Open Online Courses) are playing in post-pandemic academic world. The MOOC phenomenon has not only democratized education, but in words of Kunduru [1], it has made quality education accessible to all. As per the Coursera impact report published in November 2021, the platform has 189 million enrolments and 92 million registered learners globally (See Figs. 57.1 and 57.2).

The Covid pandemic gave a push to the growth of online courses, and the MOOC platforms are booming at unprecedented pace. Consequently, many universities and colleges across the world now integrate MOOC courses in their curriculum to enrich their students' learning experience.

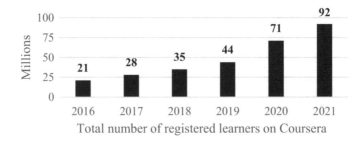

Fig. 57.1 Demand for online learning on Coursera–1 [2]

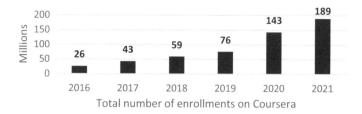

Fig. 57.2   Demand for online learning on Coursera–2 [2]

57.2   Review of Literature

Zutshi et al. [3] examined the experiences of learners who participated in MOOCs. The sample consisted of a blend of first timers and experienced MOOC participants. 21 participants were asked to reflect their experience in public blogs. Sample of blog posts was analyzed using the content analysis approach. These four categories appeared predominant in participants' blog post—assessment, instructional materials, learner interaction and engagement, and course technology. In terms of the overall participation experience, students reported mixed experiences and identified both positive and negative aspects. Students opined that they lost interest in the course wherein the assessment was poorly designed or nonauthentic. Also, it was found that students liked the platform/courses which offered relaxed and friendly atmosphere at the same time maintaining professional and serious approach.

In another study, Stracke et al. [4] reported the Global MOOC quality survey pertaining to learners' overall experiences with MOOCs. The participants in the survey included 166 MOOC learners, 68 MOOC designers, and 33 MOOC facilitators. The finding revealed that there were significant differences in MOOC learners' and designers' intentions and experiences. Therefore, the study raised a major concern whether MOOC designers understand and meet the interests and needs of the MOOC learners.

Park et al. [5] reported the MOOC experiences of three MOOC platform (Future-Learn, edX, Udacity) learners. An autoethnography method was employed by the researcher to collect empirical data in three different MOOCs. It was found that studying through MOOCs could be a challenging experience. It was found that one of the challenges for the MOOC designers is to cater to the needs to the learners who come from many different backgrounds. The researcher proposed that systematic instructional design can help to ensure quality content and more engagement in MOOCs.

A model of school learning propounded by John B. Carroll in 1963 has influenced educational researchers and developers for over three decades. Carroll expressed astonishment that his model has been well received and stated that the proposed model could still be instrumental to solve current issues in education (p. 26). Carroll's original model of school education is a formal, quasi mathematical one which states that students' academic achievement relies on five classes of variables. It also assumes

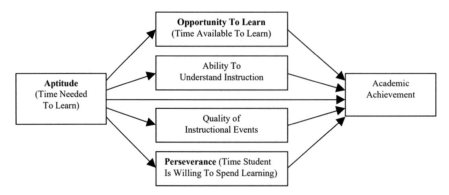

Fig. 57.3 Carroll's 'Model of School Learning'

that learners vary in the amount of time they need for learning; therefore, the teacher needs to be proficient in classroom management. One of the key factors according to Carroll in success of teaching is good planning and good instructional design. The structure of the Carroll model is represented in Figure 57.3, and each of the six factors in the model are explained below.

57.2.1 The Challenges in Education During Pandemic 2020

The covid breakout and subsequent lockdowns resulting in school closures have brought significant disruptions to education across the world. Reports (2021) suggest that from developed nations to third-world countries, the pandemic is giving rise to learning losses and increases in inequality. To give a glimpse of it, at the peak of the covid pandemic, education of over 185 million students was adversely affected when 45 countries in the Europe and Central Asia closed their academic institutions. In this unprecedented situation, academicians and administrations were caught unready for this transition and were compelled to build emergency distance learning systems almost at once. The pandemic crisis reducing the opportunities for several of the leading susceptible broods, adolescence, and adults who are lacking with the enough availability, accessibility, affordability of resources like Internet and digital equipment. For developing better understanding of the intermediations for effectiveness of online earning method with the standpoint of fish bone diagram (Fig. 57.4) .

57.3 Significance of the Study

The study on the learners' attitudes toward remote or online learning during and after COVID-19 pandemic provides inputs for various education provider to develop

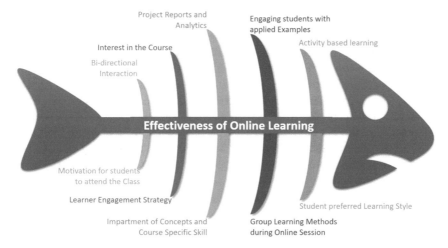

Fig. 57.4 Intermediations for effectiveness of online learning process: fishbone model

a strategy to offer better and student centric pedagogy. This study was done for the students who are studying in various colleges and universities to gain additional information regarding the contribution MOOC in teaching learning. The prime benefit to information system developers for remote education and their business operation is as, with help of digital technologies, students are maintaining the continuity of the study.

57.4   The Rationale

57.4.1   Research Objectives

1. To study the perspective of students toward their experience of MOOC learning
2. To study the MOOC design of various platforms and identify their prominent features
3. Provide inputs for information system developers for better engaging strategies for remote education and their business operation
4. Extend the Carroll's 'Model of School Learning' for MOOC design.

57.4.2   Methodology

The researched relied on both primary and secondary data for the present study. A Web-based questionnaire was developed for collecting the responses. The questionnaire consisted of Likert scale items to collect quantitative data and a few open-ended questions to collect qualitative data about participants experience of using MOOC platforms. In total, 250 filtered responses were selected to identify the preferences from the users of the MOOC users against effectiveness toward various features, facilitates and facets of the specific platforms. The responses have been collected from the students of Engineering, Pharmacy, Management, Physiotherapy, Applied Science, Computer Applications, and Nursing disciplines. In addition to this, interviews of 30 participants from the sample population were conducted to gather detailed inputs on variables not covered in structured Web-based questionnaire. The items in the survey questionnaire and interview were pertaining to following eight aspects of MOOC design/features: interactivity, content, pedagogy, motivation, assessment, learners support, collaboration, and technology.

57.5 Result and Discussion

- All the four platforms are popular among students. However, it was found that Coursera and FutureLearn are more familiar to them.
- Both students and teachers opined that Coursera has better structure of the courses and effective assessment methods.
- Concerning the partners of various platforms, students and teachers are more oriented toward Coursera and edX which are successful in hosting courses top universities of the world.
- The analysis of the data revealed that students prefer features like offline download of content, diversity of courses and graded, and effective presentation of the content.
- In terms of assessment, it was found that different platforms have adopted different methods. Courses and edX have integrated peer grading and review method where fellow learners anonymously grade peer assignments.
- For foreign language learning, most preferred platform among students was found to be edX as it offers a lot of courses in foreign languages such as French, Chinese, and Spanish.

57.6 The EYE Model for MOOC Design

See Fig. 57.5.

57.7 Conclusion

The education is undergoing massive makeover after the introduction of Web 3.0 technologies and their integration in academics. Undoubtedly, MOOCs are one of

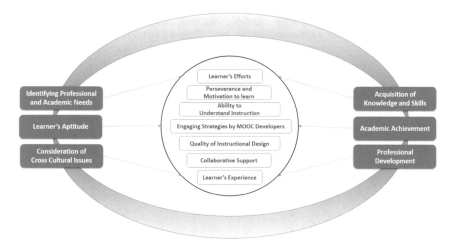

Fig. 57.5 Proposed EYE model for MOOC design (expanded from Carroll's 'model of school learning')

the most conspicuous trends in tertiary education in the last few years. In this research, we reviewed the major features of the selected MOOC platforms like the instructional design, content quality, assessment, and engagement strategies from the user perspective. Also, the research made an attempt to compare the selected platforms based on the data collected for each one. Based on the finding and the researches have proposed a model for MOOC designers. The model presented in Fig. 57.5 is intended to capture many of the complexities of inputs, processes, and outcomes involved in remote (MOOC/Online) education.

References

1. Kunduru, S.R.: The MOOC Revolution: Education on Tap (2015, November). Retrieved June 2022, from BusinessToday.in: https://www.businesstoday.in/magazine/cover-story/story/india-best-business-schools-2015-massive-open-online-courses-offers-premium-education-56525-2015-11-13
2. Coursera Impact Report. (2021, November). Retrieved from Coursera: https://about.coursera.org/press/wp-content/uploads/2021/11/2021-Coursera-Impact-Report.pdf
3. Zutshi, S., O'Hare, S., Rodafinos, A.: Experiences in MOOCs: the perspective of students. Am. J. Dist. Educ. **27**(4), 218–227 (2013). https://doi.org/10.1080/08923647.2013.838067
4. Stracke, C.M., Tan, E., Texeira, A.M., Pinto, M.D., Vassiliadis, B., Kameas, A., Sgouropoulou, C.: Gap between MOOC designers' and MOOC learners' perspectives on interaction. In: 18th International Conference on Advanced Learning Technologies, pp. 1–5 (2018)
5. Park, Y., Jung, I., Reeves, T.C.: Learning from MOOCs: a qualitative case study. Educ. Media Int. **52**(2), 72–87 (2015). https://doi.org/10.1080/09523987.2015.1053286
6. Saravanakumar, A., Ramaswami, P.: Indian higher education: issues and opportunities. J. Crit. Rev. **7**, 542–545 (2020). https://doi.org/10.31838/jcr.07.02.101

7. Sathishkumar, V., Radha, R., Mahalakshmi, K., Saravanakumar, A.: E-Learning during lockdown of Covid-19 pandemic: a global perspective. Int. J. Control Autom. **13**, 1088–1099 (2020)

Author Index

Printed in the United States
by Baker & Taylor Publisher Services